The behavior and sensory biology of elasmobranch fishes

Developments in environmental biology of fishes 20

Series Editor
EUGENE K. BALON

The behavior and sensory biology of elasmobranch fishes: an anthology in memory of Donald Richard Nelson

Guest Editors:

Timothy C. Tricas & Samuel H. Gruber

Reprinted from *Environmental biology of fishes*, Volume 60 (1–3), 2001
with addition of species and subject index

SPRINGER-SCIENCE+BUSINESS MEDIA, B.V.

A C.I.P. Catalogue record for this book is available from the library of Congress

ISBN 978-90-481-5655-9   ISBN 978-94-017-3245-1 (eBook)
DOI 10.1007/978-94-017-3245-1

Cover design by Mariko H. Gruber
using posterior portion of the porbeagle, *Lamna nasus*, from 'Fishes of the western north Atlantic' by Henry B. Bigelow & Wiliam C. Schroeder, Part 1, 1948. The posterior portion of the southern sawfish, *Pristis preotteti*, was taken from Part 2, 1953.

The logo
also designed by Mariko H. Gruber

Printed on acid-free paper

CONTENTS

Lemon shark, *Negaprion brevirostris*, in Bimini lagoon being prepared for telemetry speed trials in 1997.
Photograph by Timothy Calver.

Environmental Biology of Fishes **60:** 7–11, 2001.

Prelude to the anthology in memory of Donald Richard Nelson

There are few people on this planet who could not tell you something about the behavior of the sea creature known as the shark. Ask most people and they will likely cite voracious feeding and aggressive behaviors, but ask a shark scientist and you will get a very different answer. In truth, information on the natural behavior of sharks is known for only a few of the more than 350 species. Even less is known about their cousins, the more than 450 species of derived rays and skates. The main reason for this great void is that elasmobranch fishes are relatively large, wide-ranging, and live in an inhospitable and secretive environment. Thus they are relatively difficult to observe in the field or study under laboratory conditions. The same reasons explain why so little is known of how the sensory systems of these animals function in natural settings. As a result, much of what we know about the natural behavior of sharks and rays comes from brief field studies conducted without a premeditated paradigm or experimental design.

One extraordinary exception is the research from 1962–1997 conducted by Donald Richard Nelson, whom this anthology honors. Don was a diver, teacher, scientist and adventurer who arguably spent more time underwater observing these animals than any other human. His bold and numerous expeditions to study shark behavior and sensory biology in the field are unmatched. Perhaps most significant is that his enthusiasm and teachings have continued to stimulate many of his students and colleagues. Don's work, and that of his successors, will become even more significant as time progresses given the great academic pressures for successful grantsmanship and the current lack of funding resources for the study of shark behavior. In retrospect, Don has left a healthy legacy that will hopefully propel this field well into the future.

The symposium to honor the contributions of Don Nelson to the study of shark behavior was convened on 18 July 1998 at the annual meetings of the American Elasmobranch Society held at the University of Guelph (Figures 1–3). At the formal meeting there were 17 presentations that covered a wide range of topics on the sensory biology and behavior of sharks and rays. From the original symposium presentations, 14 scientific papers were contributed for this volume. The purpose of this publication is twofold. First, there has not been a scientific treatise published on the behavior of elasmobranch fishes for more than 20 years. In that time period, many new technologies were developed (some of which were pioneered in Don Nelson's laboratory) and applied to the study of shark behavior and sensory biology. Second, there is a new generation of young students who often find it difficult to consolidate information on elasmobranch behavior, or identify critical study questions. Thus, we asked authors of this volume to include a significant review of their study area, and also to identify at least 3 hypotheses that can be tested. It is our hope that this treatise will provide an impetus for scientists (young and old) to continue the study of shark behavior and sensory biology.

The articles that follow the introductions to Don Nelson can be separated into three categories: sensory biology, behavior and telemetry. The first paper is by symposium keynote speaker Art Myrberg, who in his own right is a founder of the study of shark behavior and sensory biology. His paper addresses the history and current status of research on how sharks hear and orient to sound sources in their natural world, which remains one of the most controversial questions in auditory biology. Karen Maruska then presents a detailed review of the lateral line system of sharks and rays, new comparative data on the poorly studied batoids, and an interpretation of this complex but poorly studied system in relation to elasmobranch natural behavior. In his paper on the electrosense, Tim Tricas argues for the 'neuroecology' approach to the study of elasmobranch sensory systems, and presents new information on how the skate and great white shark may use their respective electrosensory systems in the real world. John New's paper on the enigmatic cerebellum, a brain structure that is greatly enlarged in some sharks and batoids, identifies the intricacies of this sensorimotor interface and how it may function in the expression of behavior. John Montgomery & Mike Walker review the subject of shark posture, orientation and navigation where they emphasize the importance of integrating sensory system functions with natural behaviors. The work by Joe Sisneros & Don Nelson posthumously provides the history of the U.S. Navy's effort to develop an effective shark repellent and the subsequent discovery of pardaxin in the Moses sole. They then present new data on the irritant response of sharks to more economical

Figure 1. Participants in the Donald R. Nelson symposium sponsored by the American Elasmobranch Society at the University of Guelph on 18 July 1998. Standing in the upper picture are (from left to right) Scott Eckert, John New, John Montgomery, Sonny Gruber, Karen Maruska, Arthur Myrberg, Jr., Peter Klimley, Nancy Kohler, David Bodznick, Timothy Tricas, Fred Voegeli, Chris Lowe, Bob Heuter and in front, Joe Sisneros, Wesley Strong, Wes Pratt and Jeff Carrier. Photograph below shows Arthur Myrberg delivering his lively keynote presentation on shark hearing and acoustic orientation.

Figure 2. Ginger Nelson (upper photograph) thanks the American Elasmobranch Society for the symposium to honor the many accomplishments of her husband. Ginger was instrumental in establishing the Donald R. Nelson fund with AES, which promotes an annual student research award to encourage the study of the sensory biology and behavior of elasmobranch fishes. The Don Nelson symposium in Guelph was attended by many of his former students/advisees (lower photograph). From left to right are Joe Sisneros, John Morrissey, Ginger Nelson, Chris Lowe, William Fouts, Ron Abir and Tim Tricas.

Figure 3. Festivities at the American Elasmobranch Society banquet following the Nelson symposium. Social hour in which José Castro (left) describes the reproductive biology of sharks to unidentified graduate student (right) under the watchful eye of Art Myrberg (center). Co-organizers Tim Tricas (left) and Sonny Gruber (right) discuss the symposium and plan the publication of this volume. Mariko Gruber (seated at center) moderates.

substitute surfactants. Phil Motta & Cheryl Wilga give a much-needed update on the feeding mechanism of sharks and how modern techniques have advanced the study of the function and evolution of shark feeding behavior. Wes Pratt & Jeff Carrier's review of the reproductive behavior of sharks and rays also includes recent work on their gold mine of mating nurse sharks in the Florida Keys. The final part of the volume deals with applications of old and new technologies to monitor the movements and behavior of sharks and rays in the field. Nancy Kohler & Pat Turner provide a detailed history of how standard tagging techniques are used to study migration patterns of sharks. They then synthesize the major National Marine Fisheries effort in the United States and others conducted around the world. This is followed by a series of telemetry papers, the first of which is authored by Fred Sundström, Sonny Gruber and colleagues. They summarize past research conducted on the behavior and movements of sharks that use standard telemetry tracking techniques, and then provide new information on the movement patterns, activity, and homing behavior of juvenile lemon sharks around Bimini. The applications of ultrasonic telemetry used to study swimming physiology and endothermy in sharks is reviewed by Chris Lowe & Ken Goldman, and several timely studies in this area are proposed. This is followed by a review of the current state-of-the-art in remote monitoring equipment compiled by engineer Fred Voegeli and associates. The issue closes with two excellent field studies that demonstrate applications of new and exciting tracking techniques. The paper on archival tagging written by Grant West & John Stevens of CSIRO, Australia provides a remarkable example of how this new technology can provide unprecedented information on the vertical movements of large sharks. This impressive work is matched by that of Scott Eckert & Brent Stewart, who report their transoceanic satellite trackings of whale sharks. Clearly, new techniques applied to the field such as these will continue to reveal information that was only a dream for us a few years ago.

The guest editors would like to thank all the participants of the symposium in Guelph, and those who contributed papers in this printed volume. The symposium was made possible with the full support of the American Elasmobranch Society and its members. Special thanks are given to then AES president Phil Motta for his administrative support, Jeff Carrier for his never-ending technical and photographic assistance during the presentations, and Chris Lowe for his helpful assistance with details on this publication and the Nelson Fund. We also thank the more than 100 AES and ASIH members that formed the interactive audience at the Nelson symposium meeting. We extend our deepest appreciation to Eugene Balon for offering us pages in this journal, and his ability to bend with the wind as we tried to bring this event to press. Finally, we thank Ginger Nelson for her endless support of Don, and our efforts to crystallize his impact on the study of shark behavior.

Melbourne and Miami, Florida, 1 May 2000

Timothy C. Tricas
Samuel H. Gruber

A note on shark drawings on the blank pages of this volume

Paul Vecsei & Sarah Heaton

We searched through shark publications and found, with the rare exception, that most lateral views of these creatures are rather poorly executed. While there exist many wonderful underwater photographs of sharks in their natural environment, scientific photographs stressing the morphological characters of these animals are much more difficult to find. At the beginning of this project, José Castro sent several slides of his lateral views of sharks. His images were taken at a great enough distance from the subject as to minimize the parallax distortions that plague this sort of picture.

Our technique was as follows. The slide was projected onto a Lintech 3000 illustration board so that the image was between 40 and 70 cm in length. With a pencil, we traced the contour and all details. Letter codes were used on the emerging sketch to indicate the tone or degree of density of the stippling that is to follow (see Figure 1a).

Stippling was used to indicate shape, tone and pattern. A base coat was applied using a 0.5 mm tip of the Koh-i-Noor rapidograph. To give form and to smoothen out the rough appearance of the initial coat, a second and third coats were applied with a 0.25 mm tip. All aspects of detail were taken into account and the end result (up to 30 hours later) was a reasonable replica of a particular specimen (Figure 1b). José Castro evaluated the results during our visit to Mote Marine Laboratory. He reveiwed the illustrations, giving useful commentaries that sometimes lead to changes being made on a particular specimen.

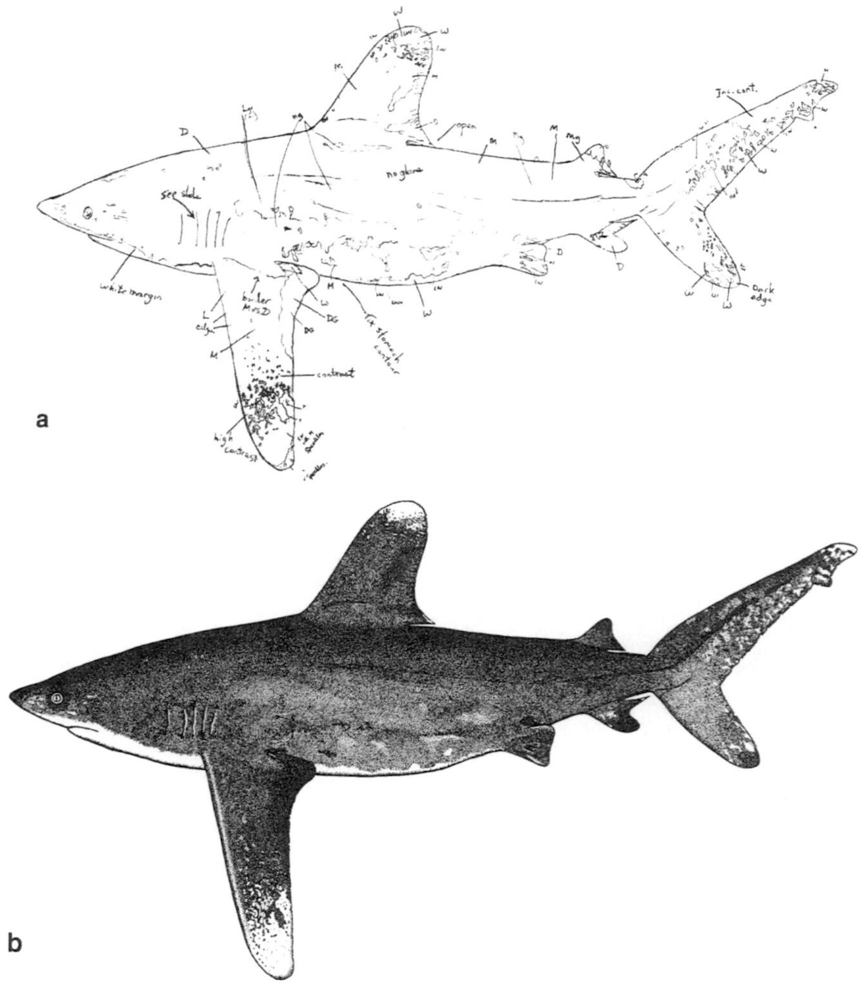

Figure 1. Female of the oceanic whitetip, *Carcharhinus longimanus*, 185 cm long. Original by Sarah Heaton.

Part 1. Biography and eulogy

Female of the longfin mako, *Isurus paucus*, 152 cm long from Fort Pierce, Florida. Original by Paul Vecsei (PV).

Male of the bigeye thresher, *Alopias superciliosus*, 157 cm long. Original by Sarah Heaton (SH).

Environmental Biology of Fishes **60:** 15–18, 2001.

Shark meets man: the research and academic life of Donald Richard Nelson (1937–1997)

Timothy C. Tricas
Department of Biological Sciences, Florida Institute of Technology, Melbourne, FL 32901, U.S.A.
Present address: Department of Zoology, University of Hawaii at Manoa, Honolulu, HI 96822, U.S.A.
(e-mail: tricas@hawaii.edu)

Key words: acoustic, behavior, ethology, ray, shark, sensory biology, telemetry

Synopsis

This symposium is dedicated to the memory of Donald Richard Nelson, a pioneer in the study of shark behavior and sensory biology, a dedicated teacher, and friend to many. Don was an ultimate marine biologist who studied these enigmatic animals submerged in their own environment. He had many talents that ranged from expert diver to ethologist to cinematographer to engineer. Persistence and a critical approach to scientific challenges were his landmarks, and these qualities touched all of his students and many of his colleagues. His pleasant nature, never-ending friendship and dependable support are missed by many.

Don was born in Plainfield, New Jersey on 21 September 1937. He received his B.A. in Biology from Rutgers University in 1958, and entered graduate school at the University of Miami under the guidance of fish behaviorist Warren Wisby. In 1962 Don received a M.S. in Fisheries for his study on the behavior of penaeid shrimp, and then immediately commenced his doctoral work. Within one year Don's first scientific paper appeared in *Science* co-authored with his graduate student colleague, Sonny Gruber. This paper was the first to demonstrate attraction of sharks to low frequency sounds, and set the stage for Don's career in shark behavior and sensory biology. Don received a Ph.D. in Marine Biology in 1965 for his novel work on hearing and acoustic orientation in sharks.

Immediately following his graduate study, Don took his first and only academic appointment in the Biology Department at California State University at Long Beach. Here he launched a highly productive research career that included 25 field expeditions to sites in the Gulf of California, Marshall Islands, Polynesia, Australia and Bahamas. Over the next 30 years his laboratory produced nearly 50 publications, a remarkable statistic given the fact that many resulted from time-consuming field expeditions. Don's great knowledge of comparative animal behavior made him a highly respected teacher and scholar. For his academic duties Nelson taught a rigorous and challenging course in animal behavior in which the details of observation and synthesis were emphasized. In his last decade, he brought the joys of marine biology to literally thousands of undergraduates in his highly regarded Marine Natural History course. In it he developed an impressive interactive, multimedia CD-ROM with a large marine species database. Don was a member of several professional societies, and a founder of the American Elasmobranch Society in which he served as president in 1996.

In 1969 the Office of Naval Research awarded Don the first of many grants to perform experimental studies of hearing and acoustic orientation in sharks. His technical expertise with electronics equipment, which he acquired empirically for laboratory studies while in graduate school, led to excellent laboratory studies on hearing thresholds and diel behavior of small sharks. At the same time he and his students took traditional 'laboratory technology' into the field, and spent countless hours in the waters of the Pacific studying the behavior and ethology of sharks in their natural habitats, often with the support of the National Geographic Society. This led to many important papers on shark sensory biology in the 1970's, perhaps most notable the study with graduate student Richard Johnson on acoustic attraction of Pacific reef sharks (Nelson & Johnson 1972).

As a young naturalist in the woods of New Jersey, Don studied bird behavior and was greatly influenced by the ethologists Konrad Lorenz and Niko Tinbergen. His interest in shark ethology is particularly evident in the paper on the agonistic display of the gray reef shark at Enewetak Atoll (Johnson & Nelson 1973). This milestone study was the first to show that shark attack behavior could be graded and involve a complex, predictable series of action patterns. It also demonstrated, in great detail, that shark attacks were not necessarily feeding motivated. While to many behaviorists this clearly appeared to be evidence for territorial behavior in sharks, Don's critical scientific mind was not convinced, and the function of the agonistic display in nature still remains to be demonstrated.

The shark research program in Nelson's laboratory at Long Beach State was known not only to the scientific world but also to the international public. Between 1968 and 1994, Don and his students were involved in the making of no less than 25 documentaries on television or educational films. His high profile in shark research was greatly complemented by his own cinematographer skills. Don spent thousands of hours filming his study subjects, and compiled a large library of 16-mm film that includes classic footage of the acoustic attraction and agonistic display experiments at Enewetak Atoll. Much of this footage still appears on television today.

Although Don's first love was to study sharks under the surface and 'tête-à-tête', he clearly recognized the limitations of directly working in the underwater world. Divers could not observe sharks for long periods of time, and almost nothing was known about shark movements under darkness. Thus, in the early 1970's he and his students embarked on a line of behavioral research that used a new technology known as ultrasonic telemetry. This occurred in the prehistoric era, before integrated circuits, lithium batteries and well-stocked telemetry vendors. There were even fewer biologists who could tackle the custom electronics designs needed for their research projects. Nevertheless, Don and his students (the present author included) spent countless weekends and holidays designing and soldering together ultrasonic transmitters from discrete transistors, capacitors and resistors. Working closely for the next two decades with his student and colleague Jim McKibben, Nelson's laboratory pioneered the use of ultrasonic techniques and produced many important theses and papers on the movement patterns and home ranges of sharks. Don's technical approach to remote sensing is apparent even in his last senior author paper in which the rare megamouth shark was tracked off of the southern California coastline (Nelson et al. 1997).

Don's applied talents are evident even during the last years of his career. The discovery that the irritant effect of the Moses sole's pardaxin secretion was due to its surfactant properties led Nelson's laboratory to study the potency of cheap alternative surfactants and ways to apply them in the field. These experiments and potential for application to diver safety are found in his last published work, which is a part of this symposium volume (Sisneros & Nelson 2001).

Don supervised 21 Masters and 1 Ph.D. student (Peter Klimley at Scripps Institute of Oceanography) while at Cal State Long Beach, with many still active in elasmobranch research and participants of this symposium. All can tell you that his favorite tools were a pencil (or pen), paper (or napkin) and a cup of coffee, and with these he produced remarkable achievements in engineering design. Perhaps the most important gift he gave to his students was the confidence to view perspectives from both sides of the fence, and that there was almost always a workable solution to a logistic problem. Each of us still carry that confidence with us today, and recognize that gift more than ever since Don's passing on 7 March 1997 after a lengthy bout with melanoma skin cancer.

Don is survived by his wonderful wife, Ginger, who fully understood and supported his fascination with his animals. She was instrumental in coordinating efforts for the Nelson symposium at Guelph, and establishing the Nelson Research Fund, which is now administered by the American Elasmobranch Society. This perpetual fund now supports student research projects in shark behavior and sensory biology, with the goal of helping young scientists to view, as did Don, the secrets of these shrouded animals.

List of publications by Donald R. Nelson

1. Nelson, D.R. 1962. An evaluation of the Columbia obstruction method for studying the behavior of the pink shrimp, *Penaeus duorarum*. Masters Thesis, University of Miami, Miami. 77 pp.
2. Nelson, D.R. & S.H. Gruber. 1963. Sharks: attraction by low-frequency sounds. Science 142: 975–977.
3. Wisby, W.J., J.D. Richard, D.R. Nelson & S.H. Gruber. 1964. Sound perception in elasmobranchs. pp. 255–268. *In:* W.N. Tavolga (ed.) Marine Bio-acoustics. Pergamon Press, New York.
4. Nelson, D.R. 1965. Hearing and acoustic orientation in the lemon shark, *Negaprion brevirostris* (Poey), and other large sharks. Ph.D. Dissertation, University of Miami, Miami. 158 pp.
5. Nelson, D.R. 1967. Cardiac responses to sounds in the lemon shark, *Negaprion brevirostris*. pp. 533–544. *In:* P.W. Gilbert, R.F. Mathewson & D.P. Rall (ed.) Sharks, Skates, and Rays. John Hopkins Press, Baltimore.
6. Nelson, D.R. 1967. Hearing thresholds, frequency discrimination, and acoustic orientation in the lemon shark, *Negaprion brevirostris* (Poey). Bull. Mar. Sci. 17: 741–768.
7. Nelson, D.R. 1969. The silent savages. Oceans 1: 8–22.
8. Nelson, D.R. 1969. Meeting the shark underwater. Underwater Naturalist 6: 13–26.
9. Nelson, D.R., R.H. Johnson & L.G. Waldrop. 1969. Responses in Bahamian sharks and groupers to low-frequency, pulsed sounds. Bull. S. Calif. Acad. Sci. 68: 131–137.
10. Nelson, D.R. & R.H. Johnson. 1970. Diel activity rhythms in the nocturnal, bottom-dwelling shark, *Heterodontus francisci* and *Cephaloscyllium ventriosum*. Copeia 1970: 732–739.
11. Nelson, D.R. & R.H. Johnson. 1972. Acoustic attraction of Pacific reef sharks: effect of pulse intermittency and variability. Comp. Biochem. Physiol. 42: 85–96.
12. Standora, E.A., T.C. Sciarrotta, D.W. Ferrel, H.C. Carter & D.R. Nelson. 1972. Development of a multichannel, ultrasonic telemetry system for the study of shark behavior at sea. Office of Naval Research, Technical Report 5. 21 pp.
13. Johnson, R.H. & D.R. Nelson. 1973. Agonistic display in the gray reef shark, *Carcharhinus menisorrah*, and its relationship to attacks on man. Copeia 1973: 76–84.
14. Ferrel, D.W., D.R. Nelson, T.C. Sciarrotta, E.A. Standora & H.C. Carter. 1974. A multichannel ultrasonic biotelemetry system for monitoring marine animal behavior at sea. Instrument Soc. Amer. Trans. 13: 120–131.
15. Nelson, D.R. 1974. Ultrasonic telemetry of shark behavior. Naval Research Reviews 27:1–21.
16. Finstad, W.O. & D.R. Nelson. 1975. Circadian activity rhythm in the horn shark, *Heterodontus francisci*: effect of light intensity. Bull. S. Calif. Acad. Sci. 74: 20–26.
17. Kelly, J.C. & D.R. Nelson. 1975. Hearing thresholds of the horn shark, *Heterodontus francisci*. J. Acoust. Soc. Amer. 58: 905–908.
18. Nelson, D.R. & R.H. Johnson. 1976. Some recent observations on acoustic attraction of Pacific reef sharks. pp. 229–239. *In:* A. Schuijf & A. Hawkins (ed.) Sound Reception in Fish. Elsevier, Amsterdam.
19. Nelson, D.R. 1977. On the field study of shark behavior. Amer. Zool. 17: 501–507.

20. Sciarrotta, T.C. & D.R. Nelson. 1977. Diel behavior of the blue shark, *Prionace glauca*, near Santa Catalina Island, California. U.S. Fish. Bull. 75: 519–528.
21. Standora, E.A. & D.R. Nelson. 1977. A telemetric study of the behavior of free-swimming Pacific angel sharks, *Squatina californica*. Bull. S. Cal. Acad. Sci. 76: 193–201.
22. Johnson, R.H. & D.R. Nelson. 1978. Copulation and possible olfaction-mediated pair formation in two species of carcharhinid sharks. Copeia 1978: 539–542.
23. Nelson, D.R. 1978. Telemetering techniques for the study of free-ranging sharks. pp. 419–482. *In:* E. Hodgson & R. Mathewson (ed.) Sensory Biology of Sharks, Skates, and Rays, Office of Naval Research, Washington, D.C.
24. Nelson, D.R. & R.H. Johnson. 1980. Behavior of the reef sharks of Rangiroa, French Polynesia. Nat. Geo. Soc. Res. Rep. 12: 479–499.
25. Nelson, D.R. 1980. The body language of sharks. NAUI News. June 1980: 6–7.
26. Klimley, A.P. & D.R. Nelson. 1981. Schooling of the scalloped hammerhead shark, *Sphyrna lewini*, in the Gulf of California. U.S. Fish. Bull. 79: 356–360.
27. Nelson, D.R. & J.N. McKibben. 1981. Timed-release, recoverable, ultrasonic/radio transmitters for tracking pelagic sharks. pp. 90–104. *In:* F.M. Long (ed.) Proceedings of the Third International Conference on Wildlife Biotelemetry, ICWB, University of Wyoming, Laramie.
28. Nelson, D.R. 1981. Aggression in sharks: is the gray reef shark different? Oceanus 24: 45–55.
29. Nelson, D.R. 1983. Shark attack and repellency research: an overview. pp. 11–74. *In:* B. Zahuranec (ed.) Shark Repellents from the Sea: New Perspectives, AAAS Selected Symposium Series, Westview Press, Boulder, CO.
30. Gruber, S.H., E. Zlotkin & D.R. Nelson. 1984. Shark repellents: behavioral bioassays in laboratory and field. pp. 26–42. *In:* L. Bolis, J. Zadunaisky & R. Gilles (ed.) Toxins, Drugs, and Pollutants in Marine Animals, Springer-Verlag, Berlin.
31. Klimley, A.P. & D.R. Nelson. 1984. Diel movement patterns of the scalloped hammerhead shark, *Sphyrna lewini*, in relation to El Bajo Espritiu Santo: a refuging central-position social system. Behav. Ecol. Sociobiol. 15: 45–54.
32. Klimley, A.P. & D.R. Nelson. 1985. Functional analysis of schooling in the scalloped hammerhead, *Sphyrna lewini*. Nat. Geo. Soc. Res. Rep. 21: 227–229.
33. McKibben, J.N., D.R. Nelson & S.H. Gruber. 1985. A microprocessor data-logging monitor for aquatic biotelemetry. pp. 90–96. *In:* R.W. Weeks & F.M. Long (ed.) Proceedings of the Fifth International Conference on Wildlife Biotelemetry, ICWB, University of Wyoming, Laramie.
34. Gruber, S.H., D.R. Nelson & J.N. McKibben. 1986. Tracking sharks in Bimini Lagoon: one phase of a bioenergetic study. Newsletter of Int. Assoc. Fish Ethologists 9: 5–8.
35. Nelson, D.R., R.R. Johnson, J.N. McKibben & G.G. Pittenger. 1986. Agonistic attacks on divers and submersibles by gray reef sharks, *Carcharhinus amblyrhynchos*: antipredatory or competitive? Bull. Mar. Sci. 38: 68–88.
36. Nelson, D.R. 1987. On the use of ultrasonic tracking in orientation studies. pp. 118–129. *In:* W.F. Herrnkind & A.B. Thistle (ed.) Signposts in the Sea, Florida State Univ., Tallahassee.
37. Sydlik, M.A., R.B. Barlow, A.T. Stull, D.R. Nelson & C. Kammire. 1987. Migratory behavior of individual horseshoe crabs. Biol. Bull. 173: 438.
38. Gruber, S.H., D.R. Nelson & J.F. Morrissey. 1988. Patterns of activity and space utilization of lemon sharks, *Negaprion brevirostris*, in a shallow Bahamian lagoon. Bull. Mar. Sci. 43: 61–76.
39. Klimley, A.P., S.B. Butler, D.R. Nelson & A.T. Stull. 1988. Diurnal movements of scalloped hammerhead sharks (*Sphyrna lewini*) to and from a seamount in the Gulf of California. J. Fish Biol. 33: 751–761.
40. Nelson, D.R. 1990. Telemetry studies of sharks: a review, with applications in resource management. pp. 245–262. *In:* H.L. Pratt, T. Taniuchi & S.H. Gruber (ed.) Elasmobranchs as Living Resources, NOAA Tech. Report 90.
41. Nelson, D.R. 1991. Shark repellents: how effective, how needed? pp. 106–108. *In:* S.H. Gruber (ed.) Discovering Sharks, Special Pub. 14, Amer. Littoral Soc., Highlands.
42. Myrberg, A.A. & D.R. Nelson. 1991. Shark behavior: what have we learned? pp. 92–100. *In:* S.H. Gruber (ed.) Discovering Sharks, Special Pub. 14, Amer. Littoral Soc., Highlands.
43. Strong, W.R., Jr., R.C. Murphy, B.D. Bruce & D.R. Nelson. 1992. Movements and associated observations of bait-attracted white sharks, *Carcharodon carcharias*: a preliminary report. Aust. J. Mar. Freshwater Res. 43: 13–20.
44. Lowe, C.G., R.N. Bray & D.R. Nelson. 1994. Feeding and associated electrical behavior of the Pacific electric ray, *Torpedo californica*, in the field. Mar. Biol. 120: 161–169.
45. Nelson, D.R. & W.R. Strong, Jr. 1993. Chemical repellent tests on white sharks, *Carcharodon carcharias*, with comments on repellent delivery methods. pp. 471–475. *In:* A.P. Klimley & D.G. Ainley (ed.) Biology of the White Shark, Academic Press, San Diego.
46. Nelson, D.R., J.N. McKibben, W.R. Strong, Jr., C.G. Lowe, J.A. Sisneros, D.M. Schroeder & R.J. Lavenberg. 1997. An acoustic tracking of a megamouth shark, *Megachasma pelagios*: a crepuscular vertical migrator. Env. Biol. Fish. 49: 389–399.
47. Fouts, W.R & D.R. Nelson. 1999. Prey capture by the Pacific angel shark, *Squatina californica*: visually mediated strikes and ambush-site characteristics. Copeia 1999: 304–312.
48. Sisneros, J.S. & D.R. Nelson. 2001. Surfactants as chemical shark repellents: past, present, and future. Env. Biol. Fish. 60: 117–129 (this volume).

Environmental Biology of Fishes **60**: 19–27, 2001.

Donald Richard Nelson: the most unforgettable character I ever met

Samuel H. Gruber
Bimini Biological Field Station, Rosenstiel School of Marine and Atmospheric Sciences, University of Miami, Miami, FL 33149, U.S.A. (e-mail: sgruber@rsmas.miami.edu)

Key words: ornithology, ichthyology, spear fishing, elasmobranch, shark, ethology, bioacoustics, agonistic display, underwater telemetry, shark repellent, boat building

Synopsis

This volume is dedicated to the work, life and memory of Donald Richard Nelson. The present article is based primarily on the author's 38-year friendship and personal experience with the man. The text chronicles his life and times from budding naturalist to champion spearfisher through graduate school and on to becoming scientist, teacher and mentor to dozens of graduate students. The author perceives Nelson as a unique character, creative, introspective, kind, wise and gentle. His influence in the field of contemporary elasmobranch biology, especially ethology and sensory physiology was enormous.

I first met Don Nelson in the fall of 1961, 38 years ago. Through the decades, Don has influenced my career in manifold ways. He was a unique character – perhaps the most unique character I have had the pleasure to know as a close friend. In writing this small tribute to the man, I want to do justice to his memory ... somehow distill the essence of Don into these few paltry lines so that those who did not know him well can experience a feeling of what he was like. But, can I achieve this goal? With this in mind, herewith follows my photographic essay ... a distillation of the life and times of Donald Richard Nelson seen through the lens of one of his greatest admirers (Figure 1).

Don Nelson was many things to many people. He was friend, colleague, scientist, teacher, logician, innovator, builder, wildlife photographer, athlete, husband, father and grandfather. Don greatly enriched the lives of the people lucky enough to know him – including and especially me! He discovered things about sharks and ways to study them new to science, he taught hundreds of students the wonders of animal behavior – he *inspired* dozens of graduate students and indoctrinated them into the ways of the scientific method. He appeared in the media before literally hundreds of millions of people to explain his passion for and understanding of sharks. In our discipline, Donald Richard Nelson will be sorely missed.

Here are some of the milestones of Don's life and career: Donald Nelson was born on 21 September 1937 in Plainfield, New Jersey, the only child of Anna Marie and Leonard Nelson. Don grew up during World War 2 in a caring, nuclear family, near the eastern shore of New Jersey (Figure 2). One of his earliest memories was of a trip out west during the forties. That experience stayed with him for a lifetime and even a half-century later, Don's wife Ginger often referred to him as 'Her Cowboy'. Don was influenced by the sea at an early age. He was interested in boats and fishing and was also very proud of his accomplishments as a boy scout. He earned a plethora of merit badges that served him well through his life.

Don went to Clara Barton elementary school in Edison, New Jersey, then on to high school in Perth Amboy. He became interested simultaneously in photography and ornithology. When the other boys were playing baseball Don was out in the woods or at the shore birding, taking pictures or adding to his amazing collection of bird nests. He related early on that his goal in life was to photograph every living species of bird. That might have been the end of the story for Marine Biology and us shark-types, except that the pull of the sea was strong.

Owing to his athletic ability, he soon became adept at spear fishing – the first of many interests that would

Figure 1. Don Nelson in a classic pose before the self-designed, personally constructed, wet submersible he named 'SOS' (Shark Observation Submersible). The photograph was made during the 1978 expedition to Enewetak Atoll in the central Pacific (photograph by James McKibben).

connect my life and career to his. Don told me of the incredible situation in the cold, turbid, temperate waters off the Jersey coast. He said that the visibility was often less than the length of a spear gun. Growing up in South Florida, it seemed inconceivable to me that one could ever get a spear into a fish under those conditions. But he did – and it was these early experiences that would later mature to the level of a champion athlete and spearfisher.

Don's interest in ornithology and ichthyology led him to take a degree in the biological sciences at Rutgers University in 1958. In the fall of 1958, he relocated to South Florida to begin his graduate work at the Marine Laboratory of the University of Miami – later to become the Rosenstiel School of Marine and Atmospheric Sciences. But he also came to South Florida to test his spear fishing skill in the clear waters of the tropical reefs. At that time, there were still plenty of fish on the reefs and Miami was perhaps the spear fishing capital of the world.

By 1959, Don was ready to make his mark. One day on the reef, he encountered a serious underwater challenge: here was this 4 m tiger shark looking at Don as if he were a somewhat impaired, large, marine organism – perhaps a strange sea turtle. And there was Don looking at the big tiger as a trophy. The poor shark never had a chance! Social and conservation expectations were very different in the 1950s. We really had no idea, given its reputation, whether a tiger shark would attack on the spot. Nor did anyone really consider that

Figure 2. a – Don with mother Nelson and grandmother Nelson at their home in Plainfield, New Jersey. Don was about 2 1/2 years old when the picture was taken in 1940. b – Cowboy Don's big trip out west. He was about eight years old at the time in 1945. c – Fisherman Don at about nine years with a good-sized striper, *Morone saxatilis*.

killing such a magnificent creature as a trophy might present a moral dilemma. Yet, that encounter on the reefs in 1959 was both to change Don's life and eventually much of what we know about shark behavior (Figure 3a). It would be three more years before I met Don, an encounter that would change my life.

At the time in 1961, I was searching for a professor who would take on a shark fanatic like me as a graduate student. I found Warren Wisby, a young, very sharp Ph.D. out of Professor Arthur Hassler's laboratory at University Wisconsin. W. Wisby had done some extremely creative behavioral experiments with salmon, putting the finishing touches on the theory of olfactory imprinting, the mechanism that brings salmon back to their home stream. At the time, Dr. Wisby, a new professor at the University of Miami's Marine Laboratory was Don's MS thesis chairman. Don's thesis project involved trying to determine the physiological strength of an olfactory attractant by measuring the intensity of an aversive electric field just strong enough to keep the commercially important pink shrimp from reaching a food source. For someone used to dominating 4 m tiger sharks, working with a 15 gram shrimp hardly presented the kind of challenge that Don was looking for. However, it did pay the bills.

By 1961 when I arrived at the newly named Institute of Marine Sciences, there was a concerted effort by the US Office of Naval Research to begin to understand the behavior of sharks. W. Wisby won his first-ever grant from the Navy to look into shark's senses – primarily hearing. Thus it was that Don and I were put on to the same project. He sort of took me under his wing as I was just starting out and he had been at the Institute some time and knew the ropes. In addition, we became fast friends because of our all-consuming interest in sharks and our athletic orientation to diving, especially spear fishing.

I was a fair spearfisher but Don was a master! He belonged to a competitive team called the 'Glug Glugs' and they were indeed world champions in 1961 (Figure 3b,c). Through most of his graduate career, Don kept up with spear fishing. Eventually he bought into a fast skiff, a 7 m Prowler, known to be the best in those early days. Every weekend Don and two others would blast

a
b
c
d
12
STATES NATIONAL
CHAMPIONSHIP

across the Straits of Florida to the reefs off Bimini and spear a hundred kilo of grouper, working for hours on end, free diving in 20 m of water. In those days, gasoline was 7 cents a liter and filleted grouper went for 50 cents a kilo. His spear fishing efforts, paid off the boat, kept Don in beer and covered a good bit of his graduate education. But it was too good to last!! In 1964, frustrated at his apparent lack of progress Wisby gave Don an ultimatum: 'Get rid of your interest in the Prowler and finish up your dissertation or find a new project.' Characteristically, Don sold out and indeed finished that very year (Figure 3d).

However, I'm getting ahead of myself. In the summer of 1961, I was plugging away at the anatomy of the shark's inner ear. One day, Don and I were musing over the possible use of hearing to an animal that produces no sounds. We both had experienced the rapid appearance of sharks while spear fishing. Don was absolutely convinced that sounds, produced by speared fishes attracted shark from long distances. So we set out to test this hypothesis. With the blessing of W. Wisby, we started making and analyzing the sound spectrum of signals produced by various fishes that we had speared. By this time, Don's laboratory experiments suggested that lemon sharks could only hear low frequency sounds well below 1000 Hz. To our delight, the sounds of speared fish were right in the best portion of the shark's hearing range. Thus, we made some synthetic, struggling fish sounds and recorded them on audio tape. Using a sophisticated Navy underwater speaker called a J-9 and a tape player we set up field tests with quiet, control periods and test periods with the fish sounds on. Here's the way Don put it in an article written in 1970: 'Sam and I were apprehensive on the first day of playback. We knew there was an element of risk for it was necessary to observe from in the water and we had no shark cage. We chose a spot on the leeward slope of the reef and lowered the sound projector to 12 m. Sam entered the water for the first 15 min control period. He saw no sharks during this period. We then began playing sounds. After 12 minutes Sam saw a large shark approaching – it was a tiger – about $2\frac{1}{2}$ meters long. It circled the speaker, left and returned again. Sam climbed into the boat and we congratulated ourselves on our first sonic attraction.'

I remember it a little differently. On seeing the huge tiger shark, I literally flew out of the water and landed on the deck upright, scared and shaking but ecstatic. In the initial playback series, we saw a total of 22 sharks, none during the control periods and all during the 18 sound playback periods (Figure 4). This led to our first-ever scientific publication, which was accepted in 1963 in the journal Science. Two years later, Don Nelson or Reef, as his friends liked to call him, left Miami heading out to California where he accepted a tenure-track teaching position in the Department of Biology at California State University, Long Beach. In the ensuing 32 years, Don broke new ground in the study of shark behavior, as you have learned from Don's former student and first editor of this volume, Tim Tricas.

Working under a Navy grant, which ran for 17 years, and then later with me on an NSF grant, Don was mentor to 23 graduate students, and published over 50 scientific papers. Don and I were co-authors on six of these. His main research interests focused on the behavior of sharks. Specifically he worked on shark hearing, shark repellents, movement patterns and social behavior. Between 1968 and 1989, Don organized and/or participated in 24 research expeditions, ranging from the Bahamas to the Sea of Cortez to Polynesia. Many of these were on research vessels of his own design and construction! I had the pleasure of working with Don on several research cruises.

Looking back at some of the important milestones, Don's research career can be summed up as follows: (1) He performed the first laboratory analysis of shark hearing, conclusively proving that wild sharks have directional hearing and are naturally attracted to low frequency, pulsed sounds. (2) He described for the first time in detail, field behavior and diel activity patterns of large, free ranging sharks such as the gray reef, reef whitetip, blue, angel, leopard and megamouth sharks. This included a description of the unique and dangerous agonistic display of gray reef shark as well as an explanation of the schooling behavior of scalloped

←

Figure 3. a – Don's first interest in sharks was for sport. He single-handedly bagged this 4 m tiger *Galeocerdo cuvier* with a spear. At the time, he would have been considered an amazing, fearless hero for dispatching this oceanic menace. How times have changed! b – World champion spearfisher, Nelson receives his award at the 1961 championships in Miami, Florida. c – A typical weekend catch; Don and colleagues would spear 3 to 4 dozen such groupers, free diving for up to 10 h a day in 25 m of water. In the 1960s, such grouper would fetch \$ 0.50 per 1 kg filleted. d – Don in his academic regalia being hooded by his 'father' professor, Warren J. Wisby immediately prior to receiving his doctorate in 1965.

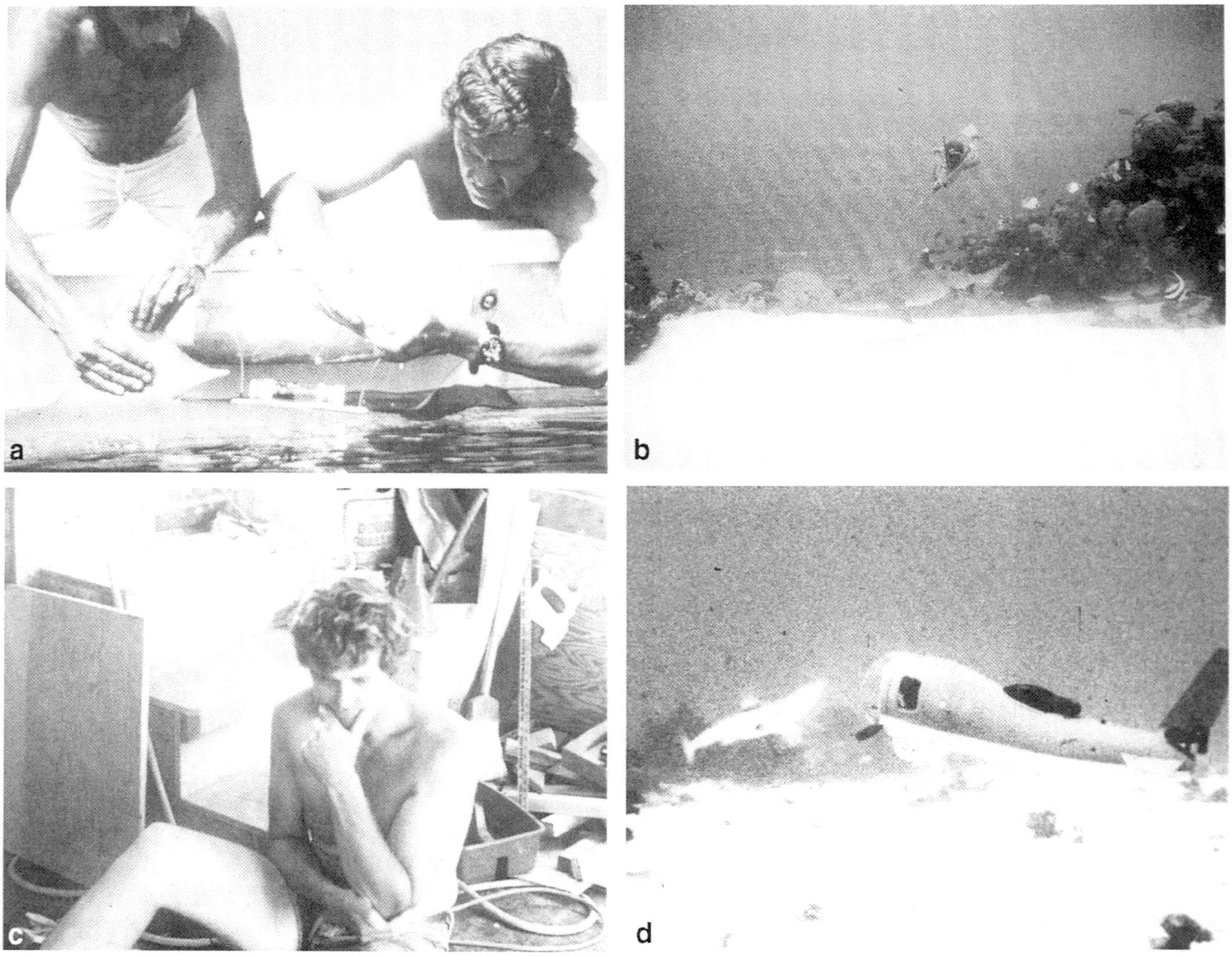

Figure 4. a – Don and author attaching an acoustic transmitter to the back of a lemon shark, *Negaprion brevirostris*. During this 1979 cruise of the R/V Calanus, Don taught the author the fundamentals of shark tracking. b – Don's 'Kamakazi' method of marking or intimidating a shark. In this case, taken during the 1974 expedition to Rangiroa, a whitetip reef shark, *Triaenodon obesus*, receives a spaghetti tag. In similar encounters with gray reef sharks, *Carcharhinus ambyrhynchos*, Don would press in on the shark to evoke an agonistic display. c – Don in deep thought during construction of the saloon of his research trawler R/V Discovery. Don often adopted this 'pondering' pose when faced with a series of choices requiring a decision. d - SOS in action during the 1979 expedition to Rangiroa. Photograph taken from a 16 mm frame shows a grey reef shark, *Carcharhinus ambyrhynchos*, in full agonistic display just prior to violently attacking the submersible.

hammerheads. (3) He developed practical underwater telemetry devices and pioneered the use of submersibles to study natural behavior of sharks in the wild. This included a unique study of the only known living specimen of the megamouth shark. In his final paper, Nelson described the amazing, several hundred meters, diurnal, vertical migration of a 5 m specimen!!! Finally, Don led the search for an effective shark repellent, one that would repel but not harm sharks (Figure 5a).

In addition to his long and fruitful research career, Don was an academician in the true sense of the word, competent in many areas of teaching. He taught seven courses ranging from general zoology to animal behavior to marine biology and served on no less than 15 University committees. Don belonged to 11 learned societies and was frequently called on to evaluate National Science Foundation proposals as well as to review and judge scientific articles for no less than 8 scientific journals.

These cold facts do not begin to explain the depth, character, kindness and creativity of the man. A quick non-scientific survey of Don's students and colleagues confirmed my impressions: Don marched to the beat of a different drummer. First, he was calm and unflappable. He never got flustered, angry or spiteful. He truly approached life with a child-like equanimity and fascination with discovery. Two apocryphal shark stories come to mind: One was a chance encounter in 1960 with a great white shark on a reef off Grassy Key, Florida. Don had been spear fishing alone in about 13 m. Here are Don's own words from a 1970 article: 'To me the shark seemed enormous, 12 feet long and 1000 lb. It swam stiffly and had a lunate tail. I realized that it was probably a white shark. My mind abruptly switched from identification to survival. The shark came so close, I had to withdraw my out-stretched spear for fear of touching or inciting the shark. For a very long minute, the shark and I swam parallel, eye-to-eye. On the fourth pass, I was able to gain my wits and look for identifying features like fin shape, body markings and tooth shape. Unlike all other sharks I have encountered, this one seemed interested in me. I had the distinct impression that he was visually examining me as a possible meal.' This is a perfect example of what I mean when I say that Don was unflappable – calmly looking at the tooth shape of the monster preparing to consume him!

Don had been working for several years at Eniwetak Atoll, in the Pacific, trying to understand an unusual and frightening behavior pattern performed by gray reef sharks whenever divers threatened them. In the early days, Don used what became known as the 'Kamikaze Technique'. He would simply free dive on a reef shark minding its own business and chase it along the reef face until it became rather trapped. At that moment, the shark would either flee or begin the sinister underwater ballet Don called an agonistic display. One day in 1976, working with our editor, then graduate student Tim Tricas, Don speared a couple of fish to attract some sharks. However, this time a 2 m long shark swam directly up off the bottom straight at Don and Tim. At a distance of only 1 m, the shark went into an intense agonistic display.

According to Tim, who was 'riding shot gun', the shark became increasingly excited and pressed in on them. But Tim was reluctant to hit the shark with his explosive bangstick because he knew that quite a few of the shells were duds due to water seepage; and he didn't want to further anger the shark! Push was rapidly coming to shove. The shark was literally on the verge of attacking. Then, over Tim's shoulder came the unflappable Don, like the Cavalry to the rescue. He poked the excited shark in the side with his spear and drove it off. Back on the boat, a shaking Tim Tricas was treated to a classical Nelson line: 'Well', said Don, 'that fish was rather pugnacious'.

This incident brings up several facets of Don's unique and wondrous personality: He was a kind of genius. He would consider his experiences, ponder and ponder, then come up with a hypothesis to answer a problem. Don was the consummate devil's advocate and, as a teacher, forced us to think clearly and discard all our assumptions. In the case of the gray reef shark, Don knew from that day on, a new technique would have to be found to elicit the agonistic display. So with absolutely no engineering background, over dinner with his students, Don got out his pencil and on the back of a napkin, his favorite teaching device, he sketched out a design for a wet, one man sub. He named it the SOS – Shark Observation Submersible (Figures 1,4d).

This was not the first vessel he designed, built, maintained and operated. Which brings up another facet of his personality: once Don decided to move on a project, nothing could stop him! He would find a way. If he didn't have access to a research platform it just didn't matter. He personally built and financed the following vessels and was working on his last – an ROV. (1) The *Skokende Hai*, a ferrocement ketch he built in the 70s and sailed to Tahiti with absolutely no maritime construction, sailing or open ocean navigation experience. (2) The *Prionace*, a small fiberglass tracking boat he built and used for years in the Catalina channel. (3) And of course SOS, (4) and finally, his true pride and joy the 60' diesel trawler *Discovery*, which Don built from the keel up (Figures 5a,b).

Throughout his life, Don was an athlete: yes, he was a champion free diver, but later he took up tennis, skiing and running (Figure 6). He often competed and was to meet his wife Ginger during a 10 km run. Don also loved music: as a graduate student in Miami, Don was forced to do his shark hearing experiments in the early hours of the morning when traffic was minimal. This

a
LONG
BEACH,CA
b
c
d
e
f

was because the low frequency rumble of trucks on the road would interfere with his acoustic testing. So, in the long, dark hours, alone with his sharks before sunrise, Don taught himself to play guitar and banjo.

Don was also a master of public relations. We started doing films together in 1963, working on the blockbuster-thriller 007 movie *Thunderball*. Don was also scientific advisor on the award winning Spielberg film *Jaws* (Figures 5c,d).

On a personal note, I want everyone to know how much I owe Don and how much I relied on his clear thinking and wisdom. Don stuck with me through the good times and the bad. He visited me in Miami in 1988, when I was at death's door suffering from cancer. How ironic it is that I am writing the article he should be writing for me! Don was always ready to offer a helpful hand, good advice or a shoulder to cry on. I truly relied on his clear thought and intellect. If I became stuck on a difficult scientific problem or a tricky interpretation, I would pick up the phone. I will dearly miss these telephone calls, none of which lasted for under 2 h. I will miss his enduring friendship.

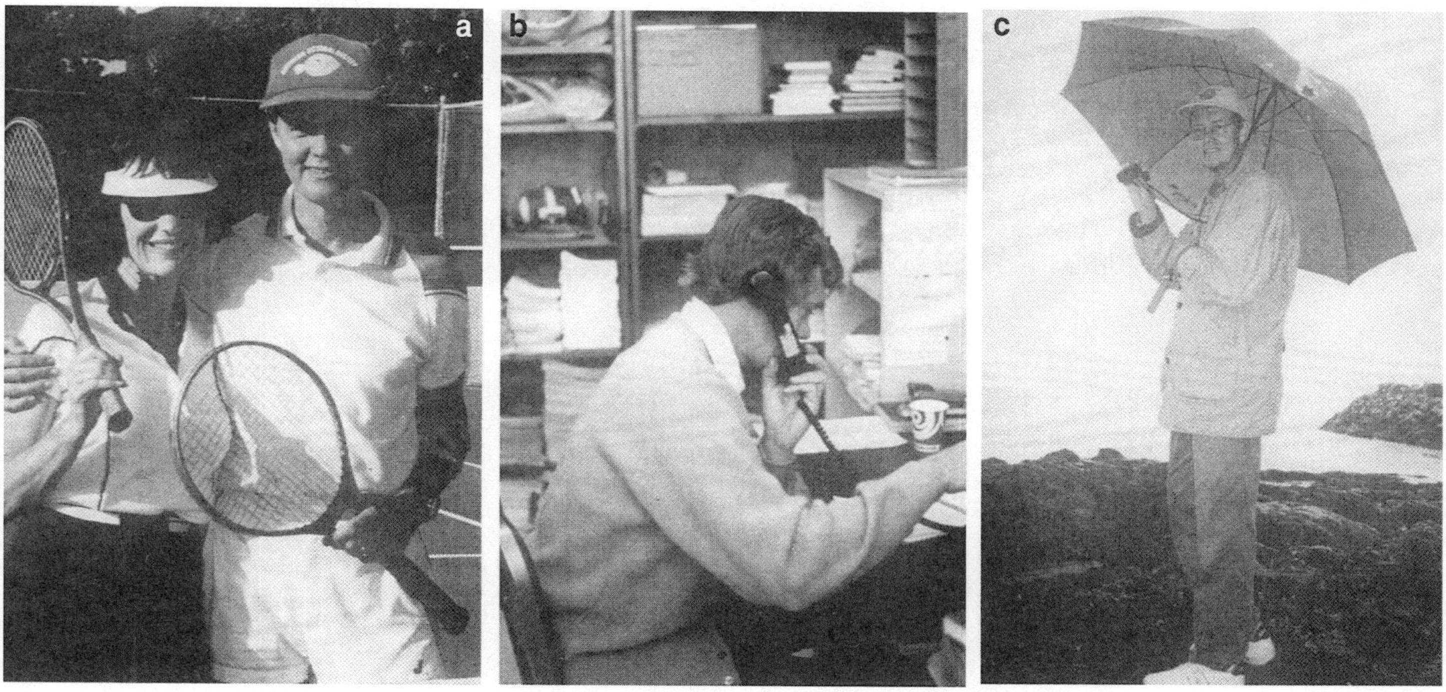

Figure 6. a – Don and wife Ginger on the courts in the 1980s. Don was an active athlete throughout his life and often used his athletic skills in his research. b – Don at the telephone in his office at California State University, Long Beach in the early 1990s. The author and Don often had long and fruitful telephone conversations throughout their careers. c – Don in the mid 1990s at the rocky shore working on his final, monumental project. He aimed to photograph every marine organism in southern California for inclusion in his CD-ROM on marine biology. This project took him back to his early interest in ornithology as he photographed many species of marine birds in the area.

←

Figure 5. a – Don's greatest achievement in vessels, the diesel trawler R/V 'Discovery' shown with biologists carrying out shark repellent tests in the late 1980s. b – Don playing guitar in the saloon of R/V Discovery during one of his popular Christmas parties aboard the vessel. He taught himself to play as a graduate student a quarter century earlier in Miami. c – Don and other 'bad guys' on the set of the James Bond 007 thriller 'Thunderball' in 1962 off Lyford Cay, New Providence, Bahamas. As extras, Don and author handled live tiger sharks and participated in the elaborately choreographed, underwater battle between good (Bond) and evil (Spectre). d – Don consults for the 1975 blockbuster film 'Jaws'. e – Don (right), author (left) and Office of Naval Research's Bernie Zahuranec (reclining) at Fullerton, California during the 1983 symposium on the great white shark, *Carcharodon carcharias*. During this poolside meeting the groundwork was laid for establishing the American Elasmobranch Society of which Nelson, Leonard Compagno and the author were founders. f – Author's last photograph of Don, wife Ginger and the author taken at the 1995 AES meeting in New Orleans. Don approached his affliction with the same optimism, equanimity and intelligence that he displayed in all aspects of his life.

Male of a longfin mako, *Isurus paucus*, 160 cm long, captured off Florida (PV).

Female of a shortfin mako, *Isurus oxyrinchus*, 170 cm long (SH).

Part 2. Sensory biology and behavior

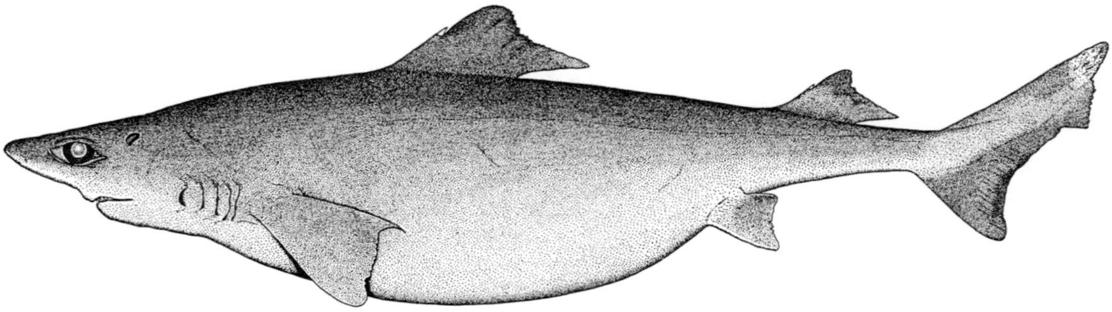
A gravid female of the little gulper shark, *Centrophorus uyato*, 85 cm long from the Gulf of Mexico (PV).

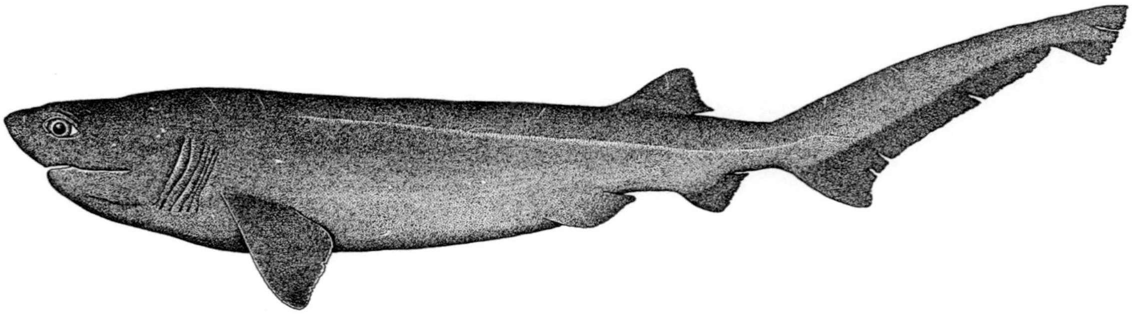
Male of the sixgill shark, *Hexanchus griseus*, 99 cm long, from Puget Sound (PV).

Environmental Biology of Fishes **60:** 31–45, 2001.

The acoustical biology of elasmobranchs

Arthur A. Myrberg, Jr.
Rosenstiel School of Marine and Atmospheric Science, University of Miami, 4600 Rickenbacker Causeway, Miami, FL 33149, U.S.A. (e-mail: amyrberg@rsmas.miami.edu)

Received 5 December 1998 Accepted 6 January 2000

Key words: hearing, underwater sound, ear, sensory biology, sharks, rays

Synopsis

A report on the auditory capabilities and their associated functions of elasmobranch fishes along with a brief review of the physics of underwater sound as it relates to hearing by fishes is provided. The inner ears of elasmobranchs possess structures that are no different from other fishes, except for an enlarged macula neglecta, which is unique among fishes. Hearing abilities among sharks (the only elasmobranchs examined) have demonstrated highest sensitivity to low frequency sound (40 Hz to approximately 800 Hz), which is sensed solely through the particle-motion component of an acoustical field. Free-ranging sharks are attracted to sounds possessing specific characteristics: irregularly pulsed, broad-band (most attractive frequencies: below 80 Hz), and transmitted without a sudden increase in intensity. Such sounds are reminiscent of those produced by struggling prey. A sound, even an attractive one, can also result in immediate withdrawal by sharks from a source, if its intensity suddenly increases 20 dB [10 times] or more above a previous transmission. Present evidence also shows that the lateral line system does not respond to normal acoustical stimuli. Morphological and physiological evidence suggest that the maculae neglecta possess acoustical relevance and may explain directionality of response despite physical principles that provide still other hypotheses for acoustical directionality. Brain centers controlling acoustical response, particularly among sharks, must be explored in the near future with careful consideration of the habits of subjects based on indications of species-differences regarding the importance of acoustical stimuli to their activities. Numerous facts and ideas about the acoustical biology of elasmobranch fishes make certain that future research will be most rewarding.

Introduction

The ear of elasmobranchs, like the ear of all vertebrates, consists of three quite different functional systems, which provide information on different kinds of movement resulting from gravitational forces, linear and angular accelerations of the head and acoustical stimuli. This review deals, specifically, with the auditory system, which concerns acoustical stimuli. Earlier reviews of the auditory system of elasmobranchs include Wisby et al. (1964), Popper & Fay (1977) and Corwin (1981a). All forms of mechanoreception by elasmobranchs were reviewed by Roberts (1978). This paper honors the memory of Donald R. Nelson, whose studies are fundamental to much of our present knowledge about the physiology and the function of hearing by elasmobranchs.

A brief review of the physics of underwater sound

Sound obeys the same physical principles when traveling through air or water, but quantitative differences in its structure and behavior occur due to differences in the density and compressibility of the two media (see Urick 1975). These differences include speed (4.8 times faster in water than in air) and wavelength of any given frequency (4.8 times longer in water than in air).

As sound passes through water, its pressure component is accompanied by oscillations (successive compressions and rarefactions) of water particles along the axis of propagation. Pressure, being a scalar quantity, provides information about sound-intensity, but no information about the direction of propagation. Particle-motion, on the other hand, has three interrelated vector quantities: displacement (distance each water particle moves), velocity (speed of such movement), and acceleration (rate of change in velocity), each of which provides information about the intensity of a sound as well as its direction. Sound pressure and the resulting particle-motion are related to each other and to the distance from the sound-source. Under idealized conditions, when the sound is near the source, the sound pressure falls off as $1/r$ where $r =$ distance from the source, while particle motion falls off as $1/r^2$ for a pulsating source (volumetric change) or $1/r^3$ for a vibrating source (no volumetric change). Sound, as it travels away from the source, conforms, initially, to a spherical wave. When the sound is far from the source, its pressure and particle-motion both fall off as $1/r$ and the sound is considered a plane-wave. As the source of a spherical wave is approached, particle-motion increases at a much faster rate than its corresponding pressure, since additional particle-motion is caused by the motion of the source itself (the acoustical near-field). At a distance where the ratio of pressure and particle motion remain constant, i.e., in a plane-wave, the acoustical far-field exists. The distance where the spherical wave becomes a plane-wave is termed the near-field/far-field boundary. The boundary is frequency dependent; the lower the frequency, the larger the near-field. Although the near-field has little importance in air, its importance to fishes is due to the relative, non-compressibilty of water. For example, the near-field for a sound of 500 Hz extends about 3 m from the source in water, while in air, that distance is only about 0.6 m.

Although experimental evidence has demonstrated that fishes can indeed orient to sound in the far-field (Schuijf et al. 1972, Schuijf & Siemelink 1974), the near-field is of primary importance to fishes because low frequency sounds, which most species attend to, have near-fields that can extend for many meters from their sources. Also, when considering vibrating sources (dipoles), such as a fish tail, sounds are likely restricted to the near field, since they haven't much of a compressional wave and therefore little sound propagation. Even pulsating sources, e.g., swimbladders (monoples), require considerable energy to produce compressional waves capable of reaching the far field. Finally, although little incident energy passes across the water–air interface, the ratio between pressure and particle motion will change at that interface as well as at the water–bottom interface and that includes complex structures such as the coral reef (Hawkins & Myrberg 1983). Thus, underwater sound can form acoustical fields that can easily become complex; but it also obeys physical laws that can provide relative ease in predicting, to a reasonable extent, its actions so long as one is sufficiently aware of these laws and their limitations. Those physical phenomena associated with sound propagation in water, as it applies to hearing in fishes, have been discussed at several levels of complexity (e.g., Hawkins & Myrberg 1983, Schellert & Popper 1992, Kalmijn 1988a, 1997) and the interested reader is encouraged to seek out these references and their sources.

The elasmobranch hearing organ

Elasmobranchs possess only inner ear labyrinths. They have no accessory organs of hearing often found in bony fishes, e.g., a swimbladder, bony connections between the swimbladder and the labyrinths. Each labyrinth includes three membranous semicircular canals and three membranous sacs, the latter of which are located below and slightly posterior to the brain (Figure 1). Each labyrinth consists of two relatively large chambers, the utriculus and the sacculus with the latter expanding at one end, forming a small, third chamber, the lagena. Each chamber possesses a sensory epithelium, termed a macula, housing the receptors of the hearing organ, the hair cells. Also, each has a mucilaginous cupula, containing numerous calcium carbonate granules (the otoconia), which covers major portions of the maculae. Additional sensory epithelia (two patches in sharks, one in batoids), the macula neglecta, are housed in a small duct, extending from the base of the parietal fossa to the saccular chamber of each inner ear. Unlike the other maculae, the macula neglecta has no otoconial mass.

Each macula contains hair cells; each cell has an apical ciliary bundle, consisting of a single, long

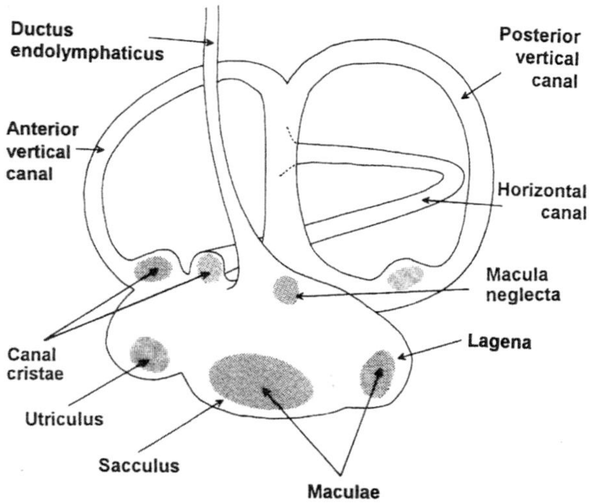

Figure 1. Simplified diagram of the elasmobranch labyrinth. View is from medial aspect, rostral on left (modified from Roberts 1978).

Figure 2. Scanning electron micrograph of hair cell and supporting cell surfaces in the macula neglecta of a carcharhinid shark. The ciliary complex projecting from the apical surface of each hair cell contains a size-graded series of stereocilia with a single kinocilium next to the longest stereocilium. Scale bar = 3μm (Corwin 1981c with permission).

kinocilium and up to 60 stereocilia (Figure 2). The latter are graded in size, with the longest adjacent to the kinocilium. Such hair cells are numerous in sharks, e.g., the saccular macula of the oceanic whitetip *Carcharhinus longimanus*, 1.6 m in standard length (SL), has approximately 4 200 000 and that same macula in the blacktip *C. limbatus*, 1.7 m SL, has approximately 3 200 000 (Corwin 1985). Response of the hair cell to an acoustical stimulus occurs when associated particle motions cause the cilia to bend based on the inertial differences between components having different densities, i.e., otoconia, basal tissues. Movement of the stereocilia of a given hair cell towards its kinocilium results in an increased number of action potentials per unit time relative to the spontaneous activity of the associated neuron of the eighth cranial nerve; the opposite movement, i.e., the kinocilium moving towards the stereocilia, results in a decrease in the number of action potentials per unit time sent by its neuron to the brain (Flock 1967).

Lowenstein & Roberts (1951) showed that the macula of the lagena and most of the utricular macula of *Raja clavata* did not respond to vibrations, but to gravitational stimuli, e.g., tilting. Vibration sensitivity was provided by a small part of the utricular macula (the lacinia) and the anterior 2/3 of the saccular macula. Greatest sensitivity to vibration was found at the macula neglecta. Thus, the authors suggested that the latter macula might well be an important organ for sound detection in elasmobranchs, a role suggested twenty years later by both Tester et al. (1972) and Fay et al. (1974) based on anatomical and neurophysiological grounds, respectively. Corwin (1981b) examined click-evoked, eighth nerve potentials of the lemon shark, *Negaprion brevirostris*, after electrocautery lesioning of the saccular macula, followed by lesioning the macula neglecta. Severe reduction of neural response after the first lesion, followed by total loss of response following the second lesion indicated that only these macula held the primary auditory receptors in the ear of the lemon shark, essentially that shown by Lowenstein & Roberts for the skate 30 years earlier. Corwin (op. cit.) also extended the findings by Fay et al. (1974), demonstrating auditory sensitivity by the macula neglecta with acoustical clicks, replacing vibration of the skin overlying the parietal fossa, used in the earlier work. It is noteworthy that also Bullock & Corwin (1979) showed that the largest averaged evoked potentials in specific regions of the brain of the blacktip reef shark, *Carcharhinus melanopterus*, were obtained if a sound source was directly over the parietal fossa, while such potentials were usually suppressed just lateral to or behind the fossa. Such findings supported the view that the parietal fossa is an important portal for sound and directly connected to it via the foramen ovalis is the posterior canal duct, housing the macula neglecta.

Hearing abilities of sharks

Although Vilstrup (1951), Clark (1959), Dijkgraaf (1963a) and a few earlier workers had shown that several species of sharks could recognize different types of underwater sounds, the first hearing curve or audiogram of an elasmobranch was that of the bull shark, *Carcharhinus leucas* (Kritzler & Wood 1961). The audiogram, based on pure tones and obtained by instrumental conditioning, showed the typical 'U' shaped curve noted in the audiograms of vertebrates, indicating that the subject could hear acoustical frequencies from 100 Hz to 1500 Hz, with greatest sensitivity, relative to ambient noise, between 400 and 600 Hz.

The next major publication on hearing by elasmobranchs was that by Nelson & Gruber (1963), reporting that free-ranging sharks (bulls, hammerheads, lemons, tigers) could be attracted to underwater sounds representing the vibrations caused by the struggling actions of fishes (grouper), when speared. Such sounds were a strong attractant if they were rapidly and irregularly pulsed and possessed only low frequencies, i.e., below 60 Hz. A sound, when possessing only higher frequencies, i.e., filtered band of 400–600 Hz, was far less an attractant and a low frequency, but continuous, sound attracted no sharks.

Information about the auditory sensitivity of sharks was extended to the lemon shark *Negaprion brevirostris* (Nelson 1967). The author, using instrumental conditioning, demonstrated clear frequency discrimination (40 Hz vs. 60 Hz = 0.5 octave), good acoustical orientation to a sound source (9.5° error in a pool 4 m, dia) and also established an audiogram. The audiogram varied from that for the bull shark by showing greatest sensitivity at 40 Hz and relatively poor sensitivity at 400–600 Hz. Nelson's method of recording signal intensity was the same as used by Kritzler & Wood (1961), in units of pressure, i.e., one component of sound in water and that which terrestrial animals use to detect acoustical signals in air. Nelson mentioned that since sharks do not possess a pressure to particle-motion converter, such as a swimbladder, they likely do not respond to acoustical pressure. Unfortunately, a sensor for the other component of a sound field, particle motion, was not commercially available at the time. Nelson's (1967) behavioral thresholds were later compared to neural thresholds (compound action potentials of the eighth cranial nerve) provided by members of the same species (Corwin 1981b). Both data sets showed good similarity, i.e., a relatively shallow plateau in sensitivity below 200 Hz and a steep decrease in sensitivity above that point.

Behavioral thresholds, by lemon sharks, were again established for several pure tones (20, 320, 640, 1000 Hz) by Banner (1967), using both a pressure hydrophone and a particle-motion sensor (accelerometer). Results clearly demonstrated that his subjects were attending to the particle-motion component of the sound field and not to its pressure component. Kelly & Nelson (1975), using the same sensors with horn sharks, *Heterodontus francisci*, noted that their subjects showed different pressure measurements at sensory threshold for a given frequency at two locations in their enclosure, but the same particle-motion measurement at threshold at these different locations. The results demonstrated that their subjects, like those examined by Banner (op. cit.), were attending to the particle-motion component of the sound field. Subsequently, based on the apparent lack of a pressure to particle-motion converter, particle motion has been considered to be the relevant stimulus for perceiving a sound field by other elasmobranchs. That point has been confounded, however, by van den Berg & Schuijf (1983), upon examining sound discrimination by the catshark, *Chiloscyllium griseum*. Based on phase differences existing between acoustically generated particle motion and pressure, their findings indicated that subjects were indeed attending to pressure as well as particle motion, implying that at least their subjects possessed a pressure-sensitive mechanism. Barring errors in measurement, the authors provided the suggestion, originally postulated by Corwin (1977), that the cupula-loaded macula neglecta appears well positioned to detect displacements concentrated at the membranous fenestra ovalis caused by compressions of the fluid volume through the parietal fossa.

Attraction to sound sources by sharks

Although sharks had been clearly shown to be attracted to sounds that simulate the struggling actions of speared fishes (Nelson & Gruber 1963), no study had examined if such predators respond to the natural sounds of their prey. Banner (1968), after reporting that free-ranging lemon sharks, shortly after birth, were readily attracted to instrumentally produced, low frequency, pulsed sounds, examined whether young animals react to the sounds produced by prey (and non-prey) in their natural habitat. He recorded and played back numerous sounds from animals available to these young predators, e.g., wounded and struggling fishes, accelerating and jumping fishes, and even vocalizations (catfish, i.e., non-prey of such small sharks) (Banner 1972). He found that any low frequency, natural sound, if rapidly pulsed, was extremely attractive. Such sounds included those from erratic movements, e.g., struggling, but also rapidly pulsed vocalizations, even from non-prey. Other prey sounds, possessing only one or two pulses, e.g., accelerations and jumping, were not attractive. Finally, sounds, when suddenly transmitted as a subject was slowly approaching or when near a source, resulted in rapid flight. Since a sound was played back at the intensity of recording, all were of low amplitude and, thus, the greatest distance of attraction was 4.5 m from the source. That distance was minuscule compared to that found in studies using high intensity, sonic transmissions in the open ocean (e.g., Nelson et al. 1969, Nelson & Johnson[1], 1972, Myrberg et al. 1976, 1978).

Interest in further examining the attractive nature of sounds to free-ranging sharks remained high between the late 1960s and the late 1970s due largely to three points. The first centered on the serious doubt expressed by several physicists that free-ranging sharks could orient to a sound source from the acoustical far-field, countering a report that such animals had shown precisely such behavior (Wisby & Nelson[2]). The acoustical far-field is considered a distance, exceeding one wavelength from the source, e.g., approximately 15 m for a sound having a frequency of 100 Hz, and beyond which the ratio of particle motion to pressure remains constant. The acoustical near-field, on the other hand, is considered a distance generally less than one wavelength from the source, where the ratio of particle motion to pressure increases as the source is approached. The physicists attributed the results with sharks solely to an unusual condition during tests, i.e., shallow water, whereby sound, repeatedly interacting with the surface and bottom, propagates differently than deep water and which somehow caused the unique findings. The second point involved those acoustical parameters (other than intensity) that could increase a sound's attractiveness. The final point focused on the question: could acute hearing lead to an effective shark repellent?

The first confirmation of sonic attraction by sharks from the acoustical far-field occurred at a video-acoustic installation at the deep edge of the reef line off North Bimini, Bahamas (Myrberg et al. 1969). Tests, with appropriate controls, were conducted at depths exceeding 20 m with video documentation. Sharpnose sharks, *Rhizoprionodon* sp. (probably *porosus*), were attracted to 'overdriven' sine waves (= sounds whose energy is distributed primarily in the fundamental and its odd harmonics) as well as broad-band sounds throughout the day. Subjects showed marked increase in activity as their numbers increased near the source, but they also demonstrated habituation to such sounds in the absence of positive reinforcement. In that same year, several species of sharks showed such sonic attraction to instrumental sounds (broad-band with frequencies between 50 and 200 Hz) at various locations in the Bahamas over reefs as well as over the deep (1830 m) waters of the Tongue of the Ocean (Nelson et al. 1969). The authors confirmed habituation to acoustical signals and also observed epipelagic, silky sharks, *Carcharhinus falciformis*, rapidly orienting to speared fishes and to the frenzied feeding actions of nearby fishes. The latter behavior supported an earlier hypothesis by Banner (1968) that hydroacoustic stimuli, associated with normal feeding behavior by certain fishes, were relevant stimuli for sharks, possibly more so than the struggling sounds of prey.

Sonic attraction studies continued over the deep waters of the Florida Strait and the Tongue of the Ocean, Bahamas. Controlled testing showed that broad-band sounds, reaching 500–1000 Hz, were attractive to sharks of the open ocean: silky sharks, *Carcharhinus falciformis*, and oceanic whitetips, *C. longimanus*, with attractiveness increasing as spectral frequencies were sequentially lowered to a 25–50 Hz band. The difference in effectiveness reached almost 1000 fold. Attraction from the acoustical far-field (sound-source distance, 250–400 m) was again clearly demonstrated (Myrberg et al. 1972). Subsequent tests, lowering the spectral frequency band (40–80 Hz, 20–40 Hz, 10–20 Hz) showed no further increase in attractiveness (Myrberg et al. 1976). Increasing the pulse rate of regularly pulsed sounds (1, 5, 10, 20 per sec) also increased attractiveness, but the most rapidly pulsed sounds did not attain the attractiveness of irregularly pulsed sounds, though the latter were more slowly pulsed. There was no simple relationship

[1] Nelson, D.R. & R.H. Johnson. 1970. Acoustic studies on sharks, Rangiroa Atoll, July, 1969. Off. Nav. Res. Tech. Rep. 2. 15 pp.

[2] Wisby, W.J. & D.R. Nelson. 1964. Airplane observations of acoustic orientation in sharks. Amer. Fish. Soc. Conf. (abstract).

between attractiveness and the distribution of acoustic energy within a sound. For example, a signal of 20 pulses sec^{-1}, having pulse durations of 10 msec, was more attractive than a signal of 10 pulses sec^{-1}, having pulse durations of 50 msec; but the latter was more attractive than a signal of 5 pulses sec^{-1}, having pulse durations of 50 msec.

Similar studies in the Pacific showed the same results as those obtained in the Atlantic. Lower frequency sounds (25–50 Hz) were clearly superior attractants to those containing higher frequencies (50–100 Hz) (Nelson & Johnson[1]), habituation to acoustic signals readily occurred (Nelson & Johnson 1972), and natural sounds, including fish vocalizations (*Myripristis*), were attractive (Nelson & Johnson 1976). In addition, observations indicated that although single-pulsed sounds were generally non-attractive, exceptions to that rule could occur through apparently unintentional, instrumental conditioning, such as speargun discharges, due to frequent hunting by divers in specific areas (Nelson & Johnson 1976).

Table 1 lists those species of sharks that have been attracted to underwater sound. Most are well known and often they are the more common species in their respective habitats, e.g., the coral reef, the epipelagic region, and even shallow, inshore waters. An earlier list is provided by Myrberg (1978).

Withdrawal from sound sources by sharks

The two epipelagic species *Carcharhinus falciformis* and *C. longimanus*, which had shown strong attraction to different underwater sounds also showed, interestingly, rapid withdrawal from a sound source (Myrberg et al. 1978). Such a response was seen under specific circumstances, i.e., sudden onset of transmission of an intense sound when the subject was within 10 m of the source. The same response would also occur with a sudden change in transmission of an attractive sound to that of another sound, e.g., the 'scream' of a killer whale or even another attractive sound, if the latter's intensity was considerably higher than the preceding signal. The response was similar to that previously shown by young lemon sharks, when confronted with sudden onset of transmissions (Banner 1972). This withdrawal response was further examined, using adult lemon sharks housed in a large circular channel (230 m) (Klimley & Myrberg 1979). The study established that an individual, when approaching within 10 m of a source and confronting a broad-band sound whose onset is abrupt and of high intensity (more than 20 dB [10 times] above ambient level), will suddenly turn and withdraw from such a source. Even attractive sounds became repellents when possessing such characteristics. One type of sound failed, however, to elicit withdrawal: audible pure tones (to 40 dB [100 times] above ambient noise). Interestingly, such pure tones had been found, previously, to be ineffective attractants. An important complication was also evident from the studies; there appeared to be species-typical differences in habituation rates to such signals, e.g., silky sharks would withdraw from such signals 6–8 times in a row; but oceanic whitetips would withdraw from such signals only once.

Lateral line involvement?

Over the years, the lateral line system has been considered a detector of hydrodynamic changes close to the body and, at times, also designated as a detector of low frequency sound. The latter function was strongly advanced by van Bergeijk (1967), who assailed the role of the inner ear of fishes in detecting near-field motions, assigning that role instead to the lateral line system. Although contrary evidence had been obtained and reviewed by Dijkgraaf (1963a,b), van Bergeijk's influence was great at the time. Few attempts have been made to establish the function of the inner ear under exclusion of the lateral line, but three studies are noteworthy. One dealt with an elasmobranch. Corwin & Northcutt (1982) found clear response to acoustical clicks in the ventromedial division of the lateral mesencephalic nucleus in the brain of the thornback guitarfish, *Platyrhinoidis triseriata*. Such response disappeared when the eighth cranial nerve was cut, but no change occurred when all lateral line nerves were cut. Using high intensity sound stimuli, they also found no noticeable increase in the metabolic activity of anatomically recognized CNS projections of the lateral line system. Two additional studies on teleosts also clearly demonstrated sound detection and directionality despite the complete elimination of the lateral line system (Schuijf & Siemelink 1974 for *Gadus morhua*, Schuijf et al. 1972 for *Labrus berggylta*). Görner & Kalmijn (1989), also using *P. triseriata*, have shown that the lateral line is responsive to water current oscillations near the body with good response below 10 Hz. Above that frequency, sensitivity rapidly declined until at 128 Hz, response was shown to stimuli that were 45 dB above the reference level (taken

Table 1. Sharks attracted by underwater sounds.

Region	Common name	Species	Reference
Pacific	Blacktip reef	*Carcharhinus melanopterus*	Nelson & Johnson (1970)
			Nelson & Johnson (1972)
	Gray reef	*Carcharhinus amblyrhyncos*	Brown[3]
			Nelson & Johnson (1970)
			Nelson & Johnson (1976)
	Lemon	*Negaprion fosteri*	Nelson & Johnson (1972)
	Reef whitetip	*Triaenodon obesus*	Brown[3]
			Nelson & Johnson (1970)
			Nelson & Johnson (1972)
			Nelson & Johnson (1976)
	Silvertip	*Carcharhinus albimarginatus*	Nelson & Johnson (1972)
			Nelson & Johnson (1976)
Atlantic	Bonnethead	*Sphyrna tiburo*	Nelson et al. (1969)
	Bull	*Carcharhinus leucas*	Nelson & Gruber (1963)
			A.A. Myrberg personal observation
	Lemon	*Negaprion brevirostris*	Nelson & Gruber (1963)
			Banner (1968)
			Banner (1972)
	Nurse	*Ginglymostoma cirratum*	Richard (1968)
			Myrberg et al. (1969)
			Nelson et al. (1969)
	Reef	*Carcharhinus springeri*	Richard (1968)
			A.A. Myrberg personal observation
	Sharpnose	*Rhizoprionodon porosus*	Richard (1968)
			Myrberg et al. (1969)
World-Wide	Blue	*Prionace glauca*	D.R. Nelson personal observation
	Hammerhead	*Sphyrna lewini*	Klimley & Nelson (1981)
	Oceanic whitetip	*Carcharhinus longimanus*	Myrberg et al.[4]
			Myrberg et al.[5]
			Myrberg et al. (1976)
			Myrberg et al. (1978)
	Shortfin mako	*Isurus oxyrinchus*	D.R. Nelson personal observation
	Silky	*Carcharhinus falciformis*	Evans & Gilbert[6]
			Myrberg et al. (1972)
			Myrberg et al.[4]
			Myrberg et al.[5]
			Myrberg et al. (1976)
	Thresher	*Alopias* sp.	D.R. Nelson personal observation
	Tiger	*Galeocerdo cuvieri*	Nelson & Gruber (1963)

[3]Brown, T. 1968. Shark research programme at Rangiroa, French Polynesia for the year 1968. Report to the French Polynesian Government, Papeete. 17 pp.

[4]Myrberg, A.A., Jr., C.R. Gordon & A.P. Klimley. 1975a. Attraction of free-ranging sharks by acoustic signals in the near subsonic range. Tech. Rept. to Office of Nav. Res., Contract No. N00014-67-A-0201-0008, Rosenstiel School of Marine and Atmospheric Science, TR75-4, University of Miami, Coral Gables. 42 pp.

[5]Myrberg, A.A., Jr., C.R. Gordon & A.P. Klimley. 1975b. Rapid withdrawal from a sound source by sharks under open-ocean and captive conditions. Tech. Rept. to Office of Nav. Res., Contract No. N00014-67-A-0201-0008, Rosenstiel School of Marine and Atmospheric Science, TR75-5, University of Miami, Coral Gables. 32 pp.

[6]Evans, W.B. & P.W. Gilbert. 1971. The force of bites by the silky shark (*Carcharhinus falciformis*) measured under field conditions. Naval Research and Development Center, San Diego. 8 pp.

as the acceleration of a 1 Hz stimulus of 0.5 cm^{-2} displacement amplitude of the small, moving sphere used during tests). Finally, Sand (1981) concluded from a series of carefully controlled experiments that the trunk lateral line of the European roach, *Rutilus rutilus*, although extremely sensitive to local water displacements relative to the skin surface, has drastically reduced sensitivity when the fish is free to move with the water mass. Since the latter condition occurs as a sound wave passes, he maintained that the trunk lateral line is of little use for detecting sound-induced water displacements beyond a few body lengths even at enormous sound intensities.

Functional anatomy of the ears of elasmobranchs

Further insight into the probable importance of hearing among elasmobranchs was also obtained by considering the anatomy of the inner ears of both sharks and batoids that inhabit different zones of the ocean. Corwin (1978), examining the labyrinths of various elasmobranchs, noted some clear differences between primarily piscivores living in the pelagic zone and those animals living on or near the substrate that are largely not piscivorous (Figure 3). The latter showed a small macula neglecta within a short posterior canal duct connected to a relatively small, ventrally positioned sacculus, while the former possessed a large sacculus with a moderately large, dorsally positioned posterior canal duct, containing a well developed macula neglecta. The macula neglecta of those batoids examined contained only one epithelial surface on the posterior wall of the duct, while amongst those sharks examined, all showed two epithelial surfaces, one on the posterior wall of the duct and a smaller patch on the anterior wall. The hair cells of the piscivorous species were polarized in one direction on one epithelial surface and in the opposite direction on the other epithelial surface. The hair cells of the mainly non-piscivorous species showed a more random pattern of polarization on their epithelial surface(s) (Figure 4). The different arrangements found in the labyrinths of those species examined appeared to relate more to feeding habits than to phyletics. And it is noteworthy that those species that appeared to be more euryphagous showed structural intermediacy. Further differences in the structural design of labyrinths were also noted among pelagic sharks by Roberts (1978) and by Corwin (1989), including great variation in the

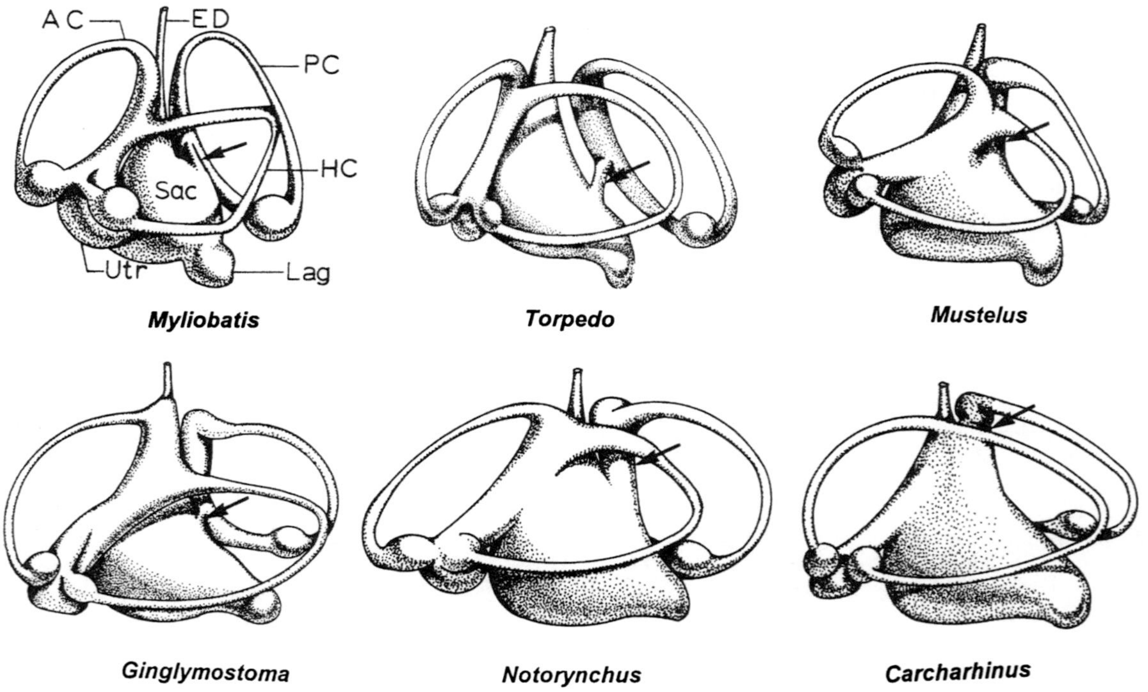

Figure 3. Drawing of elasmobranch labyrinths as they would appear in lateral view. Arrows show the posterior canal duct. Note the species-typical differences in the size and position of the ducts. In *Carcharhinus* and *Notorynchus*, they are particularly large and are closer to the fenestrae ovalis than they are in the four other species. AC = anterior semicircular canal, ED = endolymphatic duct with dorsal portion cut off, HC = horizontal semicircular canal, Lag = lagena, PC = posterior semicircular canal, Sac = sacculus, Utr = utriculus (Corwin 1978 with permission).

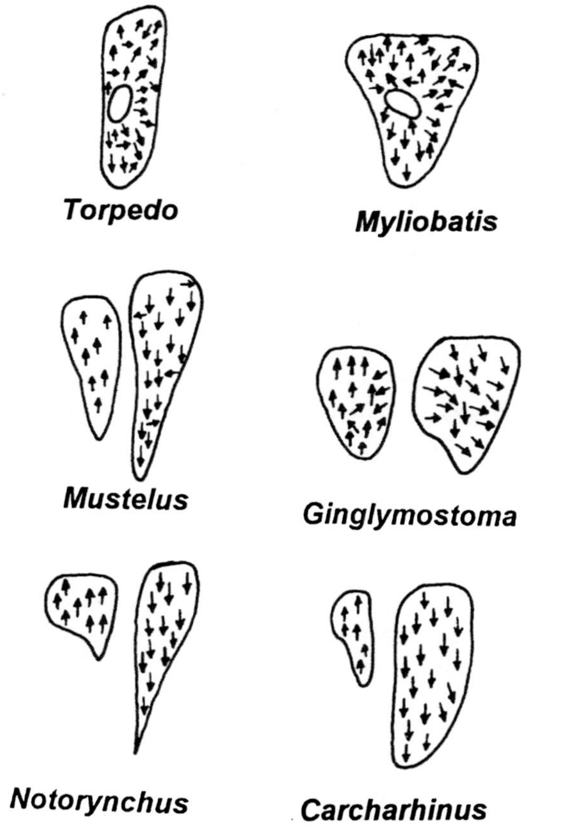

Figure 4. Hair cell orientation in the maculae neglecta from the left ears of the six elasmobranchs shown in Figure 3. The excitatory direction of representative hair cell ciliary bundles are marked by small arrows, with the head of each arrow indicating the orientation of the hair-cell kinocilium. In *Torpedo* and *Myliobatis*, the cells show no consistent orientation and their single epithelia contain central zones lacking hair cells. In *Notorynchus* and *Carcharhinus*, the cells of each epithelial surface appear to be perfectly aligned; those in left epithelial patch are oriented dorsoanteriorly while those in the right epithelial patch are oriented ventroposteriorly. In *Mustelus* and *Ginglymostoma* alignments appear intermediate with occasional orthogonal orientation in *Mustelus* and more random orintations in *Ginglymostoma* (Corwin 1978 with permission).

length of the endolymphatic duct, based apparently on the depth of the parietal fossa. Reasons for such disparity are unclear.

An excellent illustration of the macula neglecta and its anatomical relationships to nearby structures in a typical piscivore, the blacktip reef shark, *Carcharhinus melanopterus*, was provided by Fay et al. (1974) (Figure 5). The epithelial patches reside in the lumen of the posterior canal duct which contacts, anteriorly, the membranous opening at the base of the parietal fossa,

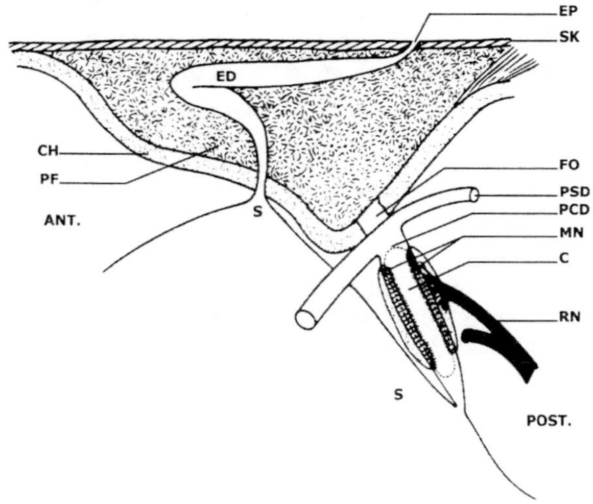

Figure 5. Drawing of macula neglecta and its anatomical relationships to nearby structures in a carcharhinid shark. C = cupula, CH = condrocranium, ED = endolymphatic duct, EP = endolymphatic pore, FO = fenestra ovalis, MN = macula neglecta, PCD = posterior canal duct, PSC = posterior semicircular canal, PF = parietal fossa, RN = ramus neglectus nerve, S = sacculus, SK = skin overlying fossa (modified from Fay et al. 1974 with permission).

the fenestra ovalis. The latter is the largest opening from the otic capsule to the connective tissue of the parietal fossa. The duct ends, posteriorly, in the dorsal aspect of the sacculus at a transparent and very thin part of the labyrinth's wall (analogous to the round window in the mammalian ear). The gelatinous cupula, resting on the hair cells, has a density nearly matching that of water. No otoconial mass exists.

The epithelia of the maculae neglecta of the shark *Carcharhinus melanopterus* and the skate *Raja clavata* grow with age and, as expected, so too do the number of hair cells (Corwin 1981c, 1983). The ramus neglectus (a branch of the eighth cranial nerve, leading to the macula neglecta) of the skate also increases in size with age as does also threshold sensitivity, when recorded from that nerve. Increasing sensitivity reached 500 fold between a skate, 0.6 year old and one, 6 years of age (Corwin 1983). This particular macula, in the skate, *Raja ocellata*, also shows gender differences: adult females have more hair cells in the epithelium and a larger ramus neglectus than found in conspecific males (Barber et al. 1985). Such findings indicate sexual differences in auditory function. Along similar lines, Carrier et al. (1994) have provided the first indication of the auditory system playing a role in sexual matters among elasmobranchs, i.e., sounds of thrashing by interacting male and female nurse sharks, *Ginglymostoma cirratum*, on the breeding ground, appear to be attractive to other males in the vicinity.

Current hypotheses for directional hearing in sharks

Three hypotheses have been proposed to explain how sharks can rapidly orient to distant sound sources. The first, proposed by Schuijf (1974, 1975), involves a timing analysis between the direct and indirect sound waves from a source after they reach a shark's inner ears. A sound reaches the inner ear along two paths: the direct path, lying between the source and the ear, and the indirect path, reaching the ear after reflecting from the water-surface (Figure 6). That 'soft' surface results in the reversal of phase of the particle-motion component (Lloyd mirror effect, see Urick 1975) relative to that component of the direct wave as noted by the arrows in the insets of Figure 6 (note the mirror symmetry existing at the vertical [dashed line] above the shark). The two waves stimulate differently oriented hair cells in one or more maculae and since the indirect path is longer, the oscillation of the surface-reflected wave possesses a time delay relative to that of the direct wave. It can be shown that the resultant particle-motion vector describes a rotational trajectory,

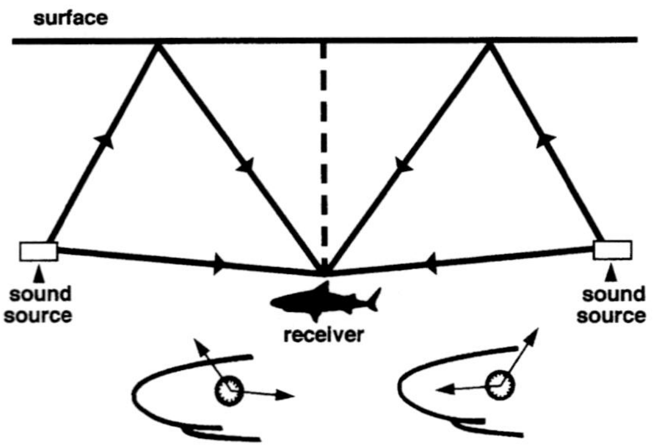

Figure 6. Phase analysis hypothesis of sound directionality in sharks. The insets show the contributions of the direct wave and the surface-reflected wave to the particle displacements at the labyrinths of the fish. The indirect dispacements in the insets are not directed in the propagation direction at the illustrated instant, as the indirect wave undergoes an inversion in the compression phase on reflection at the soft surface. For further clarification, see text (modified from Schuijf 1981 with permission).

whose rotational sense is opposite for sounds 180° apart (Schuijf & Buwalda 1980, Schuijf 1981). This model solves the crucial 180° ambiguity in particle-motion oscillation, but it has a rather shallow depth restriction based on the differing path lengths of the direct and indirect sound waves. Reliability is lost in the case of sine waves if the reflected wave is delayed by more than one-fourth of a period. That depth is reached when the sound source is more distant than one-quarter of the wavelength from both the surface or the bottom (Schuijf & Buwalda 1980). Thus, orienting to a sound source, producing a signal of 20 Hz, would be limited to depths shallower than approximately 20 m or within 20 m from the bottom. An additional problem also arises when phase relations are used to resolve 180° ambiguities when dealing with fish sounds. Phase implies sinusoidal signals and the sounds from fishes are notoriously nonsinusoidal. In such cases, resolution of the 180° ambiguity must involve a timing analysis apart from a strict phase analysis. How this is achieved is unknown, but ideas as to a possible answer are provided by Schuijf (1974).

The second hypothesis, proposed by Corwin (1981a, b), centers on the macula neglecta. The author maintains that in piscivorous species, the unique polarization patterns of hair cells of the macula neglecta, paralleling the long axis of the posterior canal duct, which extends from the fenestra ovalis (Figure 5), would show greatest sensitivity anteriodorsomedially (Figure 7). Such detection, when functioning in parallel with the symmetrically responsive saccular detection, is expected to resolve the problem of 180° ambiguity. If a sound comes from the front, the highly polarized hair cells of the macula neglecta will move strongly due to the relatively unobstructed canal duct via the foramen ovalis, while their response will be less to a sound coming from the rear because of the more obstructed pathway from that direction (as implied from previous recordings from the ramus neglectus of two species of sharks (see above). Since the size of response by saccular hair cells will be symmetrical regardless of source location, the combined response from the two different regions would be greatest from the zone depicted in Figure 7 and clearly less from the opposite direction. Thus, the 180° ambiguity is solved. This hypothesis is confronted by an additional hypothesis, i.e., the mechanism generating the shearing force needed to bend the hair cells within the macula neglecta. The author maintains that the cupula of the non-mass loaded macula, with its low density, is not attached to the epithelium,

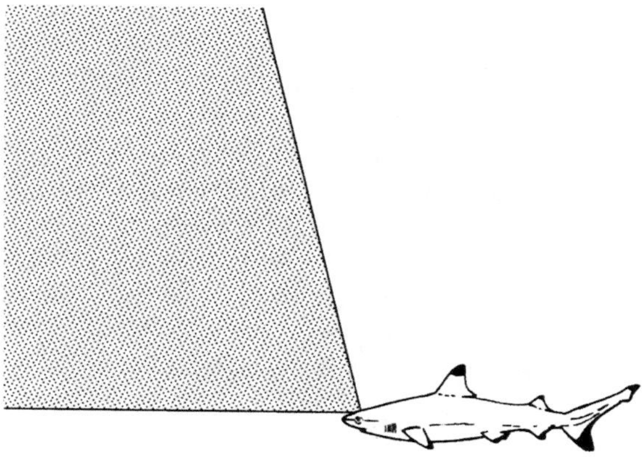

Figure 7. Representation of the zone predicted as most effective for acoustic stimulation (waves in reception zone) of the macula neglecta via the parietal sound path in a piscivorous shark. The hypothesis is based on the anatomical descriptions and physiological recordings that relate to a forward-biased sensitivity. For further clarification, see text (modified from Corwin 1981a with permission).

while the latter is firmly attached to the calcified cartilage of the skull. Since the density of the skull is greater than that of the cupula, the epithelia should move with lower amplitude and a phase lag relative to the movement by the cupula. Thus, a shearing motion should be generated at the level of the hair-cell cilia that extends between the acoustically well coupled cupula and the more poorly coupled sensory epithelium (see Corwin 1981b, figure 13A–D). This argument still runs into the problem posed by physicists, i.e., the body of a fish moves in sympathy with the passing sound wave except for the more highly dense otoconia that possess greater inertia. Corwin's view is based on the highly organized structures of the macula neglecta and the response of the organ to sound stimuli from different directions.

Although Corwin does not restrict sound directionality to the near-field, the structural organization of the macula neglecta suggests that it is a near-field acoustic detector. Present evidence for its sensitivity does not suggest otherwise (Fay et al. 1974, Bullock & Corwin 1979, Corwin 1981b). An interesting issue, regarding Corwin's orientation-hypothesis, relates to the study by van den Berg & Schuijf (1983). The latter authors found acoustic pressure perception in the catshark, *Chiloscyllium griseum*. That fact is difficult to resolve, since elasmobranchs have long been considered as not possessing a pressure to particle-motion converter. Perhaps, the foramen ovalis is such a converter (see above). The study awaits confirmation.

The final hypothesis proposed by Kalmijn (1988a, 1997), emphasizes solely the near-field (for a sound of 10 Hz, that distance reaches approximately 150 m). Kalmijn's interest has been to examine directional hearing as it relates to a common acoustical stimulus of a prey item, e.g., struggling or thrashing. The latter have acoustical fields (essentially acceleration fields) consisting primarily of a dipole term, which is directly analogous to a bioelectric field, i.e., changes in the direction of accelerations, the units of mass, take the place of changes in the direction of the units of charge within an electric field. The similarities of such fields were noted by Kalmijn based on his previous studies of electroreception in sharks (Kalmijn 1988b). At some distance, where the acoustical dipole field is perceived, the shark does not need to know the precise location of the prey source; it needs only turn in such a fashion as to keep a constant angle between the accelerations that the inner ear receives and its body axis to arrive in the vicinity of the source (Figure 8). Note in the figure that the shark does not move on a bee-line to the source. Rather, by following the angular rate of change, experienced over time, within the acceleration field of the near-field source, the shark's movements should prove as effective in guiding the predator to its target as movements by the weakly electric mormyrid, *Brienomyrus brachyistius*, when following current lines in an electric field (Schluger & Hopkins 1987, Hopkins 1988).

The above hypotheses do not invoke a mechanism for sound localization. Those by Schuijf and Corwin attempt to explain how subjects will resolve the

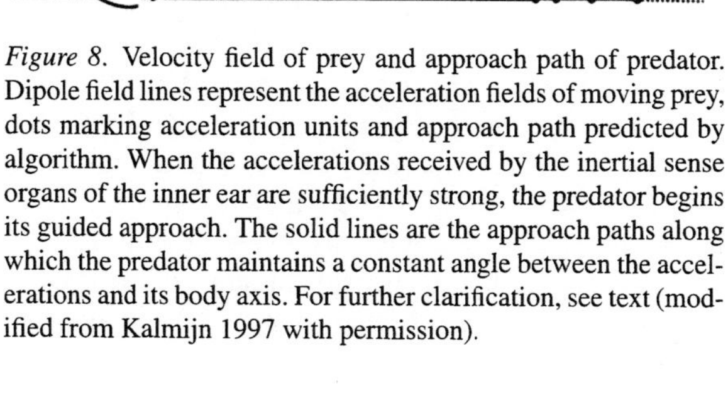

Figure 8. Velocity field of prey and approach path of predator. Dipole field lines represent the acceleration fields of moving prey, dots marking acceleration units and approach path predicted by algorithm. When the accelerations received by the inertial sense organs of the inner ear are sufficiently strong, the predator begins its guided approach. The solid lines are the approach paths along which the predator maintains a constant angle between the accelerations and its body axis. For further clarification, see text (modified from Kalmijn 1997 with permission).

180° ambiguity resulting from particle-motion oscillations; that, by Kalmijn, invokes a mechanism that brings the shark sufficiently close to the source so that more precise sensory systems (electrical, lateral line, vision) will be able to pinpoint the location of the source.

Auditory centers in the brains of elasmobranchs

Little information is available on auditory processing in the brains of elasmobranchs. Bullock & Corwin (1979) reported that auditory stimuli produced evoked potentials in the cerebellum, the mesencephalon, and even the telencephalon of carcharhinid and triakid sharks, suggesting distinct anatomical pathways for auditory inputs to telencephalic levels. With few exceptions, information focuses on two species: the thornback guitarfish, *Platyrhinoidis triseriata*, and the clear nose skate, *Raja eglanteria*. Findings, centering on *P. triseriata*, are reviewed by Northcutt (1980). The species possesses an octavolateralis area of the rhombencephalon, composed of three columns of which the ventral column contains four octaval nuclei rostrocaudally: anterior, magnocellular, descending, and posterior nuclei. Primary lateral line afferents do not terminate on octaval nuclei, with the possible exception of the magnocellular nucleus. Descending fibers of the ventral column terminate on neurons of the magnocellular, descending, and posterior nuclei as well as on dendrites from the reticular formation. Ascending fibers of that column terminate throughout the anterior nucleus and the lower leaf of the cerebellar auricle. Further work (Corwin & Northcutt 1982) on the same species, using acoustic stimulation and employing both deoxyglucose autoradiography and evoked potentials, has provided additional information. Equilibrium information appears to be processed primarily in the magnocellular, descending, and posterior nuclei, while acoustic centers in the rhombencephalon include the anterior nucleus, the medial octavolateralis nucleus, a nucleus of the lateral lemniscus, and the reticular formation and in the midbrain: the lateral mesencephalic nucleus. Equilibrium and acoustic processing are apparently at least partially separate in the medulla. Uncertainty remained, however, since equilibrium detectors were not stimulated. Also, interpretation was difficult because of multiendorgan input into each nucleus. Nevertheless, evidence suggests that the lateral line system does not contribute to acoustical sensitivity in that elasmobranch. Projections of the eighth cranial nerve from the labyrinth of *R. eglanteria* have been examined by Barry (1987). He provided an anatomical description of the organization of afferents from the individual endorgans, including a newly described nucleus for ascending auditory and vestibular afferents, the periventricular octaval nucleus. The paper also discusses the auditory and vestibular functions of the various octaval nuclei relative to the known functions of the labyrhinthine endorgans and compares its findings with those obtained from other fishes.

Considerations for future research

The dipole hypothesis

All information on the hearing abilities of free-ranging sharks has been obtained by using monopole sources. Sounds produced by such sources, e.g., J-series speakers, though clearly heard and homed in on by sharks, are actually not mimicking biologically based vibrations that are due to a dipole source, e.g., thrashing, struggling. Let it, first, be noted that Banner (1972) and also Nelson & Johnson (1976) showed that monopole sources (i.e., fish calls) do attract free-ranging sharks. There is also no question that the rapid and direct orientation to distant sound sources by sharks created the early, major stumbling block for the widely held theory by van Bergeijk (1964) that fishes cannot orient to sound from the acoustic far-field. Such studies aided in the early demise of that theory. Yet, when it comes to a strictly biological meaning, a problem exists with equating the characteristics of sounds produced by a monopole and a dipole source. A dipole sound doesn't have much of a compressional wave and, therefore, little sound propagation. Consequently, the region of effectiveness for such a source must be largely restricted to the acoustic near-field, which, depending on frequency, may reach many meters from the source. On the other hand, a monopole source, such as an underwater speaker, when provided sufficient energy, can readily produce a good compressional wave and associated water movements that can reach well into the far-field. I think it's past due that a dipole source, truly mimicking the thrashing actions of a prey item, be presented to free-ranging or range-limited sharks so as to determine effective distances of attraction by such a biologically relevant stimulus.

Directional hearing hypotheses

The three hypotheses, explaining possible mechanisms for directional hearing by sharks, need to be tested in the field with appropriate controls. Schuijf's hypothesis needs to be tested at depths that preclude the information necessary to recognize the direction of the source, i.e., transmitting a tonal sound at depths deeper than one quarter of the wavelength from either the surface or the bottom. Since tonal sounds have never attracted unconditioned sharks (personal observation), the hypothesis, theoretically, could be examined by instrumental conditioning of free-ranging sharks to tonal sounds. Serious consideration must be exercised, however, when contemplating the paradigm, using the attraction-response, with such animals. Corwin's hypothesis should be examined by neurosurgical techniques, rendering specific maculae insensitive to directional information. Also, sheltering the parietal fossa from sound must be considered. Kalmijn's hypothesis requires subjects to be precisely tracked to a dipole source, while not being able to orient, visually, to it. The hypothesis also predicts poor attraction from locations orthogonal to the axis of vibration.

Sex-based hearing capability hypothesis

The gender differences found in the macula neglecta of the skate, *Raja ocellata*, by Barber et al. (1985) and the indication noted by Carrier et al. (1994) that certain sounds serve a sexual function pose a most intriguing question. Are there other gender related sonic effects to be found among elasmobranchs? Elasmobranchs are clearly aware of much of the acoustical environment that surrounds them and there is no reason to assume that sexual differences in attending to that portion of the environment have no relevance.

Taxon-specific differences in the auditory central nervous system

Although there exists good evidence of structurally and physiologically definable acoustic centers in the elasmobranch brain, response properties of the various centers and associated pathways need further study, particularly among sharks. Since what we presently know suggests that elasmobranchs may well vary in their auditory capabilities, it seems that the choice of species may be critical as to demonstrating specific functions of centers and terminal fields.

Finally, despite the relatively few scientists who have examined either directly or indirectly the sense of hearing in the elasmobranchs, there are presently data and ideas which strongly indicate that numerous lines of rewarding research are available to those who choose to further our knowledge of such a remarkable sensory system.

Acknowledgements

I thank Samuel Gruber and Timothy Tricas for inviting me to participate in the symposium honoring Donald Nelson. The American Elasmobranch Society and Samuel Gruber provided the funds to participate. Gratitude is expressed to Jeffrey Corwin, Richard Fay, Carl Hopkins, Adrianus Kalmijn, Peter Klimley, and Arthur Popper for providing information necessary for this review and to Joseph Serafy and two anonymous reviewers for constructive comments on the manuscript.

References cited

Banner, A. 1967. Evidence of sensitivity to acoustic displacements in the lemon shark, *Negaprion brevirostris* (Poey). pp. 265–273. *In:* P.H. Cahn (ed.) Lateral Line Detectors, Indiana University Press, Bloomington.

Banner, A. 1968. Attraction of young lemon sharks, *Negaprion brevirostris*, by sound. Copeia 1968: 871–872.

Banner, A. 1972. Use of sound in predation by young lemon sharks, *Negaprion brevirostris* (Poey). Bull. Mar. Sci. 22: 251–283.

Barber, V.C., K.I. Yake, V.F. Clark & J. Pungur. 1985. Quantitative analyses of sex and size differences in the macula neglecta and ramus neglectus in the inner ear of the skate, *Raja ocellata*. Cell Tissue Res. 241: 597–605.

Barry, M.A. 1987. Afferent and efferent connections of the primary octaval nuclei in the clearnose skate, *Raja eglanteria*. J. Comp. Neurol. 266: 457–477.

Bergeijk, W.A. van. 1964. Directional and nondirectional hearing in fish. pp. 281–299. *In:* W.N. Tavolga (ed.) Marine Bioacoustics, Pergamon Press, New York.

Bergeijk, W.A. van. 1967. The evolution of vertebrate hearing. pp. 1–49. *In:* W.D. Neff (ed.) Contributions to Sensory Physiology, Academic Press, New York.

Berg, A.V. van den & A. Schuijf. 1983. Discrimination of sounds based on phase difference between particle motion and acoustic pressure in the shark *Chiloscyllium griseum*. Proc. R. Soc. London B. 218: 127–134.

Bullock, T.H. & J.T. Corwin. 1979. Acoustic evoked activity in the brain in sharks. J. Comp. Physiol. 129: 223–234.

Carrier, J.C., H.L. Pratt, Jr. & L.K. Martin. 1994. Group reproductive behaviors in free-living nurse sharks, *Ginglymostoma cirratum*. Copeia 1994: 646–656.

Clark, E. 1959. Instrumental conditioning of lemon sharks. Science 130: 127–128.

Corwin, J.T. 1977. Morphology of the macula neglecta in the shark genus *Carcharhinus*. J. Morph. 152: 341–361.

Corwin, J.T. 1978. The relation of inner ear structure to the feeding behavior in sharks and rays. Scanning Electron Microsc. 1978; II: 1105–1112.

Corwin, J.T. 1981a. Audition in elasmobranchs. pp. 81–102. *In:* W.N. Tavolga, A.N. Popper & R.R. Fay (ed.) Hearing and Sound Communication in Fishes, Springer-Verlag, New York.

Corwin, J.T. 1981b. Peripheral auditory physiology in the lemon shark: evidence of parallel otolithic and non-otolithic sound detection. J. Comp. Physiol. 142A: 379–390.

Corwin, J.T. 1981c. Postembryonic production and aging of inner ear hair cells in sharks. J. Comp. Neurol. 201: 541–553.

Corwin, J.T. 1983. Postembryonic growth of the macula neglecta auditory detector in the ray, *Raja clavata*: continual increases in hair cell number, neural convergence, and physiological sensitivity. J. Comp. Neurol. 217: 345–356.

Corwin, J.T. 1985. Over two million hair cells in a single auditory epithelium. Assoc. Res. Otolaryngol. Abstr. 8: 3–4.

Corwin, J.T. 1989. Functional anatomy of the auditory system in sharks and rays. J. Exp. Zool., Suppl. 2: 62–74.

Corwin, J.T. & R.G. Northcutt. 1982. Auditory centers in the elasmobranch brain stem: deoxyglucose autoradiography and evoked potential recording. Brain Res. 236: 261–273.

Dijkgraaf, S. 1963a. Sound reception in the dogfish. Nature 197: 93–84.

Dijkgraaf, S. 1963b. The functioning and the significance of the lateral line organs. Biol. Rev. 38: 51–105.

Fay, R.R., J.I. Kendall, A.N. Popper & A.L. Tester. 1974. Vibration detection by the macula neglecta of sharks. Comp. Biochem. Physiol. 47A: 1235–1240.

Flock, Å. 1967. Transducing mechanisms in the lateral line canal organ receptors. Cold Spring Harbor Symp. Quant. Biol. 30: 133–144.

Görner, P. & A.J. Kalmijn. 1989. Frequency response of lateral line neuromasts in the thornback ray (*Platyrhinoidis triseriata*). Proc. 2nd Intern. Cong. Neuroethology: 82.

Hawkins, A.D. & A.A. Myrberg, Jr. 1983. Hearing and sound communication underwater. pp. 347–405. *In:* B. Lewis (ed.) Bioacoustics, A Comparative Approach, Academic Press, New York.

Hopkins, C.D. 1988. Social communication in the aquatic environment. pp. 233–268. *In:* J. Atema, R.R. Fay, A.N. Popper & W.N. Tavolga (ed.) Sensory Biology of Aquatic Animals, Springer-Verlag, New York.

Kalmijn, A.J. 1988a. Hydrodynamic and acoustic field detection. pp. 83–130. *In:* J. Atema, R.R. Fay, A.N. Popper & W.N. Tavolga (ed.) Sensory Biology of Aquatic Animals, Springer-Verlag, New York.

Kalmijn, A.J. 1988b. Detection of weak electric fields. pp. 151–186. *In:* J. Atema, R.R. Fay, A.N. Popper & W.N. Tavolga (ed.) Sensory Biology of Aquatic Animals, Springer-Verlag, New York.

Kalmijn, A.J. 1997. Electric and near-field acoustic detection, a comparative study. Acta Physiol. Scand. 161 Suppl.: 25–38.

Kelly, J.C. and D.R. Nelson. 1975. Hearing thresholds of the horn shark, *Heterodontus francisci*. J. Acoust. Soc. Am. 58: 905–909.

Klimley, A.P. & A.A. Myrberg, Jr. 1979. Acoustic stimuli underlying withdrawal from a sound source by adult lemon sharks, *Negaprion brevirostris* (Poey). Bull. Mar. Sci. 29: 447–458.

Klimley, A.P. & D.R. Nelson. 1981. Schooling of the scalloped hammerhead shark, *Sphyrna lewini*, in the Gulf of California. U.S. Fish. Bull. 79: 356–360.

Kritzler, H. & L. Wood. 1961. Provisional audiogram for the shark, *Carcharhinus leucas*. Science 133: 1480–1482.

Lowenstein, O. & T.D.M. Roberts. 1951. The localization and analysis of the responses to vibration from the isolated elasmobranch labyrinth. A contribution to the problem of the evolution of hearing in vertebrates. J. Physiol. 114: 471–489.

Myrberg, A.A., Jr. 1978. Underwater sound – its effect on the behavior of sharks. pp. 391–417. *In:* E.S. Hodgson & R.F. Mathewson (ed.) Sensory Biology of Sharks, Skates and Rays, U.S. Government Printing Office, Washington D.C.

Myrberg, A.A., Jr., A. Banner & J.D. Richard. 1969. Shark attraction using a video-acoustic system. Mar. Biol. 2: 264–276.

Myrberg, A.A., Jr., S.J. Ha, S. Walewski & J.C. Banbury. 1972. Effectiveness of acoustic signals in attracting epipelagic sharks to an underwater sound source. Bull. Mar. Sci. 22: 926–949.

Myrberg, A.A., Jr., C.R. Gordon & A.P. Klimley. 1976. Attraction of free ranging sharks by low frequency sound, with comments on its biological significance. pp. 205–228. *In:* A. Schuijf & A.D. Hawkins (ed.) Sound Reception in Fish, Elsevier, Amsterdam.

Myrberg, A.A., Jr., C.R. Gordon & A.P. Klimley. 1978. Rapid withdrawal from a sound source by open-ocean sharks. J. Acoust. Soc. Am. 64: 1289–1297.

Nelson, D.R. 1967. Hearing thresholds, frequency discrimination, and acoustic orientation in the lemon shark, *Negaprion brevirostris* (Poey). Bull. Mar. Sci. 17: 741–768.

Nelson, D.R. & S.H. Gruber. 1963. Sharks: attraction by low frequency sounds. Science 142: 975–977.

Nelson, D.R. & R.H. Johnson. 1972. Acoustic attraction of Pacific reef sharks: Effect of pulse intermittency and variability. Comp. Biochem. Physiol. 42A: 85–89.

Nelson, D.R. & R.H. Johnson. 1976. Some recent observations on acoustic attraction of Pacific reef sharks. pp. 229–239. *In:* A. Schuijf & A.D. Hawkins (ed.) Sound Reception in Fish, Elsevier, Amsterdam.

Nelson, D.R., R.H. Johnson & L.G. Waldrop. 1969. Responses in Bahamian sharks and groupers to low-frequency, pulsed sounds. Bull. Southern Calif. Acad. Sciences 68: 131–137.

Northcutt, R.G. 1980. Central auditory pathways in anamniotic vertebrates. pp. 79–118. *In:* A.N. Popper & R.R. Fay (ed.) Comparative Studies of Hearing in Vertebrates, Springer-Verlag, New York.

Popper, A.N. & R.R. Fay. 1977. Structure and function of the elasmobranch auditory system. Amer. Zool. 17: 443–452.

Richard, J.D. 1968. Fish attracted with pulsed low-frequency sound. J. Fish. Res. Board Can. 25: 1441–1452.

Roberts, B.L. 1978. Mechanoreception and the behaviour of elasmobranch fishes with special reference to the acoustico-lateralis system. pp. 331–390. *In:* E.S. Hodgson & R.R. Mathewson (ed.) Sensory Biology of Sharks, Skates and Rays, U.S. Government Printing Office, Washington D.C.

Sand, O. 1981. The lateral line and sound reception. pp. 459–478. *In:* W.N. Tavolga, A.N. Popper & R.R. Fay (ed.) Hearing and Sound Communication in Fishes, Springer-Verlag, New York.

Schellart, N.A.M. & A.N. Popper. 1992. Functional aspects of the evolution of the auditory system of actinopterygian fish. pp. 295–322. *In:* D.B. Webster, R.R. Fay & A.N. Popper (ed.) The Evoluionary Biology of Hearing, Springer-Verlag, New York.

Schluger, J. & C.D. Hopkins. 1987. Electric fish approach stationary signal sources by following electric current lines. J. Exp. Biol. 130: 359–367.

Schuijf, A. 1974. Field studies of directional hearing in marine teleosts. Ph.D. Dissertation, University of Utrecht, Utrecht. 119 pp.

Schuijf, A. 1975. Directional hearing of cod (*Gadus morhua*) under approximate free field conditions. J. Comp. Physiol. 98: 307–332.

Schuijf, A. 1981. Models of acoustic localization. pp. 267–310. *In:* W.N. Tavolga, A.N. Popper & R.R. Fay (ed.) Hearing and Sound Communication in Fishes, Springer-Verlag, New York.

Schuijf, A. & R.J.A. Buwalda. 1980. Underwater localization – a major problem in fish acoustics. pp. 43–77. *In:* A.N. Popper & R.R. Fay (ed.) Comparative Studies of Hearing in Vertebrates, Springer-Verlag, New York.

Schuijf, A. & M.E. Siemelink. 1974. The ability of cod (*Gadus morhua*) to orient towards a sound source under approximate free field conditions. Experientia 30: 773–775.

Schuijf, A., J.W. Baretta & J.T. Wildschut. 1972. A field investigation on the discrimination of sound direction in *Labrus berggylta*. Neth. J. Zool. 22: 81–104.

Tester, A.L., J.I. Kendall & W.B. Milisen. 1972. Morphology of the ear of the shark genus *Carcharhinus* with particular reference to the macula neglecta. Pacif. Sci. 26: 264–274.

Urick, R.J. 1975. Principles of underwater sound. McGraw-Hill Book Co., New York. 384 pp.

Vilstrup, T. 1951. Structure and function of the membranous sacs of the labyrinth in *Acanthurus vulgaris*. Ejnar Munksgaard, Copenhagen. 134 pp.

Wisby, W.J., J.D. Richard, D.R. Nelson & S.H. Gruber. 1964. Sound perception in elasmobranchs. pp. 255–267. *In:* W.N. Tavolga (ed.) Marine Bio-Acoustics, Pergamon Press, New York.

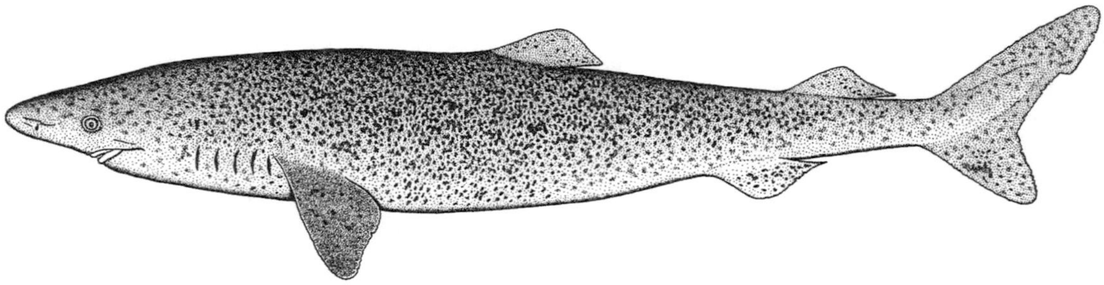

Female of the Greenland shark, *Somniosus microcephalus*, from a mounted specimen at Harvard Museum of Comparative Zoology, 2 m long (PV).

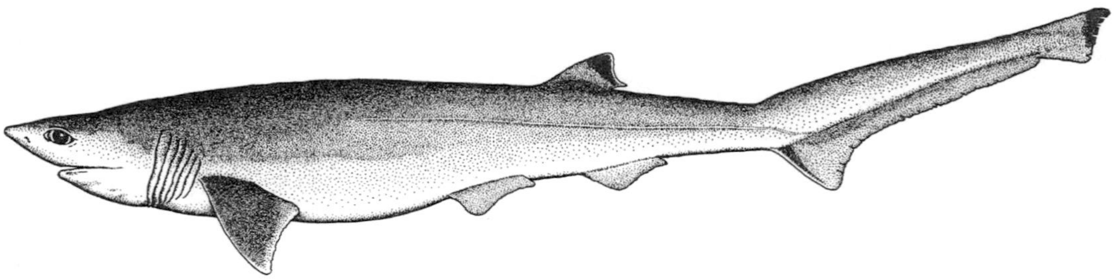

A female of the sharpnose sevengill shark, *Heptranchias perlo*, 80 cm long from Japan (PV).

Environmental Biology of Fishes **60:** 47–75, 2001.

Morphology of the mechanosensory lateral line system in elasmobranch fishes: ecological and behavioral considerations

Karen P. Maruska
Department of Biological Sciences, Florida Institute of Technology, 150 West University Blvd., Melbourne, FL 32901, U.S.A.
Present address: Department of Zoology, University of Hawaii at Manoa, Honolulu, HI 96822, U.S.A.
(e-mail: maruska@hawaii.edu)

Received 11 February 2000 Accepted 19 April 2000

Key words: batoid, canal, hair cell, mechanotactile, neuromast

Synopsis

The relationship between morphology of the mechanosensory lateral line system and behavior is essentially unknown in elasmobranch fishes. Gross anatomy and spatial distribution of different peripheral lateral line components were examined in several batoids (*Raja eglanteria*, *Narcine brasiliensis*, *Gymnura micrura*, and *Dasyatis sabina*) and a bonnethead shark, *Sphyrna tiburo*, and are interpreted to infer possible behavioral functions for superficial neuromasts, canals, and vesicles of Savi in these species. *Narcine brasiliensis* has canals on the dorsal surface with 1 pore per tubule branch, lacks a ventral canal system, and has 8–10 vesicles of Savi in bilateral rows on the dorsal rostrum and numerous vesicles ($\bar{x} = 65 \pm 6$ SD per side) on the ventral rostrum. *Raja eglanteria* has superficial neuromasts in bilateral rows along the dorsal body midline and tail, a pair anterior to each endolymphatic pore, and a row of 5–6 between the infraorbital canal and eye. *Raja eglanteria* also has dorsal canals with 1 pore per tubule branch, pored and non-pored canals on the ventral surface, and lacks a ventral subpleural loop. *Gymnura micrura* has a pored dorsal canal system with extensive branch patterns, a pored ventral hyomandibular canal, and non-pored canal sections around the mouth. *Dasyatis sabina* has more canal pores on the dorsal body surface, but more canal neuromasts and greater diameter canals on the ventral surface. *Sphyrna tiburo* has primarily pored canals on both the dorsal and ventral surfaces of the head, as well as the posterior lateral line canal along the lateral body surface. Based upon these morphological data, pored canals on the dorsal body and tail of elasmobranchs are best positioned to detect water movements across the body surface generated by currents, predators, conspecifics, or distortions in the animal's flow field while swimming. In addition, pored canals on the ventral surface likely also detect water movements generated by prey. Superficial neuromasts are protected from stimulation caused by forward swimming motion by their position at the base of papillar grooves, and may detect water flow produced by currents, prey, predators, or conspecifics. Ventral non-pored canals and vesicles of Savi, which are found in benthic batoids, likely function as tactile or vibration receptors that encode displacements of the skin surface caused by prey, the substrate, or conspecifics. This mechanotactile mechanism is supported by the presence of compliant canal walls, neuromasts that are enclosed in wide diameter canals, and the presence of hair cells in neuromasts that are polarized both parallel to and nearly perpendicular to the canal axis in *D. sabina*. The mechanotactile, schooling, and mechanosensory parallel processing hypotheses are proposed as future directions to address the relationships between morphology and physiology of the mechanosensory lateral line system and behavior in elasmobranch fishes.

Introduction

The mechanosensory lateral line system is present in all fishes and aquatic amphibians (Dijkgraaf 1962, Lannoo 1987, Northcutt 1992), and detects near field water movements relative to the skin surface (Harris & Van Bergejik 1962, Kalmijn 1989, Coombs 1994). The functional unit of the lateral line system is the neuromast receptor organ, which consists of sensory hair cells and support cells covered by a gelatinous cupula. Displacement of the cupula by viscous drag due to water movements causes modulation of the spontaneous primary afferent discharges sent to the mechanosensory processing centers in the hindbrain (Hoagland 1933, Münz 1985, Denton & Gray 1988).

Chondrichthyan fishes (holocephalans and elasmobranchs) have several types of mechanosensory lateral line end organs that are classified by morphology and location. Superficial neuromasts (or pit organs) are located on the skin surface either in grooves (batoids) or between modified scales (sharks) with their cupulae directly exposed to the water (Figure 1). In contrast, canal neuromasts are situated in sub-epidermal canals that are either in contact with the external environment via tubules that terminate in pores, or are isolated from the environment in non-pored canals. In chimaerid fishes, canal neuromasts are also located within a system of open grooves (Cole 1896, Ekström von Lubitz 1981). Vesicles of Savi consist of neuromasts enclosed in sub-epidermal pouches found primarily on the ventral surface of some torpedinid, narcinid, and dasyatid batoids (Savi 1844, Nickel & Fuchs 1974, Chu & Wen 1979). Spiracular organs are located in diverticula of the first visceral pouch in several fish taxa including elasmobranchs, and are stimulated by flexion of the cranial-hyomandibular joint (Barry et al. 1988a,b, Barry & Bennett 1989). The morphology and spatial distribution of these mechanoreceptors determines the distance range of the lateral line system, extent of the receptive field, frequency response properties, and which component of water motion is encoded (Denton & Gray 1983, 1988, Hoekstra & Janssen 1986, Münz 1989, Kroese & Schellart 1992). Elasmobranchs are among the few fish taxa to possess multiple types of mechanosensory receptors, and therefore provide an excellent opportunity to study structure-function-behavior relationships of different mechanoreceptors within a single group of fishes.

The mechanosensory lateral line system shows considerable morphological diversity among chondrichthyan species, but the functional significance of this diversity has received little attention. Current knowledge of lateral line function is based primarily on behavioral and physiological studies in a limited number of amphibian and teleost species, which do not possess many of the morphological specializations found in chondrichthyan taxa. Therefore, in the absence of these types of experiments in chondrichthyans, comparative studies on the peripheral organization and morphology of the lateral line system can provide valuable data to infer behavioral functions for the lateral line system in this group of fishes. For example, neuromast morphology determines which type of information (e.g. velocity, acceleration, or displacement) is encoded (Kroese & Schellart 1992), spatial distributions of lateral line components define the receptive field of the system (Denton & Gray 1983), and position of lateral line mechanoreceptors on the head and body can have behavioral implications. Therefore, the purpose of this study is to integrate data on the morphology and peripheral organization of the lateral line system in several elasmobranch species (*Narcine brasiliensis*,

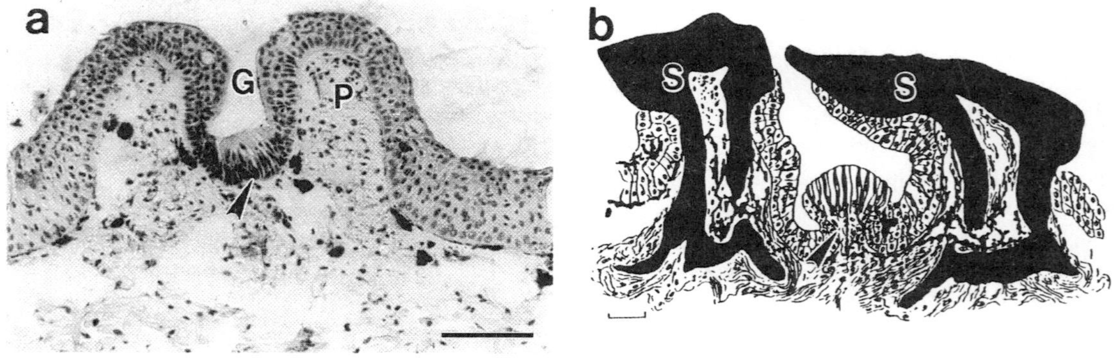

Figure 1. Morphology of superficial neuromasts in batoids and sharks. a– Transverse section of a single superficial neuromast papilla in the skate, *Raja eglanteria*. Superficial neuromasts are located on raised papillae (P) on the skin surface with a sensory epithelium (arrowhead) positioned at the base of a central groove (G). Scale bar = 100 μm. b– Schematic transverse section of a single superficial neuromast in the nurse shark, *Ginglymostoma cirratum*. Sensory epithelia (arrowhead) of shark superficial neuromasts are located between the bases of modified scales (S). Scale bar = 50 μm. Cupulae are not shown (modified from Budker 1958).

Raja eglanteria, *Gymnura micrura*, *Dasyatis sabina*, and *Sphyrna tiburo*) with their known ecology and behavior to infer possible biological functions for different mechanoreceptor types.

Past studies

The lateral line canal system was first discovered in chondrichthyans in the 1600's and was originally thought to be a mucus-producing organ in all fishes (Stenonis 1664). Direct histological evidence of the sensory nature of the lateral line canal neuromasts was provided by Leydig (1850), and subsequent physiological studies in teleost fishes determined that the lateral line was capable of detecting water movements at a range of approximately 1–2 body lengths (Hofer 1908, Hoagland 1933, Dijkgraaf 1934). Despite the number of studies on lateral line structure and function in teleost fishes and amphibians (Flock 1965a,b, Coombs et al. 1988, Coombs & Janssen 1989, Münz 1989, Görner & Mohr 1989), relatively few examine the system in cartilaginous fishes.

General morphology and organization of the canal portion of the lateral line system received the most attention from early anatomists, who used detailed descriptive terminology to label each section of the canal system based on anatomical landmarks and innervation patterns (e.g. Garman 1888, Johnson 1917). However, in subsequent years much of this comprehensive terminology was simplified so that only several canals (e.g. hyomandibular, infraorbital, supraorbital, mandibular, and posterior lateral line) are now recognized. The holocephalan fishes have a lateral line canal system that consists of canals and open grooves, which were suggested to represent an intermediate morphology between the free neuromasts of cyclostomes and the sub-epidermal canals of elasmobranchs and teleosts (Cole 1896, Garman 1888, Ekström von Lubitz 1981). Early studies in elasmobranch fishes described the lateral line canal system as a series of sub-epidermal canals that contained neuromasts arranged in an almost continuous sensory epithelium (Ewart 1892, Ewart & Mitchell 1892, Garman 1888, Johnson 1917). In addition to pure descriptions of canal distribution, several studies also provided data on the innervation and histology of the lateral line system in species such as the Greenland shark, *Somniosus microcephalus* (Ewart 1892), the skate, *Raja batis* (Ewart & Mitchell 1892), and others (Johnson 1917). These studies were followed by several other anatomical, histological, and finally ultrastructural examinations of elasmobranch lateral line canal systems (Tester & Kendall 1969, Roberts & Ryan 1971, Hama & Yamada 1977, Boord & Campbell 1977, Chu & Wen 1979), but references to its biological function in these fishes were few.

Many of these studies on the morphology and spatial distributions of elasmobranch lateral line systems also showed that the canal system in elasmobranchs differed from that of bony fishes, and therefore may have different response properties and biological functions. For example, elasmobranch lateral line systems differ from that in bony fishes due to the presence of non-pored canals, which contain innervated neuromasts. Early researchers had observed non-pored canals in elasmobranch fishes, but the functional or behavioral significance of these canals was not discussed (Garman 1888, Ewart & Mitchell 1892). Further, unlike the canals in most bony fishes (Webb & Northcutt 1997), pored lateral line canals in elasmobranchs contain multiple neuromasts between adjacent pores and the sensory epithelium is nearly continuous (Johnson 1917, Tester & Kendall 1969, Hama & Yamada 1977). This is the pattern present in many adult elasmobranch species, as well as in the primary canals of the ratfish *Chimaera monstrosa* (Ekström von Lubitz 1981), but its functional significance also remains unknown.

Distribution, innervation and morphology of elasmobranch superficial neuromasts (or pit organs) were first described in the skate *R. batis* (Ewart & Mitchell 1892). Elasmobranch pit organs were initially thought to function as chemoreceptors because of their structural similarity to taste buds and their sensitivity to monovalent cations such as potassium and sodium (Katsuki & Hashimoto 1969, Katsuki et al. 1969). However, other studies demonstrated that pit organs contain sensory hair cells, an overlying cupula, and are innervated by branches of lateral line nerves (Ewart & Mitchell 1892, Tester & Kendall 1967, Tester & Nelson 1969, Maruska & Tricas 1998). Therefore, pit organs in elasmobranchs are now considered to be mechanoreceptors comparable to the superficial (or free) neuromasts of teleost fishes.

Vesicles of Savi are another type of elasmobranch mechanoreceptor, which were first described in torpedo rays (Savi 1844). Subsequent morphological studies on the vesicles of Savi in *Torpedo* and *Narcine* revealed their close association with cartilaginous skeletal elements, the lack of connection with the external environment, their innervation by branches of lateral line nerves, and the presence of one large central and two smaller peripheral neuromasts within each vesicular pouch (Norris 1932, Nickel & Fuchs 1974, Barry & Bennett 1989). To date, vesicles of Savi are found only in some species of torpedinid, narcinid, and dasyatid batoids (Norris 1932, Nickel & Fuchs, 1974, Chu & Wen 1979, Barry & Bennett 1989, Maruska & Tricas 1998), but the function of these mechanoreceptors remains to be unequivocally demonstrated.

Thus, past studies on the morphology and distribution of lateral line canals and mechanoreceptors in chondrichthyan taxa are largely descriptive, and generally do not include complete descriptions of the entire mechanosensory system (e.g. canals, canal neuromasts, superficial neuromasts, vesicles of Savi) within a single species. More importantly, the majority of past studies fail to integrate morphological data with the ecology and behavior of each species in order to infer biological functions of the lateral line system in chondrichthyan fishes.

Materials and methods

General organization, distribution, and morphology of the mechanosensory lateral line system was examined in the lesser electric ray, *Narcine brasiliensis*, clearnose skate, *Raja eglanteria*, butterfly ray, *Gymnura micrura*, Atlantic stingray, *Dasyatis sabina*, and the bonnethead shark, *Sphyrna tiburo*. These species are all common inshore or estuarine elasmobranchs along the southeastern United States that feed in part on benthic invertebrate or teleost prey, and represent a range of body shapes. Canal and neuromast distributions were examined by gross dissection and standard histological techniques. Spiracular organs were not examined in this study and will not be discussed. Morphology and spatial distribution of the lateral line system was integrated with known ecology and behavior of each species in order to propose biological functions of different receptor types in these elasmobranch fishes.

Peripheral organization of the lateral line system

Adult and juvenile *R. eglanteria*, *G. micrura*, *N. brasiliensis* and *S. tiburo* of both sexes were collected in near shore waters off Cape Canaveral, Florida by bottom otter trawl at a depth of approximately 15 m. Adult *D. sabina* were collected by dip or seine net in the shallow waters (0.2–1 m) of the Banana River estuary, Florida. Atlantic stingrays were used as a representative batoid to quantify differences between dorsal and ventral lateral line canal systems, and for scanning electron microscopy (see below). All specimens were euthanized on ice, fixed in 10% formalin in seawater, and stored in 50% isopropyl alcohol. The distribution of canals on the dorsal and ventral surface was mapped following pressure injection of an aqueous 0.5% methylene blue or a 0.5% toluidine blue solution (in 20% isopropyl alcohol) into the canals and visualized with a dissecting microscope. Multiple injections were performed at different locations and then overlying or underlying epidermal tissue was removed to visualize and map the distribution of all of the canals and tubules on the head and body in each species. Tubules are defined as extensions of the main canal which lack sensory neuromasts, and terminate in pores at the skin surface. A total of at least five individuals of each species were used to determine canal distributions. Terminology for classification of lateral line canals as hyomandibular (including the subpleural loop), infraorbital, supraorbital, scapular, and mandibular follows that of Ewart & Mitchell (1892), Tester & Kendall (1969), Chu & Wen (1979), and Maruska & Tricas (1998). The term 'posterior lateral line' which is based on innervation of this canal by the posterior lateral line nerve is used instead of 'lateral canal' used by earlier researchers (e.g. Ewart & Mitchell 1892, Tester & Kendall 1969, Chu & Wen 1979). Distribution of the vesicles of Savi in *N. brasiliensis* was determined by removal of epidermal tissue, application of an aqueous 0.5% methylene blue solution, and subsequent dissection to expose the neuromasts within the vesicular pouches. The number of vesicles of Savi on the dorsal and ventral surfaces was also counted in 6 individuals (disk widths 9.0–13.0 cm).

The distribution of superficial neuromasts was examined only in the clearnose skate, *R. eglanteria*. Superficial neuromasts in batoids are located on raised papillae that contain a central groove with a sensory epithelium at the base of this groove. These papillae were easily visualized with a dissecting microscope in freshly euthanized animals. All superficial neuromasts were marked with colored latex paint to map their distribution. The orientation of grooves that contain superficial neuromasts were recorded as the angular deviation from the rostrocaudal body axis and midsagittal plane.

Canal neuromasts, superficial neuromasts and vesicles of Savi along with surrounding tissue were removed from 10% formalin-fixed, isopropyl alcohol preserved specimens and prepared for histological analysis. Tissue was dehydrated in a graded ethanol series, cleared in toluene, and infiltrated with and embedded in paraffin (Paraplast®). Tissue was sectioned in the transverse or sagittal plane at 10 μm, mounted on chrom-alum coated slides, and stained with Mayer's or Ehrlich's hematoxylin and eosin (Presnell & Schreibman 1997). The anatomy of the canal neuromasts, superficial neuromasts and vesicles of Savi was examined with a compound microscope.

The morphology of the lateral line canals on the dorsal and ventral surface was quantitatively compared in a representative batoid species, *D. sabina*. Lateral line canal pores were marked with toluidine blue solution or latex paint and viewed with a dissecting microscope to perform counts. In addition, each lateral line canal was injected with a 0.5% toluidine blue solution in order to stain the cupulae and neuromasts. The canals were then cut open and individual neuromasts were counted. The total number of lateral line canal pores and canal neuromasts was counted on the left side of each ray and doubled (assuming bilateral symmetry) on both the dorsal and ventral surface of five individuals (disk width 22–26 cm). Enumeration did not include the portion of the posterior lateral line canal on the tail. Diameter of the pored hyomandibular canal on the dorsal surface and non-pored hyomandibular canal on the ventral surface was also measured (neuromast base to canal roof) with an ocular micrometer from transverse histological sections in five stingrays (disk width 22–26 cm). Number of pores, neuromasts, and canal diameters were compared between the dorsal and ventral surface with an unpaired Student's t-test.

Hair cell morphology

Neuromasts from the non-pored hyomandibular and infraorbital canals on the ventral surface of *Dasyatis sabina* were examined with scanning electron microscopy (SEM) to describe neuromast and hair cell morphology and to determine hair cell polarity. The hyomandibular canal lateral to the gill slits or infraorbital canal near the mouth was dissected open in freshly euthanized rays to expose the canal neuromasts. Cupulae were then mechanically removed with a water jet. A 1.0–1.5 cm section of canal including neuromasts was removed, fixed in 2% glutaraldehyde in Millonig's buffer, rinsed in Millonig's buffer, rinsed in 0.1 M phosphate buffer, post-fixed in 1% osmium tetroxide, rinsed in 0.1 M phosphate buffer, and dehydrated in an ethanol series. Tissue was dried in an LADD critical point dryer with carbon dioxide as a transitional fluid and sputter coated with gold-palladium alloy. Neuromasts were viewed with a Hitachi S-2700 SEM at an accelerating voltage of 8–10 kV and images were recorded on VHS tape for analysis. Each hair cell has many stereocilia in a stepwise arrangement of increasing height towards a single kinocilium. The polarity of each hair cell was measured as the angular deviation of the axis of maximum excitation of the hair cell (towards the stepwise stereocilia and single kinocilium) from the longitudinal axis of the canal. Therefore, 0° and 180° represent hair cells oriented in opposite directions along the longitudinal canal axis. Measurement error was estimated to be ±15°. Hair cell polarities were determined for 150 randomly selected hair cells from 15 neuromasts in the hyomandibular canal in 6 stingrays (disk width 22–26 cm).

Results

Peripheral organization of the lateral line system

The mechanosensory lateral line system of elasmobranch fishes consists of superficial neuromasts on the skin surface, a sub-epidermal canal system, and the sub-epidermal vesicles of Savi. The presence, number and distribution of each mechanoreceptor type varies among species. Therefore, the peripheral organization of the lateral line system in *Narcine brasiliensis*, *Raja eglanteria*, *Gymnura micrura*, *Dasyatis sabina* and *Sphyrna tiburo* will be described separately below. Several morphological features of the lateral line system in each species are summarized in Table 1.

Narcine brasiliensis (Batoidea: Narcinidae)
The lesser electric ray, *N. brasiliensis*, has both vesicles of Savi and a lateral line canal system on the dorsal surface (Figure 2). Vesicles of Savi are located in bilateral rows on the rostrum from the anterior edge of the eye to the edge of the rostrum. There are approximately 8–10 vesicles in each row just lateral to the rostral cartilage. Vesicle pouches are oval with their long axis oriented approximately 45° to the rostrocaudal body axis. Vesicles within a rostrocaudal row have a common orientation.

All canals on the dorsal body surface have pores except for a short section of the hyomandibular that

Table 1. Summary of morphological features of the lateral line system in *Narcine, Raja, Gymnura, Dasyatis*, and *Sphyrna*.

Species	Surface	SN location (#)	VS location (#)	Pored canals	# pores per tubule branch*	Non-pored canals
Narcine brasiliensis	Dorsal	ND	Bilateral rows along rostrum midline (8–10 per row)	HYO, IO, SO, PLL	1	None
	Ventral	ND	Rostrum and anterior to electric organs ($\bar{x}$ = 65 ± 6 SD per side)	None	NA	None
Raja eglanteria	Dorsal	Bilateral medial rows to end of tail, anterior to EP (2), between IO canal and eye (5–6)	None	HYO, IO, SO, PLL	1	SO
	Ventral	None	None	HYO, IO (SPL absent)	1	IO, MAN, SO
Gymnura micrura	Dorsal	ND	None	HYO, IO, SO, PLL	1–35	SO
	Ventral	ND	None	HYO, IO (SPL present)	1–3	HYO, IO, MAN, SO
*Dasyatis sabina***	Dorsal	Bilateral medial rows to end of tail (~100 per side)	None	HYO, IO, SO, PLL	1–10	SO
	Ventral	None	Bilateral rows along rostrum midline (6–10 per row)	HYO, MAN (SPL present)	1	HYO, IO, SO
Sphyrna tiburo	Dorsal	Lateral along PLL and on dorsolateral body (> 400 per side), anterior to EP (2)†	None	HYO, IO, SO, PLL	1–20	IO, HYO
	Ventral	Mandibular row, umbilical row†	None	HYO, IO, SO (SPL NA)	1–20	IO, MAN

Some canals contain pored and non-pored sections and are listed under both categories. *numbers represent the range of average values from all canals; **from Maruska & Tricas (1998); †from Tester & Nelson (1969); EP = endolymphatic pore, HYO = hyomandibular, IO = infraorbital, MAN = mandibular, ND = not determined, NA = not applicable, SO = supraorbital, SPL = subpleural loop, SN = superficial neuromast, VS = vesicles of Savi.

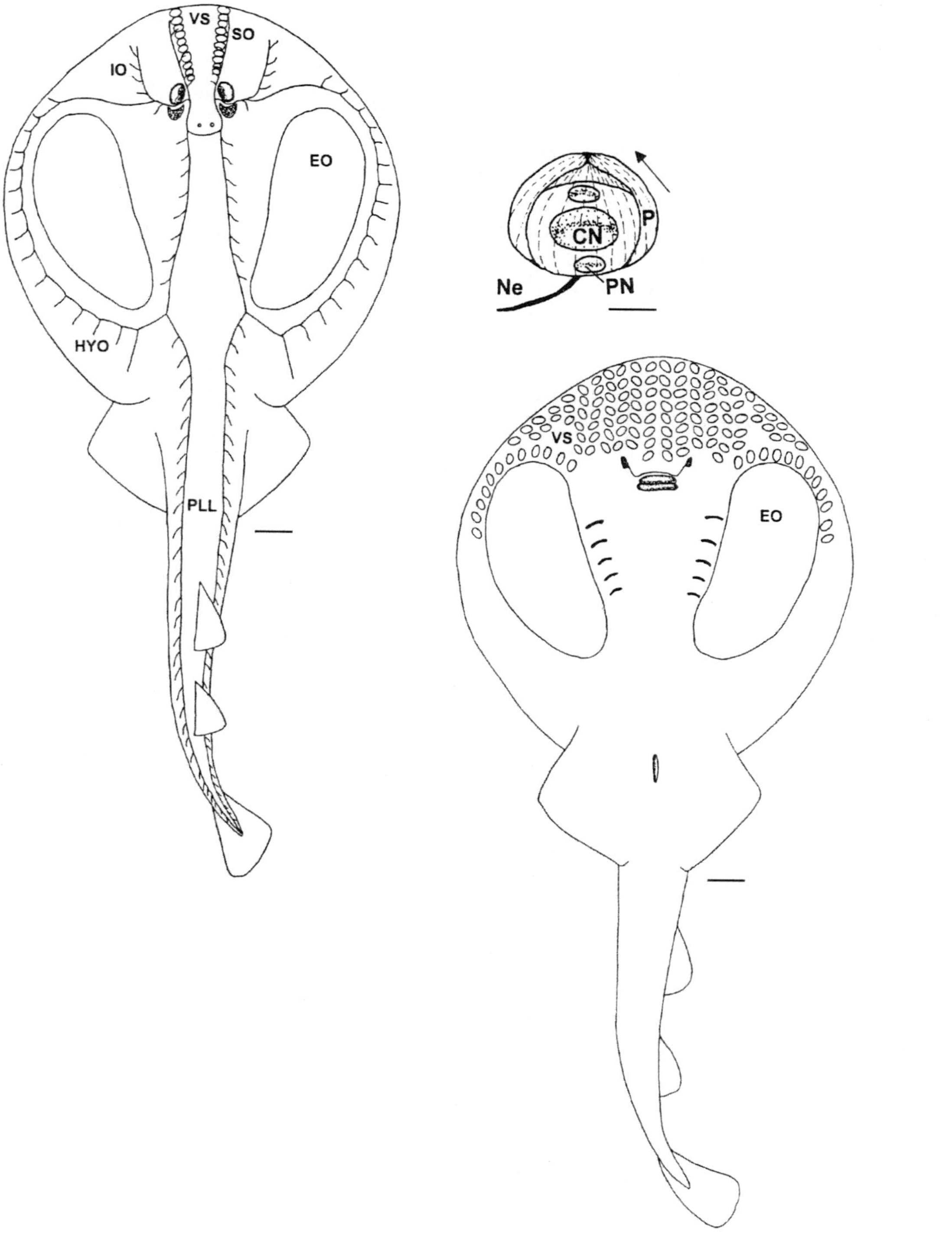

Figure 2. Distribution of the lateral line canal system and vesicles of Savi on the dorsal (upper) and ventral (lower) surface of the lesser electric ray, *Narcine brasiliensis*. Canals on the dorsal surface are bilateral, interconnected and pored, while the ventral surface lacks a canal system. Vesicles of Savi (VS) are arranged in a single row just medial to the supraorbital canal (SO) on the dorsal surface, and in several rows on the rostrum and along the anterior edge of the electric organ (EO) on the ventral surface. Scale bars = 1 cm. Schematic diagram of a single vesicle of Savi in *N. brasiliensis* (upper right). Each vesicle contains a large central neuromast (CN) and two smaller peripheral neuromasts (PN) covered by separate cupulae (not shown) all enclosed in a thin-walled pouch (dashed lines). A connective tissue pedicle (P) supports the vesicle on either side of the neuromasts. Arrow lies along the rostrocaudal body axis. Scale bar = 400 μm (HYO = hyomandibular canal, IO = infraorbital canal, Ne = nerve, PLL = posterior lateral line canal).

is connected to the infraorbital canal just anterior to the electric organ (Figure 2). The main hyomandibular canal extends from the infraorbital canal near the spiracle, to the disk margin where it then caudally circumscribes the electric organ. This canal then joins the posterior lateral line canal near the midline. The hyomandibular canal has lateral tubules (0.5–1.5 cm in length) that terminate in single pores near the disk margin. Tubules, which lack sensory neuromasts, are extensions of the main canal that terminate in pores at the skin surface. The dorsal infraorbital canal extends from the supraorbital canal on the head, between the eye and spiracle, and then terminates near the disk margin. The dorsal supraorbital canal is located on the cranium and extends from near the endolymphatic pores, where it is connected across the midline by a commissural canal, to the rostrum tip just lateral to the vesicles of Savi. Several (3–5) short tubules extend medially from the supraorbital canal and terminate in pores directly dorsal to the vesicles of Savi. The posterior lateral line canal begins at the supraorbital canal near the endolymphatic pores along the cranium. This canal extends to the tip of the tail and has numerous lateral tubules that terminate in pores ventral to the main canal along its entire length.

The ventral lateral line system completely lacks canals, but has a complex of vesicles of Savi (Figure 2). Vesicles of Savi are located on the ventral rostrum anterior and lateral to the mouth, and along the anterior edge of the electric organs. Vesicles on the rostrum are arranged in 6–7 rostrocaudal rows per side, each of which contains 7–10 vesicles. Each side of the ventral body surface contains approximately 65 vesicles ($\bar{x} = 65 \pm 6$ SD, $n = 6$). Each vesicle is oval in shape and oriented approximately 45° to the rostrocaudal body axis. All vesicles within an individual row have a common orientation, but are oriented orthogonal to vesicles in the adjacent row. Individual vesicles of Savi have no connection to the external environment and are located in pouches approximately 0.5–2.0 mm below compliant epidermal, dermal and connective tissue layers (Figure 3a–d). In addition, there is no evidence that the compartments of adjacent vesicles are contiguous. Each vesicle consists of a large central neuromast and two smaller adjacent neuromasts, all three of which are innervated (Figure 3b). The cupula of the central neuromast appears more dense than the cupulae of the smaller adjacent neuromasts (observed in fresh tissue). The walls of each individual pouch are thin and the roof over the large central neuromast is supported by an arch formed by connective tissue pedicles that extend from the floor of the vesicle (Figure 2). The vesicles are also attached at their bases to a connective tissue band, which is closely associated with the overlying cartilage (Figure 3c).

Raja eglanteria (Batoidea: Rajidae)

The clearnose skate, *R. eglanteria*, has superficial neuromasts located on raised papillae with the sensory neuromast at the base of a central groove. The sensory epithelium is likely covered by a cupula that is in contact with the external environment, but cupulae were not observed in this species possibly because this structure is often lost during histological processing. Superficial neuromasts occur in bilateral rows along the dorsal midline from the suprabranchial region to the caudal fin, and are located on the dorsolateral surface of the tail near the median spines (Figure 4). In addition, a group of 5–6 superficial neuromasts are found between the infraorbital canal and the eye, and a single pair of superficial neuromasts is located anterior to each endolymphatic pore. Superficial neuromasts near the eye have a central groove oriented almost parallel to the rostrocaudal body axis while those anterior to the endolymphatic pores are oriented approximately 135° to the rostrocaudal body axis. Central grooves located on the body and tail are oriented between 90° and 135° to the rostrocaudal body axis in the transverse plane.

The canals on the dorsal surface of *R. eglanteria* are bilateral, interconnected and independently penetrate the disk to join the canals on the ventral surface (Figure 5). The hyomandibular canal extends caudally and laterally from the infraorbital canal on the rostrum, along the pectoral fin margin where it then extends medially to join the anterior branch of the scapular canal. A posterior branch of the scapular canal extends from the posterior lateral line canal to the caudal disk edge several centimeters posterior to the anterior branch of the scapular canal. The hyomandibular and scapular canals contain straight tubules, which extend anteriorly to terminate in single pores. The infraorbital canal is contiguous with the supraorbital canal between the eye and spiracle, and extends rostrally to the edge where it penetrates the disk to join the infraorbital canal on the ventral surface. The infraorbital canal also contains short tubules that extend laterally or rostrally from the main canal. The supraorbital canal on the head extends anteriorly to the tip of the rostrum where it joins the supraorbital canal on the ventral surface. A deep commissural canal posterior to the endolymphatic pores connects the dorsal supraorbital canals on the left and right sides of the body. Several tubules extend laterally from the supraorbital canal and terminate in single pores in the region rostral to the eyes. However, there is a section of supraorbital canal on the rostrum that is non-pored. The posterior lateral line canal begins near the endolymphatic pores and extends along the midline to the tip of the tail. Tubules extend ventral to the posterior lateral line canal from the scapular canal junction to the caudal edge of the pelvic fins, and both dorsal and ventral to the posterior lateral line canal from the pelvic fins to the tip of the tail.

The canals on the ventral surface are also interconnected (except mandibular) and independently penetrate the disk to join the canals on the dorsal surface (Figure 5). The hyomandibular canal extends rostrally from near the caudal edge of the pectoral fin, around the gill slits, where it joins the infraorbital canal just lateral to the nares. Several tubules extend laterally from this canal in the vicinity of the gill slits, and medially in the area caudal to the gill slits to terminate in single pores. The clearnose skate has no subpleural loop of the ventral hyomandibular canal as seen in most other batoid species (Chu & Wen 1979). The infraorbital canal extends from the hyomandibular canal, medially to the posterior edge of the naris and along the rostrum midline. There is a single pore in the canal section between the hyomandibular canal and naris, while the infraorbital canal that extends along the midline of the rostrum contains several tubules that terminate in single pores medially. Another section of infraorbital canal is contiguous with the hyomandibular canal lateral to the naris, and extends to the anterior pectoral fin margin where it penetrates the disk to join the infraorbital canal on the dorsal surface. The ventral infraorbital canals are joined across the midline by a commissural connection anterior to the mouth. The ventral supraorbital canal extends from the infraorbital canal just anterior to the infraorbital–hyomandibular canal junction, extends to the anterior disk margin just lateral to the infraorbital canal, and penetrates the disk at the rostrum tip to join the supraorbital canal on the dorsal surface. A short mandibular canal is located posterior to the lower jaw and extends across the midline as a single canal that lacks pores. Vesicles of Savi were not observed in this species.

Gymnura micrura (Batoidea: Gymnuridae)

The canal system on the dorsal surface of the butterfly ray, *G. micrura*, consists of the same main canal pattern found in other batoid species. However, this species differs due to the profuse number of tubules that terminate in pores over the entire body surface (Figure 6). The hyomandibular canal begins at the rostrum, extends along the entire disk margin, and joins the posterior lateral line canal near the caudal edge of the pectoral fin. Tubules extend from this canal, often branch several times, and terminate in approximately 1–10 pores near the disk margin. The dorsal infraorbital canal is contiguous with the supraorbital canal on the head, extends between the eye and spiracle, and then bifurcates into two divisions just below the eye. The caudal division branches extensively in the mid-disk region, whereas the anterior division branches over the anterior disk and then penetrates the rostrum to join the infraorbital canal on the ventral surface. The dorsal supraorbital canal is located on top of the cranium and has numerous tubules that are medial and caudal to the eye and spiracle. The supraorbital canal then extends anteriorly to the tip of the rostrum where it penetrates the disk to join the supraorbital canal on the ventral surface. The posterior lateral line canal extends from the supraorbital canal on the head, along the dorsal midline, to the tip of the short tail. A scapular canal loop is also present in the mid-disk region near the dorsal midline. All the canals on the dorsal surface have numerous tubules that terminate in multiple pores per tubule over the entire body surface, with the exception of a short segment of the supraorbital canal on the rostrum that lacks pores.

The lateral line system on the ventral surface consists of both pored and non-pored canals that are bilateral, interconnected (except mandibular) and independently penetrate the disk to join the canals on the dorsal surface (Figure 6). The pored hyomandibular canal extends laterally along the pectoral fin margin from the midline of the rostrum to the widest point of the pectoral fin, where it then extends medially. The hyomandibular canal then shows a caudal change in position before transition to the non-pored section of this canal on the caudal region of the pectoral fin. Many tubules extend from this canal and terminate in pores at the disk margin. Each hyomandibular canal tubule on the anterior disk generally branches at a point several millimeters from the disk edge and terminates in 2–3 pores. In contrast, each tubule on the caudal portion of the subpleural loop of the hyomandibular canal generally terminates in only a single pore. The non-pored hyomandibular canal section then extends anteriorly from the caudal region of the pectoral fin along the

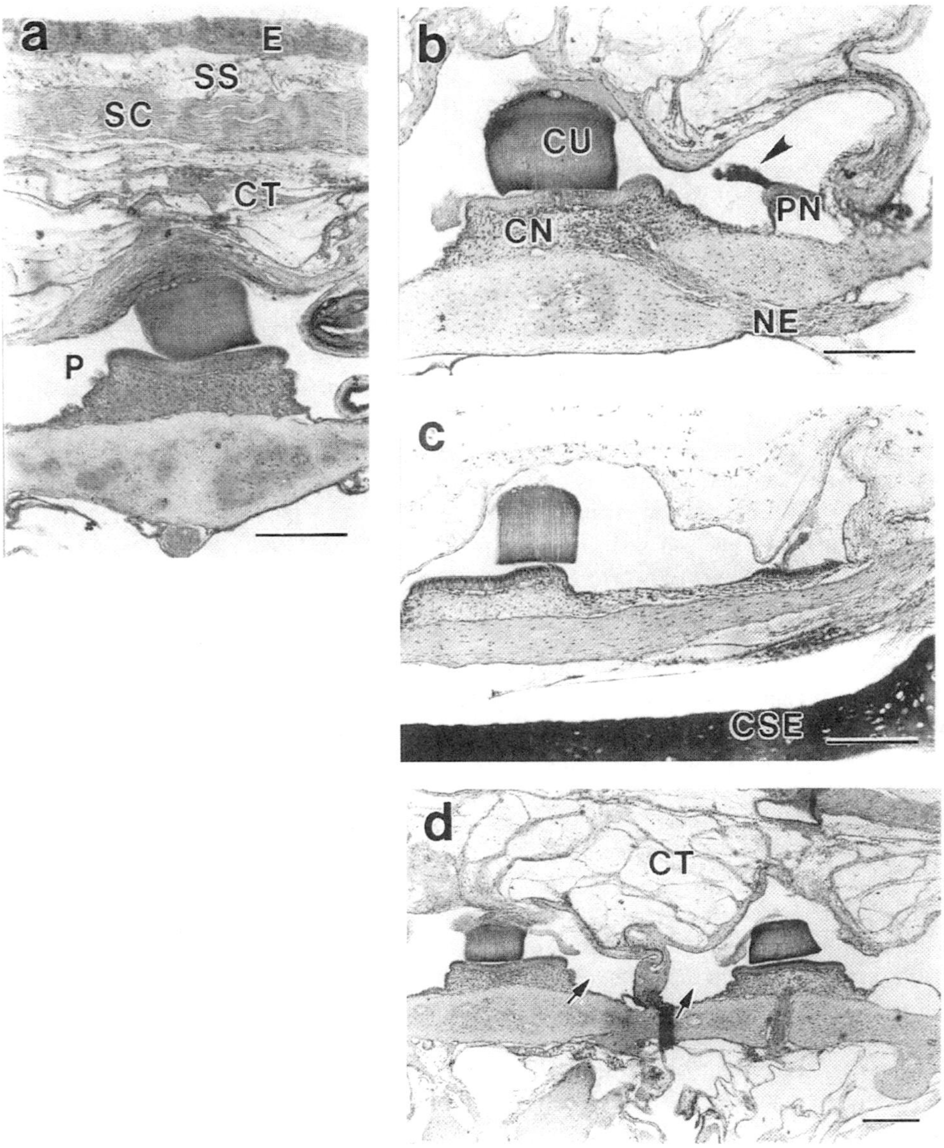

Figure 3. Histological sections of the vesicles of Savi on the ventral surface of the lesser electric ray, *Narcine brasiliensis*. a – Vesicles of Savi are located in sub-epidermal pouches (P) below epidermal (E), stratum spongiosum (SS), stratum compactum (SC) and loose connective tissue (CT) skin layers. Scale bar = 200 μm. b – Each vesicle of Savi consists of a large central neuromast (CN) and two smaller peripheral neuromasts (PN), all innervated by nerve fibers (NE). The cupula (CU) of the central neuromast is dense with a striated appearance, while the cupula (arrowhead) of the smaller peripheral neuromasts is gelatinous and similar in structure to canal neuromast cupulae. Scale bar = 200 μm. c – Vesicles of Savi are closely associated with cartilaginous skeletal elements (CSE) at their base. Scale bar = 200 μm. d – Vesicles of Savi are located in distinct rows along the ventral rostrum. Lumina of adjacent vesicles (arrows) do not appear connected but lie approximately 100 μm apart below loose connective tissue (CT). Scale bar = 200 μm. Ventral side is up.

midline just lateral to the gill slits, where it joins the non-pored sections of the infraorbital and supraorbital canals near the mouth. The infraorbital canal extends laterally from this point to form a large loop that contains straight tubules, which overlie the hyomandibular canal tubules along the disk margin. The infraorbital canal also extends medially and penetrates the disk adjacent to the hyomandibular canal on the rostrum

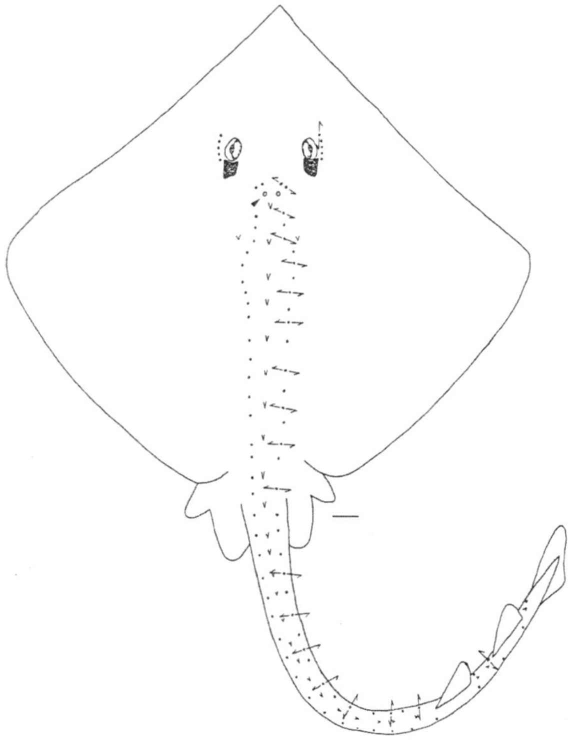

Figure 4. Distribution of superficial neuromasts on the dorsal surface of the clearnose skate, *Raja eglanteria*. Each dot represents a single superficial neuromast papilla and arrows (right side) show the groove orientation on every other neuromast. A single pair of superficial neuromasts is located anterior to each endolymphatic pore (arrowhead), a row of 5–6 neuromasts is positioned beneath each eye within the infraorbital canal, and the remainder of superficial neuromasts extend from the suprabranchial region along the midline to the end of the tail. Superficial neuromast grooves near the endolymphatic pores are positioned at approximately 135°, grooves beneath the eye at 0° (parallel), and grooves along the midline at 90°–135° to the rostrocaudal body axis. Scale bar = 1 cm.

to join the infraorbital canal on the dorsal surface. The supraorbital canal also extends rostrally from this point, forms a small non-pored loop, and then shows a medial change in position to join an additional section of the supraorbital canal on the rostrum. A short mandibular canal is located posterior to the lower jaw and extends across the midline as a single non-pored canal. Vesicles of Savi were not observed in this species.

Dasyatis sabina (Batoidea: Dasyatidae)

The distribution of lateral line canals, superficial neuromasts, and vesicles of Savi in the Atlantic stingray, *D. sabina*, was previously described in detail by Maruska & Tricas (1998). Dorsal canals are primarily pored and include the supraorbital on the cranium, infraorbital beneath the eye, hyomandibular on the pectoral fins, scapular loop on the caudal trunk and posterior lateral line along the dorsal midline and tail (Figure 7). The ventral lateral line system of the stingray consists of both pored and non-pored canals as well as vesicles of Savi. Vesicles of Savi are located in bilateral rows along the midline of the rostrum and are contiguous with the ventral supraorbital canal. Non-pored canals on the ventral surface include the supraorbital and infraorbital around the mouth, nares and rostrum, and a section of hyomandibular canal along the midline just lateral to the gill slits. Pored canals on the ventral surface include the hyomandibular canal that forms the subpleural loop, and the short mandibular canal posterior to the lower jaw.

The total number of canal pores and neuromasts was counted on both the dorsal and ventral surface, and the diameter of the dorsal pored hyomandibular and ventral non-pored hyomandibular canal was measured in the stingray to quantitatively compare dorsal and ventral canal organization. There are over 250 pores on the dorsal surface and less than 90 on the ventral surface (Figure 8a) of the stingray. However, there are about twice as many neuromasts on the ventral surface ($\bar{x} = 1028 \pm 51$ SE) than on the dorsal surface ($\bar{x} = 570 \pm 32$ SE) (Figure 8b). The diameter of the non-pored hyomandibular canal on the ventral surface is on average 2.5 times larger than the diameter of the hyomandibular canal on the dorsal surface (Figure 8c). There is a difference between the dorsal and ventral surface in *D. sabina* for all comparisons tested (unpaired t-test, $p < 0.001$). Although not quantitatively examined, these relative differences between dorsal and ventral lateral line systems are also present in *R. eglanteria*, *N. brasiliensis* and *G. micrura*. In addition, the morphology and structure of the canal system on the ventral surface differs from the dorsal in all batoids examined. Generally, ventral canals are large in diameter, have compliant canal walls, and lie deep to loose, pliable dermal and connective tissue layers (Figure 9a–c). In contrast, the dorsal canals are smaller in diameter, have rigid canal walls, and lie deep to dense dermal and connective tissue layers (Figure 9d). However, further quantitative histological evaluations of the skin and canal wall structures are needed to confirm relative stiffness of the canals on the dorsal versus ventral surface in batoids.

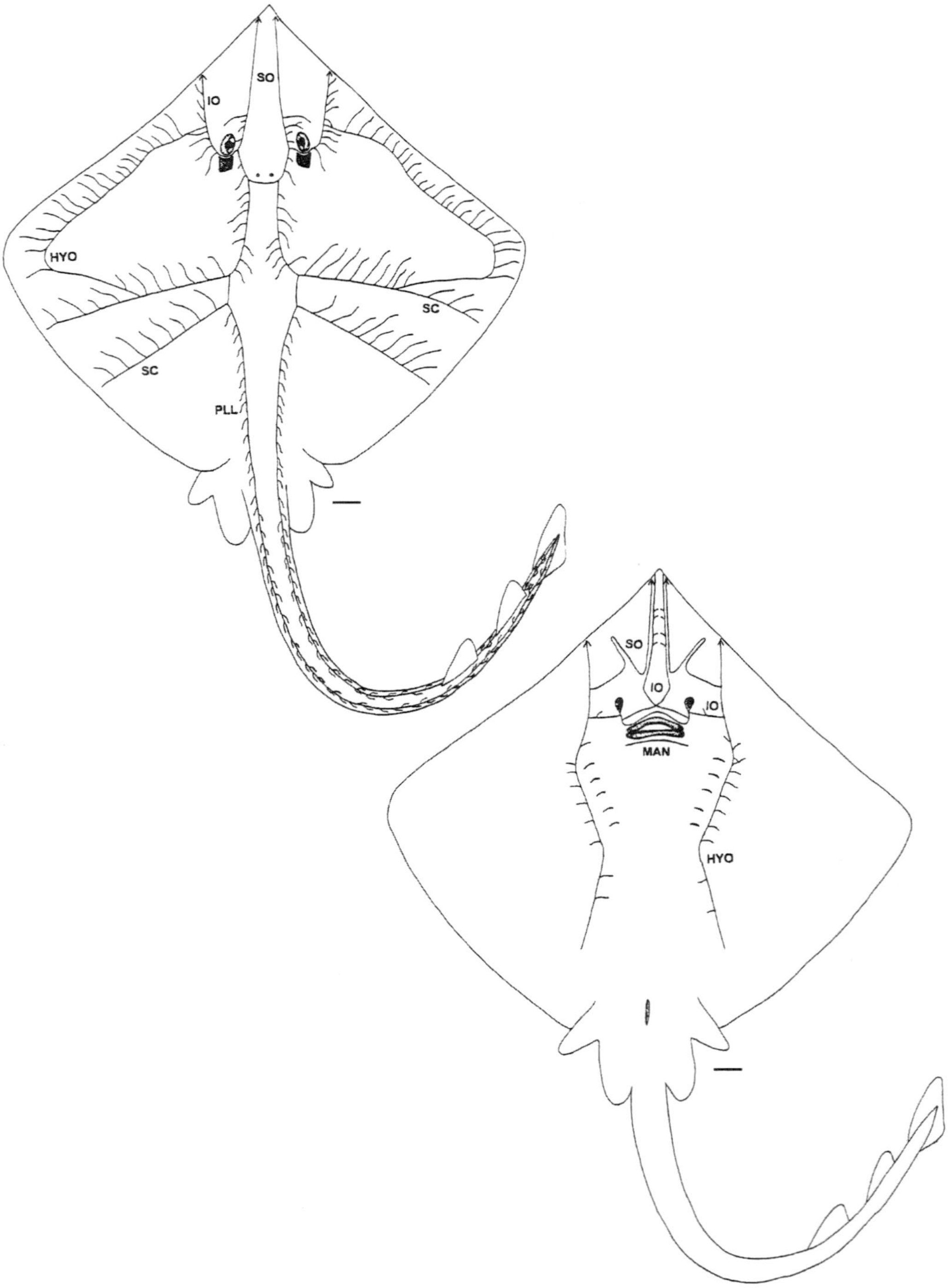

Figure 5. Distribution of the lateral line canal system on the dorsal (upper) and ventral (lower) surface of the clearnose skate, *Raja eglanteria*. All canals (except mandibular) are interconnected both among and within sides. Straight tubules extend from the main canals to terminate in single pores. The ventral hyomandibular canal does not contain a subpleural loop in this species. Scale bars = 1 cm (HYO = hyomandibular canal, IO = infraorbital canal, MAN = mandibular canal, PLL = posterior lateral line canal, SC = scapular canal, SO = supraorbital canal).

Figure 6. Distribution of the lateral line canal system on the dorsal (upper) and ventral (lower) surface of the butterfly ray, *Gymnura micrura*. All canals (except mandibular) are interconnected both among and within sides with extensive tubule branching on the dorsal surface. The ventral system consists of both pored canals, and non-pored canals along the midline and around the mouth. Scale bars = 1 cm (HYO = hyomandibular canal, IO = infraorbital canal, MAN = mandibular canal, PLL = posterior lateral line canal, SC = scapular canal, SO = supraorbital canal, SPL = subpleural loop).

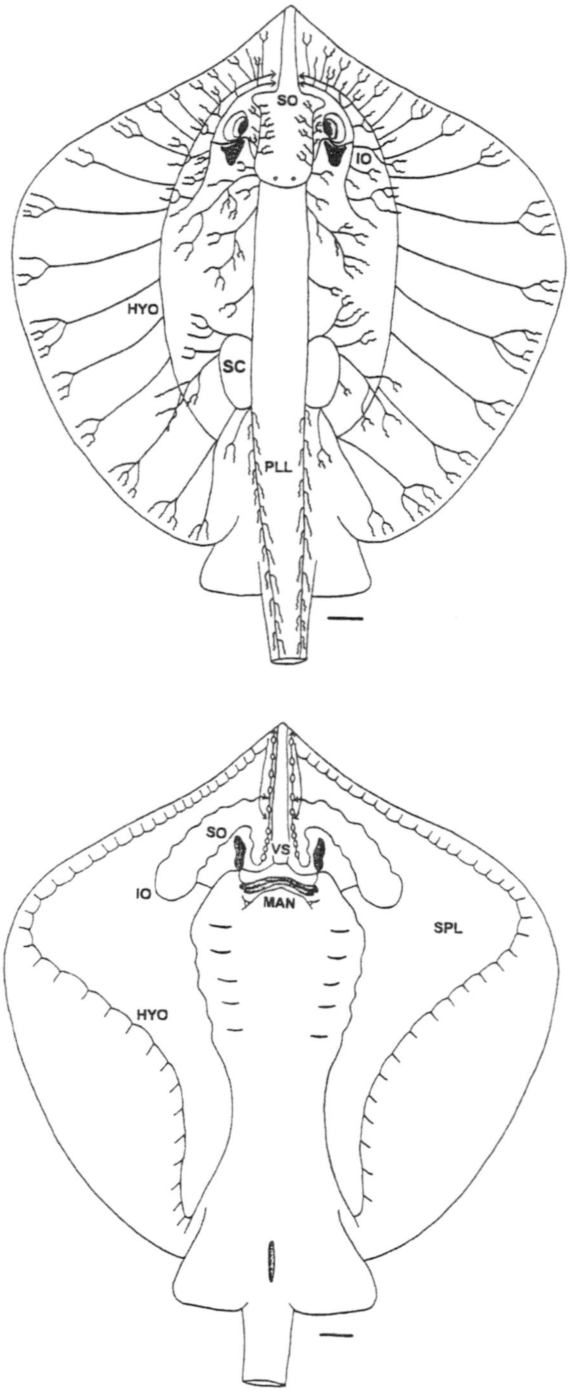

Figure 7. Distribution of the lateral line canal system and vesicles of Savi on the dorsal (upper) and ventral (lower) surface of the Atlantic stingray, *Dasyatis sabina*. Dorsal canals contain numerous lateral tubules that terminate in pores across the entire body surface. The infraorbital, supraorbital, and sections of the hyomandibular canal near the mouth, rostrum, and along the ventral midline, lack pores but do contain innervated neuromasts. Vesicles of Savi (VS) are located in bilateral rows on the ventral rostrum midline and are isolated from the surrounding water, but lumina of adjacent vesicles are connected via tubules. Scale bars = 1 cm (HYO = hyomandibular canal, IO = infraorbital canal, MAN = mandibular canal, PLL = posterior lateral line canal, SC = scapular canal, SO = supraorbital canal, SPL = subpleural loop).

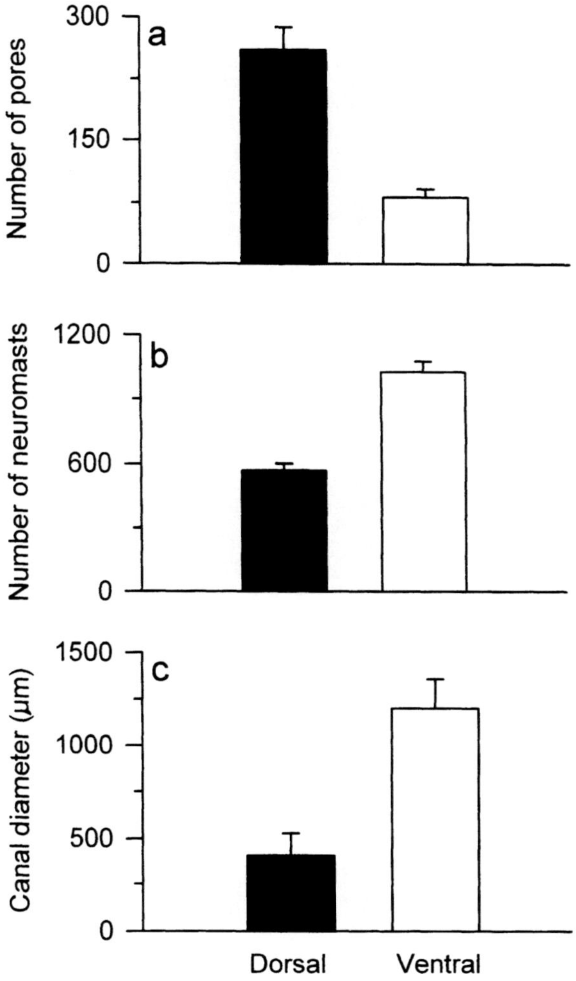

Figure 8. a – Mean number of lateral line canal pores on the entire dorsal (dark) and ventral (open) surface of adult Atlantic stingrays, *Dasyatis sabina*. There are approximately three times as many pores on the dorsal surface ($\bar{x} = 260 \pm 28$ SE) compared to the ventral surface ($\bar{x} = 82 \pm 10$ SE). b – Mean number of lateral line neuromasts on the dorsal (dark) and ventral (open) surface of *D. sabina*. There are approximately twice as many neuromasts on the ventral surface ($\bar{x} = 1028 \pm 51$ SE) compared to the dorsal surface ($\bar{x} = 570 \pm 32$ SE). c – Mean diameter of the pored dorsal hyomandibular canal (dark) and non-pored ventral hyomandibular canal (open) in *D. sabina*. The ventral non-pored canal diameter ($\bar{x} = 1200 \pm 158\,\mu m$ SE) is approximately 2.5 times greater than its dorsal counterpart ($\bar{x} = 405 \pm 125\,\mu m$ SE). The number of pores and neuromasts were counted on one side of the animal and doubled to represent the entire body. Counts do not include the portion of the posterior lateral line on the tail. Error bars represent standard error of the mean. There is a difference between the dorsal and ventral surface in all comparisons (unpaired t-test, $p < 0.001$). n = 5 rays, disk width = 22–26 cm.

Hair cell morphology

Canal neuromasts in the non-pored canals on the ventral surface of the Atlantic stingray, *D. sabina*, consist of a sensory epithelium surrounded by a population of large mantle cells. The sensory epithelium is a central strip, which extends along the length of the neuromast but does not span its entire width. The sensory epithelium consists of sensory hair cells interspersed with support cells that contain numerous microvilli on the apical surface (Figure 10a). Each hair cell has about 60–100 stereocilia with the typical staircase arrangement towards the single kinocilium (2–8 µm long). Some kinocilia have bulbous endings (Figure 10b) which are approximately 1.5–2.0 µm in diameter, but the percentage of hair cells with bulbous endings could not be accurately assessed due to frequent damage of kinocilia during tissue processing. Each hair cell is approximately 2–6 µm in diameter, but distances between adjacent hair cells in the sensory epithelium are variable (range = 2–25 µm) (Figure 10a). Smaller cells with only a few stereocilia of equal height and a single long kinocilium (7–10 µm long) were abundant at the ends of the sensory epithelium. However, it is unknown whether these cells are innervated, are another hair cell type, are support cells, or are precursor hair cells.

Hair cells in the neuromasts of the hyomandibular and infraorbital canals did not follow the organized arrangement of adjacent hair cells with opposite polarity oriented parallel to the canal axis (0° or 180°) commonly observed in teleosts. In the stingray, the majority (~ 76%) of individual hair cells within a neuromast were oriented within 45° of the main canal axis (Figure 11), and a small number of hair cells (~ 24%) were also oriented nearly perpendicular (90° or 270° ± 45°) to the canal axis (Figure 10a). However, it should be noted that measurement error was estimated at 15°. Adjacent hair cells of opposite polarity were observed throughout the length of the sensory epithelium, but they were interspersed with hair cells of varying polarities.

Sphyrna tiburo (Galeomorphii: Sphyrnidae)

Lateral line canals on the head of the bonnethead shark, *S. tiburo*, consist of the bilateral and interconnected hyomandibular, infraorbital, and supraorbital canals, as well as the short mandibular canals on the lower jaw, and the posterior lateral line canal found longitudinally along the trunk (Figures 12 and 13). The canal system on the dorsal surface includes numerous branched

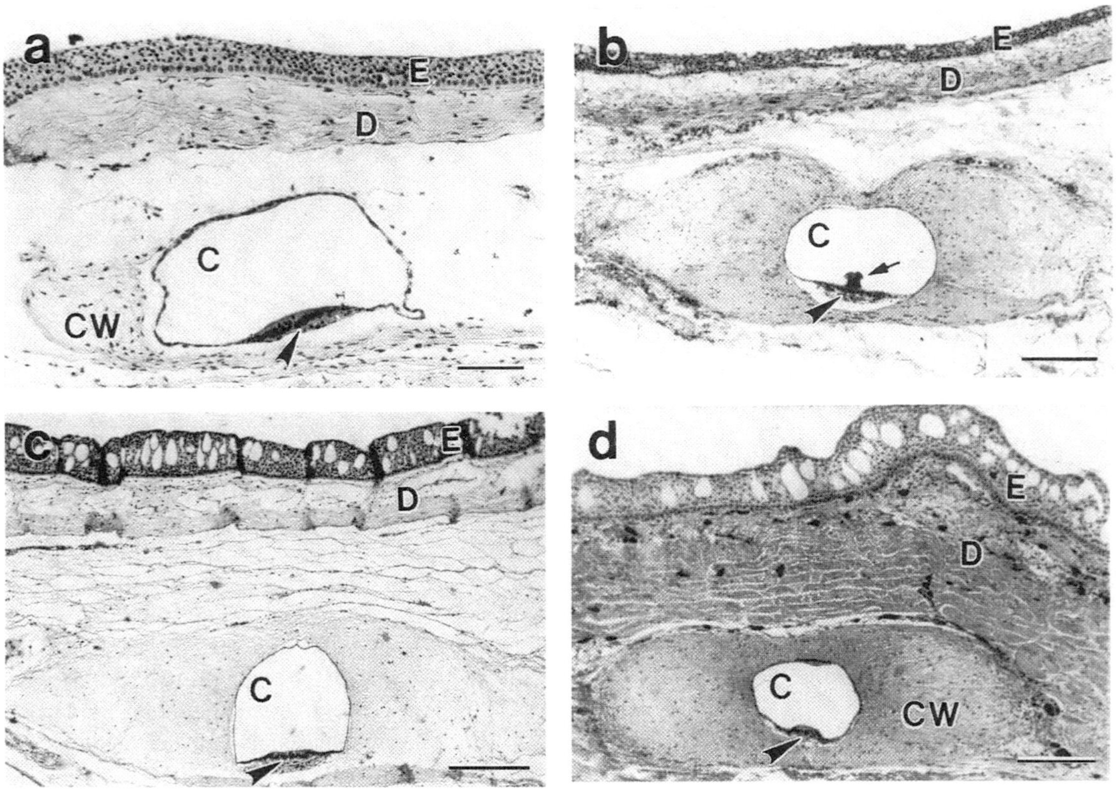

Figure 9. Histological cross sections of ventral non-pored hyomandibular lateral line canals in *Raja eglanteria* (a), *Gymnura micrura* (b), and *Dasyatis sabina* (c) and dorsal posterior lateral line canal in *Dasyatis sabina* (d) for comparison. a–c Ventral canal neuromasts (arrowheads) are located in canals (C) deep to epidermal (E) and loose, compliant dermal (D) skin layers (stratum spongiosum and compactum). Notice the large canal diameters, and loose organization of the canal walls (CW) and dermal skin layers in the ventral canals (a–c) compared to the dorsal canal (d). A portion of the cupula (arrow) is shown in (b) but is absent in a, c, d. Ventral surface of the animal is up in a, b, c and dorsal is up in d. Scale bars = (a) 100 μm, (b) 200 μm, (c) 200 μm, (d) 100 μm.

tubules that terminate in many pores primarily along the rostral edge of the head, and in a horizontal band between the eyes (infraorbital canal) (Figure 12). The hyomandibular canal on the dorsal surface of the head is reduced compared to the batoid species examined and consists of a small section that extends from the infraorbital canal medial to the eye, across the midline near the endolymphatic pores, where it joins the posterior lateral line canal. There is a deep hyomandibular canal section located lateral to the endolymphatic pores, which penetrates the head to the ventral side. The infraorbital canal consists of a loop that extends from the eye rostromedially to the dorsal midline. The most caudal portion of this loop has many small diameter tubules that branch just dorsal to the canal, while the most rostral portion of the loop has no tubules. The infraorbital canal is contiguous with the supraorbital canal near the eye. The supraorbital canal is displaced medially towards the midline of the snout, and has branched tubules that terminate in pores along the margin of the head. The supraorbital canal also penetrates the snout to join the ventral supraorbital canal near the nares. The posterior lateral line canal extends from the endolymphatic pores on the head, dorsolaterally along the body to the tip of the upper lobe of the caudal fin (Figure 13). Tubules extend in a posterior direction both dorsal to and ventral to the main canal along the entire length. However, the majority of tubules extend ventral to the posterior lateral line canal, especially on the caudal fin.

Canals on the ventral surface of the head are pored, with the exception of the mandibular canal and short sections of the infraorbital canal anterior to the mouth (Figure 12). The hyomandibular canal has numerous

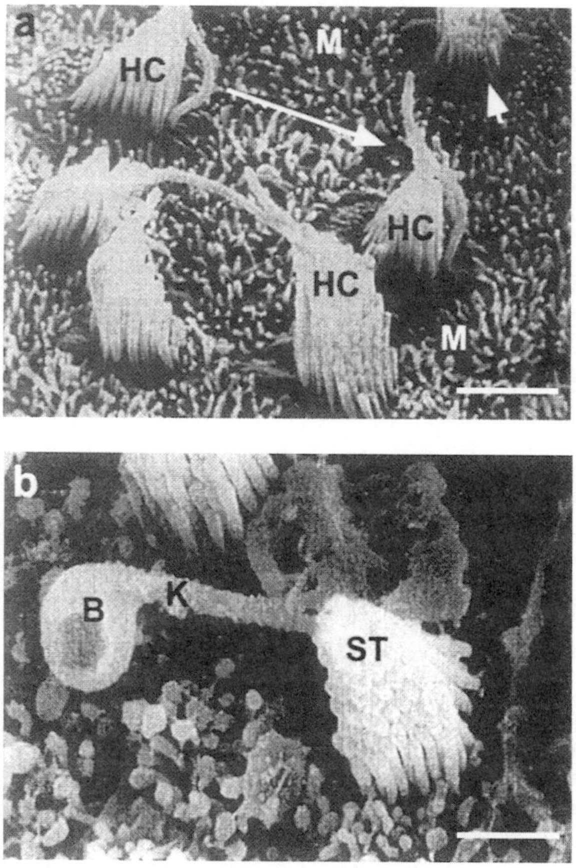

Figure 10. Scanning electron micrographs of neuromasts from the ventral non-pored hyomandibular (a) and infraorbital (b) canals in the Atlantic stingray, *Dasyatis sabina*. a – The sensory epithelium is composed of sensory hair cells (HC) and support cells with numerous microvilli (M) at the apical surface. The majority of hair cells are oriented parallel to or within 45° of the main canal axis. However, a small number of hair cells is oriented off the canal axis (small arrow). Large arrow shows the longitudinal canal axis. Scale bar = 2.0 μm. b – Individual hair cells contain 60–100 stereocilia (ST) in a staircase arrangement of increasing height towards a single kinocilium (K). Some hair cells contain kinocilia with bulbous endings (B). Scale bar = 1.0 μm.

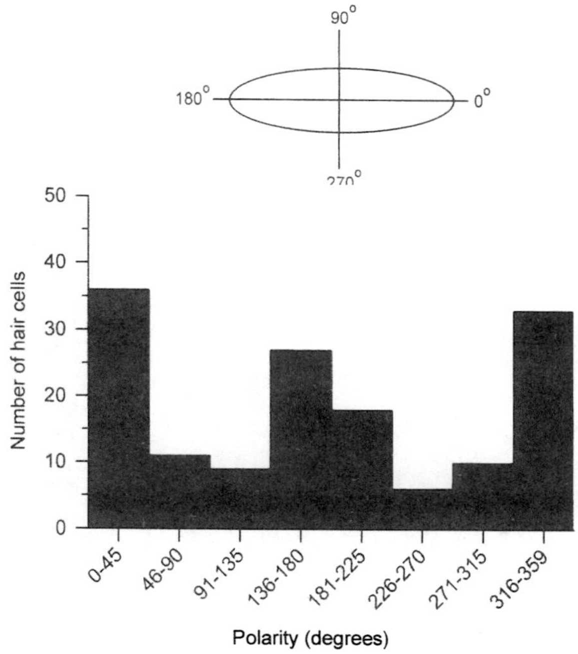

Figure 11. Frequency distribution of hair cell polarities in the neuromasts of the non-pored hyomandibular lateral line canal on the ventral surface of the Atlantic stingray, *Dasyatis sabina*. The neuromast and canal axis lies along the 0°–180° line (0° and 180° directions were chosen arbitrarily) (inset). The majority (> 75%) of hair cells are oriented within 45° of the canal axis with some oriented nearly perpendicular (90° or 270° ± 45°) to the canal axis. Bars represent a total of 150 hair cells measured from 15 neuromasts in 6 fish.

lateral tubules that terminate in pores rostral and caudal to the eye, and is connected to both the mandibular canal and the infraorbital canal. The hyomandibular canal also penetrates the head lateral to the mouth to join the dorsal canals. The infraorbital canal consists of a loop that extends from a non-pored segment of the infraorbital canal, out to the naris. Another portion of the infraorbital canal extends directly from the non-pored segment just below the loop, laterally to the edge of the head, and has numerous tubules that terminate in pores ventral to the main canal. The supraorbital canal also extends from the non-pored infraorbital canal segment and bifurcates to either side of the ventral midline. Each division extends to the edge of the snout and then caudally along the margin of the head, to the naris where it penetrates to join the supraorbital canal on the dorsal surface. The short non-pored mandibular canal joins the hyomandibular canal on either side of the lower jaw, but does not appear joined across the midline.

Discussion

This study describes the morphology and spatial distribution of the lateral line system in *Raja eglanteria*, *Narcine brasiliensis*, *Gymnura micrura*, *Dasyatis sabina*, and *Sphyrna tiburo*. Superficial neuromasts in

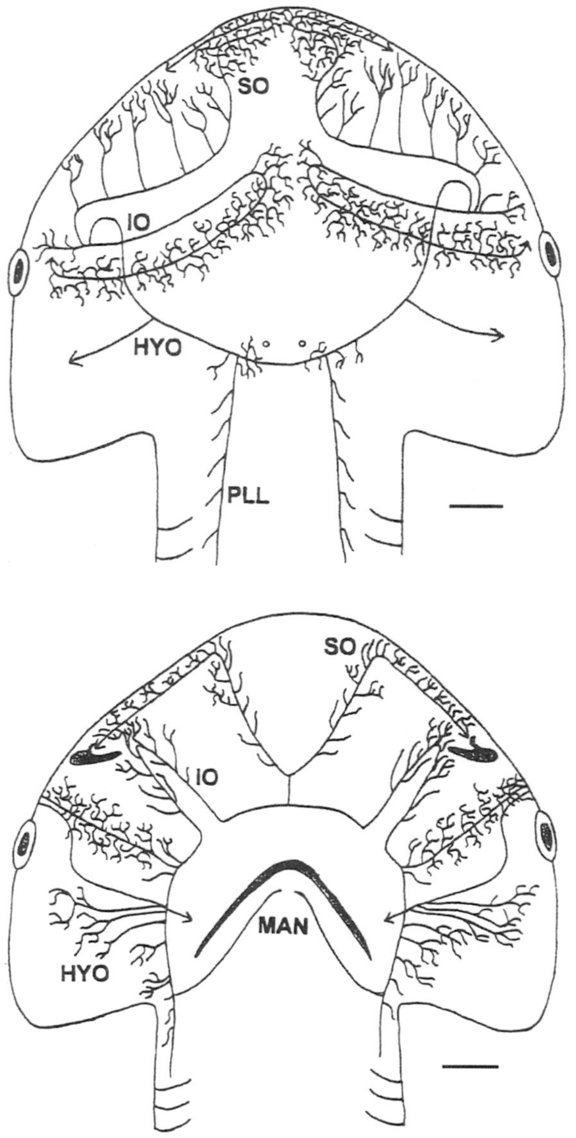

Figure 12. Distribution of the lateral line canal system on the dorsal (upper) and ventral (lower) surface of the head of the bonnethead shark, *Sphyrna tiburo*. Canals are bilateral and interconnected both among and within sides. Numerous branched tubules extend from most canals and terminate in pores at the surface. Scale bar = 0.5 cm (HYO = hyomandibular canal, IO = infraorbital canal, SO = supraorbital canal, MAN = mandibular canal, PLL = posterior lateral line canal.

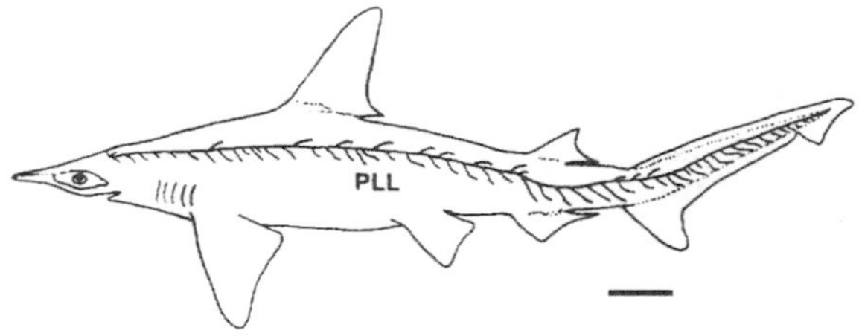

Figure 13. Distribution of the posterior lateral line canal in the bonnethead shark, *Sphyrna tiburo* (lateral view). The posterior lateral line (PLL) extends from the endolymphatic pores on the head to the tip of the upper caudal fin lobe and contains lateral tubules that terminate in pores along its length. Scale bar = 0.5 cm.

R. eglanteria are located in papillar grooves that are positioned along the dorsal midline, a pair anterior to each endolymphatic pore, and a row between the infraorbital canal and the eye. Pored canals are found on the dorsal surface of all species, but the number of pores per tubule branch increases from less (*N. brasiliensis*) to more (*G. micrura*) recently derived batoids. Pored canals are also found along the margin of the pectoral fins on the ventral surface of all species (except *N. brasiliensis*, which lacks a ventral canal system). Non-pored canals are found primarily on the ventral surface of all batoids examined (except *N. brasiliensis*) and are located on the rostrum, around the mouth, and along the midline just lateral to the gill slits. In contrast to the batoids, the bonnethead shark *S. tiburo* has primarily pored canals on both the dorsal and ventral surfaces of the head. Vesicles of Savi are found on the dorsal and ventral surface of *N. brasiliensis*, but only on the ventral surface of *D. sabina*. Quantitative comparison of the dorsal and ventral canal system in *D. sabina* shows that the ventral surface has fewer pores, but more sensory neuromasts and larger diameter canals than the dorsal surface. These morphological data are discussed below and are integrated with known ecology and behavior of elasmobranchs to propose biological functions for each mechanoreceptor type.

Superficial neuromast morphology and distribution

Morphology and distribution of superficial neuromasts is described in relatively few elasmobranch species (Ewart & Mitchell 1892, Tester & Nelson 1969, Maruska & Tricas 1998). In all batoids studied thus far, superficial neuromasts are located on papillae with sensory epithelia positioned at the base of central grooves. These papillae are raised approximately 0.3–0.5 mm above the surrounding skin and cupulae are directly exposed to the water. Superficial neuromasts in the clearnose skate, *R. eglanteria*, are located in a row along the posterior lateral line canal, a row of about 5 between the infraorbital canal and the eye, and a pair anterior to each endolymphatic pore. This organization is very similar to that observed in the skate, *R. batis* (Ewart & Mitchell 1892), and the stingray, *D. sabina*, with the exception of the pair anterior to the endolymphatic pores and the distinct row near the eyes, which were not observed in the stingray (Maruska & Tricas 1998). The main axis of superficial neuromast grooves on the body and tail in both *R. eglanteria* and *D. sabina* range from 90°–135° to the rostrocaudal body axis. However, the orientation of superficial neuromasts near the spiracle in the stingray and endolymphatic pores in the skate is at a slightly greater degree range (100°–160°).

The orientation of superficial neuromast grooves provides insight into their possible biological functions. The position of the neuromast and cupula at the base of a well-developed groove oriented perpendicular to the rostrocaudal body axis enhances water flow parallel to the cupula especially when the animal lies motionless on the substrate, and minimizes stimulation of these receptors during forward swimming motion. This superficial neuromast morphology also indicates these receptors are directionally sensitive in elasmobranchs. This differs from most teleost fishes that have superficial neuromasts on the skin surface, which can be stimulated by water disturbances from any direction. The orientation of superficial neuromast grooves in batoids indicates a maximum response to water movements along the dorsal transverse body axis and a minimum response along the rostrocaudal body axis. However, superficial neuromasts near the eye in the skate have grooves oriented nearly parallel with the rostrocaudal body axis which indicates a maximum response to water movements along the longitudinal body axis from anterior or posterior to the eye. In addition, placement of superficial neuromasts along the dorsolateral aspect of the tail in *Raja* indicates a best response to water movements along the dorsal transverse tail axis. However, superficial neuromasts on the tail of *D. sabina* are positioned laterally which indicates a best response to vertical water movements parallel to the dorsoventral axis (Maruska & Tricas 1998). This difference in position may be explained by subtle differences in habitat and behavior of the skate compared to the stingray. For example, water movements in the dorsoventral axis along the tail would be more common for a ray swimming above the substrate than a skate that often propels itself forward with modified pelvic fins directly on the substrate. However, these proposed axes of best sensitivities are based on gross morphology and assume the hair cells are oriented parallel to the groove axis, but this requires electron microscopic confirmation.

In contrast to the batoids, superficial neuromasts in sharks generally lie between the bases of modified scales and are recessed below the skin surface. Superficial neuromasts in sharks follow the generalized plan of a line posterior to the mouth (mandibular row), a line between the pectoral fins (umbilical row), lines on the dorsolateral portion of the body and caudal fin, and a pair anterior to each endolymphatic pore (Budker 1958, Tester & Nelson 1969) (Figure 14a). Although the number and distribution of superficial neuromasts differs among species, most superficial neuromasts in sharks are associated with or positioned dorsal to the posterior lateral line canal (Tester & Nelson 1969) (Figure 14b–e). The number of superficial neuromasts in the shark species examined thus far ranges from 77 per side in the spiny dogfish, *Squalus acanthias*, to over 600 per side in the scalloped hammerhead, *Sphyrna lewini* (Tester & Nelson 1969) (Figure 14b–e).

A relationship appears to exist between the number of superficial neuromasts and the habitat and activity level of the shark species. Benthic dwelling and relatively slow swimming shark species such as the nurse shark, *Ginglymostoma cirratum*, and spiny dogfish, *Squalus acanthias*, have few superficial neuromasts with the majority located in a row just dorsal to the posterior lateral line canal along the trunk (Figure 14b,c). In contrast, more coastal-pelagic or active sharks such as the scalloped hammerhead, *S. lewini*, and bonnethead, *S. tiburo*, contain numerous superficial neuromasts distributed both dorsal to and ventral to the posterior lateral line canal along the body (Figure 14d,e). This relationship contradicts generalizations made in teleost fishes where sluggish fish or those that live in quiet waters have an increased number of superficial neuromasts, while active fish or those that live in turbulent waters have few superficial neuromasts. The position and number of superficial neuromasts in both sharks and batoids may either have some ecological significance, or represent functional specializations driven by evolution, but conclusions cannot be made until additional taxa are examined.

Based upon morphology and distribution, superficial neuromasts in both sharks and batoids likely function to detect water movements generated by predators, prey, or conspecifics, or may mediate hydrodynamic imaging

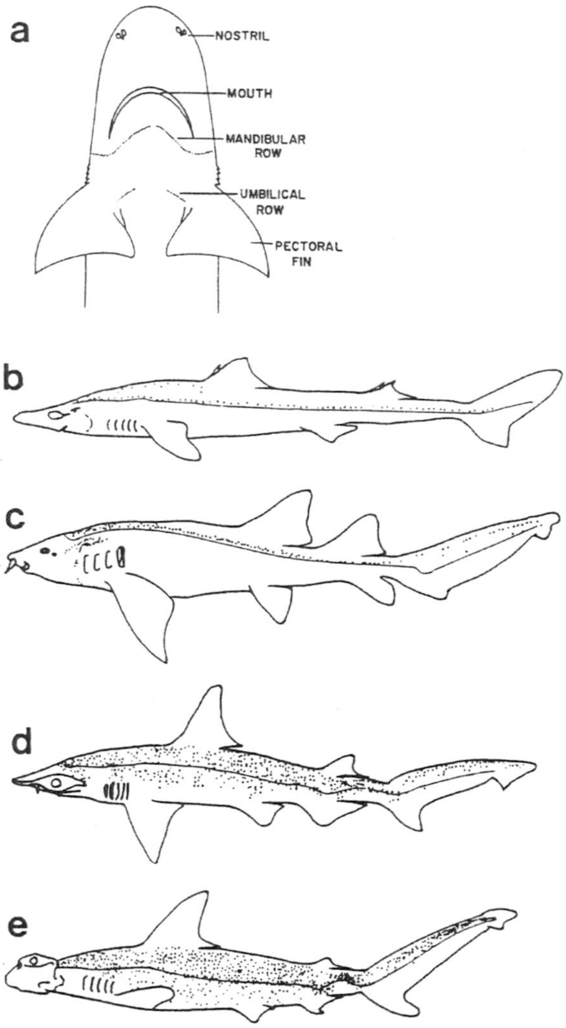

Figure 14. Distribution of superficial neuromasts (pit organs) in sharks. a – Ventral surface of the lemon shark, *Negaprion brevirostris*, shows the row of mandibular neuromasts beneath the lower jaw and the umbilical neuromasts between the pectoral fins. b – Superficial neuromasts in the spiny dogfish, *Squalus acanthias*, are few (approximately 77 per side) and located along the dorsal aspect of the posterior lateral line canal. c – Superficial neuromasts in the nurse shark, *Ginglymostoma cirratum*, are also few in number with the majority located above the posterior lateral line canal (line along dorsolateral portion of body). d – Superficial neuromasts in the bonnethead shark, *Sphyrna tiburo*, are numerous (> 400 per side) and distributed both dorsal and ventral to the posterior lateral line canal. e – Superficial neuromasts in the scalloped hammerhead, *Sphyrna lewini*, are more numerous (> 600 per side) and extend both dorsal and ventral to the posterior lateral line canal. Each dot represents a single superficial neuromast (modified from Tester & Nelson 1969).

Table 2. Summary of hypothesized behavioral functions of mechanosensory lateral line systems found in elasmobranch fishes.

Behavior	Superficial neuromasts	Vesicles of Savi*	Pored canals	Non-pored canals
Rheotaxis	+		+	
Predator avoidance	+		+	
Schooling†	+		+	
Social communication	+	+	+	+
Prey detection	+**	+	+	+
Hydrodynamic imaging	+		+	

The six general behaviors that are known to be mediated by the lateral line system in bony fishes are listed, and a + indicates a hypothesized function for each specific system in elasmobranchs. Use of each system for specific behaviors may differ between elasmobranch species (e.g. sharks versus batoids) due to both variations in lateral line systems and ecology or behaviors. *vesicles of Savi do not apply to shark species; **superficial neuromasts are likely not used for prey detection by benthic feeding batoids; †facilitation of schooling behavior would only apply to those species of sharks and batoids that are known to school.

(Table 2), but conclusions on biological function cannot be made until the adequate stimulus for these mechanoreceptors is physiologically demonstrated. In addition, superficial neuromasts in elasmobranchs may mediate rheotactic behavior (orientation to water currents) as demonstrated in teleosts (Montgomery et al. 1997), but this also remains to be tested.

Dorsal canal system morphology and organization

Differences in the spatial distribution of lateral line canals are often related to variations in body shape (Coombs et al. 1988). For example, the dorsoventrally flattened body of batoids is correlated with the extension of both the hyomandibular canal onto the enlarged pectoral fins, and the supraorbital canal onto the elongated rostrum (Chu & Wen 1979). Thus, the reduced tubule branching on the dorsal surface of *N. brasiliensis* may be related to the decreased size (width) of the disk. However, lateral line canals on the dorsal surface of both *N. brasiliensis* and *R. eglanteria* also contain only straight lateral tubules that terminate in single pores. This differs from many of the myliobatid rays, which show extensive branching of lateral tubules that terminate in many pores over the entire dorsal body surface (Chu & Wen 1979). The elaborate organization of tubule branching in the dorsoventrally flattened

G. micrura may serve to place the receptive field over the greatly expanded pectoral fins and likely increases sensitivity to water movements across the dorsal surface. However, the number and complexity of tubule branching is also greater in the more recently derived batoids such as *Rhinoptera, Aetobatus*, and *Manta* spp. (Garman 1888, Chu & Wen 1979), and may be a result of phylogeny. In addition, the number and complexity of tubule branches often increases with size of the animal which is likely an adaptation to maintain or increase sampling area as the fish grows. Thus, future examinations of lateral line characteristics across taxa should address whether differences in spatial distribution are correlated with body shape, ecology, ontogeny, or are a result of phylogenetic relationships.

Chondrichthyan lateral line canals differ from those in most bony fishes by the presence of multiple neuromasts between adjacent pores (Johnson 1917, Tester & Kendall 1969, Ekström von Lubitz 1981). The lepidosirenid lungfish, *Protopterus* (Webb & Northcutt 1997), and the oyster toadfish, *Opsanus tau* (Clapp 1898) also have multiple neuromasts between adjacent pores in their head lateral line canals. Therefore, multiple neuromasts between pores may have evolved independently in these groups of fishes. However, the presence of multiple neuromasts between pores and their organization into a nearly continuous sensory epithelium within the canal of many elasmobranch species (Johnson 1917, Tester & Kendall 1969, Boord & Campbell 1977, Maruska & Tricas 1998) indicates they likely have different response properties and possibly biological functions, but this remains to be physiologically and behaviorally tested.

The expansion of lateral line tubules and pores on the dorsal surface of all batoids examined may function to detect water movements over the disk surface and would be advantageous when a batoid lies buried or motionless on the substrate. The pored canals on the dorsal surface likely function to detect water movements across the body surface generated by predators, epifaunal prey items, or possibly conspecifics during mating (Table 2). These pored canals, as well as the superficial neuromasts, may be especially important for the detection of water movements generated by predators. Lateral line-mediated predator avoidance is likely an important behavioral response for adult and especially juvenile benthic batoids. Also, many skate species copulate on the ocean bottom, often for several hours at a time with their dorsal surface exposed to predators such as sharks. Thus, the ability to detect predators via the lateral line system while in copulo would ultimately enhance survival and reproductive success, but this hypothesis remains to be tested.

Pored canals on the dorsal surface of the head and the posterior lateral line canal on the body and tail of elasmobranchs may also serve to detect water movements generated by conspecifics to facilitate schooling behavior, and to locate objects in the environment via detection of distortions in the animal's own flow field while swimming (Table 2). Several shark (sphyrnid) and ray (myliobatid) species are known to school during certain times of the year and at certain life-history intervals (Castro 1983, Michael 1993). Therefore, the pored canals on both the dorsal and ventral surfaces of these species are likely used to detect water movements generated by conspecifics to maintain position within a school. Thus, it would be interesting to examine the correlation between schooling elasmobranch species and distribution of pored lateral line canals.

Ventral canal system morphology and organization

The lateral line canal system on the ventral surface of batoids, and ventral portion of the head of sharks generally consists of both pored and non-pored canal sections. The distribution of pored and non-pored sections of canal varies among species and this may be correlated with ecology and behavior, or explained by phylogenetic relationships among taxa. For example, the hyomandibular canal in *R. eglanteria* does not contain the subpleural loop in the mid-disk region that is characteristic of most batoids, but contains short tubules that terminate in single pores along the gill slits which is not characteristic of most other species (Garman 1888, Ewart & Mitchell 1892, Chu & Wen 1979), but is similar to the winter skate, *R. ocellata* (Garman 1888). Therefore, differences in canal structure among species within a single family (e.g. Rajidae) may be related to ecology or behavior. However, until a greater number of species are examined, conclusions on whether these differences reflect ecological correlations or phylogenetic trends cannot be made.

Neuromasts in the non-pored hyomandibular canal of *D. sabina* have sensory hair cells with numerous stereocilia in a stepwise arrangement of increasing height towards a single kinocilium. This morphology was also observed in the canal neuromasts of the spotted shark, *Mustelus manazo* (Hama & Yamada

1977) and in the vesicles of Savi of *Torpedo* spp. (Nickel & Fuchs 1974), but not in the posterior lateral line canal of the catshark, *Scyliorhinus canicula* (Roberts & Ryan 1971). Smaller cells with a single kinocilium and stereocilia of equal length were observed near the ends of the neuromast in *D. sabina*, but it is unknown whether these cells are innervated, or possibly represent a developmental interval of hair cells as described in *S. canicula* (Roberts & Ryan 1971). It is possible that these are supporting cells and that their position at the end of the neuromast and their long kinocilia both function to stabilize and support the cupula between adjacent neuromasts, but ultrastructural examination is required to test this hypothesis. In addition, many hair cells in the canal neuromasts of the stingray have kinocilia with bulbous endings similar to those observed in the lateral line system of the ratfish, *Chimaera monstrosa* (Ekström von Lubitz 1981) and saccular organ of the frog, *Rana catesbeiana* (Hillman & Lewis 1971). In the frog, the apex of this bulbous kinocilium contacts the otolithic membrane, while the base rests on a portion of the hair cell that lacks the rigid cuticular plate of the stereocilia. Movement of the stereocilia and kinocilium produces a deformation in the hair cell membrane at the base of the kinocilium. It was suggested that this specialized anatomical arrangement functions as a mechanical coupling system for the transformation of shearing motion of the otolithic membrane into a generator potential (Hillman & Lewis 1971). A similar mechanism may be operating in the stingray where bulbous kinocilia in the neuromasts of non-pored canal sections are anchored to the cupula and couple small cupular movements in any direction to hair cell stimulation. This organization may be advantageous in non-pored canals that are hypothesized to encode local displacements of the ventral skin surface in batoids (Sand 1937, Maruska & Tricas 1998). However, ultrastructural examination of the hair cell, kinocilium, and cuticular plate are needed to determine the presence or absence of a similar system in the elasmobranch lateral line.

Directional sensitivity of canal neuromasts in teleost fishes generally results from pairs of hair cells with opposite polarities oriented parallel to the longitudinal canal axis. Hair cells in canal neuromasts in some chondrichthyan species do not follow this strict arrangement of adjacent cells of opposing polarities and often contain hair cells oriented nearly perpendicular to the canal axis (Roberts 1969, Roberts & Ryan 1971, Ekström von Lubitz 1981), while other species do have this primarily parallel arrangement (Hama & Yamada 1977). However, relatively few studies examine the polarity of hair cells in cartilaginous fishes, thus the taxonomic distribution of this character is not well known. The majority of hair cells in the non-pored hyomandibular canal on the ventral surface of *D. sabina* are oriented within 45° of the main canal axis, with a small number oriented nearly perpendicular to this axis. This indicates that some hair cells will be stimulated by cupular movements in any direction. Presence of hair cells oriented at various angles within neuromasts may broaden the directionality to tactile stimuli at different locations on the skin. For example, small displacements of the skin directly beneath the ventral canals would cause displacement of the cupula inside the canal and stimulation of hair cells oriented near 0° or 180° along the canal axis. However, skin displacements at some lateral distance from the canal may also cause cupular movements and stimulation of hair cells oriented orthogonal to the longitudinal canal axis. This hypothesis is supported by the fact that these non-pored ventral canals have a large diameter which covers a greater area of underlying skin surface, and they are located above loose, compliant skin layers. Therefore, it is likely that these ventral non-pored canals function as displacement detectors (or mechanotactile receptors) rather than detectors of water movement. Further, the position of these non-pored canals around the mouth and rostrum indicates that they may be used in the detection and guidance of prey to the mouth as suggested by the mechanotactile hypothesis (Maruska & Tricas 1998). However, the functional significance of hair cell orientations both parallel and perpendicular to the canal axis in ventral non-pored canal neuromasts cannot be determined until they are compared with those of pored canal neuromasts. In addition, the response properties of these canal neuromasts should be determined and the mechanotactile receptive field on the skin surface beneath these canals requires investigation before hypotheses on biological function can be tested.

Non-pored canals on the ventral surface of elasmobranch species may also function as sensitive tactile receptors to facilitate body positioning during copulation (Table 2). Elasmobranchs have internal fertilization that often requires the male to bite the pectoral fin of the female in order to help position his body either along side, or underneath the female for insertion of the clasper (intromittent organ) into the cloaca. Therefore, the non-pored canals on the rostrum, around the mouth, and along the midline may function as tactile receptors to facilitate body positioning between males and females during copulation. It would be interesting to examine seasonal changes in sensitivity of the neuromasts in the non-pored canals on the ventral surface in an elasmobranch species that undergoes a distinct annual reproductive cycle such as the Atlantic stingray, *D. sabina* (Maruska et al. 1996), to test the hypothesis that the lateral line plays a role in the coordination of copulatory behavior.

Vesicles of Savi

Vesicles of Savi are found primarily on the ventral surface of torpedinid, narcinid, and dasyatid batoids (Norris 1932, Szabo 1958, 1968, Nickel & Fuchs 1974, Chu & Wen 1979, Barry & Bennett 1989, Maruska & Tricas 1998), but the function of these receptors still remains unknown. *Narcine brasiliensis* has neuromasts that are located in enclosed vesicular pouches, which sit approximately 0.5–2.0 mm below a relatively compliant skin surface. Morphology of vesicles of Savi is similar in *Narcine* and *Torpedo* where each vesicle contains 3 neuromasts (one large central and 2 smaller peripheral), each of which has its own cupula and hair cells polarized parallel to the center of the neuromast (Szabo 1968, Barry & Bennett 1989). However, the cupula of the central neuromast in *N. brasiliensis* appears more dense than the cupulae of the peripheral neuromasts (observed in fresh tissue). Therefore, it is possible that the cupula of the central neuromast is weighted, and may be similar to the situation found in the otolithic organs of teleost fishes. Therefore, these electric rays may encode linear accelerations of the body in multiple directions because the vesicles of Savi are arranged in rows oriented 45° from the rostrocaudal body axis on both the dorsal and ventral surfaces. However, further analysis of cupular composition, hair cell polarization, and neuromast-cupular coupling is needed before conclusions can be made on the significance of variation in cupular density within the vesicles of Savi.

Vesicles of Savi in the dasyatid rays differ slightly in morphology and distribution from *Torpedo* and *Narcine* (Chu & Wen 1979, Maruska & Tricas 1998). In dasyatids, vesicles of Savi are located only on the ventral surface, contain only a single neuromast, are contiguous with the lumen of the supraorbital canal on the rostrum, and adjacent vesicles are connected by tubules (Maruska & Tricas 1998). Garman (1888) described a change in the subrostral canal of the batoids, *Potamotrygon* and *Paratrygon*, from a tubular structure to a row of closed rings connected by tissue, which may represent another morphological variant of the vesicles of Savi. Several researchers suggest that vesicles of Savi may represent an obsolescent canal condition, but conclusions can not be drawn until the morphological diversity of these structures is assessed. Also, the presence of vesicles of Savi only in the torpediniform and dasyatid batoids suggests that these vesicles evolved independently in these groups, but conclusions on evolutionary relationships can not be proposed until additional taxa are examined.

Vesicles of Savi are hypothesized to be receptors used to detect substrate-bourne vibrations transmitted via the skin or cartilaginous attachments in *Narcine* (Barry & Bennett 1989), or serve as specialized tactile receptors sensitive to displacement of the underlying skin caused by contact with prey, conspecifics, or the substrate (Nickel & Fuchs 1974, Maruska & Tricas 1998). Electrophysiological experiments in *Torpedo* indicate the vesicles of Savi have a peak sensitivity to vibrations of 150–200 Hz (Szabo 1968). Barry & Bennett (1989) suggest the vesicles may be high frequency vibration receptors due to their isolation and protection of the central neuromast by an arch in *Narcine*, but this remains to be experimentally tested. The concentration of vesicles on the rostrum around the mouth and their rostrocaudally arranged rows in *Narcine* puts them in an ideal location to aid in the localization and guidance of the mouth over prey items. Vesicles of Savi would not respond to pressure differences across the skin surface caused by water movements because they are not connected to the external environment, but should be sensitive to direct displacement of the compliant underlying skin. The specific orientation of the rows of vesicles (45° to the rostrocaudal axis) in *Narcine* may also provide the animal with directional information on the location of small prey items. However, comparisons of the morphology, distribution, and physiology of vesicles of Savi among batoid taxa warrants further investigation to examine the functional specialization and evolution of these mechanoreceptors.

Feeding neuroecology of the lateral line system

Feeding ecology and behavior can often be correlated with the peripheral distribution of the lateral line system in fishes (Dijkgraaf 1962, Hensel 1978). The batoids examined in this study generally feed on infaunal or epifaunal organisms that often require excavation from the substrate and are often outside of the animal's visual field. The lesser electric ray, *N. brasiliensis*, feeds predominantly at night on burrowing polychaete worms with some amphipods, decapod shrimp, sipunculid worms, and anguilliform eels also reported in the diet (Funicelli 1975, Rudloe 1989). The clearnose skate, *R. eglanteria*, is a deep-water benthic batoid reported to feed on invertebrates such as mollusks and small crustaceans as well as the benthic tonguefish, *Symphurus plagiusa*, and other teleosts (Hildebrand & Schroeder 1928, Fitz & Daiber 1963, Schwartz 1996). Butterfly rays, *Gymnura* spp., are primarily piscivorous feeders that actively prey on teleost fishes such as spot, *Leostomous xanthurus*, and pinfish, *Lagodon rhomboides*. In addition, some crustaceans, gastropods and cephalopods were also found in the diet of smaller *Gymnura*, but teleosts become more important in the diet as the ray grows (Daiber & Booth 1960). The Atlantic stingray, *D. sabina*, feeds day and night almost exclusively on small benthic invertebrates such as amphipods, isopods, ophiuroids and polychaetes that they excavate from the substrate (Cook 1994, Bradley 1996). Integration of the feeding ecology and behavior of each of these species with the morphology and spatial distribution of the lateral line system supports the hypothesis that the lateral line system functions in predation in elasmobranch fishes.

Prey detection in elasmobranchs is mediated by multiple sensory systems, but the ventral lateral line system in batoids probably serves to locate prey and guide the mouth over it during the final stages of prey capture. The pored canal system on the ventral surface may function to detect water movements generated by locomotion, respiration, and filter-feeding activities of prey (Montgomery & Skipworth 1997), and allow the batoid to reposition its body to orient the non-pored canals and mouth directly over the prey. Non-pored canals are often located on the ventral rostrum and around the mouth, and may function as specialized tactile receptors stimulated by prey contact with the skin surface as proposed by the mechanotactile hypothesis in the stingray (Maruska & Tricas 1998). Similarly, the vesicles of Savi distributed around the mouth and on the rostrum in batoids may also serve as mechanotactile receptors involved in prey localization. The spatial separation of vesicles of Savi would allow them to serve as point source detectors when prey are just rostral or lateral to the mouth.

In addition to the correlation between peripheral lateral line organization and feeding ecology and behavior of batoids, other evidence supports the mechanotactile hypothesis of lateral line function. First, although the number of lateral line pores is greater in the canals on the dorsal surface of *D. sabina*, the canals on the ventral surface often contain twice as many sensory neuromasts. Second, the average diameter of non-pored canals on the ventral surface is 2.5× greater than that of the dorsal canals in *D. sabina*. This indicates there is a wide tactile receptive field on the ventral skin surface beneath the canal that will move canal fluid and stimulate the neuromast when it is displaced. For any given displacement of the skin surface, fluid flow inside a larger diameter canal will attenuate at shorter distances from the stimulus than in a smaller diameter canal, permitting the ray to better localize a prey item because fewer neuromasts will be stimulated. The compliant nature of the canal walls and dermal layers superficial to the ventral canal would also facilitate movement of the cupula and canal fluid in response to skin displacement. Third, cutaneous sensory endings (putative tactile receptors) in elasmobranchs are stimulated by 20 μm displacements of the skin surface (Murray 1961), but a skin displacement of less than 20 μm should stimulate the canal neuromasts of the lateral line system, making it more sensitive to tactile stimulation than the general cutaneous tactile system. Therefore, these data support the hypothesis that the non-pored canals and vesicles of Savi on the ventral surface in the stingray and other batoids function as mechanotactile receptors that likely play a role in the localization and capture of prey. However, the mechanotactile hypothesis remains to be physiologically and behaviorally tested.

The bonnethead, *S. tiburo*, was used in this study to compare the organization of the lateral line system in batoids to a shark species with similar food habits. The bonnethead is a small abundant coastal shark species that feeds primarily at night on motile invertebrates such as crabs, shrimp, bivalves, and cephalopods that often reside in seagrass beds (Cortés et al. 1996). Although the bonnethead often feeds on benthic invertebrates, it does not have the extensive non-pored canal system or vesicles of Savi present in benthic feeding batoids. Therefore, it is likely that prey localization is mediated by the detection of water motions caused by movements of prey via a pored canal system rather than a mechanotactile mechanism mediated by the non-pored canals as suggested in batoids (Maruska & Tricas 1998). However, many shark species do have sections of non-pored canals on the head (Garman 1888, Chu & Wen 1979). Non-pored canals on the head of sharks may also serve as mechanotactile receptors to facilitate prey localization during the final stages of prey capture and handling or during copulation, or may help reduce stimulation of the canal system during forward swimming movements. However, differences in morphology and spatial distribution of the lateral line system between batoids and sharks may also result from different evolutionary selective pressures. Thus, functional or phylogenetic interpretations should be treated with caution until additional taxa are examined.

Future directions

In teleosts, the mechanosensory lateral line functions to detect water flow across the skin surface to facilitate prey detection (Hoekstra & Janssen 1985, Montgomery & Saunders 1985, Montgomery et al. 1988, Montgomery 1989, Janssen et al. 1995), social communication (Satou et al. 1991, 1994), schooling (Partridge & Pitcher 1980), predator avoidance (Blaxter & Fuiman 1990, Fuiman 1993), rheotaxis (Montgomery et al. 1997), and object localization or hydrodynamic imaging (Campenhausen et al. 1981, Hassan 1989). Several of these functions were shown behaviorally or physiologically to be mediated by a specific class of mechanoreceptor organ. For example, rheotaxis and predator avoidance in fish larvae is mediated primarily by superficial neuromasts (Montgomery et al. 1997, Blaxter & Fuiman 1990), while schooling and localization of objects is often mediated by the canal system (Partridge & Pitcher 1980, Hassan 1989). Elasmobranch lateral lines differ from teleosts by the placement of superficial neuromasts within grooves that enhance a bidirectional sensitivity, and canal neuromasts that are organized as a nearly continuous sensory epithelium with multiple neuromasts between pores. In addition, specialized non-pored canals are common in most species, and vesicles of Savi are found in some rays. Thus, it cannot be assumed that class-specific mechanoreceptor-mediated behaviors in teleosts apply to elasmobranchs. Also, the biological function of the specialized non-pored canal system and vesicles of Savi remain to be demonstrated. Testing of mechanoreceptor-specific functions requires quantitative comparisons of response properties of different mechanoreceptors as well as direct behavioral experimentation. Several of these mechanoreceptor-specific function hypotheses were mentioned in previous sections and summarized in Table 2, while a few are discussed below.

Mechanotactile hypothesis

The mechanotactile hypothesis of lateral line function in batoids states that ventral non-pored canals likely function as specialized tactile receptors used to facilitate prey capture (Maruska & Tricas 1998). Detection of weak water jets by the short-tailed stingray, which simulated water movements generated by prey, provided behavioral evidence for lateral line-mediated prey detection in elasmobranch fishes (Montgomery & Skipworth 1997). However, the relative roles of the pored and non-pored canal systems on the ventral surface of batoids during prey detection and localization is unknown. The hypothesis that non-pored canals encode displacement of the skin can be tested by electrophysiological determination of response properties of primary afferents that innervate the neuromasts in these canals. Some evidence already exists for stimulation of lateral line canals across the skin in both teleosts and elasmobranchs, but characterization of sensitivity and receptive field organization remains unknown (Sand 1937, Denton & Gray 1983, 1988). Also, the question of whether or not a natural source (e.g. prey) is able to stimulate these receptors and elicit behaviors will require behavioral testing. Many of the lateral line-mediated behaviors in teleosts were demonstrated by studies that sequentially eliminated each sensory system, including pharmacological techniques used to block lateral line receptors (e.g. cobalt chloride, gentamicin sulfate). However, elasmobranchs are often large and difficult to deal with in captivity, standard pharmacological methods (e.g. cobalt chloride blockers) do not work in salt water, and elimination of the lateral line system via surgical transection of nerves without damage to the electrosensory system is difficult. Therefore, the logistics of conducting lateral line behavioral experiments in elasmobranch fishes must be tediously resolved before conclusions on biological function can be demonstrated.

Schooling hypothesis

Teleost fishes use their lateral line systems in conjunction with vision to maintain position within a school (Partridge & Pitcher 1980). Individual fish detect short-term changes in the velocity and direction of their nearest neighbors primarily via the pored lateral line canal along the trunk. Several elasmobranch fishes are also known to form aggregations at certain times of the year or day for reasons such as mating and parturition, feeding, or predator avoidance. These aggregations range from large schools with hundreds of individuals (e.g. *Sphyrna lewini* and *Rhinoptera bonasus*) to smaller groups of only a few individuals (e.g. *Carcharhinus amblyrhynchos*). However, relatively little is known about the organization and function of elasmobranch schools and whether the lateral line system plays a role in this behavior. Individuals within a school would detect water movements produced by swimming neighbors primarily via pored lateral line canals along the body in sharks (i.e. posterior lateral line canal) and on both the dorsal and ventral pectoral fins in batoids. Therefore, it is possible that elasmobranch species that form aggregations have lateral line specializations such as increased canal branching to expand the receptive field, increased number of pores, or increased numbers of superficial neuromasts. Morphological studies across taxa would test the hypothesis that lateral line organization is correlated with schooling behavior in elasmobranchs. In addition, behavioral experiments which measure the ability to maintain position within a school in fish with different portions of the lateral line system ablated can reveal the relative importance of the mechanosensory system in elasmobranch schooling behavior.

Mechanosensory parallel processing hypothesis

Lateral line canals on the head are innervated by the anterior lateral line nerve complex, and those of the body and tail by the posterior lateral line nerve. Both branches enter the brain and terminate somatotopically around cell plates within the medial octavolateralis nucleus of the medulla (Bodznick & Northcutt 1980, Puzdrowski & Leonard 1993). However, these nerves contain neurons that innervate both superficial and canal neuromasts. Therefore, the hypothesis that superficial neuromasts that encode lower frequency velocity information, and canal neuromasts that encode higher frequency acceleration information have separate parallel processing pathways should be tested via neuroanatomical and neurophysiological techniques. Central physiological and neuroanatomical studies in elasmobranchs have shown mechanosensory lateral line regions from the medulla to the telencephalon (Bleckmann et al. 1987, 1989, Bleckmann & Bullock 1989, Boord & Montgomery 1989), but identification of distinct cell populations that process velocity versus acceleration or displacement information received from different receptor classes has received little attention. Also, central processing of mechanosensory information from the distinct dorsal and ventral surfaces of batoids requires investigation in a behavioral context. It is currently unknown how different types of mechanosensory information are processed and integrated in the elasmobranch brain to elicit specific lateral line-mediated behaviors.

The importance of the mechanosensory lateral line system in the coordination of behaviors such as feeding, schooling, predator avoidance, hydrodynamic imaging, and courtship remain to be investigated in elasmobranch fishes. However, it is only after basic questions on lateral line structure and function are answered that more complex questions such as how the central nervous system processes and integrates lateral line information with other sensory cues to elicit behaviors can be logically approached.

Acknowledgements

I thank Timothy C. Tricas for his endless guidance, support and helpful comments on this research and manuscript, Captain Rich Gurlek of the R/V Delphinus and student volunteers for help with animal collections, Michael Helmstetter and Raynor Howard of the Brevard Teaching and Research Laboratories for SEM assistance, the American Elasmobranch Society for student travel funds, and J.F. Webb and an anonymous reviewer for helpful comments on the manuscript. I also thank T.C. Tricas and S. Gruber for organizing this symposium and for the invitation to participate.

References cited

Barry, M.A., D.H. Hall & M.V.L. Bennett. 1988a. The elasmobranch spiracular organ I. Morphological studies. J. Comp. Physiol. A 163: 85–92.

Barry, M.A., D.H. Hall & M.V.L. Bennett. 1988b. The elasmobranch spiracular organ II. Physiological studies. J. Comp. Physiol. A 163: 93–98.

Barry, M.A. & M.V.L. Bennett. 1989. Specialized lateral line receptor systems in elasmobranchs: the spiracular organs and vesicles of Savi. pp. 591–606. *In*: S. Coombs, P. Görner & H. Münz (ed.) The Mechanosensory Lateral Line-Neurobiology and Evolution, Springer-Verlag, New York.

Blaxter, J.H.S. & L. A. Fuiman. 1990. The role of the sensory systems of herring larvae in evading predatory fishes. J. Mar. Biol. Ass. U.K. 70: 413–427.

Bleckmann, H. & T.H. Bullock. 1989. Central nervous physiology of the lateral line, with special reference to cartilaginous fishes. pp. 387–408. *In*: S. Coombs, P. Görner & H. Münz (ed.) The Mechanosensory Lateral Line – Neurobiology and Evolution, Springer-Verlag, New York.

Bleckmann, H., T.H. Bullock & J.M. Jorgensen. 1987. The lateral line mechanoreceptive mesencephalic, diencephalic, and telencephalic regions in the thornback ray, *Platyrhinoidis triseriata* (Elasmobranchii). J. Comp. Physiol. A 161: 67–84.

Bleckmann, H., O. Weiss & T.H. Bullock. 1989. Physiology of lateral line mechanoreceptive regions in the elasmobranch brain. J. Comp. Physiol. A 164: 459–474.

Bodznick, D. & R.G. Northcutt. 1980. Segregation of electro- and mechanoreceptive inputs to the elasmobranch medulla. Brain Res. 195: 313–321.

Boord, R.L. & C.B.G. Campbell. 1977. Structural and functional organization of the lateral line system of sharks. Amer. Zool. 17: 431–441.

Boord, R.L. & J.C. Montgomery. 1989. Central mechanosensory lateral line centers and pathways among the elasmobranchs. pp. 323–340. *In*: S. Coombs, P. Görner & H. Münz (ed.) The Mechanosensory Lateral Line – Neurobiology and Evolution, Springer-Verlag, New York.

Bradley, J.L. 1996. Prey energy content and selection, habitat use and daily ration of the Atlantic stingray, *Dasyatis sabina*. M.S. Thesis, Florida Institute of Technology, Melbourne. 49 pp.

Budker, P. 1958. Les organes sensoriels cutanés des sélaciens. pp. 1033–1062. *In*: P.-P. Grassé (ed.) Traité de Zoologie, Vol. 13, fasc. 2, Masson et Cie, Paris.

Campenhausen, C. von, I. Riess & R. Weissert. 1981. Detection of stationary objects by the blind cave fish *Anoptichthys jordani* (Characidae). J. Comp. Physiol. 143: 369–374.

Castro, J.I. 1983. The sharks of north American waters. Texas A&M University Press, College Station. 180 pp.

Chu, Y.T. & M.C. Wen. 1979. A study of the lateral-line canal system and that of Lorenzini ampullae and tubules of elasmobranchiate fishes of China. Monograph of Fishes of China, Academic Press, Shanghai. 132 pp.

Clapp, C.M. 1898. The lateral line system of *Batrachus tau*. J. Morph. 15: 223–265.

Cole, F.J. 1896. On the cranial nerves of *Chimaera monstrosa*, with a discussion of the lateral line system and of the morphology of the chorda tympani. Trans. R. Soc. Edin. 38: 631–680.

Cook, D.A. 1994. Temporal patterns of food habits of the Atlantic stingray, *Dasyatis sabina* (LeSeur, 1824), from the Banana River Lagoon, Florida. M.S. Thesis, Florida Institute of Technology, Melbourne. 45 pp.

Coombs, S. 1994. Nearfield detection of dipole sources by the goldfish (*Carassius auratus*) and the mottled sculpin (*Cottus bairdi*). J. Exp. Biol. 190: 109–129.

Coombs, S. & J. Janssen. 1989. Peripheral processing by the lateral line system of the mottled sculpin (*Cottus bairdi*). pp. 299–319. *In*: S. Coombs, P. Görner & H. Münz (ed.) The Mechanosensory Lateral Line – Neurobiology and Evolution, Springer-Verlag, New York.

Coombs, S., J. Janssen & J.F. Webb. 1988. Diversity of lateral line systems: evolutionary and functional considerations. pp. 553–593. *In*: J. Atema, R.R. Fay, A.N. Popper & W.N. Tavolga (ed.) Sensory Biology of Aquatic Animals, Springer-Verlag, Heidelberg.

Cortés, E., C. Manire & R.E. Hueter. 1996. Diet, feeding habits, and diel feeding chronology of the bonnethead shark, *Sphyrna tiburo*, in Southwest Florida. Bull. Mar. Sci. 58: 353–367.

Daiber, F.C. & R.A. Booth. 1960. Notes on the biology of the butterfly rays, *Gymnura altavela* and *Gymnura micrura*. Copeia 1960: 137–139.

Denton, E.J. & J.A.B. Gray. 1983. Mechanical factors in the excitation of clupeid lateral lines. Proc. R. Soc. Lond. B 218: 1–26.

Denton, E.J. & J.A.B. Gray. 1988. Mechanical factors in the excitation of the lateral line of fishes. pp. 595–593. *In*: J. Atema, R.R. Fay, A.N. Popper & W.N. Tavolga (ed.) Sensory Biology of Aquatic Animals, Springer-Verlag, Heidelberg.

Dijkgraaf, S. 1934. Untersuchungen über die Funktion der Seitenorgane an Fischen. Zeitschrift fur vergleichende Physiologie 20: 162–214.

Dijkgraaf, S. 1962. The functioning and significance of the lateral line organs. Biol. Rev. 38: 51–106.

Ekström von Lubitz, D.K.J. 1981. Ultrastructure of the lateral-line sense organs of the ratfish, *Chimaera monstrosa*. Cell Tiss. Res. 215: 651–665.

Ewart, J.C. 1892. The lateral sense organs of elasmobranchs. I. The sensory canals of *Laemargus*. Trans. R. Soc. Edinb. 37: 59–85.

Ewart, J.C. & H.C. Mitchell. 1892. On the lateral sense organs of elasmobranchs. II. The sensory canals of the common skate (*Raja batis*). Trans. R. Soc. Edinb. 37: 87–105.

Fitz, E.S. & F.C. Daiber. 1963. An introduction to the biology of *Raja eglanteria* Bosc 1802 and *Raja erinacea* Mitchell 1825, as they occur in Delaware Bay. Bull. Bingham. Oceanogr. Coll. 18: 69–97.

Flock, A. 1965a. Electronmicroscopic and electrophysiologic studies on the lateral line canal organ. Acta Oto-laryngol. Suppl. 199: 1–90.

Flock, A. 1965b. Transducing mechanisms in the lateral line canal organ receptors. Cold Spring Harbor Symp. Quant. Biol. 30: 133–145.

Fuiman, L.A. 1993. Development of predator evasion in Atlantic herring, *Clupea harengus* L. Anim. Behav. 45: 1101–1116.

Funicelli, N.A. 1975. Taxonomy, feeding, limiting factors and sex ratios of *Dasyatis sabina*, *Dasyatis americana*, *Dasyatis say*, and *Narcine brasiliensis*. Ph.D. Dissertation, University of Southern Mississippi, Hattiesburg. 259 pp.

Garman, S. 1888. On the lateral canal system of Selachia and Holocephala. Bull. Mus. Comp. Zool. 17: 57–119.

Görner, P. & C. Mohr. 1989. Stimulus localization in *Xenopus*: role of directional sensitivity of lateral line stitches. pp. 543–560. *In*: S. Coombs, P. Görner & H. Münz (ed.) The Mechanosensory Lateral Line – Neurobiology and Evolution, Springer-Verlag, New York.

Hama, K. & Y. Yamada. 1977. Fine structure of the ordinary lateral line organ II. The lateral line canal organ of the spotted shark, *Mustelus manazo*. Cell Tiss. Res. 176: 23–36.

Harris, G.G. & W.A. Van Bergeijk. 1962. Evidence that the lateral-line organ responds to near-field displacements of sound sources in water. J. Acoust. Soc. Amer. 34: 1831–1841.

Hassan, E.S. 1989. Hydrodynamic imaging of the surroundings by the lateral line of the blind cave fish, *Anoptichthys jordani*. pp. 217–227. *In*: S. Coombs, P. Görner & H. Münz (ed.) The Mechanosensory Lateral Line – Neurobiology and Evolution, Springer-Verlag, New York.

Hensel, K. 1978. Morphology of the lateral-line canal system of the genera *Abramis, Blicca*, and *Vimba* with regard to their ecology and systematic position. Acta. Univ. Carol. Biol. 12: 105–197.

Hildebrand, S.F. & W.C. Schroeder. 1928. Fishes of the Chesapeake Bay. Bull. U.S. Bur. Fish. 43: 1–366.

Hillman, D.E. & E.R. Lewis. 1971. Morphological basis for a mechanical linkage in otolithic receptor transduction in the frog. Science 174: 416–419.

Hoagland, H. 1933. Quantitative analysis of responses from lateral-line nerves of fishes. II. J. Gen. Physiol. 16: 715–732.

Hoekstra, D. & J. Janssen. 1985. Non-visual feeding behavior of the mottled sculpin, *Cottus bairdi*, in Lake Michigan. Env. Biol. Fish. 12: 111–117.

Hoekstra, D. & J. Janssen. 1986. Lateral line receptivity in the mottled sculpin (*Cottus bairdi*). Copeia 1986: 91–96.

Hofer, B. 1908. Studien über die Hautsinnesorgane der Fische. I. Die Funktion der Seitenorgane bei den Fischen. Berichte der Königlich Bayerischen Biologischen Veruchsstation München 1: 115–164.

Janssen, J., W.R. Jones, A. Whang & P.E. Oshel. 1995. Use of the lateral line in particulate feeding in the dark by juvenile alewife (*Alosa pseudoharengus*). Can. J. Fish. Aquat. Sci. 52: 358–363.

Johnson, S.E. 1917. Structure and development of the sense organs of the lateral canal system of selachians (*Mustelus canis* and *Squalus acanthias*). J. Comp. Neurol. 28: 1–74.

Kalmijn, A.J. 1989. Functional evolution of lateral line and inner ear sensory systems. pp. 187–215. *In*: S. Coombs, P. Görner & H. Münz (ed.) The Mechanosensory Lateral Line – Neurobiology and Evolution, Springer-Verlag, New York.

Katsuki, Y. & T. Hashimoto. 1969. Shark pit organs: enhancement of mechanosensitivity by potassium ions. Science 180: 1287–1289.

Katsuki, Y., K. Yanagisawa, A.L. Tester & J.I. Kendall. 1969. Shark pit organs: response to chemicals. Science 163: 405–407.

Kroese, A.B. & N.A.M. Schellart. 1992. Velocity- and acceleration-sensitive units in the trunk lateral line of the trout. J. Neurophys. 68: 2212–2221.

Lannoo, M.J. 1987. Neuromast topography in urodele amphibians. J. Morphol. 191: 247–263.

Leydig, F. 1850. Über die Schleimkanale der Knochenfische. Arch. Anat. Physiol. Wiss. Med. 1850: 170–181.

Maruska, K.P., E.G. Cowie & T.C. Tricas. 1996. Periodic gonadal activity and protracted mating in elasmobranch fishes. J. Exp. Zool. 276: 219–232.

Maruska, K.P. & T.C. Tricas. 1998. Morphology of the mechanosensory lateral line system in the Atlantic stingray, *Dasyatis sabina*: the mechanotactile hypothesis. J. Morph. 238: 1–22.

Michael, S.W. 1993. Reef sharks and rays of the world: a guide to their identification, behavior, and ecology. Sea Challengers, Monterey. 107 pp.

Montgomery, J.C. 1989. Lateral line detection of planktonic prey. pp. 561–574. *In*: S. Coombs, P. Görner & H. Münz (ed.) The Mechanosensory Lateral Line – Neurobiology and Evolution, Springer-Verlag, New York.

Montgomery, J.C., C.F. Baker & A.G. Carton. 1997. The lateral line can mediate rheotaxis in fish. Nature 389: 960–963.

Montgomery, J.C., J.A. Macdonald & G.D. Housley. 1988. Lateral line function in an Antarctic fish related to the signals produced by planktonic prey. J. Comp. Physiol. A 163: 827–833.

Montgomery, J.C. & A.J. Saunders. 1985. Functional morphology of the piper *Hyporhamphus ihi* with reference to the role of the lateral line in feeding. Proc. R. Soc. Lond. B 224: 197–208.

Montgomery, J.C. & E. Skipworth. 1997. Detection of weak water jets by the short-tailed stingray *Dasyatis brevicaudata* (Pisces: Dasyatidae). Copeia 1997: 881–883.

Münz, H. 1985. Single unit activity in the peripheral lateral line system of the cichlid fish *Sarotherodon niloticus* L. J. Comp. Physiol. A 157: 555–568.

Münz, H. 1989. Functional organization of the lateral line periphery. pp. 285–297. *In*: S. Coombs, P. Görner & H. Münz (ed.) The Mechanosensory Lateral Line – Neurobiology and Evolution, Springer-Verlag, New York.

Murray, R.W. 1961. The initiation of cutaneous nerve impulses in elasmobranch fishes. J. Physiol. 159: 546–570.

Nickel, E. & S. Fuchs. 1974. Organization and ultrastructure of mechanoreceptors (Savi vesicles) in the elasmobranch *Torpedo*. J. Neurocytol. 3: 161–177.

Norris, H.W. 1932. The laterosensory system of *Torpedo marmorata*, innervation and morphology. J. Comp. Neurol. 56: 169–178.

Northcutt, R.G. 1992. The phylogeny of octavolateralis ontogenies: a reaffirmation of Garstang's phylogenetic hypothesis. pp. 21–47. *In*: D.B. Webster, R.R. Fay & A.N. Popper (ed.) The Evolutionary Biology of Hearing, Springer-Verlag, New York.

Partridge, B.L. & T.J. Pitcher. 1980. The sensory basis of fish schools: relative roles of lateral line and vision. J. Comp. Physiol. 135: 315–325.

Presnell, J.K. & M.P. Schreibman. 1997. Humason's animal tissue techniques, fifth edition. John Hopkins University Press, Baltimore. 572 pp.

Puzdrowski, R.L. & R.B. Leonard. 1993. The ocatvolateral systems in the stingray, *Dasyatis sabina*. I. Primary projections of the octaval and lateral line nerves. J. Comp. Neurol. 332: 21–37.

Roberts, B.L. 1969. Mechanoreceptors and the behavior of elasmobranch fishes with special reference to the acoustico-lateralis system. pp. 331–390. *In*: E.S. Hodgeson & R.F. Mathewson (ed.) Sensory Biology of Sharks, Skates and Rays, Office of Naval Research, Department of the Navy, Arlington.

Roberts, B.L. & K.P. Ryan. 1971. The fine structure of the lateral-line sense organs of dogfish. Proc. R. Soc. Lond. B 179: 157–169.

Rudloe, A. 1989. Captive maintenance of the lesser electric ray, with observations of feeding behavior. Prog. Fish Cult. 51: 37–41.

Sand, A. 1937. The mechanism of the lateral sense organs of fishes. Proc. R. Soc. B 123: 472–495.

Satou, M., H.A. Takeuchi, S. Kitamura, Y. Kudo, J. Nishii & M. Tanabe. 1991. Involvement of lateral line sense in inter-sexual vibrational communication during spawning behaviour in the hime salmon (landlocked red salmon, *Oncorhynchus nerka*). Neurosci. Res. Suppl. 14: 15.

Satou, M., H.A. Takeuchi, J. Nishii, M. Tanabe, S. Kitamura, N. Okumoto & M. Iwata. 1994. Behavioral and electrophysiological evidence that the lateral line is involved in the inter-sexual vibrational communication of the hime salmon (landlocked red salmon, *Oncorhynchus nerka*). J. Comp. Physiol. A 174: 539–549.

Savi, P. 1844. Etudes anatomiques sur le systeme nerveux et sur l'organe electrique de la *Torpille*. pp. 272–348. *In*: C. Matteucci (ed.) Traite des Phenomenes Electrophysiologiques des Animaux, Chez L. Mechelsen, Paris.

Schwartz, F.J. 1996. Biology of the clearnose skate, *Raja eglanteria*, from North Carolina. Florida Scientist 59: 82–95.

Stenonis, N. 1664. De muscalis et glandulis observationum specimen cum duabus epistelis quarum una ad guil. Pisonum de Rajidae etc., Hafniae. 84 pp.

Szabo, T. 1958. Quelques precisions sur la morphologie de l'appareil sensoriel de Savi dans *Torpedo marmorata*. Zeitschrift für Zellforschung 48: 536–537.

Szabo, T. 1968. Analyse morphologigicque et fonctionelle de l'epithelium sensoriel d'un mechanorecepteur. Actual. Neurol. 6: 131–147.

Tester, A.L. & J.I. Kendall. 1967. Innervation of free and canal neuromasts in the sharks *Carcharhinus menisorrah* and *Sphyrna lewini*. pp. 53–69. *In*: P.H. Cahn (ed.) Lateral Line Detectors, Indiana University Press, Bloomington.

Tester, A.L. & J.I. Kendall. 1969. Morphology of the lateralis canal system in the shark genus *Carcharhinus*. Pac. Sci. 23: 1–16.

Tester, A.L. & G.J. Nelson. 1969. Free neuromasts (pit organs) in sharks. pp. 503–531. *In*: P.W. Gilbert, R.F. Mathewson & D.P. Rall (ed.) Sharks, Skates and Rays, Johns Hopkins Press, Baltimore.

Webb, J.F. & R.G. Northcutt. 1997. Morphology and distribution of pit organs and canal neuromasts in non-teleost bony fishes. Brain Behav. Evol. 50: 139–151.

Environmental Biology of Fishes **60:** 77–92, 2001.

The neuroecology of the elasmobranch electrosensory world: why peripheral morphology shapes behavior

Timothy C. Tricas
Department of Biological Sciences, Florida Institute of Technology, 150 West University Boulevard, Melbourne, FL 32901-6988, U.S.A.
Present address: Department of Zoology, University of Hawaii at Manoa, Honolulu, HI 96822, U.S.A.
(e-mail: tricas@hawaii.edu)

Received 7 April 2000 Accepted 16 May 2000

Key words: ampullae of Lorenzini, batoid, ecomorphology, electric field, ray, skate, white shark

Synopsis

The adaptations of elasmobranch sensory systems can be studied by linking the morphological structure with the natural behavior and ecology of the organism. This paper presents the first step in a 'neuroecological' approach to interpret the spatial arrangement of the electrosensory ampullary organs in elasmobranch fishes. A brief review of the structure and function of the ampullae of Lorenzini is provided for interpretation of the organ system morphology in relation to the detection of dipole and uniform electric fields. The spatial projections of canals from discrete ampullary clusters were determined for the barndoor skate, *Raja laevis*, based upon a published figure in Raschi (1986), and measured directly from the head of the white shark, *Carcharodon carcharias*. The dorsoventrally flattened body of the skate restricts the projections of long canals to the horizontal plane. There is a distinct difference between dorsal and ventral projection patterns in all groups. Notable within-cluster features include a relatively long canal subgroup in the dorsal superficial ophthalmic (SOd) and dorsal hyoid (HYOd) clusters that are oriented parallel (bidirectionally) to the longitudinal axis of the body. It is postulated that this subgroup of canals may be important for detection and orientation to weak uniform fields. Ventral canal projections in the skate are primarily lateral, with the exception of the hyoid (HYOv) that also projects medially. This wide dispersion may function for the detection of prey located below the body and pectoral fins of the skate, and may also be used for orientation behavior. The mandibular canals located near the margin of the lower jaw (of both study species) are ideally positioned for use during prey manipulation or capture, and possibly for interspecific courtship or biting. The head of the white shark, which lacks the hyoid clusters, is ovoid in cross section and thus ampullary canals can project into three-dimensional space. The SOd and superficial ophthalmic ventral (SOv) clusters show strong rostral, dorsal and lateral projection components, whereas the SOv also detects rostral fields under the snout. In the sagittal plane, the SOv and SOd have robust dorsal projections as well as ventral in the SOv. Most notable are canal projections in the white shark buccal (BUC) ampullary cluster, which has a radial turnstile configuration on the ventrolateral side of the snout. The turnstile design and tilt between orthogonal planes indicates the white shark BUC may function in detection of uniform fields, including magnetically induced electric fields that may be used in orientation behaviors. These data can be used in future neuroecology behavioral performance experiments to (1) test for possible specializations of cluster groups to different natural electric stimuli, (2) the possibility of specialized canal subgroups within a cluster, and (3) test several models of navigation that argue for the use of geomagnetically induced electric cues.

Introduction

The 'neuroecology' of elasmobranch sensory systems

Functional morphology is a tool used to study organismal form and function, and provides important information for studies of adaptation, natural selection, and evolution. An outgrowth of functional morphology is the field of ecological morphology, often referred to as 'ecomorphology', which is based on the premise that the morphological design of an organism is associated with its ecology and behavior (see Wainwright & Reilly 1994, Motta et al. 1995). This approach is used widely to test hypotheses on feeding performance, prey selection and the evolution of feeding mechanisms in fishes (Turingan & Wainwright 1993, Westneat 1995, Hernandez & Motta 1997). The ability to link morphology and ecology relies on the identification of a quantifiable behavioral measure of 'performance' for the ecological phenotype such as prey capture success or processing ability. Remarkably, ecomorphology is infrequently applied to the study of sensory systems of fishes (e.g. Hueter 1990, Barry & Hawryshyn 1999). Like the performance measures used in feeding ecomorphology, the performance measures for sensory system ecomorphology can also be linked to feeding success or prey capture. However, since sensory systems are used primarily for pre-capture behaviors such as the detection, localization and capture of prey (rather than post-capture behaviors such as prey handling or processing), the specific performance measures will differ. Thus, the 'neuroecology' approach can be a useful complement to studies of feeding ecomorphology by establishing the constraints imposed by the sensory system morphology on the detection of natural stimuli, and in this manner they can provide new insight on behavioral adaptations.

Early anatomists who described the gross anatomical features of the ampullary organs of elasmobranchs (Stenonis 1664, Lorenzini 1678, Ewart & Mitchell 1891, Lamont 1916) were unaware of its ecological function. The discovery of the ampullary electrosense (Murray 1960, 1962, Dijkgraff & Kalmijn 1966) and its role in predation (Kalmijn 1971) has since given way to many unanswered questions that concern the ecological function and evolution of this extraordinary system. In his seminal work, Raschi (1986) compared anatomical characters of the ampullary system among 40 species of skates collected from a wide range of marine habitats. In those species examined there were more pores on the ventral than dorsal surface with pore density usually greatest around the mouth. In addition, the ampullary pores of piscivorous skates were distributed over a larger area of the body than those feeding upon benthic invertebrates, and the relative density of ventral pores was inversely related to prey mobility. Species that inhabit deep water had larger ampullae than did those in shallow waters, and were suggested to have higher sensitivity. The distribution of skin pores on deep water species may compensate for reduced visual function, and the increased pore density reflects an increased spatial resolution during predation. Such correlations and inferences about the distribution of canal pores among different species provide critical information for interpretation of function, ecology and evolution of the skate electrosense.

Raschi's (op. cit.) important work emphasized ecological associations of different spatial distributions of ampullary pores on the skin, but did not address other important functions such as directional sensitivity. This paper uses the neuroecological approach to interpret the behavioral and ecological functions of the electrosensory complex in the barndoor skate, *Raja laevis*, and white shark, *Carcharodon carcharias*. The fine structure and morphology of the ampullary system is reviewed to show how biologically significant electric fields stimulate the ampulla of Lorenzini. New morphological data are then presented on interspecific and intraspecific differences in the spatial arrangement of ampullary groups. These data are used to propose testable hypotheses on the behavioral and ecological functions of different ampullary groups in multidimensional space, the organization of the brain for information processing, and evolution of the ampullary electrosensory system.

The ampullary electrosensory system

All sharks and rays possess an electrosensory system that consists of subdermal groups of functional sensory units known as the ampullae of Lorenzini. Each functional unit consists of a small ampulla chamber that is formed by small bulbous pouches known as alveoli. A canal about 1 mm wide is attached to the ampulla and leads to the surface of the skin (Figure 1a). A layer of sensory hair cells and pyramidal support cells forms the lining of the alveoli (Waltman 1966). Each sensory hair cell has an apical kinocilium that projects into the lumen of the ampulla chamber (Figure 1b). Receptor and support cells are joined by tight junctions to form

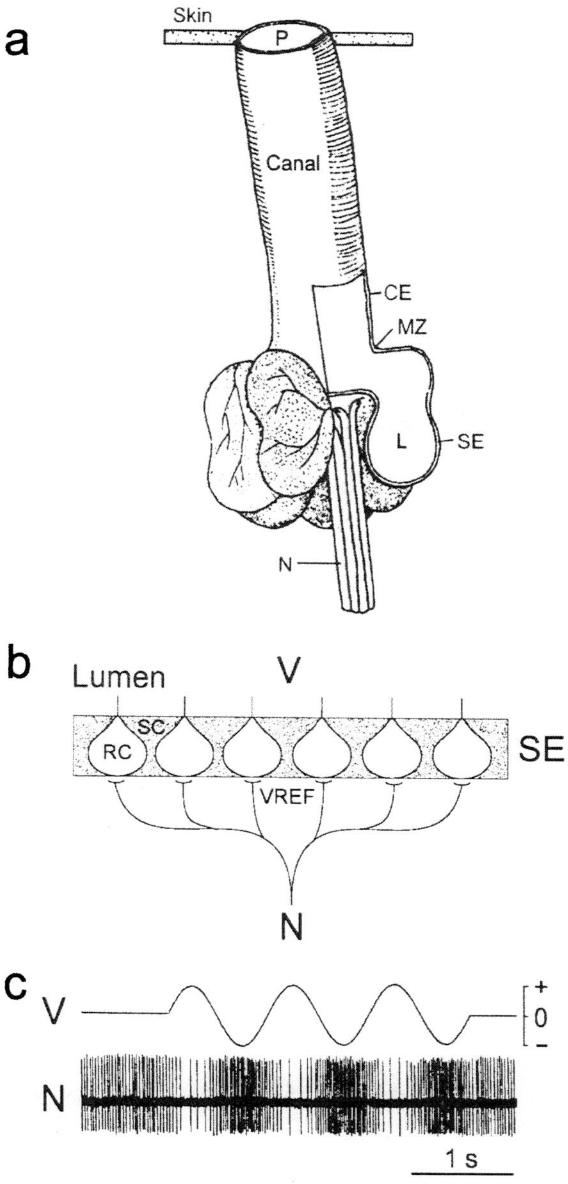

Figure 1. The ampulla of Lorenzini. a – the ampulla of the skate is formed by several alveoli that share a continuous lumen (L) and a subdermal canal that has a single pore on the skin. The sensory epithelium (SE) forms the highly resistive ampulla wall that connects with the canal epithelium (CE) at the marginal zone (MZ). The ampulla and canal are filled with a highly conductive gel. This arrangement forms an electrical 'core conductor' in which the potential within the ampulla lumen is isopotential with that at the surface pore. The sensory epithelium is innervated by primary afferent neurons (N) that conduct electrosensory information to the brain (Modified from Waltman 1966). b – in most elasmobranch species, the sensory epithelium is a layer of receptor cells (RC) and support cells (SC). Tight junctions between these cells form a high electrical resistance barrier between the lumen of the ampulla and basal portion of the receptor cells. The difference between lumen voltage (V) and reference voltage (VREF) stimulates the small apical surface of the receptor cells and controls release of neurotransmitter onto primary afferent neurons (N). c – recording from a single primary afferent neuron shows that the resting discharge pattern is modulated by a varying lumen potential. Cathodal (−) stimuli increase neural discharges, whereas anodal (+) stimuli decrease discharges (modified from Tricas & New 1998).

a high resistance electrical barrier between the apical and basal surfaces of the sensory epithelium. A double layer of squamous epithelial cells and connective tissue fibers form the canal wall and maintain the high electrical resistance between inner and outer surfaces. Both the ampulla and canal are filled with a high potassium, low-resistivity gel (Murray & Potts 1961, Doyle 1963). This arrangement forms an electrical core conductor such that the ampullary chamber is isopotential with a charge at the skin pore. Primary afferent neurons innervate the receptor cells, and encode stimulus amplitude and frequency data that are sent to the brain (Murray 1962, Montgomery 1984, Tricas & New 1998) (Figure 1c).

In marine species the ampullae are grouped into discrete subdermal clusters in the head (Figure 2) and are

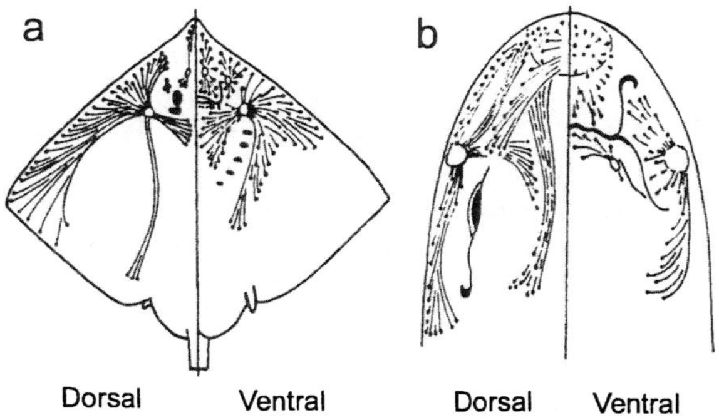

Figure 2. Diagrammatic representation of the horizontal distribution of ampullae of Lorenzini in the skate, *Raja*, and cat shark, *Scyliorhinus*. Ampullae are grouped into clusters in the head and have individual canals of different lengths that radiate in many directions. a – the body of the skate is dorsoventrally flattened, thus canals project primarily within the horizontal plane of the head and pectoral fins. b – the head of the shark is circular or elliptical in cross section, thus most canals in the shark head have projection components in multiple planes (modified from Murray 1960).

innervated by different branches of the anterior lateral line nerve (Norris 1929). Canals usually project in many directions from each cluster and their pores are distributed widely over the surface of the head (and pectoral fins in batoids). The contiguous grouping of individual ampullae into a single cluster results in a common potential at the basal region of all receptors. In contrast, all sensory cells of a single ampulla experience the same apical voltage that varies with the potential at its skin pore (Figure 3). The hair cells act as voltage detectors and release neurotransmitter onto their primary afferent neurons as a function of the difference between their apical (= pore) and basal (= internal) potentials. However, the potentials at surface pores are conserved within their respective ampullae, and the somatotopic distribution of the field is transmitted to the brain via parallel neural channels.

The morphological arrangement of the ampullary canals permits detection of both small local fields produced by biological organisms and large uniform electric fields of inanimate or animate origins (Kalmijn 1974). When a small localized dipole stimulus (such as that of a small prey) is presented at a pore that is far away from its ampulla, the potential is conducted to receptor cells within the ampulla chamber (Figure 3a). In this case, the receptor response is independent of canal length because the small spatial field does not influence the reference potential at the cluster. However, when the animal's body is within a uniform field (or at the fringe of a large dipole field) the body can admit a portion of the field that can influence the internal reference potential (Figure 3b). When the weak uniform electric field is parallel to the canal, the stimulus voltage at the apical surface of receptor cells is determined by the linear separation between the ampulla and its canal pore. Thus, long canals sample across a greater distance within the field and provide a larger potential difference for receptor cells than do ampullae with short canals. In addition, the strongest potential difference occurs when the canal is oriented parallel to the field and decreases as a cosine function as it deviates away from the direction of the field. Therefore, when an omnidirectional ampullary array is within a uniform field, the canals simultaneously sample the external potentials at different points on the body. Theoretically this can provide immediate information about the field's intensity, spatial configuration and possibly the direction of the source.

Laboratory and field studies demonstrate that the electrosense can mediate many natural behaviors of

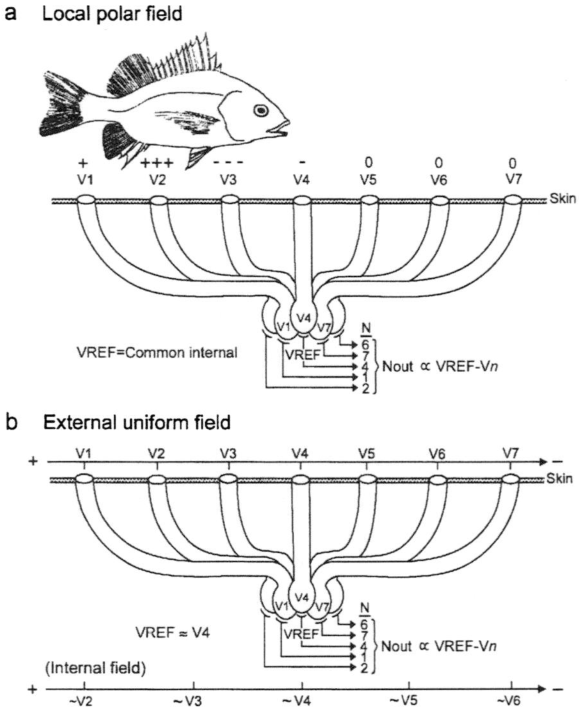

Figure 3. Simplified model for the encoding of extrinsic polar and uniform electric fields by the elasmobranch electrosensory system. a – any living prey produces a weak polar electric field formed by the differential distribution of charges on (or in) the organism. This creates weak potentials (+, −) in the water that surrounds the body of the prey. When the prey approaches, the surface pores of the ampullary system sample the field potentials (Vn) across the surface of the skin. The potentials at each pore are conducted to their individual ampullae and stimulate sensory neurons (N). In this scenario, the voltage gradient from the prey does not influence the common internal reference potential (VREF) at the ampullary cluster (and basal surfaces of receptor cells), and the effective stimulus for all ampullae is represented by the voltage drop across the skin. Neural output for each individual ampulla (Nout) is proportional to the difference between VREF and the voltage at its associated surface pore (Vn). b – when the shark or ray swims within a uniform field the field lines are parallel (and shown here as parallel to the surface of the skin). The body is more resistive than seawater but the relatively low skin resistance admits an only slightly weaker electric field through the body (Kalmijn 1974). The potential within each ampulla is isopotential with that of its pore (Vn), and is referenced to the common potential at the cluster (VREF). In this case, the stimulus within each ampulla is a function of the distance from its pore, i.e. canal length. Neural output for each ampulla (Nout) is also proportional to the difference between VREF and the potential at its associated surface pore (Vn).

rays and sharks. The ampullary receptor system can be used in the detection and capture of prey (Kalmijn 1971, 1982, Tricas 1982, Blonder & Alevizon 1988), the detection of potential predators (Sisneros et al. 1998) and in social behaviors (Tricas et al. 1995, Sisneros et al. 1998). In addition, theoretical models have proposed that geomagnetic induction of electric current could mediate orientation behavior (e.g. Kalmijn 1974, 1978, Paulin 1995). Despite these anatomical, physiological, behavioral and theoretical studies, few have interpreted the spatial arrangement of the ampullary complex in relation to the natural ecology of the animal (sensu Raschi 1986). The interpretation of specific features of the entire ampullary 'system' with the natural stimuli encountered in the wild is critical to understanding the morphological constraints of the electrosensory system, selective pressures, and specializations of different ampullary subsystems.

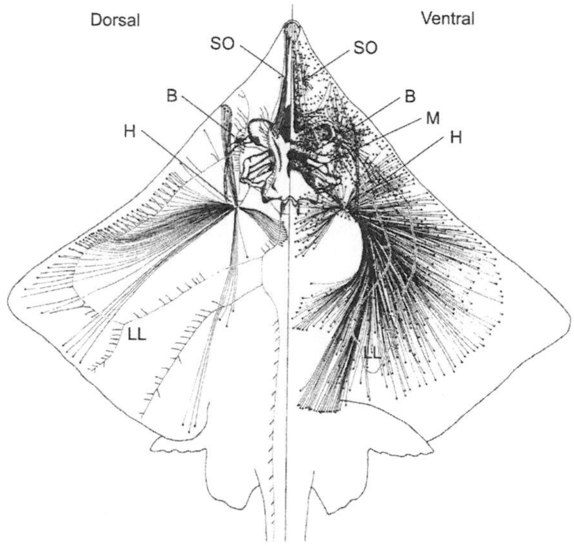

Figure 4. Ampullary clusters and canals on the dorsal and ventral surfaces of the barndoor skate, *Raja laevis*. Clusters indicated by B = buccal, H = hyoid, M = mandibular, SO = superficial ophthalmic. LL = lateral line (modified from Raschi 1986).

Materials and methods

Raja laevis

The two-dimensional spatial orientations of the ampullary canals in the barndoor skate, *Raja laevis*, were measured from the detailed morphological drawings provided by Raschi (1986) (Figure 4). Because of the dorsoventral flattening of the skate body, projection vectors for dorsal and ventral ampullary clusters were calculated only for the horizontal plane. A grid was superimposed over photocopy enlargements of the left dorsal and left ventral surfaces, and aligned with the longitudinal and transverse body axes. The center of the mandibular, buccal, hyoid and superficial ophthalmic clusters were marked as the origin for each canal projection. Angular projections from each cluster (rostral = 0°) were measured to the nearest degree with a protractor and followed a counterclockwise rotation as viewed dorsally. The length of each canal was measured as the distance from the center of its cluster to the surface pore. Canal length measurements were converted to absolute values (nearest 1 mm) relative to the mouth width of 9.5% × 700 mm TL = 67 mm as indicated by Raschi (1986). The vector angles for the complementary projections for the right side were calculated by subtraction of the measured angles from 360°. Vector data were then plotted in polar and Cartesian coordinates.

Carcharodon carcharias

The three-dimensional spatial orientation of ampullary canals was measured for the head of a juvenile white shark approximately 2.5 m total length. The entire head was fixed in a 4% formaldehyde solution and preserved in 50% isopropanol. The head was then clamped in a rigid stereotactic frame with the tip of the snout and the center of the eyes positioned in the horizontal plane to provide the longitudinal reference axis for the body. The skin was carefully removed on the left side to expose the ampullary canals of the mandibular, buccal and superficial ophthalmic clusters. The x (longitudinal), y (transverse) and z (vertical) coordinates for each surface pore and cluster center were measured with a caliper and ruler to a resolution of 1 mm. The measured coordinates for canals in each cluster were then converted to a common reference for the estimated center of each cluster (0, 0, 0). Data were plotted in polar coordinates. As discussed above, canal length is the major character that determines the sensitivity of an ampullary unit in a uniform field. Furthermore, an ampulla encodes the strongest signal when its canal is oriented parallel to the direction of the uniform field. Therefore, the spatial orientation for each canal was expressed as the vector within the horizontal, transverse and sagittal planes. Vectors for canals were determined

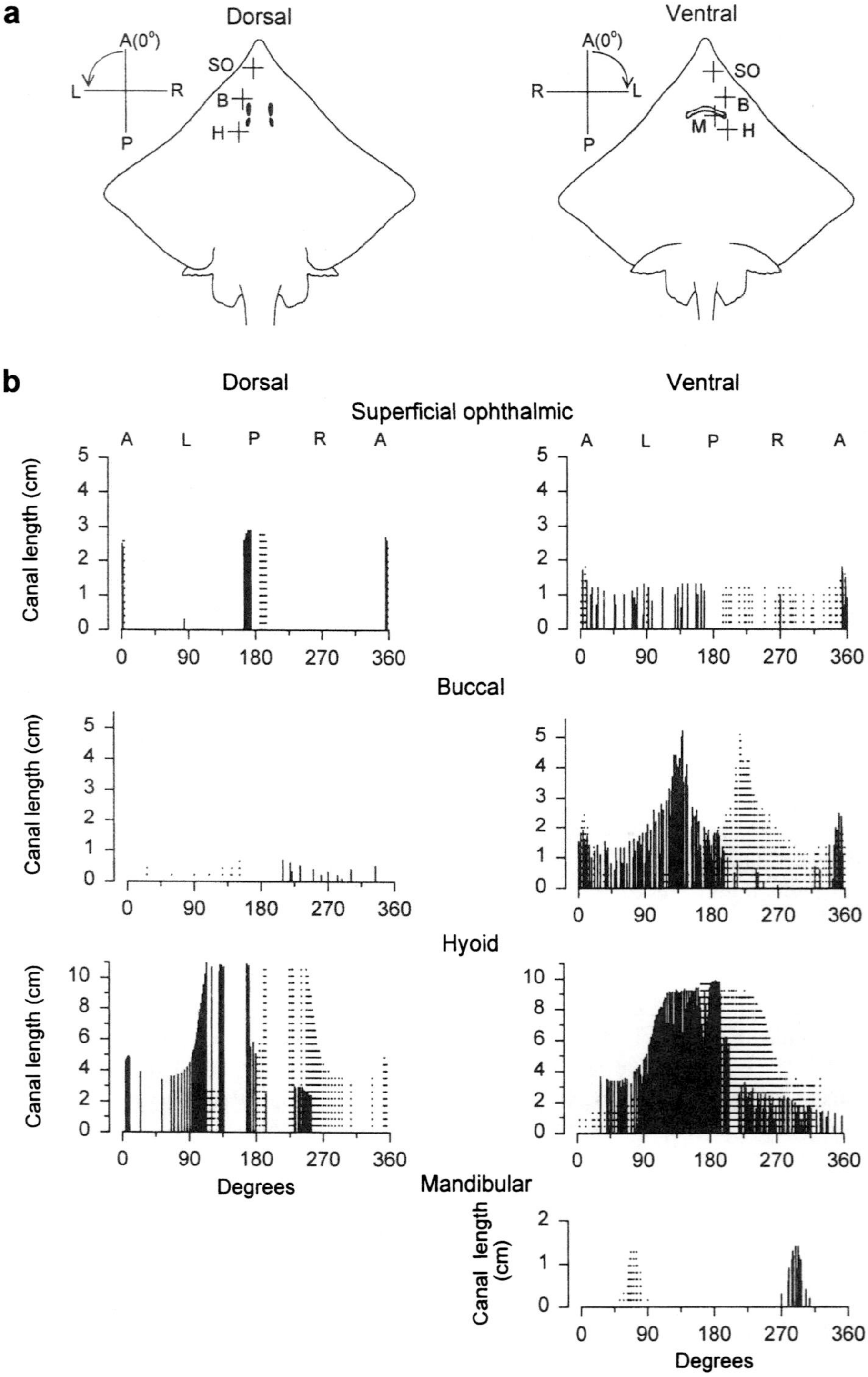

Figure 5. Projection vectors of ampullary canals in the barndoor skate, *Raja laevis*. a – dorsal and ventral locations of the four ampullary clusters on the left side of the body are indicated by crosses. All projections are referenced to 0° for the anterior (A) direction, 90° left (L), 180° posterior (P), and 270° right (R), and follow a left side rotation pattern (B = buccal, H = hyoid, M = mandibular, SO = superficial ophthalmic). b – plot of projection angle and length for individual canals from left (solid lines) and right (dotted lines) cluster pairs

for the three body planes by resolving the absolute coordinates into their respective canal lengths and projection angles. The vector angle for complementary canal projections for the contralateral side was calculated by subtraction of the measured angle from 360° and then adjusted to a 0° reference for rostral in the horizontal and sagittal planes, and a 0° reference for dorsal in the transverse plane. The bilateral distribution of canal vectors for each ampullary group was determined following a common angular rotation and then plotted in Cartesian coordinates. This analysis permits examination of angles of maximum sensitivity, null vectors where no projections occur, and possible sites for common mode rejection of electric field stimuli.

Results

The vector analysis of the ampullary canals provides information on the direction and lengths of ampullary canals in the separate ampullary clusters. The orientation data for each ampullary cluster are all presented relative to the major axes of the body, to allow comparison of relative sensitivities among ampullary groups.

The skate

The canal projections in *R. laevis* from the four discrete pairs of ampullary clusters can be reduced to the horizontal plane because of the dorsoventrally flattened body. The superficial ophthalmic (SO), buccal (BUC) and hyoid (HYO) clusters have canal projections to both the dorsal and ventral surfaces. These are subdivided into dorsal and ventral subgroups for analysis. The mandibular (MAN) projections are all concentrated on the lower jaw near the mouth.

The SO ampullary cluster is the most rostral and is located just anterior to the nasal capsule at the edge of the rostral cartilage (Figure 4,5a). Posterior to SO is the BUC cluster (= outer buccal of Raschi 1986) which is located lateral to the nasal capsule and anterior to the antorbital cartilage. The large HYO cluster is located immediately behind the hyomandibular cartilage and rostral to the branchial chambers.

There are major differences in dorsal projection patterns among clusters. The canals of the superficial ophthalmic dorsal cluster (SOd) are of approximately uniform length and have a pronounced longitudinal alignment in rostral and caudal projection groups (Figure 5b). This is in contrast to the buccal dorsal (BUCd) cluster, which has very short and medially directed canals. The canals of the dorsal hyoid (HYOd) group form a broad cruciform pattern and covers a wide area of the head and pectoral fins. The HYOd cluster also has seven canals that make a distinct caudal projection near 180°. These form two pore groups located near and approximately half the distance to the tip of the caudal disk margin. This is complemented by about 10 canals that project rostrally near 0° to the rostral margin of the disk. The lateral canals of the HYOd are the most abundant group and span an arc of about 45° around the tip of the pectoral fin. The ampullary pores along the anterior margin of the pectoral disk are positioned close to pores of the mechanosensory lateral line. A medial HYOd projection is formed by a group of the 12 shortest canals, which terminate in pores near the dorsal midline.

The ventral pores in *R. laevis* are more numerous than the dorsal surface, have a more profuse and omnidirectional projection pattern, and show distinct projection patterns among cluster groups (Table 1). The ventral superficial ophthalmic (SOv) canals are relatively uniform in length with widely spaced lateral

Table 1. Distribution of ampullae of Lorenzini within clusters in the barndoor skate, *Raja laevis*, and white shark, *Carcharodon carcharias*. Values indicate number of projections for a single ampullary cluster. BUCd = dorsal buccal, BUCv = ventral buccal, HYOd = dorsal hyoid, HYOv = ventral hyoid, MAN = mandibular, SOd = dorsal superficial ophthalmic, SOv = ventral superficial ophthalmic. In the white shark there is no hyoid cluster. Also, the white shark BUC is not divided into separate dorsal and ventral groups, thus total number of canals are given as BUCd. Canal numbers for *R. laevis* are estimated as counts taken from Figure 3 in Raschi (1986). np = not present.

Cluster	*Raja laevis*	*Carcharodon carcharias*
SOd	16	98
SOv	43	159
BUCd	12	33
BUCv	191	np
HYOd	66	np
HYOv	353	np
MAN	24	13

on the dorsal and ventral surfaces of the body. The direction of projections as well as canal lengths differ greatly both among clusters and between dorsal and ventral surfaces. There is no mandibular projection to the dorsal surface. Note that plots for ventral canals are referenced to a dorsal view and angle rotation.

projections (Figure 5b). The ventral buccal (BUCv) canals also project in a broad lateral distribution but at a higher density than SOv. In addition, the BUCv shows longer canals (and higher sensitivity to uniform fields) in the posterior lateral quadrants at 135° and 215°. The most rostral canals of the ventral hyoid (HYOv) start their projection at about 45° from the midline and increase in length by more than 2-fold as they approach and pass the tip of the pectoral fin. The longest HYOv canals project caudally and share common projection angles with shorter canals. There is a prominent HYOv caudal subgroup made of approximately 20 canals with projections near 180° that extends to the posterior margin of the disk. Much shorter canals project rostrally and medially towards the snout to complete the nearly omnidirectional canal radiation pattern of the HYOv. The mandibular (MAN) ampullae arise from clusters located on the ventral surface of the lower jaw and have very short projections along the margin of the lower jaw.

The white shark

Unlike the dorsoventrally flattened batoids, the head of most sharks is cylindrical or ovoid in cross section, thus most canals of individual ampullary clusters show projection components in at least two planes. These vectors in *C. carcharias* are resolved into three planes in order to compare the relative sensitivity and directionality of individual ampullary clusters to uniform fields. The SO cluster shows a separation into dorsal and ventral projections, and therefore is subdivided into superficial ophthalmic dorsal (SOd) and superficial ophthalmic ventral (SOv) groups.

The SO cluster is positioned approximately midway between the eye and the tip of the snout, with the SOd ampullae dorsal to those of the SOv. The polar plot of the SOd ampullary pores shows a strong rostral, lateral and dorsal projection (Figure 6a). This cluster contains the longest canals (4–4.5 cm) that project rostrally and form a small group of skin pores just lateral to the tip of the snout (Figure 6b). The remaining canals project more laterally but not medially. In the sagittal plane, the canals project back towards the eye and follow the slope of the head to the tip of the snout. In the transverse plane the dorsolateral component is seen. This arrangement indicates a predominately rostral and dorsolateral sensitivity for the SOd that is aligned with a portion of the forward and lateral visual field.

The SOv cluster is located below the SOd (Figure 7a), and like its dorsal counterpart shows strong rostral and dorsal projections. As seen in the horizontal plane, these follow the lateral margin of the snout in front of the eye (Figure 7b) and a small area on the ventral surface of the snout. Canals in the sagittal plane project omnidirectionally, with the exception of the ventrocaudal quadrant because of the position of olfactory nares.

The BUC cluster is located near the ventral surface of the snout, behind the nares, rostral to the eye and dorsal to the upper jaw (Figure 8a). The canals of this cluster project in an unique omnidirectional 'turnstile' pattern that is situated on the ventrolateral surface of the snout. The longest BUC canals are about 4 cm long and project along a rostromedial-posterolateral line as seen in the horizontal plane (Figures 8a,b). A strong downward-directed vertical sagittal plane. In the transverse plane, the major projections are dorsolateral and ventromedial. This 'turnstile' arrangement and alignment at about 45° between the sagittal-horizontal planes and also the sagittal-transverse planes indicates that the BUC cluster may be sensitive to uniform fields in more than one direction.

The MAN cluster is located beneath the lower jaw near the corner of the mouth and consists of only 13 canals in our specimen (Figure 9a). Canals project rostrally and dorsally along the margin of the lower jaw up to about 3 cm from the center of the cluster (Figure 9b). This spatial arrangement is consistent with detection of electric stimuli near the lower jaw.

Discussion

The neuroecology of the skate electrosensory system

Living elasmobranch fishes are represented by more than 800 species, of which approximately 240 species are skates (McEachran & Miyake 1990). The body form and spatial arrangement of the ampullary system of these fishes set important functional constraints on the detection of natural electric stimuli. The dorsoventral compression of the body in batoids restricts the distribution of ampullary pores to the dorsal and ventral surfaces. In addition, the projections of long canals from ampullary clusters are essentially compressed in the horizontal plane. Thus the skates and rays can detect

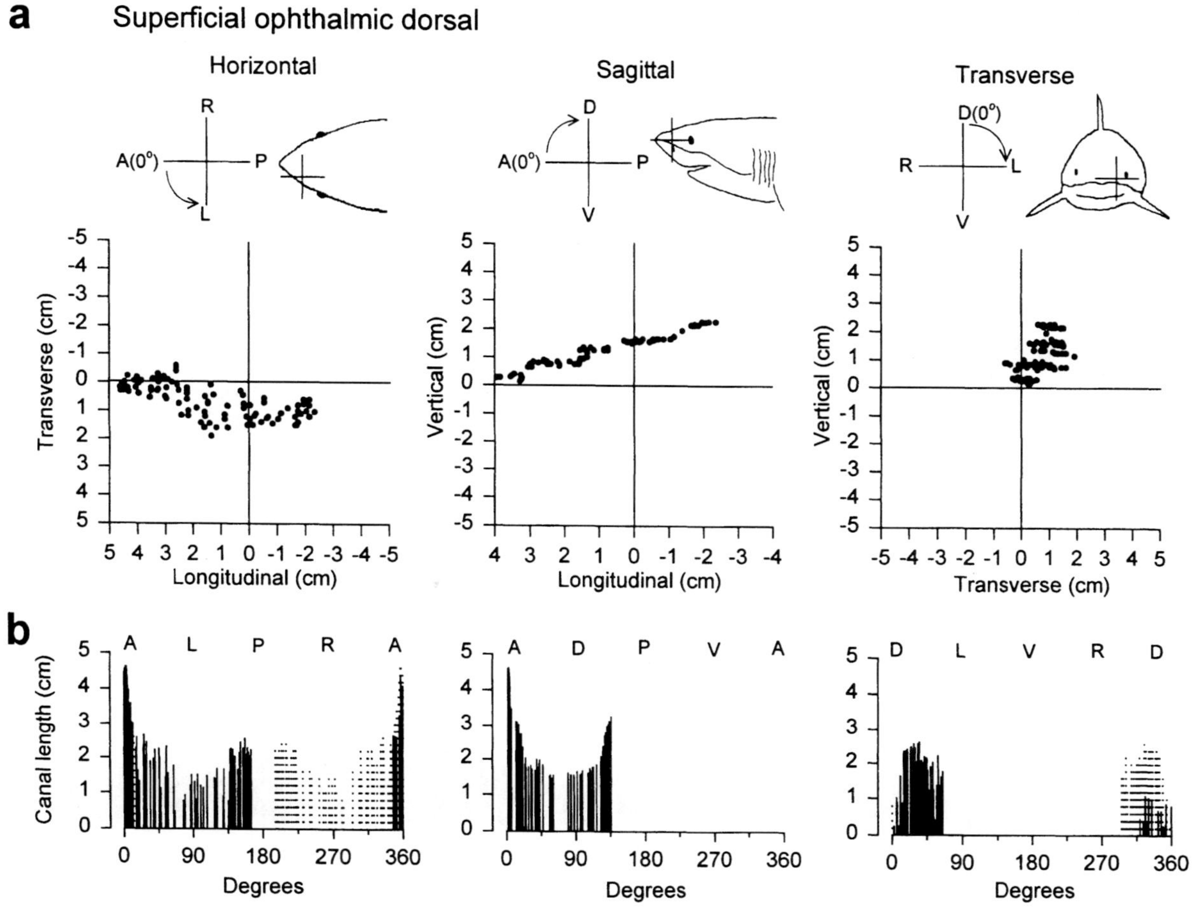

Figure 6. Projection vectors of the dorsal superficial ophthalmic ampullary group in the white shark, *Carcharodon carcharias*: a – polar coordinates for projections of the left ampullary cluster in horizontal, sagittal and transverse planes. Start of rotation (0°) is anterior for horizontal and sagittal planes, and dorsal in transverse plane. All projections are referenced to a rotation direction indicated by arrow. b – plots of projections for individual canals in horizontal, sagittal and transverse planes. Projection vectors are shown for both left (solid lines) and right (dotted lines) ampullary clusters. Note that the vertical projections are identical in the sagittal plane. Also, there are prominent rostral, dorsal and lateral projections in this group (A = anterior, D = dorsal, L = left, P = posterior, R = right, V = ventral).

small dipole fields, and are most sensitive to the horizontal components of uniform fields. Like their batoid relatives, the ampullary pores of most sharks are distributed over the surface of the head and can also detect small external polar fields such as those produced by prey. However, the head of most sharks is conical or slightly dorsoventrally flattened. As a result, ampullary canals can project omnidirectionally from clusters into three-dimensional space rather than only the horizontal plane. This complex spatial arrangement provides a potential sensitivity for a single canal to components of external uniform fields within three orthogonal planes.

One of the primary functions of the skate electrosensory system is to detect and locate bioelectric fields produced by their prey. Skates have a benthic lifestyle, a broad depth distribution and feed on a wide variety of prey primarily by a suction-grasping action of the mouth (Moss 1977). Raschi (1986) compared the ampullae of Lorenzini among 40 species of skates and showed that there were more pores on the ventral surface of the body, a feature related to their benthic feeding habits. In species that fed primarily upon benthic invertebrates, the ventral pores were more concentrated around the mouth and ventral pore density (not necessarily number of pores) was greater. Skates that fed upon more mobile fish prey had a large body size and lower pore density. The higher density of pores on the ventral surface of benthic invertebrate feeders

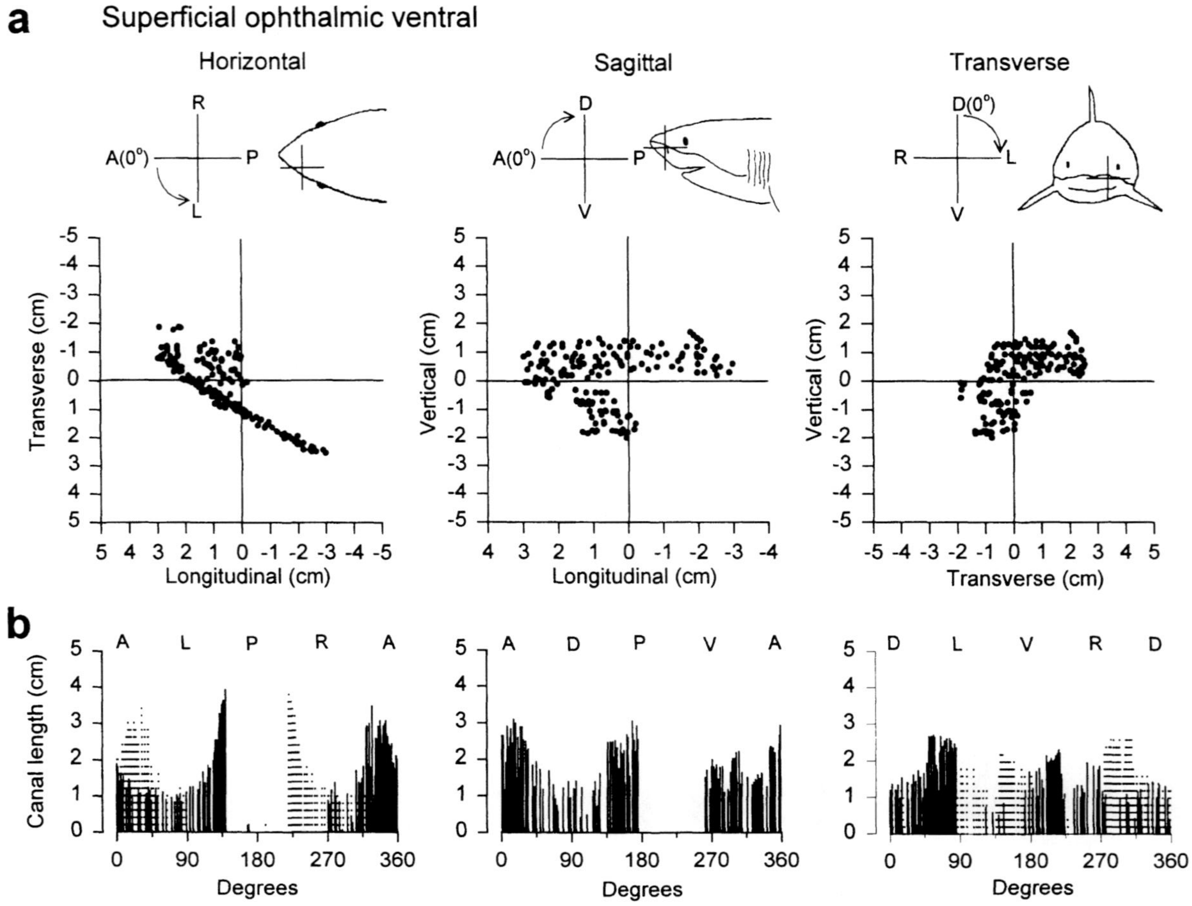

Figure 7. Projection vectors of the ventral superficial ophthalmic ampullary group in the white shark, *Carcharodon carcharias*: a – polar coordinates for projections of the left ampullary cluster in horizontal, sagittal and transverse planes. Start of rotation (0°) is anterior for horizontal and sagittal planes, and dorsal in transverse plane. All projections are referenced to a rotation direction indicated by arrow. b – plots of projections for individual canals in horizontal, sagittal and transverse planes. Projection vectors are shown for both left (solid lines) and right (dotted lines) ampullary clusters. Note that the vertical projections are identical in the sagittal plane. Also, there are prominent rostral and lateral projections in this group, and a near absence of projections due to the presence of the olfactory nares behind and below this cluster (A = anterior, D = dorsal, L = left, P = posterior, R = right, V = ventral).

provides a greater resolution for locating, manipulating and ingesting invertebrate prey that are excavated from the substrate.

The analysis of canal projection vectors for *Raja laevis* in this study provides further insight into biological functions of the skate electrosense. When a small prey field is presented at a canal pore and is small enough to not influence the reference potential at the basal region of the ampulla, sensitivity is independent of canal length. Thus ampullae with either long or short canals can provide information on the location of small prey relative to the surface of the skin. On the ventral surface of *R. laevis* the bilateral pairs of all ampullary clusters form an omnidirectional projection pattern (Figure 5) and pores cover a high proportion of the ventral body surface. This distribution would provide an excellent somatotopic map of small dipole fields near the skin surface.

In addition to bioelectric fields produced by prey, the ampullae can detect larger fields such as those produced by predators or conspecifics that are not positioned beneath the body. Weak standing bioelectric fields such as those that are produced by buried conspecific stingrays (Tricas et al. 1995) can stimulate electroreceptors from a distance, guide swimming movement towards the source and ultimately influence

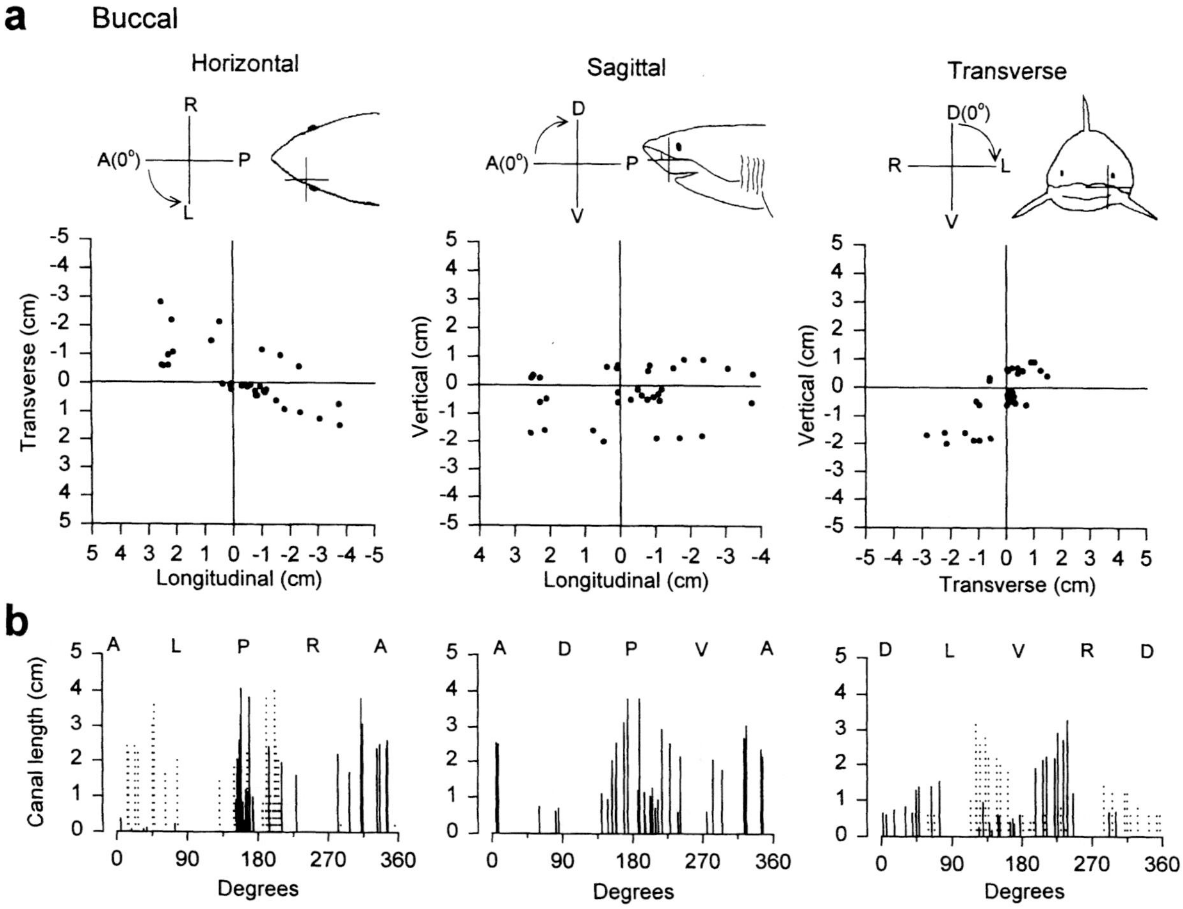

Figure 8. Projection vectors of the buccal ampullary group in the white shark, *Carcharodon carcharias*: a – polar coordinates for projections of the left ampullary cluster in horizontal, sagittal and transverse planes. Start of rotation (0°) is rostral for horizontal and sagittal planes, and dorsal in transverse plane. b – plots of projections for individual canals in horizontal, sagittal and transverse planes. Projection vectors are shown for both left (solid lines) and right (dotted lines) ampullary clusters. Note that the vertical projections are identical in the sagittal plane. The canal projections from this cluster project in a radial 'turnstile' pattern that follows the ventrolateral surface of the snout (A = anterior, D = dorsal, L = left, P = posterior, R = right, V = ventral).

both the voltage at the skin pore and the reference voltage at the ampulla. The voltage drop across the length of multiple canals could provide good directional information about the precise location of the source. In *R. laevis* a sensitivity bias based upon canal length is seen in the posterior quadrants of the BUCv and HYOv, whereas the SOv ampullae are approximately equally sensitive in the lateral direction. On the dorsal surface of *R. laevis* the SO canals are aligned with the longitudinal body axis and may provide an excellent sensory channel for orientation behavior. A similar longitudinal (and lateral) directionality exists in the HYOd group. The short medial projections of the BUCd canals indicates they may best detect strong fields above the body and between the eyes, probably when the animal is close to the source. The weakly electric organ found in the skates may serve important communication functions during social interactions (Bratton & Ayers 1987, Sisneros et al. 1998). The subgroup of the BUC could mediate inspection or approach behaviors towards the tail of a discharging conspecific. Mapping of the spatial configuration and strengths of these active discharges are needed to interpret the possible co-evolution of the electrosensory and electrogenic organ communication systems. In addition, ethological analyses of orientation behavior used in natural social settings are needed to determine any special functions for ampullary subgroups.

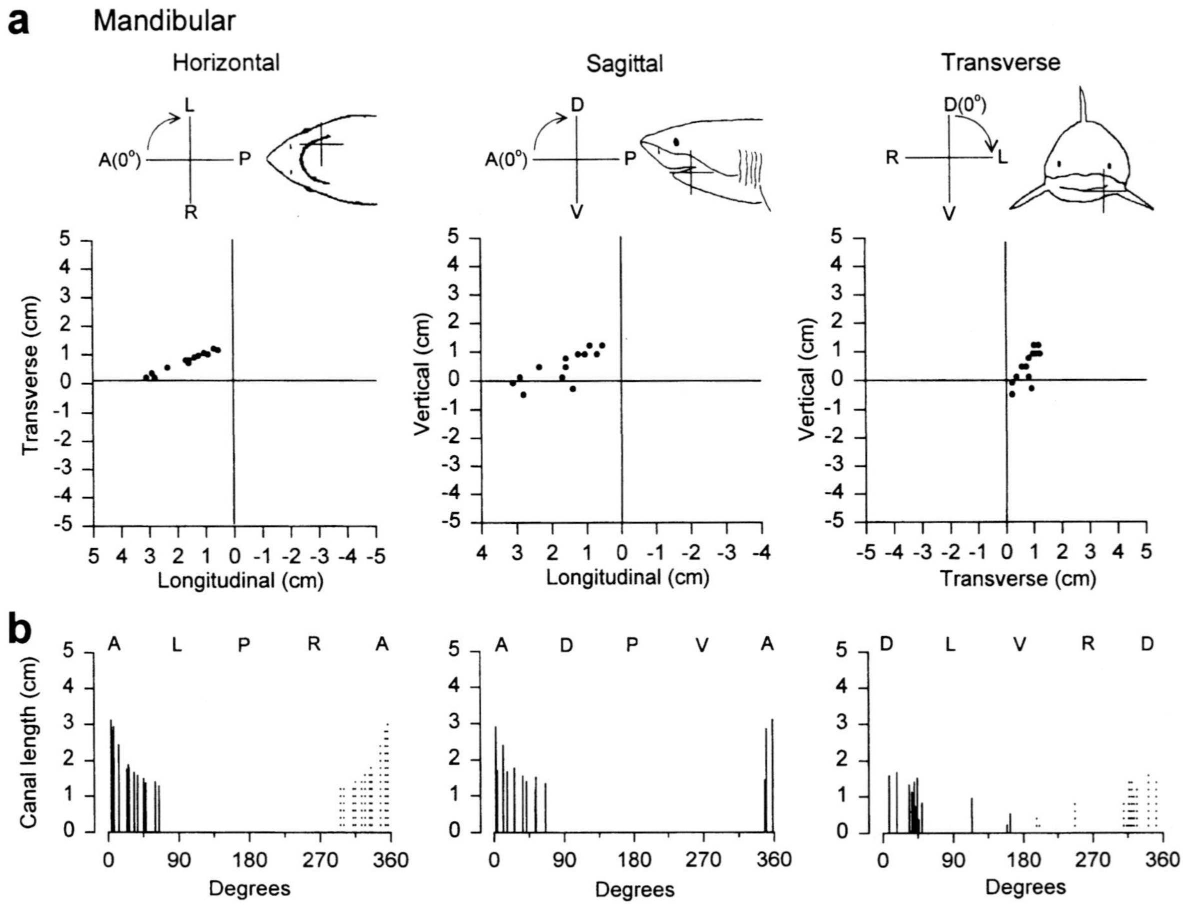

Figure 9. Projection vectors of the mandibular ampullary group in the white shark, *Carcharodon carcharias*: a – polar coordinates for projections of the left ampullary cluster in horizontal, sagittal and transverse planes. Start of rotation (0°) is rostral for horizontal and sagittal planes, and dorsal in transverse plane. b – plot of projections for individual canals in horizontal, sagittal and transverse planes. Projection vectors are shown for both left (solid lines) and right (broken lines) ampullary clusters. Note that the vertical projections are identical in the sagittal plane. These canals project along the margin of the lower jaw (A = anterior, D = dorsal, L = left, P = posterior, R = right, V = ventral).

The features of canal length and projection angle are also important when the animal encounters externally generated uniform fields. Kalmijn (1974, 1984) proposed the 'passive mode' model in which a horizontal electric field is induced by the movement of surface waters through the vertical component of the earth's magnetic field. With this system, charges of opposite polarity accumulate along the sides of a drifting shark and stimulate the ampullae. The restriction of ampullary subsystems to the horizontal plane in the skate could make an excellent comparator of charges across the body under such conditions. Differences in charge strength among the rostrally and caudally directed canals in the SOd and HYOd could sum at higher processing centers and serve as null detectors to indicate an orientation perpendicular to the flow of the stream. Alternatively, the two opposite inputs could be maximized for orientation parallel to the stream flow.

The 'active mode' model (Kalmijn 1974, 1984) proposes that a vertical electromotive field is induced as a shark swims through the horizontal component of the earth's magnetic field in an east or west direction. This results in a differential distribution of positive and negative charges on the dorsal and ventral surfaces of the body. The lack of vertically directed ampullary canals in the skate would not permit efficient detection of electric field gradients, but the charge stimuli on the two surfaces would be of opposite polarity and intensity would vary as a function of swimming speed and direction. Like the passive mode model, the ampullary groups responsible for detection of either of these stimuli would require specific organizations of the neural network in the brain. The somatotopic arrangement of ampullary clusters are conserved in the input pathways to the brainstem (e.g. Bodznick & Schmidt 1984), but the details of neural connectivity to support these models remain to be identified.

The neuroecology of the white shark electrosensory system

The barndoor skate and white shark share similar ampullary systems (white sharks lack the hyoid group) but differ greatly in body form, habitats and life style. The white shark is widely distributed in temperate coastal waters of the world. Adults are a major predator of marine mammals, especially pinnipeds (Le Boeuf et al. 1982, Ainley et al. 1985) and cetaceans (reviewed by Long & Jones 1996). Many studies have documented adult attack behavior on pinnipeds that are at rest on the surface as involving a visually-mediated search behavior followed by a rapid ambush attack from below (Tricas & McCosker 1984, McCosker 1985, Klimley et al. 1996). Just before the rapid attack, the eyes are rolled back into the orbit to reduce injury and the animal is momentarily without visual input. The electrosensory system was postulated to possibly provide information on the location of the prey during the final moments of the attack, during post-attack manipulation or pursuit of prey, or to detect changes in the disposition of the prey such as bleeding (Tricas & McCosker 1984). Many of the SO pores on the dorsal, ventral and lateral regions of the head and snout are within the visual field. Their position and concentration in this region would provide good spatial resolution for detection and tracking of a nearby prey at night, when the eyes are closed or the prey is beneath the snout. In comparison, juvenile white sharks feed primarily upon small bottom dwelling elasmobranchs and bony fishes (Tricas & McCosker 1984). White sharks are known to routinely swim near the bottom (e.g. Strong et al. 1986) and may use their SOv ampullary system to detect visually cryptic prey such as the cabezon, *Scorpaenichthys marmoratus*, that is associated with rocky inshore and kelp habitats (Limbaugh 1963). In this case the long anterior projecting canals of the SOd and SOv could detect these prey as the shark's snout passes over them.

The white shark ampullary system would also be well adapted to detect extrinsic and self-generated uniform fields. Orientations to horizontally induced fields in the passive mode (see above) could be mediated by some of the BUC or SO ampullae, most notably the laterally projecting SOv canals. In the active mode, the upward projecting SOd canals could encode the polarity and magnitude of weak charges on the dorsal surface of the head, as could components of the SOv. Most notable is the BUC cluster which sends radial projections that lay on the sloped ventral surface of the snout. The round shape of the snout tilts the BUC turnstile approximately midway between the sagittal and horizontal planes, whereas the pointed snout rotates the BUC toward the tip of the snout. This dual tilt feature theoretically makes the ampullae in the BUC equally sensitive to vertical, horizontal or oblique fields. Thus, if the SO or BUC function in detection of uniform fields, it is predicted that the neural circuitry to encode them should differ among clusters because of their different spatial organization.

Paulin (1995) proposed a model for geomagnetic induced orientation in sharks that integrates input from the vestibular labyrinths with changes in the induced voltages on the head during lateral swimming motion. Similar to Kalmijn's active mode, the induced vertical fields could be detected by the SOd, SOv, and/or BUC canals with vertical projection components. More work is clearly needed to verify the encoding of horizontal and vertical uniform fields by individual ampullary groups. In addition, there must be specific central processing pathways for encoding such information, and these remain to be characterized.

This paper has presented the first step of a neuroecological analysis, which is a detailed description of the spatial organization of the canals from different ampullary groups. In the discussion above numerous proposals were made as to how different ampulla groups and subgroups could function in the life of the skate and white shark. The next step is to identify biological contexts in which the system operates such as feeding, magnetic orientation, or the detection of mates or predators. An appropriate performance measure, such as distance of orientation from prey, can be chosen and experimental tests performed to demonstrate differences in performance within or among species.

Future direction

As outlined in the discussion, there exists numerous hypotheses on the function of the elasmobranch electrosensory system in natural settings. However, there are only a few studies that have attempted to test these models. Suggestions for future work with a 'neuroecology' emphasis are described below.

The stimulus specialization hypothesis

An efficient electrosensory system can enhance fitness in many biological contexts. This implies that selection may exist for a particular spatial configuration of the ampullary receiver system. For example, in order to determine how a sensory system imparts selective advantages during feeding, it is first critical to determine the prey species that make up the diet set. Live prey could then be brought into the laboratory and the spatial and intensity maps of their bioelectric fields constructed as done initially by Kalmijn (1974). These bioelectric configurations can be used to assess whether they differ enough so that elasmobranchs could use electric signatures to discriminate between prey species. A similar approach could be used to answer questions on electrosensory function in social behaviors or predator avoidance. Information about the spatial arrangement and intensity of these biologically important fields can be compared to the spatial features of ampullary clusters so that possible evolution of the ampullary receiver system to specific signals can be inferred and tested.

The functional subunit hypothesis

A broad range of theoretical and experimentally demonstrated functions for the electrosensory system are known, but there is almost no information on how the ampullary system may be subdivided for different behavioral functions. Anatomical studies clearly show differences in sizes, locations, pore distributions and projection patterns among elasmobranch species (Chu & Wen 1979). For example, the superficial ophthalmic cluster of the barndoor skate has a small but prominent number of long canals that are aligned with the main body axis. Do these form a functional subunit that serves a specific function such as orientation to horizontal uniform fields? Experimental studies where subunits are manipulated and the effects on behavior recorded can be conducted to address these questions. Differences in behavioral function can be confirmed by identifying different projection patterns of neurons from such subgroups.

Geomagnetic induction orientation hypotheses

There are currently a number of models which propose that elasmobranch fishes can use the ampullary electrosense to detect induced fields derived from the earth's magnetic field (Kalmijn 1974, 1984, 1997, Paulin 1995). Extensive field work has shown that sharks in open waters can make large scale directed movements in the absence of visual bottom landmarks (Sciarrotta & Nelson 1977, Carey & Scharrold 1990, Klimley 1993). However, well-designed, replicated and reviewed experiments are needed to demonstrate that these theories actually model what elasmobranchs do in the field. New technologies for in situ monitoring, recording and transmitting information about ambient electric fields and associated movements of the animal will make this possible. Furthermore, technical advances in the monitoring of sensory processing pathways in the brain of behaving animals are needed. Such field studies are a critical complement to laboratory studies in which the animal is isolated in the lab and specific variables are manipulated.

Acknowledgements

I thank Brook Swanson and Russ Brodie for their assistance in data collection, Mike Thursby for his discussions on antenna theory, David Bodznick and John New for their reading of the manuscript, and John Montgomery and Mike Paulin for discussion on biological matters. I also thank Ralph Turingan, Phil Motta and Peter Wainwright for their comments and annual thumping of ecomorphology at the F.I.S.H. meetings. This work is dedicated to the memory of Don Nelson who mentored me to doubt the obvious, swam by me in the face of danger and introduced me to bullshot.

References cited

Ainley, D.G., R.P. Henderson, H.R. Huber, R.J. Boekelheide, S.G. Allen & T.L. McElroy. 1985. Dynamics of white shark/pinniped interactions in the Gulf of Farallones. S. Calif. Acad. Sci. Mem. 9: 109–122.

Barry, K.L. & C.W. Hawryshyn. 1999. Spectral sensitivity of the Hawaiian saddle wrasse, *Thalassoma duperrey*, and implications for visually mediated behaviour on coral reefs. Env. Biol. Fish. 56: 429–442.

Blonder, B. I. & W. S. Alevizon. 1988. Prey discrimination and electroreception in the stingray *Dasyatis sabina*. Copeia 1988: 33–36.

Bodznick, D. & A.W. Schmidt. 1984. Somatotopy within the medullary electrosensory nucleus of the skate, *Raja erinacea*. J. Comp. Neurol. 225: 581–590.

Bratton, B.O. & J.L. Ayers. 1987. Observations on the electric discharge of two skate species (Chondrichthyes: Rajidae) and its relationship to behavior. Env. Biol. Fish. 20: 241–254.

Carey, F.G. & J. Scharrold. 1990. Movements of blue sharks (*Prionace glauca*) in course and depth. Mar. Biol. 106: 329–342.

Chu, Y.T. & Q.W. Wen. 1979. A study of the lateral line canals system and that of Lorenzini ampullae and tubules of elasmobranchiate fishes of China. Monograph of Fishes of China, 2, Shanghai Science Technology Press, Shanghai. 132 pp. (in Chinese).

Dijkgraaf, S. & A.J. Kalmijn. 1966. Versuche zur biologischen Bedeutung der Lorenzinischen Ampullen bei den Elasmobranchiern. Z. Vergl. Physiol. 53: 187–194.

Doyle, J. 1963. The acid mucopolysaccharides in the glands of Lorenzini of elasmobranch fish. Biochem. J. 88: 7.

Ewart, J.C. & J.C. Mitchell. 1891. On the lateral sense organs of elasmobranchs. II. The sensory canals of the common skate (*Raja batis*). Trans. R. Soc. Edin. 37: 87–105.

Hernandez, P. & P.J. Motta. 1997. Trophic consequences of differential performance: ontogeny of oral jaw-crushing performance in the sheephead, *Archosargus probatocephalus* (Teleostei, Sparidae) J. Zool. Lond. 243: 737–756.

Hueter, R.E. 1990. Adaptations for spatial vision in sharks. J. Exp. Zool. Suppl. 5: 130–141.

Kalmijn, A.J. 1971. The electric sense of sharks and rays. J. Exp. Biol. 55: 371–383.

Kalmijn, A.J. 1974. The detection of electric fields from inanimate and animate sources other than electric organs. pp. 147–200. *In:* A. Fessard (ed.) Handbook of Sensory Physiology, Vol III/3, Springer-Verlag, Berlin.

Kalmijn, A.J. 1978. Electric and magnetic sensory world of sharks, skates, and rays. pp. 507–528. *In:* E.S. Hodgson & R.F. Mathewson (ed.) Sensory Biology of Sharks, Skates, and Rays, Office of Naval Research, U.S. Government Printing Office, Washington, D.C.

Kalmijn, A.J. 1982. Electric and magnetic field detection in elasmobranch fishes. Science 218: 916–918.

Kalmijn, A.J. 1984. Theory of electromagnetic orientation: a further analysis. pp. 525–560. *In:* L. Bolis, R.D. Keynes & S.H.P. Maddrell (ed.) Comparative Physiology of Sensory Systems, Cambridge University Press, Cambridge.

Kalmijn, A.J. 1987. Electric and near-field acoustic detection, a comparative study. Acta Physiol. Scand. 161, Suppl. 638: 25–38.

Klimley, A.P. 1993. Highly directional swimming by scalloped hammerhead shark, *Sphyrna lewini*, and subsurface irradiance, temperature, bathymetry, and geomagnetic field. Mar. Biol. 117: 1–22.

Klimley, A.P., P. Pyle & S.D. Anderson. 1996. The behavior of white sharks and their pinniped prey during predatory attacks. pp. 175–191. *In:* A.P. Klimley & D.G. Ainley (ed.) Great White Sharks, Academic Press, San Diego.

Lamont, A. 1916. The lateral sense organs of elasmobranchs: the ampullary canals of the genus Raia. Trans. R. Soc. Edin. 51: 467–492.

Le Boeuf, B.J., M. Riedman & R.S. Keyes. 1982. White shark predation on pinnipeds in California coastal waters. U.S. Fish. Bull. 80: 891–895.

Long, D.J. & R.E. Jones. 1996. White shark predation and scavenging on cetaceans in the eastern north Pacific ocean. pp. 293–307. *In:* A.P. Klimley & D.G. Ainley (ed.) Great White Sharks, Academic Press, San Diego.

Lorenzini, S. 1678, Osservazioni intorno alle Torpedini. vol 1. Firenze. 136 pp.

McCosker, J.E. 1985. White shark attack behavior: observations of and speculations about predator and prey strategies. S. Calif. Acad. Sci. Mem. 9: 123–135.

McEachran, J.D. & T. Miyake. 1990. Phylogenetic interrelationships of skates: a working hypothesis (Chondrichthyes, Rajoidei). pp. 285–326. *In:* H.L. Pratt, S.H. Gruber & T. Taniuchi (ed.) Elasmobranchs as Living Resources: Advances in the Biology, Ecology, Systematics, and the Status of the Fisheries, U.S. Department of Commerce, NOAA Technical Report NMFS 90.

Montgomery, J.C. 1984. Frequency response characteristics of primary and secondary neurons in the electrosensory system of the thornback ray. Comp. Biochem. Physiol. 79A: 189–195.

Moss, S.A. 1977. Feeding mechanisms in sharks. Amer. Zool. 17: 355–364.

Motta, P.J., S.F. Norton & J.J. Luczkovich. 1995. Perspectives on the ecomorphology of bony fishes. Env. Biol. Fish. 44: 11–20.

Murray, R.W. 1960. Electrical sensitivity of the ampullae of Lorenzini. Nature 187: 957.

Murray, R.W. 1962. The response of the ampullae of Lorenzini in elasmobranchs to electrical stimulation. J. Exp. Biol. 39: 119–128.

Murray, R.W. & T.W. Potts. 1961. The composition of endolymph and other body fluids in elasmobranchs. Comp. Biochem. Physiol. 2: 65–75.

Norris, H.W. 1929. The distribution and innervation of the ampullae of Lorenzini of the dogfish, *Squalus acanthias*. Some comparisons with conditions in other plagiostomes and corrections of prevalent errors. J. Comp. Neurol. 47: 449–465.

Paulin, M.G. 1995. Electroreception and the compass sense of sharks. J. Theor. Biol. 174: 325–339.

Raschi, W. 1986. A morphological analysis of the ampullae of Lorenzini in selected skates (Pisces, Rajoidei). J. Morph. 189: 225–247.

Sciarrotta, T.C. & D.R. Nelson. 1977. Diel behavior of the blue shark, *Prionace glauca*, near Santa Catalina, California. U.S. Fish. Bull. 75: 519–528.

Sisneros, J.A., T.C. Tricas & C.A. Luer. 1998. Response properties and biological function of the skate electrosensory system during ontogeny. Journal of Comparative Physiology, A 183: 87–99.

Stenonis, N. 1664. De musculis et glandulis observationum specimen cum duabus epistolis quarum una ad guil. Pisonum de anatome Rajae etc., Amstelodami.

Strong, W.R., Jr., B.D. Bruce, D.R. Nelson & R.D. Murphy. 1996. Population dynamics of white sharks in Spencer Gulf, South Australia. pp. 401–416. *In:* A.P. Klimley & D.G. Ainley (ed.) Great White Sharks, Academic Press, San Diego.

Tricas, T.C. 1982. Bioelectric-mediated predation by swell sharks *Cephaloscyllium ventriosum*. Copeia 1982: 948–952.

Tricas, T.C. & J.E. McCosker. 1984. Predatory behavior of the white shark, *Carcharodon carcharias*, and notes on its biology. Proc. Calif. Acad. Sci. 43: 221–238.

Tricas, T.C. & J.G. New. 1998. Sensitivity and response dynamics of electrosensory primary afferent neurons to near threshold fields in the round stingray. J. Comp. Physiol. 182: 89–101.

Tricas, T.C., S.W. Michael & J.A. Sisneros. 1995. Electrosensory optimization to conspecific phasic signals for mating. Neuroscience Letters 202: 129–132.

Turingan, R.G & P.C. Wainwright. 1993. Morphological and functional bases of durophagy in the queen triggerfish, *Balistes vetula* (Pisces, Tetraodontiformes). J. Morphol. 215: 101–118.

Wainwright, P.C. & S.M. Reilly. 1994. Ecological morphology. University of Chicago Press, Chicago. 367 pp.

Westneat, M.W. 1995. Phyologenetric systematics and biomechanics in ecomorphology. Env. Biol. Fish. 44: 263–283.

Environmental Biology of Fishes **60:** 93–108, 2001.

Comparative neurobiology of the elasmobranch cerebellum: theme and variations on a sensorimotor interface

John G. New
Department of Biology and Parmly Hearing Institute, Loyola University Chicago, 6525 N. Sheridan Road, Chicago, IL 60626, U.S.A (e-mail: jnew@luc.edu)

Received 4 March 1999 Accepted 8 April 2000

Key words: anatomy, behavior, metencephalon, shark, ray, skate

Synopsis

The organization of the vertebrate cerebellum has been thoroughly studied over the past century, but the function of this structure remains poorly understood. In elasmobranch fishes, the cerebellum displays tremendous variation in size and development although the basic and conservative nature of cerebellar circuitry as seen in other vertebrate taxa is largely retained. Large and morphologically complex cerebelli have evolved independently in both sharks and batoids, and the relative development of this structure in both taxa parallels those of birds and mammals. There are relatively few studies of the physiological role of the cerebellum in generating or shaping behaviors, however, and a convincing explanation of cerebellar hypertrophy in elasmobranchs is lacking. The purpose of this article is to review the current understanding of the structure of the cerebellum in elasmobranch fishes, the physiological responses of cerebellar neurons and the possible role of the cerebellum in behavior. I will also provide a number of hypotheses for future research directions, based upon models that have been suggested by different investigators. These hypotheses include models of cerebellar function as a sensory coincidence detector, a dynamic state estimator and/or a direct modulator of motor programs. Hypotheses concerning the possible organization of cerebellar microcomplexes, the evolution of afferent and efferent cerebellar connections paralleling those observed in mammals and the role of the cerebellum in learning are also suggested.

Introduction

The vertebrate cerebellum is an enigmatic portion of the central nervous system. In one sense it is a phylogenetically conservative structure: the neural elements and cytoarchitecture comprising the cerebellar circuitry are remarkably similar among vertebrate taxa, with limited and well-defined neural input and output pathways. Moreover, the cerebellar circuitry is composed of a relatively few types of nerve cells, and many of these elements strongly resemble each other consistently across the various vertebrate groups. However, despite the relative simplicity and conservative nature of cerebellar circuitry throughout the vertebrates, and despite the inviting nature of the cerebellum as a model for understanding neural processing in the vertebrate brain, a clear understanding of cerebellar function in vertebrates remains elusive.

In most respects, the organization of the elements of cerebellar neural circuitry in elasmobranchs is similar in morphology and organization to those of other vertebrates. However, the elasmobranch cerebellum also exhibits large variation in size and apparent complexity among different groups of sharks and batoids. Thus, although the basic pattern of cerebellar circuitry seems to be conserved, the cerebellum in certain elasmobranchs has undergone tremendous hypertrophy, evolving from a relatively simple, smooth alobate structure to a large, multilobate organ with numerous gyri and sulci. Early anatomical studies (Voorhoeve 1917) were used to argue that the hypertrophy of the elasmobranch brain in general, and the cerebellum in particular, was a function of the increasing size of the animal, with larger species possessing more elaborate cerebelli presumably to coordinate the activities of a larger body. However allometric analyses of brain : body mass ratios in elasmobranchs (Northcutt 1978, 1989, Myagkov 1990) have indicated that the mass of the brain of certain elasmobranch groups is much greater with respect to body mass than it is in others and that these are greater than that of some avian and mammalian taxa (Figure 1). Among sharks this is particularly true for members of the families Carcharhinidae and Sphyrnidae and some species of the Lamnidae. Among batoids hypertrophy of the cerebellum is particularly notable in members of the superfamilies Dasyatoidea (stingrays) and Myliobatoidea (butterfly, eagle and manta rays), although data are available for only a relatively few species (Figure 2). The latter are in contrast to the cerebellum of the skates (Rajidae) which possess similar body size but a much smaller cerebellum. The relative mass of the cerebellum also increases with increased brain mass (Figure 3) (Northcutt 1978, 1989). We are thus presented with a wide variation in the overall development of the cerebellum among the elasmobranchs, some groups of which have independently evolved cerebelli of similar mass and apparent complexity to those of more recently derived vertebrate taxa. To what end? What elements of cerebellar function are altered or enhanced by cerebellar hypertrophy and how is this reflected in the animal's behavior? The situation is further complicated by the evidence indicating that large cerebelli have evolved independently in shark and batoid lineages (Northcutt 1989). The reason for the evolution of large and metabolically expensive central nervous system structures in elasmobranchs is by no means clear and in the case of the cerebellum the uncertainty is compounded by our poor understanding of cerebellar function in vertebrates in general. The purpose of this article is to review previous and more recent studies of the comparative anatomy, physiology and behavioral role of the elasmobranch cerebellum and to suggest new functional models and venues for future research.

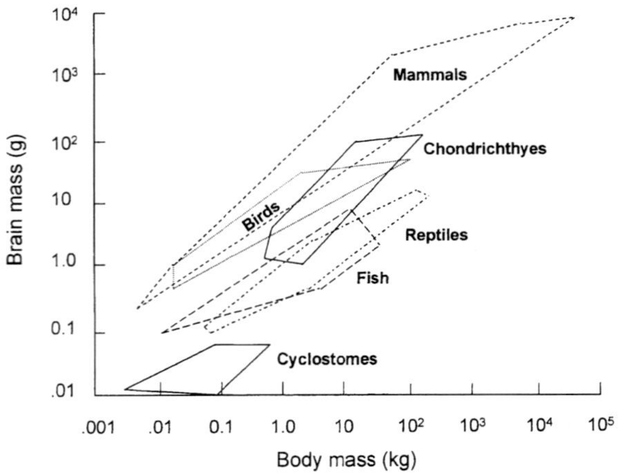

Figure 1. Brain vs. body mass in different vertebrate taxa, expressed as minimum convex polygons (from Northcutt 1977).

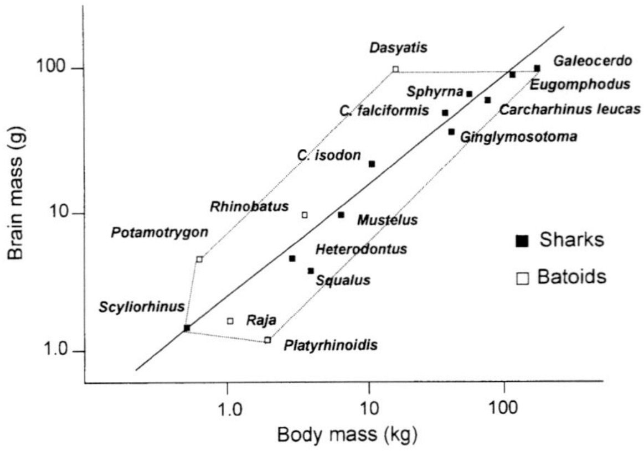

Figure 2. Distribution of brain and body masses among various elasmobranch taxa within the convex polygon illustrated in Figure 1 (from Nortcutt 1977).

Past studies of the elasmobranch cerebellum

The history of investigations of the elasmobranch central nervous system has had two distinct phases. Investigations of the comparative anatomy of the vertebrate central nervous system in general, including elasmobranchs, began in earnest in the late nineteenth and early twentieth centuries with the development of new chemical staining techniques, and in particular with the application of Camillo Golgi's silver stain. Prominent among these early studies in elasmobranchs were those of Sauerbeck (1896), Schaper (1898), Catois (1901), Houser (1901) and Vorhooeve (1917) and which culminated in the important multivolume study of Kappers et al. (1936). Following this early 'Golden Age' of comparative neuroanatomy there was a prolonged period of relatively few investigations of elasmobranch neurobiology. However, in the late sixties and seventies there was a rebirth of

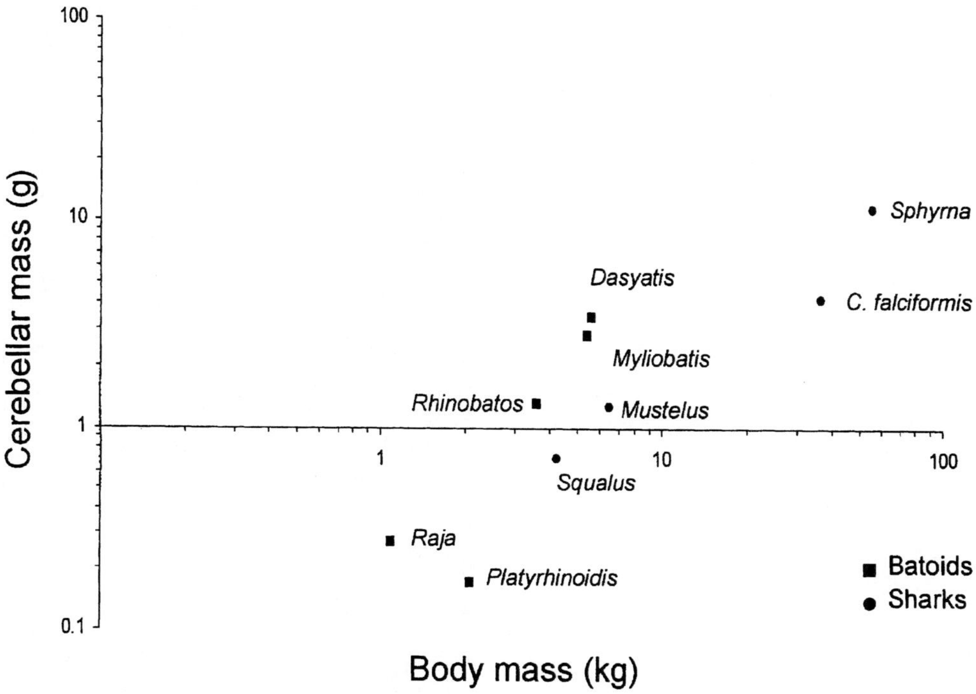

Figure 3. Relative cerebellar mass vs. body mass in a number of elasmobranch genera. Data from this study combined with data from Northcutt 1978.

interest in comparative neuroanatomy, this time with a new emphasis on the phylogeny and evolution of nervous systems. Numerous new studies of elasmobranch nervous system structure and function were conducted with the aid of newly developed tract-tracing techniques for tracing neural pathways (Northcutt 1977, 1978, 1989, Nieuwenhuys 1967, Smeets et al. 1983). However, the majority of these studies on the cerebellum in elasmobranchs were conducted on species possessing small, relatively simple cerebelli. In this paper I present new data on the anatomy of the hypertrophied cerebellum found in carcharhiniform and sphyrnid sharks, with comparisons to less complex organization in other elasmobranch species. Since sensory processing in the octavolateralis nuclei, which involves the cerebellar structures of the auricles is addressed elsewhere in this symposium, I will concentrate upon studies focussed upon the cerebellar corpus. The relatively few studies of behavior in lesioned or decerebellate animals will likewise be reviewed. Past and more recent studies on the physiology of neural elements in the cerebellum will also be presented and hypotheses to be tested by future research will be presented.

Anatomical organization of the elasmobranch cerebellum

Gross anatomy

In all elasmobranchs, the cerebellum consists of a medial, unpaired cerebellar corpus which forms the roof of the metencephalon, and bilateral auricles which are located lateral and ventral to the caudal portion of the corpus (for reviews of cerebellar anatomy and cytoarchitecture in elasmobranchs see Kappers et al. 1936, Larsell 1967, Nieuwenhuys 1967, Nicholson et al. 1969). The auricles consist of upper and lower 'leaves' or folds of tissue, separated from the corpus by a paraauricular sulcus. The folds are united rostrally, thus enclosing a lateral recess between them which is covered by the meningeal membranes and blood vessels of the tela choroidea (Figure 4). The dorsomedial edge of the upper leaf is continuous anteriorly with the ventrolateral portion of the cerebellar corpus, and the lateral portion of the lower leaf fuses with the dorsolateral portion of the rostral alar medulla, where it forms the dorsal and lateral granular ridges. Just caudal to the region in which the corpus and upper leaf fuse there are

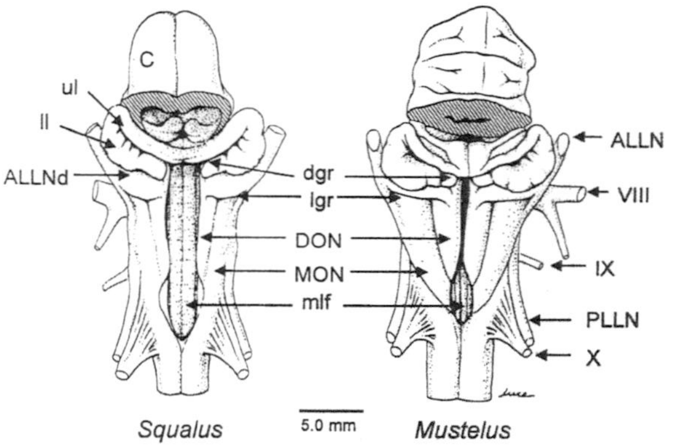

Figure 4. Comparative gross anatomy of the elasmobranch cerebellar corpus and auricles in *Squalus* and *Mustelus*. ALLN = anterior lateral line nerve; ALLNd = dorsal root of ALLN; C = cerebellar corpus; dgr = dorsal granular ridge of cerebellar auricle; DON = dorsal octavolateralis nucleus; il = inferior leaf of auricle; lgr = lateral granular ridge of cerebellar auricle; mlf = medial longitudinal fasciculus; MON = medial octavolateralis nucleus; PLLN = posterior lateral line nerve; ul = upper leaf of cerebellar auricle; VIII = octaval nerve; IX = glossopharyngeal nerve; X = vagus nerve (from Northcutt 1978).

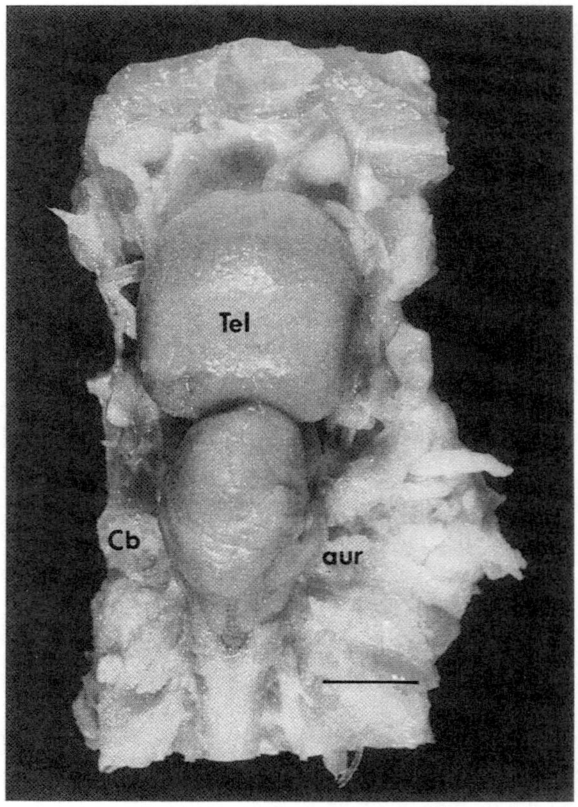

Figure 5. Brain of *Carcharhinus limbatus*. Bar scale equals 1.25 cm (aur = auricle of cerebellum; Cb = cerebellar corpus; Tel = telencephalon).

two small caudal lobules of the corpus, separated from the corpus by the sulcus posticus. Rostrally, the upper leaves fuse medially with the lower leaves. The lower leaves also fuse medially along the midline ventral to the corpus forming a bridge over the anterior fourth ventricle, the 'lower lip' of Voorhoeve (1917).

In its least complex form, as in *Scyliorhinus* and *Hexanchus*, the corpus of the elasmobranch cerebellum forms a smooth, unpaired single-lobed structure on the midline of the central nervous system with only a shallow groove along the longitudinal midline (Ariens Kappers 1936). In most elasmobranchs, however, the cerebellum is relatively larger, and a primary transverse sulcus divides the corpus into anterior and posterior lobes of approximately equal size. In galeomorph sharks such as *Carcharhinus limbatus*, a second large anterior sulcus further subdivides the anterior corpus into rostral and caudal lobes (Figure 5). Numerous secondary transverse sulci are present in all three lobes of the corpus. The cerebellum in *C. limbatus* is roughly oblong in shape and obscures most of the dorsal aspect of the midbrain tectal lobes. In the scalloped hammerhead, *Sphyrna lewini*, the cerebellum is further enlarged and assumes a roughly triangular shape, with the wide anterior edge directly adjacent to the posterior telencephalic wall and completely covering the dorsal aspect of the midbrain and diencephalon (Figure 6). The basal part of the cerebellum is connected to the rest of the central nervous system by broad cerebellar peduncles, which are connected anteriorly with the lobes of the optic tectum via the anterior medullary velum and caudally with the rostral terminus of the medulla.

Although the corpus and auricles together comprise the cerebellum, the increase in size and apparent complexity of the corpus in some taxa are not paralleled by a corresponding increase in the size of the auricles. Although comparative quantitative data are lacking, a number of studies beginning with those of Vorhooeve (1917) have observed that the corpus tends to show much greater variation and increase in size than do the auricles, thus in galeomorph sharks and myliobatoids the corpus is much larger with respect the auricles than in forms such as *Scyliorhinus* and *Raja*, indicating a functional independence of corpus and auricles, a view that is further supported by the extensive anatomical interrelationship between the auricles and the medullary octavolateralis sensory nuclei.

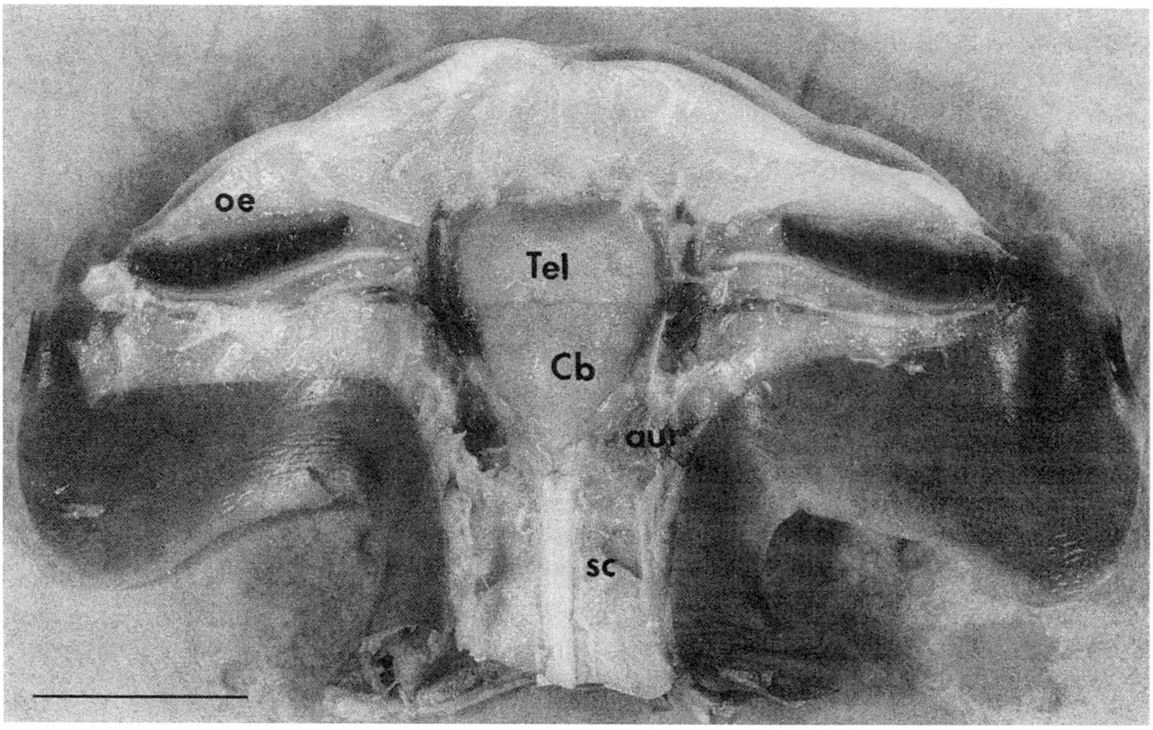

Figure 6. Dorsal view of the brain of a juvenile *Sphyrna lewini*, in situ. Bar scale equals 3.0 cm (aur = auricle of cerebellum; Cb = cerebellar corpus; oe = olfactory epithelium; Sc = spinal cord; Tel = telencephalon).

The ventricular system of the brain extends up through the cerebellar peduncles and into the central mass of the elasmobranch cerebellar corpus. In elasmobranchs with smaller cerebelli, such as *Scyliorhinus* and *Raja*, the ventricular space within the corpus is relatively large. Bilaterally within the corpus are two ridges of tissue that extends into the ventricular space from the dorsal and ventral internal walls of the corpus. These are known as the granular eminences of the corpus (the longitudinal granular folds of Larsell 1967) and consist mainly of cerebellar granule cells (see below). In elasmobranchs such as *C. limbatus* and *S. lewini*, in which the cerebellum is much enlarged and foliated, the ventricular space is greatly reduced although the bilateral eminences are still evident in *C. limbatus* (unpublished observations).

Fine structure and cytoarchitecture

The fine structure of the cerebellum in elasmobranchs is very similar to that of other gnathostome taxa. The corpus of the cerebellum consists of an external molecular layer, a Purkinje cell layer, a fibrous layer, and a deeper layer of granule cells (for reviews see Kappers et al. 1936, Larsell 1967, Nieuwenhuys 1967, Nicholson et al. 1969)

The superficial layer of the corpus, known as the molecular layer, is composed mainly of fine, unmyelinated parallel axon fibers extending transversely across the surface of the cerebellum. These parallel fibers are the axons of the granule cells located in the deeper layers of the cerebellum. In *Raja erinacea*, the approximate mean thickness of the molecular layer along the dorsal and lateral surface of the corpus varies between 600 and 800 μm. In adult individual *Squalus acanthias* and *Scyliorhinus canicula* the molecular layers have approximately the same thickness (600–900 μm). However in a juvenile specimen (approx 0.5 m TL) of *Sphyrna lewini* the molecular layer was thinner, with a mean maximum thickness of dorsal and lateral molecular layers was 400 μm or less. The total surface area of the molecular layer was doubtlessly much greater due to the extensive folding of the cerebellar surface. All of these data are from single or a limited number of specimens and rigorous quantitative analysis remains to be done.

Scattered among the parallel fibers of the molecular layer are stellate cells with numerous branching dendrites. These cells are absent near the midline where the parallel fibers enter the molecular layer. Stellate cells located in the superficial layers possess dendrites oriented primarily in the sagittal plane of the cerebellum (Schaper 1898, Larsell 1967, Nicholson et al. 1969) and display positive immunoreactivity for GABA (γ-aminobutyric acid) (Alvarez-Otero et al. 1995). Sterzi (1909) described deeper fusiform shaped cells with axons terminating near the soma of the underlying Purkinje cells, and Smeets et al. (1983) class these as a second category of stellate cells. The density of stellate cells has been analyzed in the cerebellum of only a single elasmobranch species by Alvarez & Anadon (1987). They estimate that in *Scyliorhinus* the mean stellate cell density is approximately 1.9×10^4 cells mm^{-3}. Basket cells, characteristically found in the molecular layer of tetrapods, have not been described in any elasmobranch species, although Kappers et al. (1936) suggest that the deep fusiform cells of the molecular layer may be functionally related to basket cells due to the horizontal orientation of their axons in the Purkinje cell layer.

Directly beneath the molecular layer is a prominent layer of large bodied neurons, the Purkinje cells. Purkinje cells are one of the most characterized and familiar of all vertebrate central nervous system neurons. They form a layer one or two cells thick between the molecular and granular layers. Purkinje cells are absent along the midline of the cerebellum where parallel fibers transit from the granular cell to the molecular layers. The Purkinje cells of the elasmobranch cerebellum have large round, ovoid or triangular cell somata and the size of the cell body increases with increasing thickness of the overlying molecular layer (Larsell 1967). Alvarez & Anadon (1987) estimated there were a total of 3.5×10^4 Purkinje cells in *Scyliorhinus* specimens measuring 48–58 cm. The cerebellum of *Scyliorhinus* is small and relatively smooth. The most characteristic feature of the Purkinje cells is the prominent apical dendrites extending into the molecular layer. These dendrites are prominently covered with dendritic spines and are arrayed in a plane perpendicular to the parallel fibers of the molecular layer and are oriented parallel to the sagittal plane of the cerebellum where they receive synaptic connections from the parallel fibers and stellate cells. Larsell (1967) observed that the apical dendrites of elasmobranchs are less intricately branched than are those of some teleost fishes, rarely going beyond a third order of branching. However, only a few elasmobranch species have been examined and the extent of dendritic branching may vary among elasmobranchs. The axons of the Purkinje cells form a prominent proportion of the fibrous layer directly beneath the Purkinje somata, which also consists of fibers afferent to the cerebellum from other regions of the brain.

Beneath the fibrous layer is the granule cell layer. The granule cell layer in *Scyliorhinus* forms prominent ridges, the granular eminences, on either side of the midsagittal fissure extending into the ventricular space. The granule cell layer also extends as a thin layer around the cerebellar ventricle. The granular ridges in *Scyliorhinus* occupy relatively little of the total volume of the cerebellum, much of which is taken up by the ventricular space. In *Raja*, the granular ridges are considerably expanded and occupy a greater portion of the cerebellar volume than in *Scyliorhinus*. In *Carcharhinus* and *Sphyrna* the ventricular volume is greatly reduced to form a solid layer of granule cells and granular ridges are not apparent. The granule cells themselves are small, with round, densely staining cell bodies, three or four short dendrites that terminate in claw-like branches (Larsell 1967). The fine, unmyelinated axons of the granule cells form the parallel fiber system of the molecular layer. The granule cells of the cerebellum are typically densely packed. In *Scyliorhinus* the density of granule cells is approximately 5.8×10^5 cells mm^{-3} (Alvarez & Anadon 1987). There is an increase in the density of granule cells in taxa possessing increasing cerebellar size and complexity, in *Raja* the mean granule cell density is 1.2×10^6 cells mm^{-3} and in *Sphyrna* it is 1.8×10^6 cells mm^{-3} (Figure 7). Also present in the glomerular layer are Golgi type II cells, with larger cell bodies and several sparsely branched dendrites that typically show a number of varicosities (Larsell 1967). The axons of these cells terminate deep in the granule cell layer. Golgi cells are distributed more sparsely throughout the cerebellum. The density of Golgi cells in the granule cell layer of *Scyliorhinus* is 1.8×10^3 cells mm^{-3} (Alvarez & Anadon 1987), in *Raja* 2.5×10^3 cells mm^{-3} and in *Sphyrna* 1.3×10^3 mm^{-3}. These cells also exhibit positive GABA-A immunoreactivity (Nicholson et al. 1994, Alvarez-Otero et al. 1995). Both granule cells and Golgi cells receive synaptic inputs from afferent mossy fiber synapses (Alvarez-Otero & Anadon 1992).

Cerebellar nuclei

In the cerebellar peduncle are a pair of nuclei called the medial and lateral cerebellar nuclei that may be

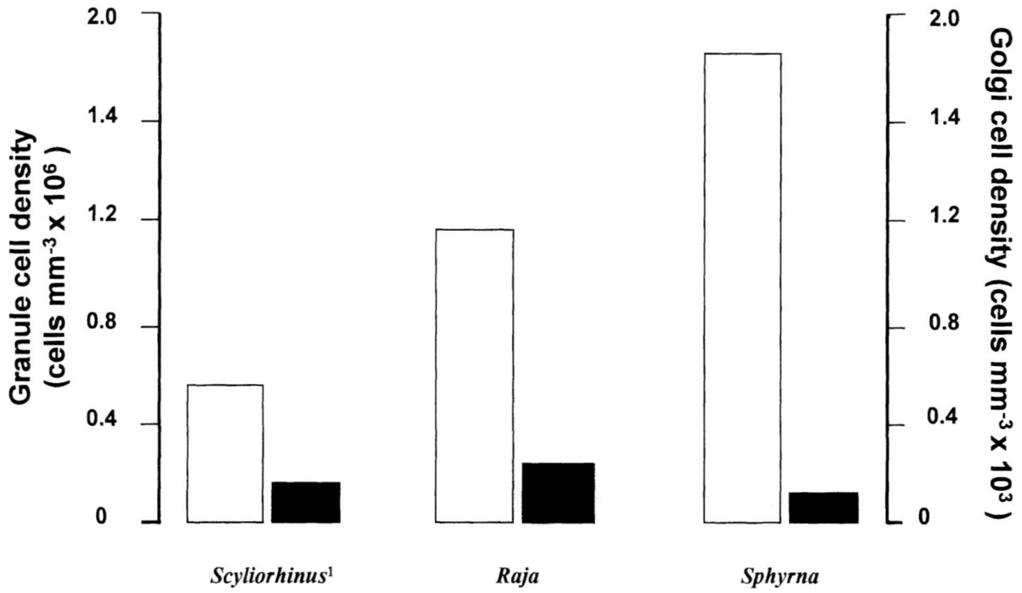

Figure 7. Relative density of granule and Golgi cell densities in the cerebellar corpus of three elasmobranch species. Dark hatched bars are granule cells, light bars are Golgi cells. *Scyliorhinus* data from Alvarez & Anadon (1987).

homologous to the deep cerebellar nuclei of tetrapods. Kappers et al. (1936) considered them to be a recipient of ascending lateral line and vestibular afferents or a portion of the octavomotor complex. Developmental studies revealed that these cells are part of the cerebellum, which develop from the cerebellar precursor, the anlage, and are not a subcerebellar structure (Rudeberg 1961). More recent neuroanatomical studies employing anterograde tracing techniques revealed that these nuclei do not receive octaval or lateralis primary afference (Koester 1983, Barry 1987), nor do they appear to be related to the oculomotor nuclei. These studies show that the cerebellar nuclei receive input from Purkinje cell axons in a manner similar to the cerebellar nuclei of tetrapod taxa, and thus may be considered a field homolog of those nuclei (Paul & Roberts 1984a, Fiebig 1988). Furthermore, Purkinje cell axons do not leave the cerebellum but terminate only within this nucleus. This implies that the evolution of vertebrate cerebellar nuclei antedates the diversion of chondrichthyan and bony fish lineages and may be a gnathostome synapomorphy (homologous character).

The cells of the cerebellar nuclei are large bipolar or triangular-shaped cells that exhibit positive immunoreactivity for glutamate (Alvarez-Otero et al. 1996). There are approximately 1000 large neuron in the cerebellar nucleus of *Scyliorhinus* that yield of Purkinje cells : cerebellar nucleus neurons convergence ratio of approximately 17 : 1 (Alvarez-Otero et al. 1996).

Cerebellar circuitry

Since there are relatively few neuronal elements in the cerebellum, and these are organized in a straightforward and regular fashion, the organization of cerebellar circuitry in elasmobranchs (and other vertebrates) is comparatively well understood. The unmyelinated axons of granule cells extend into the molecular layer, where they divide in a characteristically T-shaped fashion and proceed laterally and parallel to the surface through the molecular layer (Figure 8) (Larsell 1967, Nicholson et al. 1969). These axons cross the midline of the brain, so that the population of parallel fibers in a given region of the molecular layer arises from bilateral populations of granule cells. These parallel fibers make synaptic connections with both the apical dendrites of the Purkinje cells which are arrayed in a plane perpendicular to that of the granule cell axons and with the stellate cells of the molecular layer, which themselves project directly onto the Purkinje cell somata. As mentioned above, the Purkinje cell axons project onto the neurons of the medial and lateral cerebellar nuclei, the axons of which in turn form the output of the cerebellar corpus.

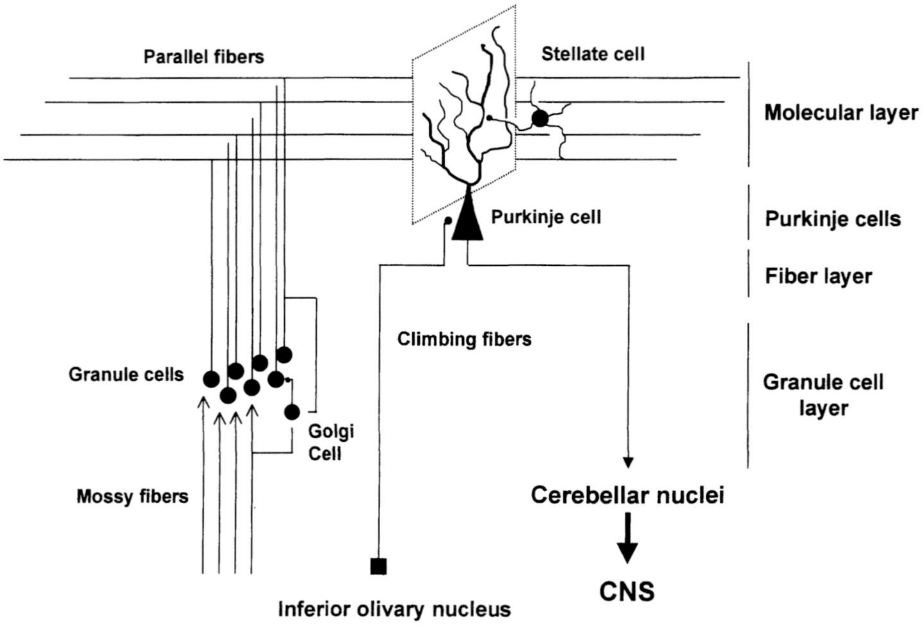

Figure 8. Schematic diagram of cerebellar circuitry.

Mossy fibers originating from the spinocerebellar axons form a principal input into the granule cell regions of the cerebellum. There has been some dispute over whether or not axons originating in the inferior olivary nucleus form climbing fibers, but experimental anatomical studies suggest that inferior olivary axons form climbing fiber-like synaptic connections onto the Purkinje cell somata, but not onto the dendrites. These fibers exhibit distinctive synapses located on the 'pincushions' of somatic spines on the Purkinje cell bodies (Alvarez-Otero et al. 1993). Similarly, although basket cells have not yet been described in any elasmobranch species, it is possible that the deep fusiform 'stellate' cells may play a similar functional role, particularly in light of the distribution of their axons. However this hypothesis remains to be tested.

Virtually all of the regions of the brainstem and midbrain which provide afferent connections to the cerebellum also receive efferent input from the cerebellar nuclei, particularly the extensive premotor circuitry of the central nervous system and the medullary and mesencephalic reticular formations. Thus the neuroanatomy of cerebellar connections suggests that the cerebellum plays a modulatory role in the activity of premotor regions in the hindbrain and midbrain, integrating information from both sensory and premotor sources and modifying the activity of the premotor reticular nuclei. This indicates that the cerebellum may modulate the motor outputs that form behaviors in elasmobranchs.

Biochemical heterogeneity and 'covert' compartmentalization

Recent advances in the understanding of the organization of the cerebellum in mammals have provided additional insight into the fine structure and possible functional role of the cerebellum. Although the cerebellum in mammals appears at first glance to be uniform in its organization (as it also does in other vertebrates) it is actually 'compartmentalized' into numerous subdivisions characterized by differing biochemical properties of cerebellar neurons, particularly the Purkinje cells. Some of these markers include acetylcholinesterase, 5′-nucleotidase, glutamate decarboxylase, zebrin I and II, synaptophysin and cytochrome oxidase (for review see Wassef et al. 1992). The heterogeneous distribution of these markers throughout the cerebellum suggest different functional subdivisions within the cerebellum. Furthermore, since many of these markers are present perinatally, prior to afferent innervation of the

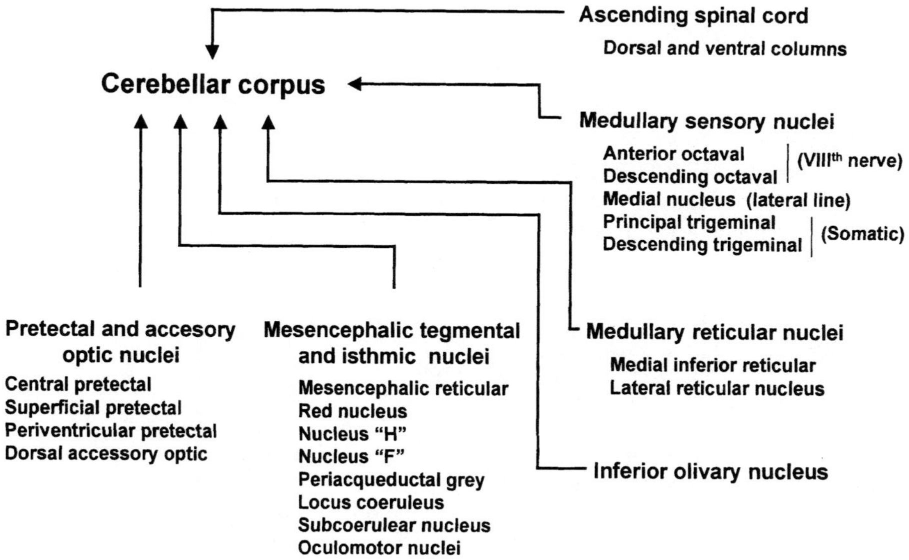

Figure 9. Afferent pathways to the elasmobranch cerebellum.

cerebellum by climbing and mossy fibers, the biochemically heterogeneous compartments may serve as organizational units in the construction of neural maps within the cerebellum. Afferent fibers invading the cerebellum are also compartmentalized in many cases, and this compartmentalization may be directly or indirectly organized by the heterogeneous expression of biochemical markers by cerebellar neurons (Hawkes et al. 1992, 1993). Immunoreactivity against zebrin-II, a 36 kDa polypeptide has also been demonstrated in other vertebrates, including lampreys, bony fishes and elasmobranchs. In lampreys, zebrin-II is expressed only in the ammocoetes in the anterior octaval nucleus of the medulla, but not in the cerebellum (Lannoo & Hawkes 1997). In most bony fishes, zebrin-II is expressed by all Purkinje cells of the cerebellar corpus. However, in the greatly hypertrophied 'gigantocerebellum' of mormyrid fishes there exists a heterogeneous distribution of zebrin-II immunoreactivity (Lannoo et al. 1991, Meek et al. 1992, Hoggat & Lannoo 1994). Zebrin-II immunoreactivity has also been reported in the cerebellum of two species of stingray of the genus *Dasyatis* (Puzdrowski 1997). Although these elasmobranch fishes possess comparatively large cerebelli, all Purkinje cells of the cerebellar corpus express positive zebrin-II immunoreactivity and there is no apparent compartmentalization or zonation of the cerebellum as observed in mammals. Thus despite the apparently 'conservative' nature of cerebellar organization in vertebrates, structural, or at least biochemical, variations in cerebellar organization in various vertebrate taxa do exist and may play differing roles in organizational or developmental mechanisms.

Afferent and efferent connections of the cerebellum

Afferent fibers to the cerebellum arise via at least eight pathways (Figure 9) (Smeets et al. 1983, Northcutt & Brunken 1984, Fiebig 1988). These include (1) fibers of the octaval and mechanosensory lateral line nerves (but not electrosensory fibers) to the granular ridges and neighboring areas of the lower leaf of the cerebellar auricle, (2) ascending projections from the medullary octaval (anterior and descending) and mechanosensory (intermediate) lateral line nuclei, (3) a trigeminocerebellar projection from cells of the mesencephalic trigeminal and also the medullary (principal and descending) trigeminal nuclei, (4) ascending spinocerebellar and reticulocerebellar axons from the spinal cord and hindbrain reticular formations, respectively, which form mossy fiber projections to the cerebellar granule cell regions, (5) axons originating from cells of the inferior olivary nucleus in the basal hindbrain which form climbing fiber input to the

Purkinje cells, (6) descending connections arising from pretectal nuclei (superficial, central and periventricular pretectal and dorsal and ventral accessory optic nuclei), (7) axons from the tegmental reticular and oculomotor nuclei and nucleus ruber and, (8) fibers from the locus coeruleus. A reciprocal connection also exists between the corpus of the cerebellum and the cerebellar nuclei. A number of investigators have suggested the presence of a projection from the inferior lobe of the hypothalamus, but more recent investigations have rendered this connection doubtful.

The sole recipient of efferent fibers from the corpus of the cerebellum are the ipsilateral medial and lateral cerebellar nuclei (Paul & Roberts 1984a, Fiebig 1988). Purkinje cell axons terminate exclusively upon these cells and do not leave the cerebellum. There are three principal output pathways from the cerebellar nuclei: (1) a cerebellobulbar tract that projects principally to the nuclei of the brainstem reticular formation, (2) the brachium conjunctivum which forms ascending and descending branches, the former to the nucleus ruber and oculomotor nuclei, the latter to the medullary reticular formation and inferior olivary nucleus and, (3) a suggested cerebellomotor tract associated with the medial longitudinal fasciculus projecting to reticular and oculomotor nuclei, which has not been verified experimentally (Figure 10). Additionally, a projection from the cerebellar nuclei to the tectum has been established (Smeets 1982). However, little is known concerning the relative weights of cerebellar afferents from different nuclei and efferent connections from the cerebellum in different elasmobranch taxa.

Behavioral role of the cerebellum and physiological correlates

Behavioral role of the cerebellum

Several studies have investigated the role of the cerebellum in locomotion, posture and other motor elements of behavior. Most have used lesions or ablation techniques on the cerebellar corpus, with varying results. Lesions of the cerebellar corpus in *Squalus* do not result in locomotor impairment but if the lesions involve the cerebellar nuclei the animal's swimming becomes uncoordinated and the animals cannot swim in a straight line (ten Cate 1930, Karamyan 1962). Lesions of the auricles result in severe motor impairments consistent with vestibular dysfunction. In *Raja* lesions involving the corpus and cerebellar nuclei do not result in severe locomotor or postural abnormalities, although overall motor activity increases and a few postural abnormalities of the fins were observed (Karamyan 1962). Paul & Roberts (1979) observed that ablation of the cerebellar corpus also attenuated the reflexive elevation of the pectoral fin following cutaneous stimulation (see below).

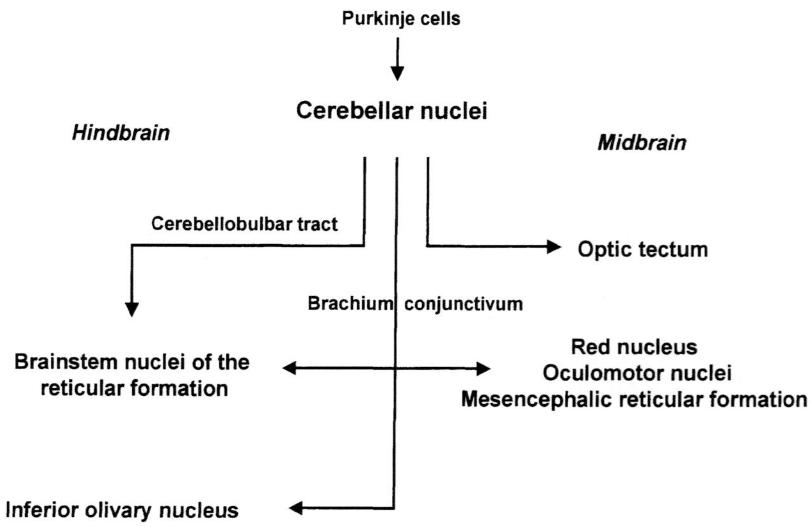

Figure 10. Efferent cerebellar pathways in elasmobranchs.

Physiological responses of the cerebellar corpus

Relatively few studies exist of the physiological responses of the cerebellum in elasmobranchs. This is unfortunate, because the organ is highly accessible to neurophysiologists particularly in the smaller shark and batoid species such as *Scyliorhinus*, *Platyrhinoidis* and *Raja*. These animals are typically hardy and tolerate the rigors of prolonged recording sessions very well.

Field potentials recorded from the surface of the molecular layer in *Scyliorhinus* (Paul 1969) and *Raja* (Young 1980) demonstrated that the parallel fiber axons of the granule cells can be activated directly via stimulation of the surface or via stimulation of the cerebellar peduncles activating the mossy fiber-granule cell pathway. These responses moved laterally over the surface of the cerebellum at a rate of 0.2 m sec^{-1} within a relatively narrow 'beam' of less than 110 μm (the maximal stimulus being 40 V, 50 μV delivered across < 0.5 mm). Nicholson et al. (1969) recorded from Purkinje cells in the cerebellar corpus of *Platyrhinoidis* using extracellular field potentials and intracellular recordings and demonstrated that Purkinje cells generate multiple action potentials following parallel fiber activation, climbing fiber stimulation, or direct injection of a depolarizing current. The multiple discharges were the result of an intrinsic rhythm within the cell, which appears to be a unique feature of elasmobranch Purkinje cells. However, repetitive discharges of the Purkinje cells were suppressed by stellate cell inhibition following direct multiple activations of the parallel fiber pathway.

The role of the cerebellum in a reflex behavior has been studied by Paul & Roberts (1979). They demonstrated that a reflexive elevation of the pectoral fin elicited via electrical stimulation was suppressed following cerebellar ablation, this effect which was reversed following transection of the spinal cord at the level of the obex. These results led the authors to suggest that medullary premotor systems, particularly the neurons of the reticular formation have a tonic inhibitory effect upon spinal cord motor reflexes, and that this inhibitory influence was modulated by the activity of the cerebellum. In subsequent studies, the same authors recorded the activity of units in the cerebellum of *Scyliorhinus* evoked by eliciting the pectoral fin elevation reflex (Paul & Roberts 1981,1983). Although the responses of Purkinje and stellate cells were variable among cells, certain common response properties emerged. With the exception of one cell, the cerebellar neuron responses had a long latency, indicating that the cerebellum did not initiate the movement of the cell, but acted as a modulator of the motor output of the central nervous system. The long latency appeared to be due in part to activation of the mossy fiber-granule cell-parallel fiber pathway, and appeared after the initiation of the reflexive movement of the fin. Further studies of the neurons of the cerebellar nucleus by the same authors demonstrated that the majority of cerebellar nucleus neurons also display a long latency (> 100 ms) excitatory response to stimuli which drive the pectoral fin reflex (Paul & Roberts 1983). The long latency of the cerebellar response suggested that the initial phasic portion of the response was unaffected by cerebellar activity. Only the longer-latency phasic portion of the response was modulated by cerebellar circuitry, an observation that matched the behavioral results observed following cerebellar ablation (see above). The regulatory role of cerebellar nuclear neurons on the reticular formation is strengthened by the extensive projections of these neurons onto the reticular neurons. Paul & Roberts (1975) estimate that greater than 90% of reticular neurons receive cerebellar nuclear afference.

Similarly, the role of the cerebellum in swimming has been studied in *Scyliorhinus* by Paul & Roberts (1984b). Spontaneous swimming movement resulted in the spontaneous rhythmic discharge of Purkinje cells and some stellate cells which was in phase with the swimming movements. These discharges persisted in animals immobilized with the neuromuscular blocker curare, in which 'fictive' swimming could be detected via rhythmic discharges of the ventral spinal roots. Burst discharges of Purkinje cells could also be elicited in these animals via cutaneous stimulation. These results indicate that the cerebellar neurons are discharging in phase with the output of the spinal locomotory pattern generators.

Sensory responses have been observed in the granule cell layer of the cerebellum by Fiebig (1988), New & Bullock (1989) and New & Baron (unpublished data). Fiebig (1988) observed multiunit activity responses to visual, tactile and electrosensory stimulation in *Platyrhinoidis* and observed that visual responses predominate in the anterior lobe and tactile responses in the posterior, with small scattered electrosensory regions in the posterior lobe as well. Visual input to the cerebellum is likely to originate from afferent fibers originating in the pretectal and accessory optic nuclei which receive direct retinal input, whereas tactile input may originate in spinocerebellar and trigeminocerebellar tracts. A source for electrosensory inputs has not been described, but previous studies on ascending electrosensory pathways have bypassed the cerebellar corpus (Boord & Northcutt 1982, Schmidt & Bodznick 1987). Studies on electrosensory responses in the granule cell layer have recorded multiunit activity from granule cells and single unit responses presumably from Golgi cells. Different electrosensory regions in the cerebellum exhibit different responses to consistently presented DC-step stimuli. The anterior region in the posterior lobe responds in a polarity dependent manner as does much of the rest of the electrosensory system. However, single cells and multiunit activity in the posterior region respond in a polarity independent manner, responding physically to a change in the direction of the change in voltage. Despite relatively strong sensory responses in the granule cell layers, sensory stimulation was generally unsuccessful at eliciting responses from Purkinje cells in immobilized animals (unpublished observations).

Future directions

The coincidence detector hypothesis

Traditionally, the cerebellum has been viewed as a region of the CNS devoted to the organization of motor commands. However, recent studies and reviews of cerebellar anatomy and physiology have challenged this notion and have suggested a larger role in sensory processing and sensorimotor integration for the cerebellum than was previously thought. Sensory representations have been recorded in the elasmobranch cerebellum (Fiebig, 1988, New & Bullock 1989), but it is not currently known whether or not the sensory organization of the elasmobranch cerebellum is similar to that observed in mammals. In the latter, the cutaneous somatosensory inputs to the granule cell layer are not organized in a continuous somatotopic fashion, but rather as a 'fractured mosaic' much as if a somatotopic map had been cut into numerous puzzle pieces and rearranged (Welker 1987). The apparently homogeneous biochemical organization of the elasmobranch cerebellum may argue against this sort of organization. However the limited physiological data available suggests that some sensory compartmentalization may exist.

Meek (1992) has argued that the structure of cerebellar circuitry in vertebrates suggests a sensory role, with Purkinje cells acting as coincidence detectors of separate sensory events. In this model, parallel fiber activity originating from different populations of granule cells coincides at different locations in the unicellular Purkinje cell array depending upon the intervals between the onset of the activity in each granule cell group. In this model, the parallel fibers act as delay lines and the Purkinje cells as coincidence detectors, responding to the temporal summation of parallel fiber inputs. Meek has argued that in the cerebellum of teleost fishes, such as the weakly electric mormyrids, such circuitry would be optimal for the detection of small temporal differences (up to 4 ms) between the inputs to different granule cell masses. Such an arrangement would be optimal for the analysis of single or low-frequency repetitive inputs with a complex and noisy shape or pattern. This possible function remains to be empirically tested in elasmobranchs, which are ideal subjects for testing this hypothesis. Since large cerebelli have evolved independently in sharks, batoids and mammals, there exists potential for comparative studies of the organization of cerebelli in these lineages and the development of testable hypotheses regarding the 'rules' of cerebellar development and evolution.

The cerebellar microcomplex hypothesis

Ito (1993, 1997) has recently suggested that the compartmentalization of the cerebellum may reflect a fundamental functional property, with cerebellar 'microcomplexes' acting as a basic functional unit of circuitry. Ito argues that such basic functional units have evolved to play numerous roles in central nervous system processing as brains have evolved including regulation of reflexes, voluntary movements and cognitive processes. The biochemical compartmentalization of the cerebellum and the presence of fractured mosaic somatotopic maps argue for a more diverse division of cerebellar tasks and roles than hitherto thought. Clearly, however, the designation of the cerebellum as a 'motor' or 'sensory' organ is less tenable than previously thought since both functions seem to find an interface in the cerebellum. Little is known about the possible structure of cerebellar microcomplexes in the elasmobranch cerebellum, however, the reports of cerebellar regions that are respond to differentially sensory stimulation (Fiebig 1988, New & Bullock 1989) may indicate that such complexes also exist in the cerebellum of elasmobranchs.

Does cerebellar connectivity in elasmobranchs parallel that of mammals?

Relatively little is known about the weight of afferent and efferent cerebellar connections in different elasmobranch taxa. Comparative studies of these connections in elasmobranchs possessing small or large cerebelli present important opportunities for new investigations. The relative importance of locus coeruleus as a source of afferents in mammals and elasmobranchs could be interesting and fuel for speculation, especially comparisons between species possessing small, smooth cerebelli with those possessing large, convoluted ones. The 'classical' view (Kappers et al. 1936) that extensive connections between cerebral cortex and cerebellum in mammals are features of 'neo'-cortex and 'neo'-cerebellum and represent unique mammalian features may need to be revised, particularly with respect to multisynaptic pathways. An example challenging this assumption has been provided in a teleost fish by Lee & Bullock (1990a,b). These studies demonstrated complex cerebellar responses following telencephalic stimuation, suggesting a wider distribution of descending telencephalic-cerebellar pathways than previously thought. The possible presence of such a pathway in elasmobanchs has not yet been investigated. Additionally there is great need to look for possible sagittal zones of afferent terminations as appear in mammals; there has not yet been any such description in elasmobranchs.

The motor program modulation hypothesis

The anatomy and physiology of the cerebellum in elasmobranchs (and in other vertebrates) indicates that it is an important center for the integration of sensory input into motor programs. The cerebellum receives inputs directly from first and second order sensory nuclei from a number of different sensory modalities, and forms extensive reciprocal connections with premotor areas of the midbrain and hindbrain, particularly the reticular formation. The cerebellum does not seem to be a direct controller of locomotor activity, normal swimming occurs following cerebellar ablation. It is clear, however, that the cerebellum plays a modulatory role, regulating the amount of inhibition that premotor regions exert upon intrinsic central motor pattern generators (Paul & Roberts 1979, 1984b). This would permit the cerebellum to alter the stereotypical activity of such pattern generators to account for changes in the environment detected by the sensory systems. One such possible modulatory activity might be in the organization of orienting responses. Such behaviors have typically been considered the domain of the midbrain tectum, but any such behavior must rely upon the alteration of existing motor outputs, particularly in an animal already in motion. Furthermore, it is known that the cerebellar nuclei have prominent projections to the midbrain tectum. The multimodal sensory input of the cerebellum and the strong reciprocal connections with the premotor regions of the neuraxis make it a likely linchpin structure for the alteration of standing motor programs in the face of novel presented stimuli. There have been a number of studies examining the role of cerebellar ablation on stereotyped locomotor outputs such as posture and normal swimming. However there is still a need for extensive studies of elasmobranch behavior in the absence or blocking of cerebellar influences, particularly with respect to responses to behaviorally relevant sensory cues such as visual, auditory or electrosensory stimuli. Chronic recording of cerebellar neurons in freely behaving or orienting elasmboranchs may prove a useful approach to understanding the modulatory role of the cerebellum in motor behaviors.

The dynamic state estimator hypothesis

Neural network modeling and the development of control algorithms have also led to new opportunities for understanding cerebellar function. Paulin (1993, 1997) has challenged traditional thinking that the cerebellum is involved in fine motor control and that 'large cerebellumed' animals display higher levels of motor dexterity and agility. Paulin argues that the large sensory inputs to the cerebellum, and the range of deficits in perceptual and cognitive, as well as motor, functions resulting from cerebellar lesions indicate a more important sensory role for the cerebellum. Paulin likens the function of cerebellar circuitry to a 'dynamic state estimator'. Classical control theory indicates that the feedback produced by a multivariate system's output is not sufficient to control the system output itself. Such theory thus predicts that sensory input produced during motor activity is insufficient to control the trajectory of the movement. Rather, a dynamical state vector must be constructed from system outputs and passed to the controller circuitry. Lacking a dynamical state vector, system output may be unstable and oscillatory, malfunctions similar to those observed following cerebellar damage. Paulin likens the function of the cerebellum to that of a

Kalman filter (Kalman 1960), a method for tracking and predicting states of dynamical systems and useful in tracking, trajectory prediction, noise rejection and navigation (Bucy & Joseph 1968). Paulin does not argue that the cerebellar circuitry performs the same algorithm as a Kalman filter, but rather that the functions of the two are parallel, biological and artificial systems approaches to solving the same problem. An important part of Paulin's argument is the relative hypertrophy of the cerebellum in animals that are 'sensory specialists'. These are animals with highly developed sensory systems that perform complex localization and orientation tasks, such as echolocation in insectivorous bats and odontocete cetaceans and electrolocation by weakly electric mormyrid fishes. Such an argument would posit that more developed sensory systems are found in, and more complex sensory localization and analytic tasks are performed by, large-cerebellumed elasmobranchs such as carcharhinid sharks and dasyatid batoids than by elasmobranchs exhibiting less cerebellar development, such as squaliforms and rajiforms. Such is not obviously the case, although much remains to be learned about the relative sensory capabilities of elasmobranch taxa. However, it may be that larger-cerebellumed elasmobranchs use sensory information to perform more complex motor tasks than do smaller-cerebellumed species, or that the former possess more complex or more extensive behavioral repertoires. A larger interface between sensory input and motor output might also be required in the latter case, which might explain relative levels of cerebellar development. The current lack of understanding of many facets of elasmobranch behavior and the absence of comparative ethological studies of the lifestyle of shark and rays with large vs. small cerebelli makes such analyses difficult. There is a clear need for more field work on the ethology of elasmobranchs.

The cerebellar learning hypothesis

Classic studies by Brogden & Gantt (1937, 1942) demonstrated that movements elicited by direct electrical stimulation of certain regions of the mammalian cerebellum could be trained to respond to neutral auditory or visual conditioned stimuli with appropriate pairing. These studies have been reproduced in detail by Thompson et al. (1998) who have presented evidence that the reinforcing pathway activated for the learning of discrete movements in mammals is in the inferior olive-climbing fiber projection to the cerebellum. This projection is an integral portion of all vertebrate cerebelli and the learning function of the cerebellum may therefore be a fundamental functional feature of cerebellar circuitry in all vertebrate taxa. An investigation of such a role in the elasmobranch cerebellum has not yet been attempted but is certainly feasible and would provide an important comparative viewpoint from which to understand the function of cerebellar circuitry.

New techniques and approaches

New telemetry and recording technologies have made increasingly feasible the marriage of physiological and behavioral approaches to neuroethology. The large size and surgical accessibility of the elasmobranch cerebellum make it an inviting target for the study of neural activity chronically recorded in behaving animals. The responses of cerebellar neurons recorded during free field behaviors may yield profitable insight into the role of the cerebellum in both stereotyped behaviors and responses such as orienting behaviors, search activity, target tracking and feeding strikes to presented stimuli. Purkinje cells are particularly inviting in this respect, due to their large size, complex spikes and regular arrangement within the cerebellum they are easily identified and recorded, and they are the principal output neurons of the cerebellar circuitry onto the cerebellar nuclei. A further puzzle concerning the Purkinje cells themselves is the relative rarity of complex spikes in elasmobranchs compared to the relatively high frequency seen in mammals.

The regularity of organization of the cerebellum also makes it a logical candidate for tissue slice investigation of cerebellar neuronal function. Comparison of the activity and physiology of cerebellar neurons in vitro across various taxa may provide useful insight into the variations imposed on cerebellar function by selective processes. A limited number of studies have already employed the elasmobranch cerebellum in this fashion (Konnerth et al. 1987), but the utility of the elasmobranch cerebellum as a model in this regard remains still largely unexplored.

Acknowledgements

I am grateful to Timothy C. Tricas and Samuel H. Gruber for inviting me to participate in the memorable symposium upon which this volume is based and for inviting me to contribute this paper. I also thank the editors, Theodore H. Bullock and an anonymous reviewer for their very helpful and constructive criticism of the manuscript.

References cited

Alvarez, R. & R. Anadon. 1987. The cerebellum of the dogfish, *Scyliorhinus canicula*: a quantitative study. J. Hirnforsch. 28: 133–137.

Alvarez-Otero, R. & R. Anadon. 1992. Golgi cells of the cerebellum of the dogfish, *Scyliorhinus canicula* (elasmobranchs): a Golgi and ultrastructural study. J. Hirnforsch. 33: 321–327.

Alvarez-Otero, R., S.D. Regueira & R. Anadon. 1993. New structural aspects of the synaptic contacts on Purkinje cells in an elasmobranch cerebellum. J. Anat. (Lond.) 182: 13–21.

Alvarez-Otero, R., S.E. Perez, M.A. Rodriguez, F. Adrio & R. Anadon. 1995. GABAergic neuronal circuits in the cerebellum of the dogfish *Scyliorhinus canicula* (elasmobranchs): an immunocytochemical study. Neurosci. Lett. 187: 87–90.

Alvarez-Otero, R., S.E. Perez, M.A. Rodriguez & R. Anadon. 1996. Organization of the cerebellar nucleus of the dogfish, *Scyliorhinus canicula* L.: a light microscopic, immunocytochemical, and ultrastructural study. J. Comp. Neurol. 368: 487–502.

Barry, M.A. 1987. Afferent and efferent connections of the primary octaval nuclei in the clearnose skate, *Raja eglanteria*. J. Comp. Neurol. 266: 457–477.

Brogden, W.J. & W.H. Gantt. 1937. Cerebellar coniditioned reflexes. Amer. J. Physiol. 119: 277–278.

Brogden, W.J. & W.H. Gantt. 1942. Intraneural conditioning: cerebellar conditioned reflexes. Archiv. Neurol. Psychiatry 48: 437–455.

Boord, R.L. & R.G. Northcutt. 1982. Ascending lateral line pathways to the midbrain of the clearnose skate, *Raja eglanteria*. J. Comp. Neurol. 207: 274–282.

Bucy, R.S. & P.D. Joseph. 1968. Filtering for stochastic processes with applications to guidance. Wiley, New York. 195 pp.

Catois, E.M. 1901. Recherches sur l'histologie et l'anatomie microscopique de l'encephale chez les poissons. Bull. Scient. France Belgique 36: 1.

Fiebig, E. 1988. Connections of the corpus cerebelli in the thornback guitarfish, *Platyrhinoidis triseriata* (elasmobranchii): a study with WGA-HRP and extracellular granule cell recording. J. Comp. Neurol. 286: 567–583.

Hawkes, R., S. Blyth, V. Chockkan, D. Tano, Z. Ji & C. Mascher. 1993. Structural and molecular compartmentation in the cerebellum. Can. J. Neurol. Sci. 20, Suppl. 3: S29–35.

Hawkes, R., G. Brochu, L. Dore, C. Gravel & N. Leclerc. 1992. Zebrins: molecular markers of compartmentation in the cerebellum. pp. 22–55. *In*: R. Llinas & C. Sotelo (ed.) The Cerebellum Revisited, Springer-Verlag, New York.

Hoggatt, A.M. & M.J. Lannoo. 1994. Monoclonal antibody anti-type I and anti-zebrin II labelling in siluriform fishes: the role of shared lineage versus shared function in polypeptide co-distributions. Brain Res. 665: 181–191.

Houser, G.L. 1901. The neurons and supporting elements of the brain of a selachian. J. Comp. Neurol. 11: 65–175.

Ito, M. 1993. Movement and thought: identical control mechanisms by the cerebellum. Trends Neurosci. 16: 448–450.

Ito, M. 1997. Cerebellar microcomplexes. pp. 475–489. *In*: J.D. Schmahmann (ed.) International Review of Neurobiology, Academic Press, San Diego.

Kalman, R.E. 1960. A new approach to linear prediction and filtering problems. J. Basic Eng. A.S.M.E. 82: 35–45.

Kappers, C.U.A., G.C. Huber & E. Crosby. 1936. The comparative anatomy of the nervous system of vertebrates, including man. Macmillan, New York. 1845 pp.

Karamyan, A.I. 1962. Evolution of the function of the cerebellum and cerebral hemispheres. Published for the National Science Foundation by the Israel Program for Scientific Translations, Jerusalem. 109 pp.

Koester, D.M. 1983. Central projections of the octavolateralis nerves of the clearnose skate, *Raja eglanteria*. J. Comp. Neurol. 221: 199–215.

Konnerth, A., A.L. Obaid & B.M. Salzberg. 1987. Optical recording of electrical activity from parallel fibers and other cell types in skate cerebellar slices in vitro. J. Physiol. 393: 681–702.

Lannoo, M.J. & R. Hawkes. 1997. A search for primitive Purkinje cells: zebrin II expression in sea lampreys (*Petromyzon marinus*). Neurosci. Lett. 237: 53–55.

Lannoo, M.J., L. Ross, L. Maler & R. Hawkes. 1991. Development of the cerebellum and its extracerebellar Purkinje cell projection in teleost fishes as determined by zebrin II immunohistochemistry. Prog. Neurobio. 37: 329–363.

Larsell, O. 1967. The comparative anatomy and histology of the cerebellum from myxinoids through birds. The University of Minnesota Press, Minneapolis. 291 pp.

Lee, L.T. & T.H. Bullock. 1990a. Cerebellar units show several types of early responses to telencephalic stimulation in catfish. Brain Behav. Evol. 35: 278–290.

Lee, L.T. & T.H. Bullock. 1990b. Cerebellar units show several types of long-lasting posttetanic responses to telencephalic stimulation in catfish. Brain Behav. Evol. 35: 291–301.

Meek, J. 1992. Why run parallel fibers parallel? Teleostean Purkinje cells as possible coincidence detectors, in a timing device subserving spatial coding of temporal differences. Neurosci. 48: 249–283.

Meek, J., T.G. Hafmans, L. Maler & R. Hawkes. 1992. Distribution of zebrin II in the gigantocerebellum of the mormyrid fish, *Gnathonemus petersii*, compared with other teleosts. J. Comp. Neurol. 316: 17–31.

Myagkov, N.A. 1990. The brain sizes of living Elasmobranchii as their organization level indicator. I. General analysis. J. Hirnforsch. 32: 553–561.

New, J.G. & T.H. Bullock. 1989. Electrosensory responses in the granule cell layer of the cerebellum of an elasmobranch. Soc. Neurosci. Abstr. 15: 1138.

Nicholson, C., R. Llinas & W. Precht. 1969. Neural elements of the cerebellum in elasmobranch fishes: structural and functional characteristics. pp. 215–244. *In*: R. Llinas & R.F. Mathewson (ed.) Neurobiology of Cerebellar Evolution and Development, Institute for Biomedical Research, American Medical Association, Chicago.

Nicholson, L.F.B., J.C. Montgomery & R.L.M. Faull. 1994. GABA, muscarinic cholinergic, excitatory amino acid, neurotensin and opiate binding sites in the octavolateralis column and cerebellum of the skate *Raja nasuta* (Pisces: Rajidae). Brain Res. 652: 40–48.

Nieuwenhuys, R. 1967. Comparative anatomy of the cerebellum. pp. 1–93. *In*: C.A. Fox & R.S. Snider (ed.) Progress in Brain Research, Vol. 25, Elsevier, Amsterdam.

Northcutt, R.G. 1977. Elasmobranch central nervous system organization and its possible evolutionary significance. Amer. Zool. 17: 411–429.

Northcutt, R.G. 1978. Brain organization in the cartilaginous fishes. pp. 117–194. *In*: E.S. Hodgson & R.F. Mathewson (ed.) Sensory Biology of Sharks, Skates and Rays, Office of Naval Research, Arlington.

Northcutt, R.G. 1989. Brain variation and phylogenetic trends in elasmobranch fishes. J. Exp. Zool. 2: 83–100.

Northcutt, R.G. & W.J. Brunken. 1984. Cerebellar afferents in the little skate (Batoidea). Soc. Neurosci. Abstr. 10: 853.

Paul, D.H. 1969. Electrophysiological studies on parallel fibers of the corpus cerebelli of the dogfish *Scyliorhinus canicula*. pp. 245–250. *In*: R. Llinas & C. Sotelo (ed.) Neurobiology of Cerebellar Evolution and Development, Institute for Biomedical Research/American Medical Association, Chicago.

Paul, D.H. & B.L. Roberts. 1975. Connections between the cerebellum and the reticular formation in the dogfish *Scyliorhinus canicula*. J. Physiol. 249: 62–63.

Paul, D.H. & B.L. Roberts. 1979. The significance of cerebellar function for a reflex movement of the dogfish. J. Comp. Physiol. 134: 69–74.

Paul, D.H. & B.L. Roberts. 1981. The activity of cerebellar neurones of an elasmobranch fish (*Scyliorhinus canicula*) during a reflex movement of a fin. J. Physiol. 321: 369–383.

Paul, D.H. & B.L. Roberts. 1983. The activity of cerebellar nuclear neurones in relation to stimuli which evoke a pectoral fin reflex in dogfish. J. Physiol. 342: 465–481.

Paul, D.H. & B.L. Roberts. 1984a. Projections of cerebellar Purkinje cells in the dogfish, *Scyliorhinus*. Neurosci. Lett. 44: 43–46.

Paul, D.H. & B.L. Roberts. 1984b. The activity of cerebellar neurones of the decerebrate dogfish *Scyliorhinus* during spontaneous swimming movements. J. Physiol. 352: 1–16.

Paulin, M.G. 1993. The role of the cerebellum in motor control and perception. Brain Behav. Evol. 41: 39–50.

Paulin, M.G. 1997. Neural representations of moving systems. pp. 516–535. *In*: J.D. Schmahmann (ed.) International Review of Neurobiology, Academic Press, San Diego.

Puzdrowski, R.L. 1997. Anti-zebrin II immunopositivity in the cerebellum and octavolateral nuclei in two species of stingrays. Brain Behav. Evol. 50: 358–368.

Rudeberg, S.-I. 1961. Morphogenetic studies on the cerebellar nuclei and their homologization in different vertebrates including man. Ph.D. Dissertation, University of Lund, Lund. 148 pp.

Sauerbeck, E. 1896. Beitrage zur Kenntis vom feineren Bau des Selachierhirns. Anat. Anz. B. 12: 41.

Schaper, A. 1898. The finer structure of the selachian cerebellum (*Mustelis vulgaris*) as shown by chrome silver preparations. J. Comp. Neurol. 8: 1–20.

Schmidt, A.W. & D. Bodznick. 1987. Afferent and efferent connections of the vestibulolateral cerebellum of the little skate, *Raja erinacea*. Brain Behav. Evol. 30: 282–302.

Smeets, W.J.A.J. 1982. The afferent connections of the tectum mesencephali in two chondrichthyans, the shark *Scyliorhinus canicula* and the ray *Raja clavata*. J. Comp. Neurol. 205: 139–152.

Smeets, W.J.A.J., R. Nieuwenhuys & B.L. Roberts. 1983. The central nervous system of cartilaginous fishes. Springer-Verlag, New York. 266 pp.

Sterzi, G. 1905. Sulla regio parietalis dei ciclosotomi, dei selachii e degli olocefali. Anat. Anz. 27: 346–416.

ten Cate, J. 1930. Contribution a la physiologie comparée du cervelet. III. Le cervelet des plagiostomes. Archiv. need. physiol. de l'homme et des animaux 15: 479–528.

Thompson, R.F., J.K. Thompson, J.J. Kim, D.J. Krupa & P.G. Shinkman. 1998. The nature of reinforcement in cerebellar learning. Neurobiol. Learn. Mem. 70: 150–176.

Voorhoeve. 1917. Over den Bouw van de kleine hersenen der Plagiostomen (here in English). Inaugural Dissertation, University of Amsterdam, Amsterdam. 88 pp.

Wassef, M., P. Angaut, L. Arsenio-Nunes, F. Bourrat & C. Sotelo. 1992. Purkinje cell heterogeneity: its role in organizing the topography of cerebellar cortex connections. pp. 5–21. *In*: R. Llinas & C. Sotelo (ed.) The Cerebellum Revisited, Springer-Verlag, New York.

Welker, W. 1987. Spatial organization of somatosensory projections to granule cell cerebellar cortex: functional and connectional implications of fractured somatotopy. pp. 239–280. *In*: J.S. King (ed.) New Concepts in Cerebellar Neurobiology, A.R. Liss, New York.

Young, W. 1980. Field potential analysis in elasmobranch cerebellum. Brain Res. 199: 101–112.

Environmental Biology of Fishes **60:** 109–116, 2001.

Orientation and navigation in elasmobranchs: which way forward?

John C. Montgomery & Michael M. Walker
School of Biological Sciences, University of Auckland, Private Bag 92019, Auckland, New Zealand
(e-mail: j.montgomery@auckland.ac.nz)

Received 3 December 1998 Accepted 7 January 2000

Key words: posture, sensory systems, compass sense, electroreception, lateral line

Synopsis

Elasmobranchs possess a multiplicity of mechanisms controlling posture and short distance orientation. Visual–vestibular contributions to posture and locomotion are well documented. So too, are the contributions of vision, olfaction and the octavolateralis senses to short distance orientation, particularly orientation to specific environmental stimuli such as those generated by prey. Less well understood are the mechanisms guiding orientation over longer distances. Anecdotal and systematic observations of behaviour show tidal, daily, repeat long distance, and even seasonal movement patterns. True navigation has not been demonstrated in elasmobranchs and the sensory mechanisms underlying the above movement patterns remain largely speculative. However, they are likely to include responses to water currents, and physical parameters such as temperature, pressure, and the geomagnetic field. Of particular interest in elasmobranchs is that geomagnetic orientation could be mediated directly via a magnetite based sensory system, or indirectly via the electrosensory system. Systematic studies of movement patterns and experimental studies of the underlying mechanisms of orientation are required to gain an increased understanding of orientation and navigation in this intriguing group.

Past studies

Primary orientation – posture and locomotion

Classically, orientation is divided into 'primary and secondary orientation' (Fraenkel & Gunn 1961) where 'primary orientation' is defined as the reactions that guide an animal into its normal stance. So, primary orientation covers many aspects of posture control and locomotion and, for elasmobranchs, these topics are reviewed in the book 'Physiology of elasmobranch fishes' (Roberts 1988, Montgomery 1988, Bone 1988). Since these reviews, there has been little work on elasmobranchs with respect to the precise contribution of various inputs to the control of posture and locomotion. However, there is little doubt that at the level of the sensory organ, elasmobranchs share much in common with other vertebrates, and that the sensory information provided by the visual and vestibular systems make a major contribution to posture and locomotion. Indeed, many of the pioneering studies on the vestibular labyrinth were conducted in elasmobranchs (e.g. Lowenstein & Sand 1936, Groen et al. 1952). Gravitational and angular acceleration inputs from the vestibular labyrinths (Figure 1) produce direct fin reflexes, the effect of which is to stabilise the animal (Timerick 1982). Labyrinthine input also plays a major role in the control of eye movement (Montgomery & Cotton 1984, Roberts 1988, Puzdrowski & Leonard 1994) which can in turn impact postural and locomotory control. For example, recent work on goldfish under microgravity shows that, in this species, visual inputs dominate postural control, with labyrinth inputs exerting their influence via the control of eye movement (de Jong et al. 1996). Light direction per se, also contributes strongly to primary orientation in vertebrates (e.g. Ullen et al. 1995). However, given that the normal stance of elasmobranchs is simply upright, whether they are resting

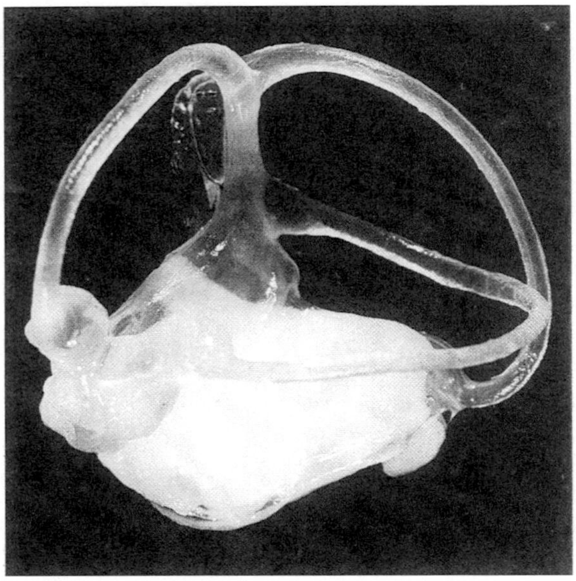

Figure 1. The left vestibular labyrinth of an elasmobranch viewed from the left side. The three semicircular canals have patches of hair cells within the ampullary swellings that respond to fluid movements produced by angular rotations of the head. The three semicircular canals, mutually at right angles, encode head rotations in three dimensional space. The dense white areas are otoliths that respond to linear accelerations and vibrations. Rostral to the left.

on the bottom or swimming, and that many are night active, or occupy low light habitats, it is to be expected that for elasmobranchs, labyrinth input probably dominates primary orientation.

Middle-scale orientation/navigation

Secondary orientation is defined as the reactions that guide an animal into its normal habitat or into other situations that are important to it (Fraenkel & Gunn 1961). For convenience, secondary orientation is often divided into 'middle' and 'long' distance orientation. Middle-scale navigation is further characterised as 'searching for food and returning home within an animal's foraging range' (Wehner 1996). Other contributions to this volume review each of the major sensory modalities of sharks and rays (with the exception of olfaction) (e.g., hearing – Myrberg 2001, mechanosense – Maruska 2001, electrosense – Tricas 2001, all this volume). In doing so, they discuss part of the domain of 'short to middle distance orientation', particularly with respect to orientation to particular environmental stimuli such as prey or conspecifics. Depending on species, and environmental circumstance, there is a distance hierarchy of senses. Under appropriate conditions, olfaction, hearing and vision operate at larger distances, whereas mechanosensory and electrosensory systems are relatively short range.

Brief mention will be made of olfactory orientation as it relates not only to finding food but also to mate localization (Johnson & Nelson 1978). This is of particular interest in elasmobranchs since the components of olfactory search behaviour are better understood in sharks than in other fish. Classic studies by Parker, Hodgson, Matthewson and Kleerekoper (reviewed in Hodgson & Matthewson 1978, Kleerekoper 1978) show that to locate an olfactory source, sharks use both a chemical gradient search and a chemically released rheotaxis (orientation to water currents). For example, nurse sharks show a true gradient search (klinotaxis). They approach the origin of the olfactory stimulus along an S-shaped track, presumably initiating turns in the direction of the nostril that receives the strongest olfactory stimulation. Evidence for this comes from Parker's (1914) early observations on *Mustelus*. Experimental animals that had one nostril blocked with a cotton plug turned predominantly towards the side of the functional nostril when activated by an olfactory stimulus. In contrast, lemon sharks encountering an olfactory stimulus turn and swim upstream, showing a chemically released rheotaxis. This behaviour would provide a quicker approach to the source than a gradient search, but depends on other sensory cues to actually identify and locate the potential food source.

The characterisation of middle-scale navigation as 'searching for food and returning home, within an animal's foraging range' (Wehner 1996) works well for the study of insects like ants and bees. However, it presupposes the existence of a 'home' and a 'home-range'. In elasmobranchs, only a few studies clearly establish a site attachment worthy of the name 'home'. Port Jackson sharks, *Heterodontus portusjacksoni*, repeatedly visit specific resting sites, and return to those sites when transferred to different localities up to 3 km away (O'Gower 1995). Juvenile hammerhead sharks, *Sphyrna lewini*, show a high fidelity to shared daytime core areas to which they return repeatedly after wider ranging nocturnal movements (Holland et al. 1993). Anecdotal evidence also supports the case for a specific site attachment in some other elasmobranchs. One individual greater spotted dogfish, *Scyliorhinus stellatus*, on the South Coast of England occupied the same small cave over an 18 month period of intermittent observations. The observations were all made during the day, and the presumption is that the animal foraged at night, but returned to the same location each day (unpublished observations).

The case for a 'home range' or 'foraging range' is better supported. Swell sharks, *Cephaloscyllium ventriosum*, rest deep within rock crevices during the day and move out onto the open bottom at night (Tricas 1982). Acoustic tracking of juvenile lemon sharks, *Negaprion brevirostris*, demonstrated established home ranges, with an activity space of 0.23–1.26 km^2 (Morrissey & Gruber 1993). Adult lemon sharks also show some degree of site attachment with activity spaces of 9–93 km^2 (Gruber et al. 1988). In addition, they showed a distinct daily pattern of movement, with their daytime location being eastward of their night-time activity spaces. It is suggested that the sharks use the sun as an orientation cue (Gruber et al. 1988).

Repeated vertical migration also comes under the heading of secondary orientation. Ironically the best example of diurnal vertical migration in an elasmobranch comes from a single specimen of one of the rarest sharks. Acoustic tracking of a megamouth shark, *Megachasma pelogios*, showed one of the most clearly defined diel patterns of movement ever obtained from an elasmobranch (Figure 2) (Nelson et al. 1997). From the point of view of the sensory cues controlling migration, it is interesting that the ascent and descent profile provided a good match to isolumes of about 0.4 lux suggesting that the shark's chosen depth was to a large degree determined by light level.

Tidal movements are evident in a number of elasmobranchs, particularly rays that forage in the intertidal. Clear evidence of their foraging activities is seen in the feeding pits that are left exposed on the sand-flats as the tide recedes (Hines et al. 1997). Large sand-flat areas that drain at low tide create a particular orientation problem – how to retreat from the flats to the channels so as not to be stranded by the falling tide. The suggestion made here is that a rule-of-thumb that would work is to always follow the current. Recent observations of movement activity in flounder show periods of random search alternating with periods of down-current movement (Moore & Montgomery unpublished observation). This single behaviour would be effective for moving an animal up onto the sand-flats as the tide comes in, and back down into the channels as the tide recedes. Such a behaviour would bear some similarities to selective tidal stream transport reported for flatfish (Harden Jones et al. 1979), however within the estuary, the importance of the behaviour would be to determine movement direction rather than assisting with the energetics of migration. Water currents are indeed an important structuring feature of acquatic environments, and most fish show distinct behavioural responses to them (called rheotaxis). Recent experimental work shows that rheotaxis at slow current flow rates is mediated by the superficial neuromasts of the mechanosensory lateral line system (Montgomery et al. 1997). Elasmobranchs too have superficial neuromasts on the surface of the body (Roberts 1978, Maruska & Tricas 1998, Maruska 2001 this volume), so they have the

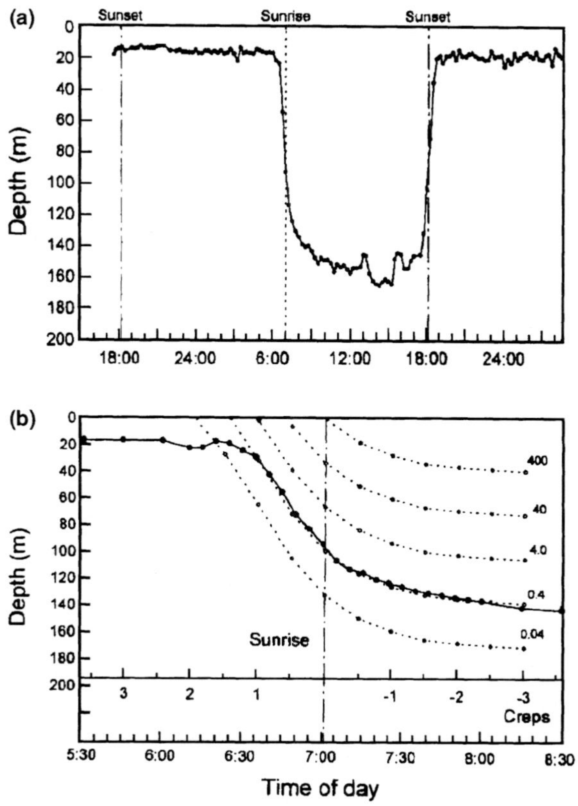

Figure 2. Vertical migration of a megamouth shark tracked off southern California: a – showing the proximity to the surface during the night, and the vertical migrations at dawn and dusk. b – showing the downward migration during the morning light transition. Dotted lines are calculated isolumes (in lux) for five different illumination levels. Note that the shark very closely follows the 0.4 lux isolume. One crep (crepuscular unit) is 25 min for the time and place of tracking (from Nelson et al. 1997).

sensory apparatus to be able to use water currents in this manner.

Downstream movement has also recently been reported in the basking shark, *Cetorhinus maximus*, (Sims & Quayle 1998). Downstream orientation is maintained until the shark encounters an increase in zooplankton density. Small-scale orientation then occurs to keep the shark within the zooplankton patch. Sims & Quayle further suggest that the basking sharks use their electrosensory system for detecting local increases in zooplankton concentration.

Long distance orientation

Elasmobranchs are generally large animals, and as such move large distances. Roving over large sections of ocean is, by itself, not a demonstration of any particular navigation ability. Selection of a particular water body, by temperature, or oxygen content, may be a sufficient explanation of distribution, as it appears to be for skipjack tuna (Barkley et al. 1978). Nakano & Nagasawa (1996) claim that six species of elasmobranchs that they analysed as by-catch in a 10-year gillnet research program, showed spatial segregation in species distribution according to oceanographic conditions such as temperature.

Griffin (1952) has provided a classification of long distance orientation which contrasts the null case (above) with compass orientation, piloting by familiar landmarks and true navigation. The simplest case of true navigation involves the use of two orthogonally arranged co-ordinates that allow the animal to know its position in relation to some specific destination. True navigation is demonstrated if an animal is experimentally displaced under these circumstances and adjusts its heading to compensate for the displacement. On the basis of these definitions, true navigation has not been demonstrated in elasmobranchs, though there are reasonable demonstrations of compass orientation and pilotage.

Circumstantial evidence for directed migrations is found in the existence of nursery areas for many elasmobranch species. Gravid females of many species of coastal sharks deliver their young in discrete geographical areas, often in shallow bays and estuaries (Castro 1993), implying an annual migration into these areas. An annual migration of whale sharks has also been reported in a north-western Australian reef system (Taylor 1996). The whale sharks arrive in autumn just after the coral has undergone mass spawning.

The best evidence for directional swimming in elasmobranchs comes from tracking studies. Sciarrotta & Nelson (1977) demonstrated that blue sharks near coastal islands spend daylight hours offshore and move inshore at night. Carey & Scharold (1990) tracked migrating blue sharks and found that they maintained straight tracks of point-to-point movements of up to 200 km over periods of up to 3 days. Klimley also used acoustic tracking methods to study scalloped hammerhead sharks, *Sphyrna lewini*. Sharks associated with the Espiritu Santo Seamount dispersed and returned to the seamount in a rhythmical fashion, moving up to a distance of 19 km away (Klimley & Nelson 1984, Klimley 1993). Individual sharks also swam in a highly directed fashion between the seamount and Las Animas Island at night, with repeated vertical excursions ranging from 100 to 450 m deep, but at all times out of view of either the sea surface or the sea floor (Klimley 1993). Headings and point-to-point movements of the scalloped hammerheads were compared with those of a transmitter floating at the surface showing that the sharks were indeed swimming in a highly oriented manner.

Persistent directional movement in the featureless middle-depths of the ocean begs the question: how is a constant heading maintained? A number of possibilities exist. For example, an inertial based system, keeping track of left and right turns would suffice for limited periods. Fine structure in the water mass may also contribute useful information (Dingle 1996). What appears to us as featureless middle-depths do in fact exhibit considerable vertical fine structure as well as a horizontal quasi-mosaic pattern in both temperature and odour pattern. The current-shear in the fine-structured gradient layers may provide a directional cue. Fish sampling this information source should make vertical movements, as do migrating salmon in Norwegian fjords (Døving et al. 1985) and the scalloped hammerheads on their visits to Las Animas. Likewise, underwater sound travels long distances and is a potential source of directional information. Rapidly accumulating evidence also shows that a wide range of animals use geomagnetic cues for orientation (Wiltschko & Wiltschko 1996, Dingle 1996), so this too should be added to the list of candidate cues. Indeed, Klimley's (1993) shark data do indicate a correlation between travel routes and magnetic topography suggesting a form of geomagnetic field based pilotage. He proposed that sharks use a topotaxis, orienting to boundaries between geomagnetic gradients in sea-floor magnetization, an ability that would require an acute sense of geomagnetic intensity.

Geomagnetic orientation

Geomagnetic orientation is well accepted, but its sensory basis is still a matter of debate. Of particular interest in elasmobranchs is that geomagnetic orientation could be mediated directly via a magnetite based sensory system, or indirectly via the electrosensory system. Critical experiments to distinguish between these two mechanisms have yet to be attempted or reported. Behavioural studies over a wide range of animal taxa show responses to both magnetic field direction (Wiltschko & Wiltschko 1995, Lohmann & Lohmann 1996), and magnetic intensity (Walker & Bitterman 1989, Walker et al. 1997). Single domain magnetite has been found in a number of species, and a magnetite based sensory detector is certainly a candidate for magnetic field detection (Walker et al. 1997).

An alternate mechanism for magnetic field detection exists in elasmobranchs. Kalmijn (1982, 1984) has suggested that geomagnetic orientation in elasmobranchs may be based on their electrosensory system (Figure 3). The suggestion is based on Faraday's law, that an electromotive force is induced in a conductor moving in a magnetic field. In the 'passive mode' horizontal electric fields generated by the movement of ocean streams within the earth's magnetic field are postulated to be used in orientation. The vector diagram indicates that the horizontal movement of the ocean stream interacting with the vertical component of the magnetic field produces the horizontal electromotive field. In the 'active mode' the horizontal velocity of the animal interacting with the horizontal component of the magnetic field produces a vertical electromotive field. A recent reanalysis (Paulin 1995) re-emphasises that electroreceptors do not have the characteristics necessary to make DC measurements. Paulin's (1995) refinement is that a compass sense could originate from the directional asymmetry of the change in induced electroreceptor voltage during turns (Figure 4). A comparison of vestibular and electrosensory inputs would provide the mechanism to determine swimming direction. The frequency response characteristics both of vestibular semicircular canals, and electrosensory receptors is given in Figure 5. Both systems have the appropriate response characteristics to encode the information generated by head turning movements. One of the attractions of Paulin's suggestion is that phase comparisons are a common feature of neuronal network processing, and it would be relatively easy to come up with a central processing algorithm to extract compass heading from the electrosensory and vestibular inputs.

Future directions

Able (1996) recognises three predominating themes of contemporary research in animal orientation and navigation. First, animal orientation systems are replete

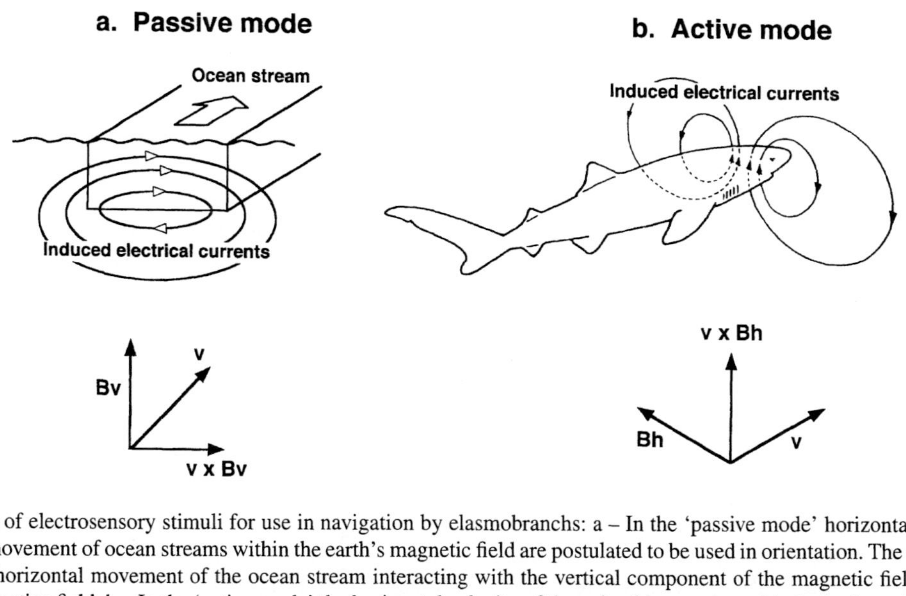

Figure 3. Sources of electrosensory stimuli for use in navigation by elasmobranchs: a – In the 'passive mode' horizontal electric fields generated by the movement of ocean streams within the earth's magnetic field are postulated to be used in orientation. The vector diagram indicates that the horizontal movement of the ocean stream interacting with the vertical component of the magnetic field produces the horizontal electromotive field. b – In the 'active mode' the horizontal velocity of the animal interacting with the horizontal component of the magnetic field produces a vertical electromotive field (from Kalmijn 1997).

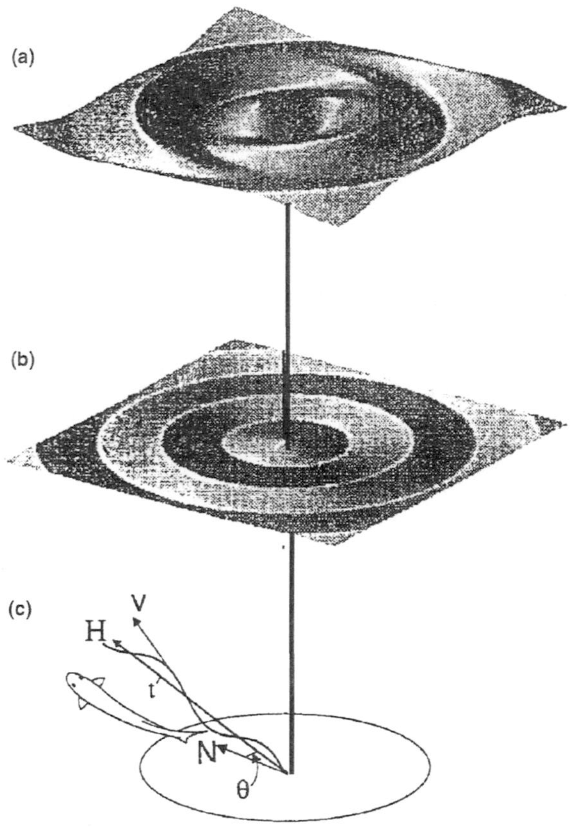

Figure 4. Surface plots of sensory afferent modulations due to head movements during swimming. a – The height of the surface at distance t from the center of the plot in direction θ represents the motion-induced receptor voltage at time t when the fish is swimming with heading θ. b – The corresponding point on this plot represents head rotational velocity at the same time. c – Swimming trajectory (N = north, H = vector in heading direction, v = velocity vector at time t, θ = heading, from Paulin 1995).

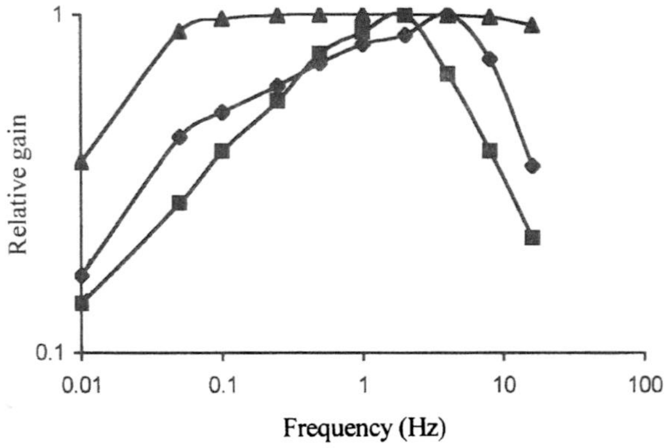

Figure 5. Frequency response curves (log log scale) of electrosensory and vestibular primary afferent neurons. Electrosensory neuron responses are shown for the round stingray (■ Tricas et al. 1995) and the thornback ray (♦ Montgomery 1984). Sinusoidal electric field stimuli were delivered at a range of frequencies and neural responses recorded as the change in firing rate of the unit in response to each stimulus. Neural responses were normalised to a gain of one. The responses of the two species are quite similar with relatively broad band pass characteristics and a peak response in the 2–4 Hz range. Also shown is the frequency range of horizontal vestibular afferents (▲). These data are the vestibular model with time constants 40 s and 0.025 s from data given in Hartmann & Klinke (1980) quoting Groen et al. (1952) for the ray labyrinth.

with interacting mechanisms and are highly flexible. Second, there is a compelling need to take studies of navigation back into the field. Finally, it is necessary to uncover, and tease apart, the 'rules of thumb' that migrators use. Studies of elasmobranch orientation and navigation have to-date been observational and correlational. This is a necessary first step to approaching a more mechanistic understanding of the sensory basis of these aspects of elasmobranch behaviour. We now require more experimental studies of orientation behaviour, showing that manipulation of sensory cues and sensory capabilities alters orientation in predictable ways. Three leading contenders for more experimental work can be suggested.

The tidal rheotaxis hypothesis

The mechanisms of orientation in intertidal areas are poorly understood. The hypothesis proposed here is that negative rheotaxis (following the current) is the 'rule of thumb' used to both find a way into shallow bays during flood tide, and also to avoid being stranded by the falling tide. As with other orientation behaviours, rheotaxis can be mediated by several sensory mechanisms. However, at low current speeds it is likely that the dominant contribution is from the superficial neuromasts of the dorsal body surface (Montgomery et al. 1997). A direct test of the above hypothesis would be to see if animals with disabled neuromasts are less proficient at negotiating this orientation task.

The horizontal electric field orientation hypothesis

Theoretical considerations permit sharks and rays to use their electrosensory systems to detect electric fields generated by ocean currents and by their own movement. Kalmijn (1988) has presented some preliminary data showing orientation responses of the stingray, *Dasyatis*, to rotation of the horizontal ambient electric field. Taken at face value, these findings indicate that the stingrays maintain their heading on the basis of the uniform horizontal field direction, rather than compass information generated from the geomagnetic field. Furthermore, the orientation response to rotation of the uniform field overrides contradictory vestibular information, which must be generated by the turning movement. Further field experiments of this kind are required to verify the finding and its implications.

The vestibulo-electrosensory compass hypothesis

Further experimental work is required to show that direct geomagnetic orientation is mediated via the electrosensory system, rather than other possible candidate mechanisms. If Paulin's (1995) analysis is correct, a constant compass heading is maintained by comparing vestibular signals with the dorso-ventral electric fields induced by the animal's movements. The prediction of this hypothesis is that appropriate modulation of a dorso-ventral electric field should produce predictable changes in orientation.

Some of these experiments will have to be in the laboratory, but it will also be important to 'ground-truth' laboratory observations in the field. Care and imagination will be required to dissect the contributing mechanisms and navigate our own way through the interacting mechanisms and behavioural flexibility to uncover the 'rules of thumb' used. 'Many fundamental and exciting things remain to be learned about the navigation behaviour of animals' (Able 1996). One has to agree with this statement for animal navigation in general and elasmobranch orientation and navigation in particular.

References cited

Able, K.P. 1996. Large scale navigation. J. Exp. Biol. 199: 1–2.

Barkley, R.A., W.H. Neill & R.M. Gooding. 1978. Skipjacktuna, *Katsuwonus pelamis*, habitat based on temperature and oxygen requirements. U.S. Fish. Bull. 76: 653–662.

Bone, Q. 1988. Muscles and locomotion. pp. 99–142. *In:* T.J. Shuttleworth (ed.) Physiology of Elasmobranch Fishes, Springer-Verlag, Berlin.

Carey, F.G. & J.V. Scharold. 1990. Movements of blue sharks (*Prionace glauca*) in depth and course. Mar. Biol. 106: 329–342.

Castro, J.I. 1993. The shark nursery of Bulls Bay, South Carolina, with a review of the shark nurseries of the southeastern coast of the United States. Env. Biol. Fish. 38: 37–48.

De Jong, H.A.A., E.N.P.M. Sondag, A. Kuipers & W.J. Oosterveld. 1996. Swimming behavior of fish during short periods of weightlessness. Aviation Space & Environmental Medicine 67: 463–466.

Dingle, H. 1996. Migration, the biology of life on the move. Oxford University Press, New York. 474 pp.

Døving, K.B., B.H. Westerberg & P.B. Johnsen. 1985. Role of olfaction in the behavioural and neuronal responses of the Atlantic salmon, *Salmo salar*, to hydrographic stratification. Can. J. Fish. Aquat. Sci. 542: 1658–1667.

Fraenkel, G.S. & D.L. Gunn. 1961. The orientation of animals, kineses, taxes and compass reactions. Dover Publications, New York. 376 pp.

Griffin, D.R. 1952. Bird navigation. Biol. Rev. Camb. Philos. Soc. 27: 359–400.

Groen, J.J., O. Lowenstein & A.J.H. Vendrick. 1952. The mechanical analysis of the responses from the end-organs of the horizontal semicircular canal in the isolated elasmobranch labyrinth. J. Physiol. 117: 54–62.

Gruber, S.H., D.R. Nelson & J.F. Morrissey. 1988. Patterns of activity and space utilization of lemon sharks *Negaprion brevirostris* in a shallow Bahamian lagoon. Bull. Mar. Sci. 43: 61–76.

Harden Jones, F.R., G.P. Arnold, M. Greer-Walker & P. Scholes. 1979. Selective tidal stream transport and the migration of plaice (*Pleuronectes platessa* L.) in the southern North Sea. J. Conseil Internat. l'Exploration de la Mer 38: 331–337.

Hartmann, R. & R. Klinke. 1980. Discharge properties of afferent fibres of the goldfish semicircular canal with high frequency stimulation. Pflugers Arch. 388: 111–121.

Hines, A.H., R.B. Whitlatch, S.F. Thrush, J.E. Hewitt, V.J. Cummings, P.K. Dayton & P. Legendre. 1997. Nonlinear foraging response of a large marine predator to benthic prey: eagle ray pits and bivalves in a New Zealand sandflat. J. Exp. Mar. Biol. Ecol. 216: 191–210.

Hodgson, E.S. & R.F. Matthewson. 1978. Electrophysiological studies of chemoreception in elasmobranchs. pp. 227–267. *In:* E.S. Hodgson & R.F. Matthewson (ed.) Sensory Biology of Sharks, Skates, and Rays, Office of Naval Research, Department of the Navy, Arlington.

Holland, K.N., B.M. Wetherbee, J.D. Peterson & C.G. Lowe. 1993. Movements and distribution of hammerhead shark pups on their natal grounds. Copeia 1993: 495–502.

Johnson, R.H. & D.R. Nelson. 1978. Copulation and possible olfaction-mediated pair formation in two species of carcharhinid sharks. Copeia 1978: 539–542.

Kalmijn, A.J. 1982. Electric and magnetic field detection in elasmobranch fishes. Science 218: 916–918.

Kalmijn, A.J. 1984. Theory of electromagnetic orientation: a further analysis. pp. 525–560. *In:* A. Bolis, R.D. Keynes & S.H.P. Madrell (ed.) Comparative Physiology of Sensory Systems, Cambridge University Press, Cambridge.

Kalmijn, A.J. 1988. Detection of weak electric fields. pp. 151–186. *In:* J. Atema, R.R. Fay, A.N. Popper & W.N. Tavolga (ed.) Sensory Biology of Aquatic Animals, Springer-Verlag, New York.

Kalmijn, A.J. 1997. Electric and near-field acoustic detection, a comparative study. Acta Physiol. Scand. 161(Suppl 638): 25–38.

Kleerekoper, H. 1978. Chemoreception and its interaction with flow and light perception in the locomotion and orientation of some elasmobranchs. pp. 269–329. *In:* E.S. Hodgson & R.F. Matthewson (ed.) Sensory Biology of Sharks, Skates, and Rays, Office of Naval Research, Department of the Navy, Arlington.

Klimley, A.P. 1993. Highly directional swimming by scalloped hammerhead sharks, *Sphyrna lewini*, and subsurface irradiance, temperature, bathymetry, and geomagnetic field. Mar. Biol. 117: 1–22.

Klimley, A.P. & D.R. Nelson. 1984. Diel movement of patterns of the scalloped hammerhead shark (*Sphyrna lewini*) in relation to El Bajo Espiritu Santo (Mexico): a refuging central-position social system. Behav. Ecol. Sociobiol. 15: 45–54.

Lohmann, K.J. & C.M.F. Lohmann. 1996. Orientation and open-sea navigation in sea turtles. J. Exp. Biol. 199: 73–81.

Lowenstein, O. & A. Sand. 1936. The mechanism of the semicircular canal: a study of responses of single fibre preparations to angular accelerations and to rotation at constant speed. Proc. R. Soc. Lond. B 129: 256–275.

Maruska, K.P. 2001. Morphology of the mechanosensory lateral line system in elasmobranch fishes: ecological and behavioral considerations. Env. Biol. Fish. 60: 47–76 (this volume).

Maruska, K.P. & T. Tricas. 1998. Morphology of the mechanosensory lateral line system in the Atlantic stingray, *Dasyatis sabina*: the mechanotactile hypothesis. J. Morph. 238: 1–22.

Montgomery, J.C. 1984. Frequency response characteristics of primary and secondary neurons in the electrosensory system of the thornback ray. Comp. Biochem. Physiol. 79A: 189–195.

Montgomery, J.C. 1988. Sensory physiology. pp. 79–98. *In:* T.J. Shuttleworth (ed.) Physiology of Elasmobranch Fishes, Springer-Verlag, Berlin.

Montgomery, J.C., C.F. Baker & A.G. Carton. 1997. The lateral line can mediate rheotaxis in fish. Nature 389: 960–963.

Montgomery, J.C. & P. Cotton. 1984. Projection of secondary vestibular neurons to the abducens nucleus in the carpet shark *Cephaloscyllium isabella*. Brain Behav. Evol. 27: 41–47.

Morrissey, J.F. & S.H. Gruber. 1993. Home range of juvenile lemon sharks *Negaprion brevirostris*. Copeia 1993: 425–434.

Myrberg, A.A., Jr. 2001. The acoustical biology of elasmobranchs. Env. Biol. Fish. 60: 31–45 (this volume).

Nakano, H. & K. Nagasawa. 1996. Distribution of pelagic elasmobranchs caught by salmon research gillnets in the North Pacific. Fisheries Sci. 62: 860–865.

Nelson, D.R., J.N. Mckibben, W.R. Strong Jr., C.G. Lowe, J.A. Sisneros, D.M. Schroeder & R.J. Lavenberg. 1997. An acoustic tracking of a megamouth shark, *Megachasma pelagios*: a crepuscular vertical migrator. Env. Biol. Fish. 49: 389–399.

O'Gower, A.K. 1995. Speculations on a spatial memory for the Port Jackson shark (*Heterodontus portusjacksoni*) (Meyer) (Heterodontidae). Mar. Freshwater Res. 46: 861–871.

Parker, G.H. 1914. The directive influence of the sense of smell in the dogfish. Bull. U.S. Bur. Fish. 33: 61–68.

Paulin, M.G. 1995. Electroreception and the compass sense of sharks. J. Theor. Biol. 174: 325–339.

Puzdrowski, R.L. & R.B. Leonard. 1994. Vestibulo-ocular connections in an elasmobranch fish the Atlantic stingray, *Dasyatis sabina*. J. Comp. Neurol. 339: 587–597.

Roberts, B.L. 1978. Mechanoreceptors and the behavior of elasmobranch fishes with special reference to the acoustico-lateralis system. pp. 331–390. *In:* E.S. Hodgson & R.F. Mathewson (ed.) Sensory Biology of Sharks, Skates and Rays, Office of Naval Research, Department of the Navy, Arlington.

Roberts, B.L. 1988. The central nervous system. pp. 99–142. *In:* T.J. Shuttleworth (ed.) Physiology of Elasmobranch Fishes, Springer-Verlag, Berlin.

Sciarrotta, T.C. & D.R. Nelson. 1977. Diel behavior of the blue shark (*Prionace glauca*) near Santa Catalina Island, California. U.S. Fish. Bull. 75: 519–528.

Sims, D.W. & V.A. Quayle. 1998. Selective foraging behaviour of basking sharks on zooplankton in a small-scale front. Nature 393: 460–464.

Taylor, J.G. 1996. Seasonal occurrence, distribution and movements of the whale shark, *Rhincodon typus*, at Ningaloo Reef, Western Australia. Mar. Freshwater Res. 47: 637–642.

Timerick, S.J.B. 1982. The response of the pectoral fin muscles of the dogfish (*Scyliorhinus canicula*) to stimulation of the labyrinth and long descending pathways. J. Physiol. (Lond.) 327: 64P.

Tricas, T.C. 1982. Bioelectric-mediated predation by swell sharks *Cephaloscyllium ventriosum*. Copeia 1982: 948–952.

Tricas, T.C. 2001. The neuroecology of the elasmobranch electrosensory world: why peripheral morphology shapes behavior. Env. Biol. Fish. 60: 77–92 (this volume).

Tricas, T.C., S.W. Michael & J.A. Sisneros. 1995. Electrosensory optimization to conspecific phasic signals for mating. Neurosci. Lett. 202: 129–132.

Ullen, F., T.G. Deliagina, G.N. Orlovsky & S. Grillner. 1995. Spatial orientation in the lamprey. J. Exp. Biol. 198: 675–681.

Walker, M.M. & M.E. Bitterman. 1989. Attached magnets disrupt magnetic field discrimination by honeybees. J. Exp. Biol. 141: 447–451.

Walker, M.M., C.E. Diebel, C.V. Haugh, P.M. Pankhurst, J.C. Montgomery & C.R. Green. 1997. Structure and function of the vertebrate magnetic sense. Nature 390: 371–376.

Wehner, R. 1996. Preface to volume Navigation Migration and Homing. J. Exp. Biol. 199: I.

Wiltschko, W. & R. Wiltschko. 1995. Magnetic orientation in animals. Springer-Verlag, Berlin. 297 pp.

Wiltschko, W. & R. Wiltschko. 1996. Magnetic orientation in birds. J. Exp. Biol. 199: 29–38.

Environmental Biology of Fishes **60:** 117–129, 2001.

Surfactants as chemical shark repellents: past, present, and future

Joseph A. Sisneros[a,c] & Donald R. Nelson[b]
[a]*Department of Biological Sciences, Florida Institute of Technology, Melbourne, FL 32901, U.S.A.*
[b]*Department of Biological Sciences, California State University, Long Beach, Long Beach, CA 90840, U.S.A.*
[c]*Present address: Department of Neurobiology and Behavior, Cornell University, Seeley G. Mudd Hall, Ithaca, NY 14853, U.S.A. (e-mail: jas226@cornell.edu)*

Received 26 January 1999 Accepted 21 February 2000

Key words: SDS, *Cephaloscyllium ventriosum*, shark chaser, pardaxin, dodecyl sulfate

Synopsis

The development of Shark Chaser by the U.S. Navy during World War II was the first serious effort to develop a chemical shark repellent. In the decade following the war reports of Shark Chaser ineffectiveness led the Office of Naval Research to search for a more efficacious shark repellent. After years without success, ONR eventually canceled the use of Shark Chaser and abandoned the search for a chemical shark repellent. In the early 1970s, interest in chemical shark repellents was renewed by the discovery of pardaxin, a natural shark repellent secreted by the Red Sea Moses sole, *Pardachirus marmoratus*. The surfactant-like nature of pardaxin led investigators to test the potential of various surfactants as repellents. Subsequent studies indicated that the shark repellent efficacy of the effective alkyl sulfate surfactants was due to their hydrophobic nature. Here we report tests conducted on juvenile swell sharks, *Cephaloscyllium ventriosum*, to determine if the noxious quality of alkyl sulfates is affected by surfactant hydrophobicity [carbon chain length and ethylene oxide (EO) groups] and counterions. Our results indicate that the aversive response of sharks to alkyl sulfate surfactants increases with carbon chain length from octyl to dodecyl, decreases with the addition of EO groups and is not affected by counterions. This study confirms that dodecyl sulfate is the most effective surfactant shark repellent, but it does not meet the Navy's potency requirement for a nondirectional surrounding-cloud type repellent of 100 parts per billion (0.1 $\mu g\ ml^{-1}$). Thus, dodecyl sulfate is only practical as a directional repellent such as in a squirt application. Future research should test the action of alkyl sulfates on cell membranes, the potential of other biotoxic agents, and semiochemicals in the search for an effective chemical shark repellent.

Introduction

Over the last 50 years various antishark measures were employed to protect humans from shark attack (Nelson 1983). Such antishark measures include electrical repellent devices (Gilbert & Springer 1963, Gilbert & Gilbert 1973), acoustical playbacks (Myrberg et al. 1978, Klimley & Myrberg 1979), visual devices (Doak 1974) and chemical repellents (Tuve 1963, Clark 1974, Gruber & Zlotkin 1982). Although none of these procedures are totally effective in preventing shark attacks, previous work on surfactant chemical shark repellents indicates some promise (Gruber et al. 1984, Zlotkin & Gruber 1984, Smith 1991, Sisneros 1993, Nelson & Strong 1996). In this paper we present a brief history of chemical shark repellents, report current work on the potential use of surfactants as shark repellents, and discuss new directions for the search of a truly effective chemical shark repellent.

Past chemical repellency research

The first serious effort to develop a chemical shark repellent did not occur until the early years of World War II. The conflict at sea and in the air in the Pacific campaign put many wounded servicemen and airmen in shark infested waters. Many deaths were attributed to shark attack as the war developed (Server 1989). The most infamous incident occurred after the 30 July 1945 Japanese torpedo attack which sunk the heavy crusier U.S.S. Indianapolis. This incident was one of the worst tragedies in U.S. naval history. In addition to the hundreds of deaths that occurred in the incident, an estimated 60–80 deaths were attributed to shark attacks on the remaining survivors who were adrift at sea for 5 days (Brown 1980). Subsequent speculation by U.S. servicemen on the threat of shark attacks became widespread enough to create a significant morale problem. In order to alleviate anxiety and renew confidence in service personnel, the U.S. Navy hurriedly began serious efforts to develop a shark repellent. The ultimate result of their rushed efforts was the development of the chemical packet known as Shark Chaser (Figure 1), sometimes more formally labeled Life Jacket Shark Repellent Compound Packet (Baldridge 1990).

Figure 1. Shark Chaser, the original U.S. Navy chemical 'shark repellent', was primarily a package of copper acetate and nigrosine dye issued to all naval personnel but later proven to be ineffective against repelling sharks. Scale bar = 5 cm.

Shark Chaser was composed of approximately 80% nigrosine-type black dye and 20% copper acetate held together by a waxy binder that allowed it to dissolve in a controlled manner over a period of 3–4 h (Tuve 1963). The special nigrosine dye component produced a black cloud visible to the user and was said to mimic the natural defensive secretions of marine mollusks such as the squid and octopus. The copper acetate component served two additional functions. First, it was recognized that copper ions inhibited the feeding of teleost fishes, and secondly, ammonium acetate, a major constituent in decomposing shark flesh, was also shown to be a feeding deterrent in the dogfish, *Mustelus canis* (Gilbert & Springer 1963, Tester 1963). It was therefore concluded that the combination of copper and acetate ions would produce an increased deterrent effect (Tuve 1963).

Due to the urgency at the time for a repellent, the testing of Shark Chaser's effectiveness was limited. Nevertheless, Shark Chaser soon became an integral component in every naval serviceman's survival gear. Following World War II, reports[1,2] of Shark Chaser's ineffectiveness began to appear. Such reports led the Office of Naval Research to reconsider the matter of chemical shark repellents in the 1950s and renew the screening and testing of possible candidate repellents (Zahuranec & Baldridge 1983). Hundreds of chemical substances were tested on sharks in an effort to find a chemical that would produce a quick and effective repellent response (Springer 1954, Gilbert & Springer 1963, Tester 1963). These chemicals included powerful toxins that could kill a shark after brief exposure but none could elicit the desired repellent response. After many fruitless attempts, research support for a chemical shark repellent ended.

Pardaxin and surfactant research

The discovery of a natural shark repellent in the early 1970s renewed interest in chemical shark repellents. Clark & Chao (1973) discovered that the milky fluid secretion from pores at the base of the dorsal and

[1] Fogelberg, J.M. 1944. Final report on the use of chemical materials as shark repellents. NRL report No. P-2373, Naval Research Laboratory, Anacostia Station, Washington, D.C. 28 pp.

[2] Fogelberg, J.M. & F.E. Brinnick. 1944. First partial report on the use of chemical materials as shark repellents. NRL report No. P-2230, Naval Research Laboratory, Anacostia Station, Washington, D.C. 35 pp.

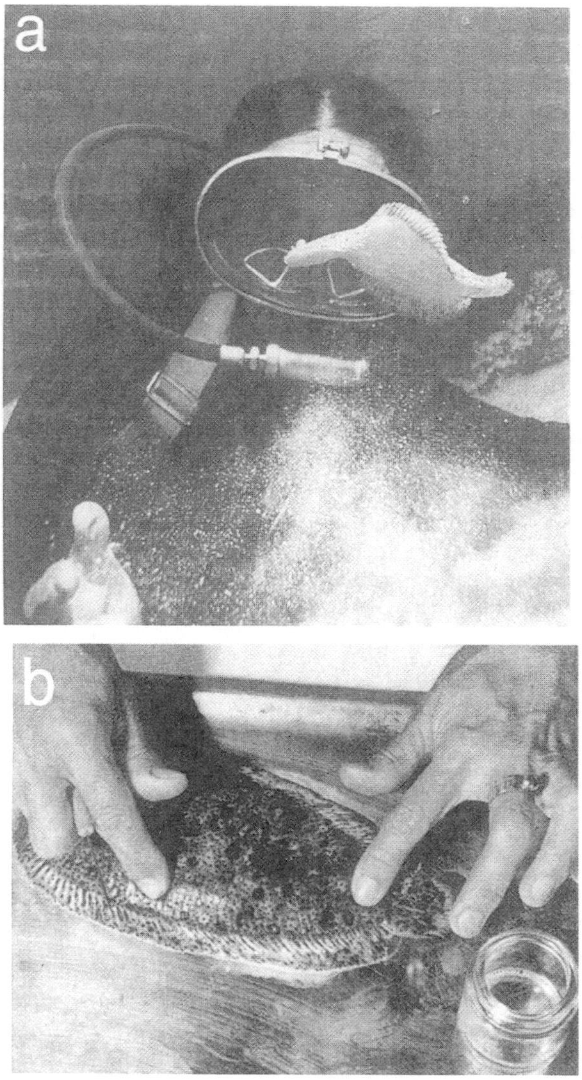

Figure 2. a – The Moses sole, *Pardachirus marmoratus*, was first demonstrated by Eugenie Clark (shown here) to produce a natural shark repellent that is both toxic and repellent to sharks. b – The Moses sole secretes a milky fluid known as pardaxin from poison glands located along the sole's anal and dorsal fins (photographs courtesy of David Doubilet).

anal fins of the Red Sea Moses sole, *Pardachirus marmoratus*, was both toxic and repellent to teleost fishes (Figure 2). Their initial observations led to a series of experiments in which Clark (1974, 1983) demonstrated the repellent effects of the secretion on captive reef whitetip sharks, *Triaenodon obesus*, and on Red Sea reef sharks, *Carcharhinus amblyrhinchus* (=*C.wheeleri*), in open water. Subsequent studies by Clark & George (1979) also revealed similar defense secretions produced by the Japanese peacock sole, *Pardachirus pavoninus*. The active ingredient of the secretion produced by *P. marmoratus* was later isolated and identified as pardaxin, an acidic protein composed of 162 amino acids with an approximate molecular weight of 17 kDa (Primor et al. 1978). Biochemical studies revealed that pardaxin affected a number of cellular targets and pathways including the neuromuscular systems of frogs (Spira et al. 1976), the gills of fishes (Primor et al. 1980) and the ion transport systems of dogfish (Primor et al. 1984).

Although pardaxin was an effective shark repellent, there were three major drawbacks to its practical use: (1) sufficient quantities were unavailable from natural sources, (2) at the time it was very difficult to synthesize, and (3) it had an extremely short shelf-life unless it was freeze-dried which reduced its potency. However, structure–activity analyses of pardaxin revealed promising alternatives to its use. Sequence analyses of the first 10 amino acids of the N-terminal primary structure of pardaxin revealed that it is extremely hydrophobic, positively charged and of amphiphatic character (Zlotkin & Barenholz 1983). Observations performed during the process of purification indicated that pardaxin possessed surface active (surfactant-like) properties by its ability to cause strong foaming and reduce water surface tension (Parness 1975). It was therefore suggested that relatively inexpensive synthetic surfactants that possess characteristics of pardaxin should be screened and tested as chemical shark repellents (Zlotkin & Barenholz 1983).

Subsequent studies revealed that certain synthetic surfactants do possess significant shark repellent properties (Gruber et al. 1984, Zlotkin & Gruber 1984, Smith 1991). In these studies, investigators tested a wide variety of different anionic, cationic and nonionic synthetic surfactants (Gruber et al. 1984, Zlotkin & Gruber 1984). Of these various surfactants, the anionic alkyl sulfate sodium dodecyl sulfate (SDS) was the most effective in repelling sharks. SDS was found to be five times more ichthyotoxic and four times more repellent to lemon sharks, *Negaprion brevirostris*, than a typical freeze-dried secretion of pardaxin (Gruber et al. 1984). The shark repellent efficacy of SDS was hypothesized to be due to the formation of large amphiphilic monolayers of surfactant alkyl chains that interact at the hydrophobic surface of lipid bilayers on buccal and gill epithelia (Tachibana & Gruber 1988). This chemical mode of action of SDS is presumably similar to pardaxin and other closely related defensive skin secretions such mosesins and pavoninins, which are also postulated to have a hydrophobic interaction with lipid membranes of buccal and gill epithelia, and ultimately elicit the repellent response (Tachibana & Gruber 1988). Other studies also suggested that the hydrophobicity of surfactants affect both the toxic and irritation effects in teleost fish (Mann 1955, Herbert et al. 1957, Pickering 1970, Parker 1971). Parker (1971) reported an increased rate of respiratory movements and air gulping in goldfish as the number of ethylene oxide (EO) groups on nonionic surfactants decreased. This increased respiratory distress was hypothesized to be due to an increase in hydrophobicity of surfactants with fewer EO groups. Thus, if shark repellency is related to the hydrophobic nature of anionic surfactants then we predict that the irritant/repellent efficacy of alkyl sulfate surfactants should increase with surfactant hydrophobicity.

Here we present the test results of seven homologous alkyl sulfate surfactants assayed as irritants/repellents on the swell shark, *Cephaloscyllium ventriosum*. The major goal of this study was to determine the effect of surfactant hydrophobicity on the shark irritant efficacy of alkyl sulfate surfactants. We investigate how carbon chain length and EO groups affect hydrophobicity and irritant efficacy. In addition, we also test whether counterions which influence surfactant solubility affect the irritant efficacy of alkyl sulfates.

Methods

Thirty-seven juvenile swell sharks, *Cephaloscyllium ventriosum*, 20.4–36.1 cm total length (TL) ($\bar{x}$ = 26.2 ± 2.9 SD cm TL) were used in tests conducted between August 1992 and April 1993. Juvenile sharks were raised from egg cases oviposted by captive adults or obtained from local collectors and public aquaria. Sharks were maintained in temperature controlled aquaria (15.0–20.0°C) and fed 2–3 times a week an *ad libitum* diet of squid, anchovies, mackerel and shrimp.

All tests were performed in a custom built roundabout test tank similar to that used by Smith (1991). The experimental tank (Figure 3) consisted of an oval channel and a removable inner holding pool built on a rectangular 1.2 × 2.4 m plywood base. The tank was covered with an inner coat of polyester laminating resin and an outer coat of light gray, gel-coat resin

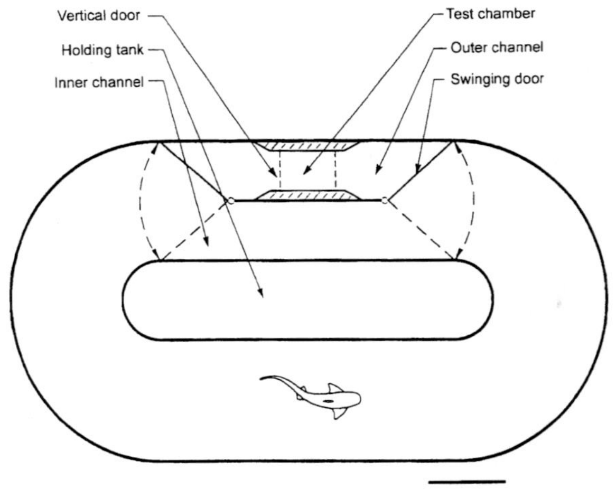

Figure 3. Experimental tank used to test surfactants as chemical shark repellents on swell sharks, *Cephaloscyllium ventriosum*. The outer channel and test chamber were used during tests while the holding tank was used prior to tests to acclimate sharks to water temperature. Swinging doors were used to direct sharks to either the inner or outer channels while the vertical doors were used to isolate chemical solutions prior to tests. Scale bar = 30 cm.

to form a smooth surface. A 5 l test chamber (20.0 cm long × 12.5 cm wide × 20.0 cm deep) made of 6.4 mm clear Plexiglass was built into the outer channel raceway (Figure 4). Sharks were directed into either the inner channel or outer channel, which contained the test chamber, by two swinging doors. The test chamber had two vertical, watertight doors which could be raised for presentations. This arrangement permitted the chamber to be isolated from the main raceway while chemicals were prepared prior to tests.

The primary test chemicals consisted of anionic surfactants sodium octyl sulfate (SOS), sodium decyl sulfate (SDecS), sodium dodecyl sulfate (SDS), sodium tetradecyl sulfate (STDS) and sodium octadecyl sulfate (SODS) (Figure 5a). SDS in 99% pure powder form (Sigma, St. Louis, MO); aqueous solutions of SDS (30% by volume), SOS (33%) and SDecS (31%) (Henkel Co., Hoboken, NJ); powdered STDS (95% pure) and SODS (93% pure) (Aldrich Chemical Co., Milwaukee, WI); and aqueous solutions of sodium dodecyl ether sulfate (SDES) (29% by volume) and magnesium dodecyl sulfate (MDS) (30%) (Figure 5b) (Lonza Inc., Long Beach, CA) were tested. Surfactants were either diluted or added to filtered seawater to produce a 1% (by volume) stock solution that was then added to the test chamber to achieve the desired concentration.

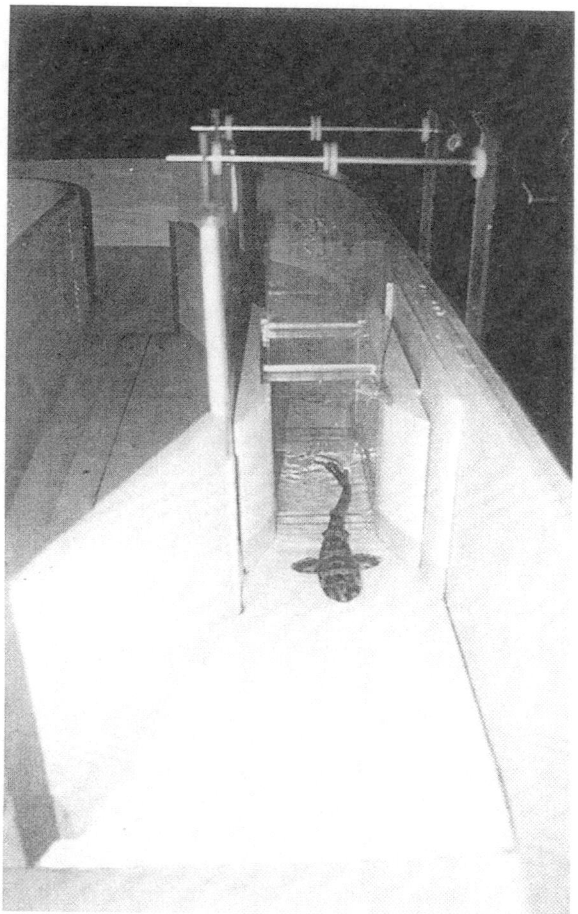

Figure 4. Test chamber used to test surfactants as chemical shark repellents on swell sharks, *Cephaloscyllium ventriosum.*

Sharks were exposed to a geometric series of chemical concentrations (Table 1) to determine the threshold concentration for each surfactant as an effective irritant/repellent. Initial concentrations for each geometric series were determined in preliminary trials. In this study sharks were exposed to only one chemical concentration per night (i.e. a single swim-through test per night) to reduce desensitization or habituation. After each test, sharks were returned to aquaria and not tested for a minimum period of at least 7 days. Due to the nocturnal nature of swell sharks, all trials were conducted at night under the illumination of dim red light.

Sharks in each night experimental session were placed in the inner pool of the tank and acclimated to the tank temperature (17.0–20.5°C) for 30 min. For each trial, a shark was randomly selected from the inner holding tank, placed in the roundabout, and allowed to swim for a minimum of 20 min before the experimental procedure was initiated. This pretest acclimation allowed the shark to adjust to a normal swimming motion.

The experimental procedure began with first isolating the test chamber by closing the vertical doors and moving the swinging doors to direct the shark through the inner channel. The appropriate amount of treatment chemical was then added to the water in the test chamber and stirred to obtain the desired concentration. After the test chemical was mixed, the test chamber solution was allowed to settle for 2–5 min before a trial was run. An experimental trial began when the swinging doors were moved to direct the shark to the outer channel. The vertical doors of the test chamber were raised when the animal approached to within approximately one body length of the chamber. Immediately after the shark exited the test chamber, the vertical doors were closed to isolate the chamber and minimize outward spread of the test chemical. Immediately following each trial, the shark was returned to its aquarium, the vertical doors of the test chamber were lifted and the entire tank emptied through a drain in the tank floor. The tank was then rinsed and refilled with fresh seawater and prepared for the next trial.

All trials were recorded on videotape using a Sony EV-C8 recorder and a CV110 Precision black-and-white video camera (0.2 lux) mounted approximately 2 m above the test chamber. Behavioral responses were classified into two levels of unconditioned aversive responses (see Results) based upon Smith (1991). The effective concentration for 50% of the test population (EC_{50}) for each chemical was determined using the Spearman–Karber method (Hubert 1980) and represents the concentration at which 50% of the population shows a categorical response. A one-way ANOVA with a Neuman–Keuls test and two-tailed *t*-tests (Sokal & Rohlf 1981) were used to test for differences between the surfactant EC_{50}s for each category.

Results

Response levels

Two distinct levels of unconditioned averse responses were identified in sharks exposed to the alkyl sulfate surfactants, and are similar to the classification used by Smith (1991). Level 1 was defined as a minimum aversion response (MAR) to the surfactant. The MAR consisted of a 'cough' or an expulsion

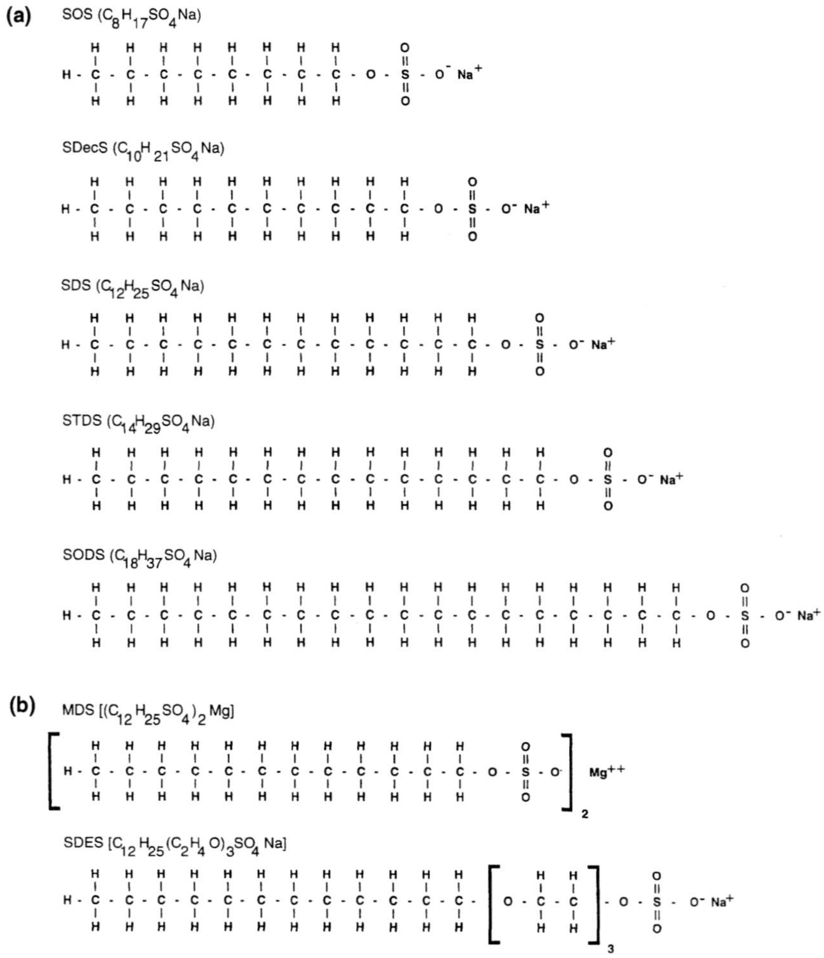

Figure 5. a – Chemical structures of the primary test chemicals: SOS, SDecS, SDS, STDS, SODS. Note that the abbreviation SDecS is used for sodium decyl sulfate to avoid confusion with the standard abbreviation of SDS for sodium dodecyl sulfate. b – Chemical structures of MDS and SDES. Note that these surfactants share the same alkyl carbon chain length of 12 carbons with SDS. MDS possess the counterion magnesium while SDES has a sodium counterion. Note that SDES has 3 ethylene oxide groups attached to the alkyl carbon chain and sulfate group.

reflex (Satchell & Maddalena 1972), and/or headshaking accompanied by a slight forward acceleration. The ‘cough’ response was considered to be a modified exhalation that forcefully expelled water from the mouth, gill slits and spiracles. Level 2, the vigorous aversion response (VAR), consisted of rapid and exaggerated headshaking, mouth gaping, and either a rapid acceleration forward or a 180° turnaround. The ‘mouth gaping’ response was characterized by a repeated opening and closing of the mouth together with large buccal and opercular expansions, presumably to flush out the buccal cavity.

Effect of alkyl carbon chain length on aversion

The irritant efficacy of SOS, SDecS, SDS, STDS and SODS was tested on 30 sharks at various chemical concentrations (Table 1) to determine the effect of surfactant hydrophobicity (which is proportional to carbon chain length) on irritant efficacy. Results show that the shark irritant efficacy of the alkyl sulfate surfactants increased as carbon chain length increased from octyl to dodecyl (Figure 6). The EC_{50}s for the MAR were 2028.1 $\mu g\,ml^{-1}$ for SOS, 175.9 $\mu g\,ml^{-1}$ for SDecS, and 36.4 $\mu g\,ml^{-1}$ for SDS. The EC_{50} of

Table 1. Concentrations of alkyl sulfate surfactants tested on juvenile swell sharks, *Cephaloscyllium ventriosum*. Note that a geometric series of concentrations was used to determine the effective dose for each surfactant.

Surfactant	Concentrations ($\mu g\,ml^{-1}$)
SOS	320, 640, 1280, 2560, 5120
SDecS	80, 160, 320, 640
SDS	20, 40, 80, 160
STDS	20, 40, 80, 160, 320, 640
SODS	20, 40, 80, 160, 320, 640
MDS	20, 40, 80, 160, 320
SDES	20, 40, 80, 160, 320

SDS was approximately 1/50th that of SOS and 1/5 that of SDecS (one-way ANOVA, Neuman–Keuls test, $p < 0.001$). The EC_{50}s for the VAR were 374.6 $\mu g\,ml^{-1}$ for SDecS, 82.6 $\mu g\,ml^{-1}$ for SDS, and inconclusive for SOS due to the lack responses at concentrations as high as 5120 $\mu g\,ml^{-1}$. The estimated 95% confidence limits (95% CL) and the number of sharks used to estimate MAR and VAR EC_{50}s are shown in Table 2. For the VAR, the EC50 of SDS was approximately 1/5 that of SDecS (two-tailed t-test, df $= 20$, $p < 0.01$). We were unsuccessful in a number of attempts to test the irritant efficacy of STDS and SODS due to the lack of solubility of these chemicals in seawater. In the 14 trials conducted using STDS and SODS no aversive responses were observed at concentrations as high as 640 $\mu g\,ml^{-1}$. In addition, no responses were observed in any of the 110 control trials performed. These results show that the intensity of the aversive response to alkyl sulfate surfactants is associated with increased carbon chain length and hydrophobicity (from octyl to dodecyl) until the surfactants become insoluble in seawater.

Effect of EO groups on aversion

The irritant efficacy of SDES was also tested on 5 sharks at 4 different chemical concentrations (Table 1) to determine the effect of hydrophobicity on the shark irritant efficacy of alkyl sulfates. SDES has the same number of carbon atoms (12) as SDS but has 3 additional EO groups that link the carbon chain to a sulfate group (Figure 5b). This arrangement of the EO group attached to the carbon chain effectively reduces the hydrophobicity of the surfactant (Satkowski et al. 1967, Kastner 1980), and the results show that SDES is less effective in producing aversion responses in sharks than SDS. The MAR EC_{50} of SDES was 67.3 $\mu g\,ml^{-1}$

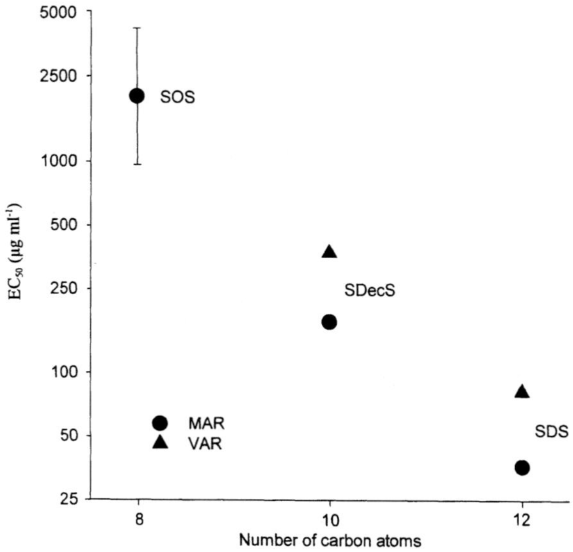

Figure 6. Relationship between surfactant EC_{50} of the MAR and VAR and alkyl carbon chain length in juvenile swell sharks, *Cephaloscyllium ventriosum*. EC_{50} represents the effective concentration for 50% of the test population for each chemical determined using the Spearman–Karber method (Hubert 1980). Note EC_{50} decreases as number of carbon atoms in the alkyl molecule increases. Standard errors for EC_{50}s are plotted but for most points are obscured by symbols.

($n = 5$, 95% CL $=$ 63.8–70.9 $\mu g\,ml^{-1}$) and for the VAR 113.1 $\mu g\,ml^{-1}$ ($n = 5$, 95% CL $=$ 106.5–120.2 $\mu g\,ml^{-1}$). For the MAR, the EC_{50} of SDES was approximately twice that of SDS (two-tailed t-test, df $= 13$, $p < 0.01$). For the VAR, the EC_{50} of SDES was approximately 1.4 times that of SDS (two-tailed t-test, df $= 13$, $p < 0.05$). No responses were observed in any control trials ($n = 25$). Thus, these results confirm that an increase in surfactant hydrophobicity does increase the shark irritant efficacy of alkyl sulfates.

Effect of counterions on aversion

The irritant efficacy of MDS was tested on 5 sharks at 4 different chemical concentrations (Table 1) to determine if the surfactant counterion affects the irritant efficacy of alkyl sulfates. MDS has the same carbon chain length (12 carbon atoms) as SDS but has the counterion magnesium instead of sodium (Figure 5b). Results show that the irritant efficacy did not differ between MDS and SDS. The EC_{50} of MDS for the MAR was 37.3 $\mu g\,ml^{-1}$ (n $= 5$, 95% CL $=$ 35.5–39.3 $\mu g\,ml^{-1}$) and 85.7 $\mu g\,ml^{-1}$ (n $= 5$, 95% CL $=$

Table 2. MAR and VAR for juvenile swell sharks, *Cephaloscyllium ventriosum*, exposed to alkyl sulfate surfactants.

Test surfactant	MAR		VAR		n
	EC_{50} ($\mu g\ ml^{-1}$)	95% CL ($\mu g\ ml^{-1}$)	EC_{50} ($\mu g\ ml^{-1}$)	95% CL ($\mu g\ ml^{-1}$)	
SOS	2028.1	970.0–4256.8	—	—	6
SDecS	175.9	168.4–183.6	374.6	363.8–385.8	11
SDS	36.4	35.2–37.6	82.6	79.5–85.7	11

EC_{50} represents the effective concentration for 50% of the test population for each chemical (Hubert 1980) and is the concentration at which 50% of the population shows a categorical response. The 95% confidence limit (95% CL) is reported for each EC_{50}. MAR is the minimal noticeable response to the surfactant while VAR is more vigorous.

81.5–90.2 $\mu g\ ml^{-1}$) for the VAR. There was no difference between the EC_{50}s of MDS and SDS for either the MAR (two-tailed t-test, df $= 14$, $p < 0.90$) or VAR (two-tailed t-test, df $= 14$, $p < 0.85$). No responses were observed in any control trials ($n = 33$). Thus, these results are consistent with the notion that counterions do not affect the irritant efficacy of alkyl sulfates.

Discussion

This study is the first to investigate the effects of surfactant hydrophobicity on the shark irritant efficacy of alkyl sulfate surfactants. The goal of this study was to determine the optimum carbon chain length in alkyl sulfate surfactant molecules that produces the greatest aversive response in sharks. Our results show that shark irritant efficacy increases as the carbon chain of alkyl sulfates increases from octyl to dodecyl. Alkyl sulfates with carbon chain lengths greater than 12 (dodecyl) were insoluble under the test conditions of this study and thus did not produce any aversive responses. Additionally, we show that other factors that can influence surfactant solubility such as counterions did not affect the shark irritant efficacy of alkyl sulfates. In this discussion we interpret our results as they relate to shark irritant efficacy and chemical structure of alkyl sulfates and discuss future directions for the search of a more effective chemical shark repellent.

The important finding of this study is that the shark irritant efficacy of alkyl sulfate surfactants increases as carbon chain length increases from 8 (octyl) to 12 (dodecyl) carbons. Alkyl sulfates that have carbon chains greater than 12 are not soluble in seawater at testable temperatures between 17°C and 27°C. Thus, the 12-carbon hydrophobic chain of SDS is the optimum chain length that produces the greatest aversion response in sharks. Similar studies that have investigated the irritating effects of alkyl sulfates (ranging from octyl to octadecyl) on skin tissue have shown that dodecyl sulfates produce the greatest irritation and potential damage (Choman 1963, Imokawa et al. 1975, Bartnik & Kunstler 1987). A similar structure–effect relationship of alkyl carbon chain length is observed with mucous membranes where maximum irritation occurs at carbon chain lengths between decyl sulfate and tetradecyl sulfate (Kastner 1980). Schott (1973) hypothesized that the maximum irritation of skin and mucous membranes by octyl to octadecyl alkyl sulfates occurs at dodecyl due to two opposing factors inherit in these surfactants: specifically the oil/water partition coefficient and the limiting monomer concentration. Together these two factors are thought to control the amount of surfactant molecules available to interact and affect the lipid components of sensitive tissues and membranes. Skin and mucous membrane irritation increases with increases in the oil/water partition coefficient (i.e. carbon chain length). However, maximum monomer (single unclustered surfactant molecule) concentration in alkyl sulfates decreases with increases in the oil/water partition coefficient. Thus alkyl sulfates with longer carbon chains should have fewer monomers in solution available to interact with hydrophobic lipid membranes. As a result of these two opposing factors, peak skin and membrane irritation is hypothesized to occur at or near the optimized carbon chain length of 12 in alkyl sulfates. Thus our results support Schott's (1973) hypothesis since the greatest shark aversion response occurs with dodecyl sulfate in the alkyl sulfates tested in this study.

Our results also show that ethylene oxide (EO) groups effectively reduce shark irritant efficacy. SDES was approximately 1.8 and 1.4 times less effective than SDS in evoking both MAR and VAR, respectively. The difference in irritant efficacy between SDES and SDS is most likely due to their differences in surfactant hydrophobicity. SDES has three EO groups attached to the 12-carbon alkyl chain while SDS lacks EO groups. The attachment of the EO groups is known to effectively reduce the hydrophobicity of the surfactant (Satkowski et al. 1967, Kastner 1980). Similar studies have also shown that increasing levels of ethoxylation decrease the skin-irritating properties of alkyl ether sulfates (Opdyke & Burnet 1965). Parker (1971) reported that goldfish showed increased respiratory distress and accelerated rates of respiratory movements as the number of EO groups decreased. Parker attributed these effects to the increase in hydrophobicity as EO groups decreased and argued that the increase in hydrophobicity increased the affinity of the alkyl carbon chain to the lipid components of gill membranes. Thus our results that increased hydrophobicity increases the shark irritant efficacy of alkyl sulfate surfactants agrees with previous studies.

One chemical characteristic that does not affect the shark irritant efficacy of alkyl sulfates is the counterion. There was no difference between MDS (counterion = magnesium) and SDS (counterion = sodium) at either the MAR or VAR. Zlotkin & Gruber (1984) reported similar results with lithium dodecyl sulfate and SDS assayed on lemon sharks, *Negaprion brevirostris*. We also observed similar results in limited tests with ammonium dodecyl sulfate and SDS which were both equally effective in evoking strong responses in adult swell sharks (unpublished data). Although the results demonstrate that counterions do not affect irritant efficacy at the temperatures (17.0–20.5°C) tested in this study, we did notice that a 1% solution of SDS would begin to form surfactant crystals and become insoluble in seawater below 17°C. In contrast, we never did observe surfactant crystal formation in a 1% solution of MDS in seawater at temperatures as low as 12°C. This difference in surfactant solubility at lower temperatures becomes significant when testing these chemicals under various field conditions, especially in the cool temperate waters of Pacific Ocean. Thus based on our results and observations, we suggest MDS may be a more practical shark irritant/repellent than SDS due the higher solubility of MDS in seawater at lower temperatures.

In summary, we have shown that the shark irritant efficacy of alkyl sulfate surfactants increases as the alkyl carbon chain increases from octyl to dodecyl, decreases with the addition of EO groups and is not affected by counterions. However, we suggest that MDS has the potential to be a more practical chemical shark repellent than SDS due to its higher solubility in seawater at lower temperatures. This study also confirms past studies that dodecyl sulfate is the most effective surfactant irritant/repellent tested to date. The MAR EC_{50} we report for SDS (36.4 $\mu g\,ml^{-1}$) is very similar to the effective concentration of SDS (35.6±3.4 SD $\mu g\,ml^{-1}$) needed to terminate tonic immobility in lemon sharks, *Negaprion brevirostris* (Gruber et al. 1984, Zlotkin & Gruber 1984). In addition, the VAR response reported for SDS and the other alkyl sulfates tested in this study was very similar to the responses observed by us in blue sharks, *Prionace glauca*, which were effectively repelled in field tests (unpublished data). However, SDS still does not meet the potency requirement for a nondirectional surrounding-cloud type repellent of 100 parts per billion (0.1 $\mu g\,ml^{-1}$) set forth by Johnson & Baldridge[3] for the U.S. Navy. The potency range of SDS reported in our study and by Smith (1991) indicates that it could be used as a directional repellent such as in a squirt application. One device that has been suggested for divers is a 'chemically enhanced' shark billy club (Figure 7) that could be used initially as a billy club, but then employ a chemical squirt of dodecyl sulfate if necessary (Nelson 1991, Nelson & Strong 1996). Thus this study supports early findings that alkyl sulfates are impractical as a nondirectional surrounding-cloud type repellent.

Future direction of chemical shark repellent research

Future candidate chemical shark repellents will need to meet the criteria set forth by Johnson & Baldridge[3] and be effective at a concentration no greater than 0.1 $\mu g\,ml^{-1}$ to be considered practical as a classical, nondirectional, surrounding-cloud type repellent. The 0.1 $\mu g\,ml^{-1}$ effective concentration represents the concentration at the boundary of a protected water volume of about 6 m^3 maintained by release from a central point source at approximately 100 $g\,h^{-1}$ over a 3.5 h period

[3] Johnson, C. S. & H. D. Baldridge. 1985. Analytic indication of the impracticability of waterborne chemicals for repelling an attacking shark. A second, confirming look. Technical document 843, Naval Oceans Systems Center, San Diego. 12 pp.

Figure 7. 'Chemical enhanced' shark billy club developed by D.R. Nelson that can deliver a rapid squirt of sodium dodecyl sulfate (SDS) to repel sharks. a – The device contains 250 ml of the SDS test solution to the right of the piston (P) and 60 psi of compressed air to the left of the piston. The device (122 cm in length) is discharged by a thumb operated valve that allows the compressed air to drive the cylinder piston to the right, thereby pushing surfactant solution out of the barrel in approximately 1 s (photograph courtesy of W.R. Strong). b – Shark researcher W.R. Strong tests the 'chemical enhanced' shark billy club on a bait-attracted white shark from the stern platform of the R/V *Alcyone* (photograph courtesy of the Cousteau Society).

under steady state conditions (Johnson & Baldridge[3]). Since surfactants are an order of magnitude less effective than the target concentration needed for a practical repellent, we suggest that other areas of research should be examined and investigated in future research. There are three potential areas of research that need to be addressed in chemical shark repellency research. These areas are the determination of the morphological target sites of chemical and natural shark repellents, further investigation of natural bioactive toxins as shark repellents, and the identification and purification of possible semiochemicals to be used as possible shark repellents. We identify and discuss three hypotheses that should be tested in future work.

The membrane lipid layer hypothesis

The site of action and biochemical mode of chemical shark repellents in elasmobranchs is one area of research that has been neglected in previous work. Various hypotheses have been proposed for the site of action and biochemical mode of pardaxin and alkyl sulfates (Primor et al. 1983 a,b, Smith 1986, Tachibana & Gruber 1988). Primor et al. (1983 a,b) hypothesized that the target sites of pardaxin were the membranal lipid components of the gills. Similarly, Tachibana & Gruber (1988) also proposed that the lipid membranes of buccal and gill epithelia and/or some unknown shark sense organ(s) were the targets affected by pardaxin and alkyl sulfates. In contrast, Smith (1986) speculated that the numerous 'pits', presumably chemoreceptors, located over the entire internal structure of the buccal cavity in swell and horn sharks were the targets of alkyl sulfates. Clearly, experimental studies that test these hypotheses will need to be performed before a truly effective chemical shark repellent can be developed. Future physiological studies that directly test the function of elasmobranch buccal 'pits' will need to be performed to determine how chemical and natural substances influence repellent behavior in sharks. The lipid membranes found in the gill and buccal cavity of elasmobranchs should also be examined to determine if the mode of action of pardaxin and alkyl sulfates is due to either a general lytic effect or due to the interaction with specific receptors that mediate sensory input for pain, olfaction, or gustation. Thus, future experiments that test this hypothesis will provide a better understanding of how shark repellent behavior is biochemically mediated and possibly lead to the better design of future chemical shark repellents.

The biotoxin hypothesis

The identification and testing of natural bioactive toxins from marine organisms as potential shark repellents is an area of research that should be further investigated. This area of research led to the discovery of the natural shark repellent pardaxin and a series of closely related ichthyotoxins known as pavoninins from the peacock sole, *Pardachirus pavoninus* (Primor et al. 1978, Clark & George 1979, Tachibana et al. 1984, Thompson et al. 1986). One ichthyotoxin that has received little attention as a potential shark repellent is saponin. Saponins also known as holothurins are ichthyotoxins that were first identified in sea cucumbers which use these toxins to protect themselves from being eaten by predators. Only crude tests regarding the toxicity of saponins have been conducted using sharks. Sobotka (1965) reported that a 1 $\mu g\,ml^{-1}$ solution of saponin (i.e. holothurin) collected from the sea cucumber *Actinopyga agassizi* killed a 22 kg lemon shark within 50 min. More recently, a naturally occurring saponin has been synthesized and identified in a class of mosesins isolated from the Red Sea Moses sole, *Pardachirus marmoratus* (Gargiulo et al. 1989). Hence, saponins are potential candidate repellents that should be screened and tested but to date have not been rigorously assayed for shark repellency. In addition to sea cucumbers, there is a plethora of other toxic marine organisms that have been tested for toxicity but have yet to be screened as shark repellents (Bakus 1983). Thus, the further investigation of the chemical defensive toxins of marine organisms may lead to the discovery of more efficacious shark repellents.

The semiochemical hypothesis

Perhaps the most encouraging area of shark repellency research is in the study of semiochemistry. This area of research was first proposed by Rasmussen & Schmidt (1992) who suggested that sharks may be chemically aware of the presence of potential danger by sensing the bodily secretions from potential predators. Semiochemicals found in the bodily secretions of predators may convey survival information to a shark and elicit rapid flight from an area that is potentially dangerous. Rasmussen & Schmidt (op. cit.) hypothesized that lemon sharks, especially juveniles, inherently recognize chemical exudates produced by the American crocodile, *Crocodylus acutus*, a known predator of sharks. Further, they demonstrated that the lemon shark, *Negaprion brevirostris*, shows aversive responses to 3 identified crocodile exudates (2-ethyl-3-methylsuccinimide, 2-ethyl-3-methylmaleimide, and 3-ethylidene-4-methylpyrrolidine-2,5-dione) produced most likely from the crocodile's chin gland, feces and blood. The concentrations needed to produce aversive responses in lemon sharks ranged from 10^{-7} to 10^{-9} M, which is near the functional limit of shark chemoreceptors (Hodgson & Mathewson 1978). Another proposed potential source for shark repellent semiochemicals may perhaps be found in decomposing shark flesh (Baldridge 1990, Rasmussen & Schmidt 1992). Anecdotal information exists from fishermen who claim that sharks will avoid areas containing decomposing carcasses of previously caught sharks. This was the reason why the U.S. Navy originally included acetate in Shark Chaser primarily due to the fact that ammonium acetate was found to be the major constituent in decomposing shark flesh. Perhaps there are semiochemicals found in extremely low concentrations in decaying shark flesh that act as alarm pheromones and provide warning signals to nearby sharks. Thus, in the search for shark repellent semiochemicals the re-examination of old ideas and the pursuit of new ones may offer the best hope for the discovery of a practical chemical shark repellent.

Acknowledgements

We thank John Reider for artwork, Michael Martinez and Jim McKibben for laboratory assistance, Shane Anderson, Jeff Landesman, John Rupp, and Mike Shaw for providing some of the animals used in this study, A.Z. Mason and R.N. Bray for comments on an earlier draft of this manuscript, and T.C. Tricas for technical comments on the final version of the manuscript. The CSULB Richard B. Loomis Memorial Fund provided partial support for this study and a student travel award from the American Elasmobranch Society (AES) enabled the author (J.A.S.) to present this paper at the AES Nelson Symposium in July 1998. This work was in partial fulfillment of a Master degree in Science for J.A.S. in the department of Biological Sciences at California State University, Long Beach.

References cited

Baldridge, Jr., H.D. 1990. Shark repellent: not yet, maybe never. Milit. Med. 155: 358–361.

Bakus, G.J. 1983. Toxicity in shallow marine waters: potential for developing shark repellents. pp. 185–199. *In*: B.J. Zahurance (ed.) Shark Repellents from the Sea: New Perspectives, Westview Press, Boulder.

Bartnik, F. & K. Kunstler. 1987. Biological effects, toxicology, and human safety. pp. 475–503. *In*: J. Falbe (ed.) Surfactants in Consumer Products: Theory, Technology, and Application, Springer-Verlag, New York.

Brown, C. 1980. Survivors of USS Indianapolis still relive their ordeal. Los Angeles Times, 27 Aug.: 9(1A).

Choman, B. 1963. Determination of the response of skin to chemical agents by an *in vitro* procedure. J. Invest. Dermatol. 40: 177–182.

Clark, E. 1974. The Red Sea's sharkproof fish. Nat. Geogr. Mag. 145: 718–727.

Clark, E. 1983. Shark repellent effect of the Red Sea Moses sole. pp. 135–150. *In*: B.J. Zahurance (ed.) Shark Repellents from the Sea: New Perspectives, Westview Press, Boulder.

Clark, E. & S. Chao. 1973. A toxic secretion from the Red Sea flatfish, *Pardachirus marmoratus* (Lacepede). Bull. Sea Fish. Res. St. (Haifa) 60: 53–56.

Clark, E. & A. George. 1979. Toxic soles, *Pardachirus marmoratus* from the Red Sea and *P. pavoninus* from Japan, with notes on other species. Env. Biol. Fish. 4: 103–123.

Doak, W. 1974. Revolutionary weapon – anti-shark wet suit. Skin Diver Mag. (June): 86–89.

Gargiulo, D., T.A. Blizzard & K. Nakanishi. 1989. Synthesis of mosesin-4, a naturally occurring steroid saponin with shark repellent activity, and its analog 7-β-galactosyl ethyl cholate. Tetrahedron 45: 5423–5432.

Gilbert, P.W. & C. Gilbert. 1973. Sharks and shark deterrents. Underwater Journal 5: 69–79.

Gilbert, P.W. & S. Springer. 1963. Testing shark repellents. pp. 477–494. *In*: P.W. Gilbert (ed.) Sharks and Survival, D.C. Heath and Company, Boston.

Gruber, S.H. & E. Zlotkin. 1982. Bioassay of surfactants as shark repellents. Nav. Res. Rev. 34: 18–27.

Gruber, S.H., E. Zlotkin & D.R. Nelson. 1984. Shark repellents. pp. 26–42. *In*: L. Bolis, J. Zadunaisky & R. Gilles (ed.) Toxins, Drugs, and Pollutants in Marine Animals, Springer-Verlag, New York.

Herbert, D.W.M., G.H. Elkins, H.T. Mann & J. Hemens. 1957. Toxicity of synthetic detergents to rainbow trout. Wat. Wast. Treat. J. 6: 394–397.

Hodgson, E.S. & R.F. Mathewson. 1978. Electrophysiological studies of chemoreception in elasmobranchs. pp. 227–267. *In*: E.S. Hodgson & R.F. Mathewson (ed.) Sensory Biology of Sharks, Skates, and Rays, Office of Naval Research, U.S. Government Printing Office, Washington, D.C.

Hubert, J.J. 1980. Bioassay. Kendall/Hunt Publishing Co., Dubuque. 164 pp.

Imokawa, G., K. Sumura & M. Katsumi. 1975. Study on skin roughness caused by surfactants: II. Correlation between protein denaturation and skin roughness. J. Am. Oil Chem. Soc. 52: 484–488.

Kastner, W. 1980. Anionic surfactants. pp. 139–150. *In*: C. Gloxhuber (ed.) Anionic Surfactants, Biochemistry, Toxicology, Dermatology, Surfactant Sci. Ser. Vol. 10, Marcel Dekker, New York.

Klimley, A.P. & A.A. Myrberg, Jr. 1979. Acoustic stimuli underlying withdrawal from a sound source by adult lemon sharks, *Negaprion brevirostris* (Poey). Bull. Mar. Sci. 29: 447–458.

Mann, H. 1955. The action of surface active washing agents on fish and animals forming fish food. Arch. Fischwiss. 6: 131–137.

Myrberg, A.A., Jr., C.R. Gordon & A.P. Klimley. 1978. Rapid withdrawal from a sound source by open-ocean sharks. J. Acoust. Soc. Am. 64: 1289–1297.

Nelson, D.R. 1983. Shark attack and repellency research: an overview. pp. 11–74. *In*: B.J. Zahuranec (ed.) Shark Repellents from the Sea: New Perspectives, Westview Press, Boulder.

Nelson, D.R. 1991. Shark repellents: how effective, how needed? pp. 106–108. *In*: S.H. Gruber (ed.) Discovering Sharks, American Littoral Society, Highlands.

Nelson, D.R. & W.R. Strong. 1996. Chemical repellent tests on white sharks, with comments on repellent delivery methods. pp. 471–475. *In*: A.P. Klimley & D.G. Ainley (ed.) Great White Sharks: The Biology of *Carcharodon carcharias*, Academic Press, San Diego.

Opdyke, D.L. & C.M. Burnett. 1965. Effect of ethoxylation on the skin irritating properties of alkyl ether sulfates. Proc. Sci. Sect. Toilet Goods Assoc. 44: 3–8.

Parker, D.O. 1971. Effect of molecular structure on the toxicity of some nonionic surfactants to the goldfish, *Carassius auratus* (Cypriniformes: Cyprinidae). Masters Thesis, California State Univ, Long Beach. 92 pp.

Parness, J. 1975. A study on the action and composition of toxic secretion of the flatfish *Pardachirus marmoratus* (Soleidae). Master Thesis, Hebrew University, Tel Aviv. 90 pp.

Pickering, Q.H. 1970. The chronic toxicity of linear alkylate sulfate (LAS) to *Pimaphales promelas* Rafinesque. J. Wat. Pollut. Contr. Fed. 42: 243–254.

Primor, N., J. Parness & E. Zlotkin. 1978. Pardaxin: the toxic factor from the skin secretion of the flatfish *Pardachirus marmoratus* (Soleidae). pp. 539-547. *In*: P. Rosenberg (ed.) Toxins: Animal, Plant, and Microbial, Pergamon Press, New York.

Primor, N., I. Sabnay, V. Lavie & E. Zlotkin. 1980. Toxicity to fish, effect on gill ATPase and gill ultrastructural changes induced by *Pardachirus* secretion and its derived toxin, Pardaxin. J. Exp. Zool. 211: 33–43.

Primor, N., E. Bonaventura & J. Bonaventura. 1983a. Effect of *Pardachirus marmoratus* secretion and its active factor, pardaxin, on elasmobranchs. pp. 151–156. *In*: B.J. Zahuranec (ed.) Shark Repellents from the Sea: New Perspectives, Westview Press, Boulder.

Primor, N., J.A. Zadunaisky & A.T. Tu. 1983b. Pardaxin, its mode of action, toxicity and structure. pp. 173–184. *In*: B.J. Zahuranec (ed.) Shark Repellents from the Sea: New Perspectives, Westview Press, Boulder.

Primor, N., J. Zadunaisky, V. Murdaugh, Jr., J. Boyer & J. Forrest, Jr. 1984. Pardaxin increases solute permeability of gills and rectal gland in the dogfish shark, *Squalus acanthias*. Comp. Biochem. Physiol. 78C: 483–490.

Rasmussen, L.E.L. & M.J. Schmidt. 1992. Are sharks chemically aware of crocodiles? pp. 335–342. *In*: R.L. Doty & D. Müller-Schwarze (ed.) Chemical Signals in Vertebrates IV, Plenum Press, New York.

Satchell, G.H. & D.J. Maddalena. 1972. The cough or expulsion reflex in the Port Jackson shark, *Heterodontus portusjacksoni*. Comp. Biochem. Physiol. 41A: 49–62.

Satkowski, W.B., S.K. Huang & R.L. Liss. 1967. Polyoxyethylene esters of fatty acids. pp. 142-174. *In*: M.J. Schick (ed.) Nonionic Surfactants, Marcel Dekker, New York.

Schott, H. 1973. Effect of chain length in homologous series of anionic surfactants on irritant action and toxicity. J. Pharmacol. Sci. 62: 341–343.

Server, L. 1989. Sharks. Crescent Books, New York. 128 pp.

Sisneros, J.A. 1993. Effect of molecular structure on the shark repellent potency of anionic surfactants. Masters Thesis, California State Univ., Long Beach. 77 pp.

Smith, L.J., Jr. 1986. Effectiveness of sodium lauryl sulfate and other chemicals as shark repellents in a laboratory swim-through test situation. Masters Thesis, California State Univ., Long Beach. 67 pp.

Smith, L.J., Jr. 1991. The effectiveness of sodium lauryl sulphate as a shark repellent in a laboratory test situation. J. Fish Biol. 38: 105–113.

Sobotka, H. 1965. Comparative biochemistry of marine animals. Bioscience 1965: 583–584

Sokal, R.R. & F.J. Rohlf. 1981. Biometry. W.H. Freeman and Company, New York. 859 pp.

Spira, M.E., M. Klein, B. Hochner, Y. Yarom & M. Castel. 1976. Ultrastructural changes accompanying the disturbances of neuromuscular transmission caused by *Pardachirus* toxin. Neuroscience 1: 117–124.

Springer, S. 1954. Laboratory experiments with shark repellents. Proc. Gulf and Caribbean Fisheries Inst. 7th Ann. Sess.: 159–163.

Tachibana, K. & S.H. Gruber. 1988. Shark repellent lipophilic constituents in the defense secretion of the Moses sole (*Pardachirus marmoratus*). Toxicon 26: 839–853.

Tachibana, K., M. Sakaitanai & K. Nakanishi. 1984. Pavoninins: shark-repelling ichthyotoxins from the defense secretion of the Pacific sole. Science 226: 703–705.

Tester, A.L. 1963. Olfaction, gustation, and the common chemical sense in sharks. pp. 255-282. *In*: P.W. Gilbert (ed.) Sharks and Survival, D.C. Heath and Co., Boston.

Thompson, S.A., K. Tachibana, K. Nakanishi & I. Kubota. 1986. Melittin-like peptides from the shark-repelling defense secretion of the sole *Pardachirus pavoninus*. Science 233: 341–343.

Tuve, R.L. 1963. Development of the U.S. Navy 'Shark Chaser' chemical shark repellent. pp. 455–464. *In*: P.W. Gilbert (ed.) Sharks and Survival, D.C. Heath and Company, Boston.

Zahuranec, B.J. & H.D. Baldridge, Jr. 1983. Shark research and the United States Navy. pp. 1–10. *In*: B.J. Zahuranec (ed.) Shark Repellents from the Sea: New Perspectives, Westview Press, Boulder.

Zlotkin, E. & Y. Barenholz. 1983. On the membranal action of pardaxin. pp. 157–171. *In*: B.J. Zahuranec (ed.) Shark Repellents from the Sea: New Perspectives, Westview Press, Boulder.

Zlotkin, E. & S.H. Gruber. 1984. Synthetic surfactants: a new approach to the development of shark repellents. Arch. Toxicol. 56: 55–58.

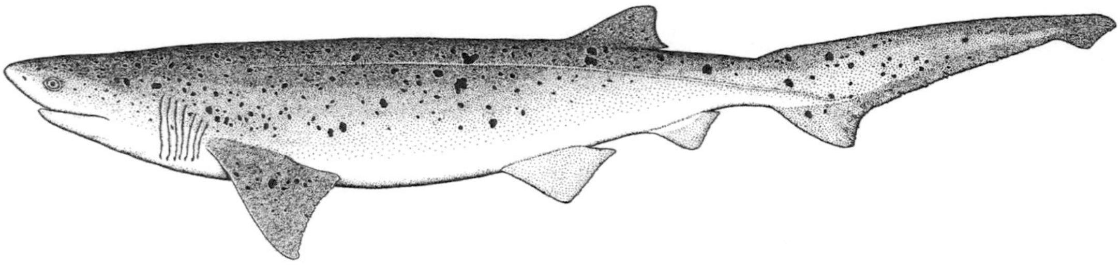

Female of the broadnose sevengill shark, *Notorychus cepedianus*, 180 cm long, captured off the coast of Japan (PV).

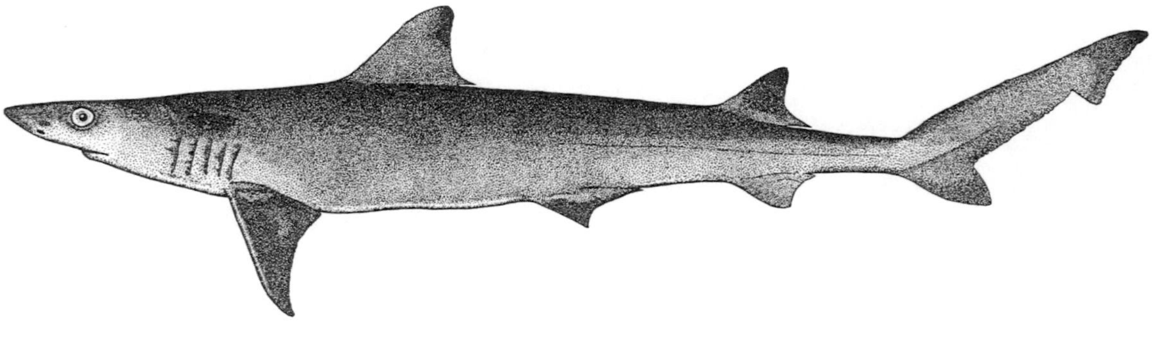

A female of the pencil shark, *Hypogaleus hyugaensis*, 62 cm long, from the Arabian Gulf (SH).

Environmental Biology of Fishes **60:** 131–156, 2001.

Advances in the study of feeding behaviors, mechanisms, and mechanics of sharks

Philip J. Motta[a] & Cheryl D. Wilga[b]
[a]*Department of Biology, University of South Florida, Tampa, FL 33612, U.S.A.*
(e-mail: motta@chuma.cas.usf.edu)
[b]*Biological Sciences, University of Rhode Island, Kingston, RI 02881, U.S.A.*
(e-mail: cwilga@uri.edu)

Received 21 May 1999   Accepted 28 October 1999

Key words: evolution, diet, prey capture, jaws, muscles, teeth, modulation, protrusion

Synopsis

Sharks as a group have a long history as highly successful predatory fishes. Although, the number of recent studies on their diet, feeding behavior, feeding mechanism, and mechanics have increased, many areas still require additional investigation. Dietary studies of sharks are generally more abundant than those on feeding activity patterns, and most of the studies are confined to relatively few species, many being carcharhiniform sharks. These studies reveal that sharks are generally asynchronous opportunistic feeders on the most abundant prey item, which are primarily other fishes. Studies of natural feeding behavior are few and many observations of feeding behavior are based on anecdotal reports. To capture their prey sharks either ram, suction, bite, filter, or use a combination of these behaviors. Foraging may be solitary or aggregate, and while cooperative foraging has been hypothesized it has not been conclusively demonstrated. Studies on the anatomy of the feeding mechanism are abundant and thorough, and far exceed the number of functional studies. Many of these studies have investigated the functional role of morphological features such as the protrusible upper jaw, but only recently have we begun to interpret the mechanics of the feeding apparatus and how it affects feeding behavior. Teeth are represented in the fossil record and are readily available in extant sharks. Therefore much is known about their morphology but again functional studies are primarily theoretical and await experimental analysis. Recent mechanistic approaches to the study of prey capture have revealed that kinematic and motor patterns are conserved in many species and that the ability to modulate feeding behavior varies greatly among taxa. In addition, the relationship of jaw suspension to feeding behavior is not as clear as was once believed, and contrary to previous interpretations upper jaw protrusibility appears to be related to the morphology of the upper jaw-chondrocranial articulation rather than the type of jaw suspension. Finally, we propose a set of specific hypotheses including: (1) *The functional specialization for suction feeding hypothesis* that morphological and functional specialization for suction feeding has repeatedly arisen in numerous elasmobranch lineages, (2) *The aquatic suction feeding functional convergence hypothesis* that similar hydrodynamic constraints in bony fishes and sharks result in convergent morphological and functional specializations for suction feeding in both groups, (3) *The feeding modulation hypothesis* that suction capture events in sharks are more stereotyped and therefore less modulated compared to ram and bite capture events, and (4) *The independence of jaw suspension and feeding behavior hypothesis* whereby the traditional categorization of jaw suspension types in sharks is not a good predictor of jaw mobility and prey capture behavior. Together with a set of questions these hypotheses help to guide future research on the feeding biology of sharks.

Introduction

The feeding behaviors, mechanisms, and mechanics of sharks are of interest to biologists for a variety of reasons. From an evolutionary and functional standpoint chondrichthyan fishes represent a basal group of jawed fishes that share a common ancestor with bony fishes (Schaeffer & Williams 1977, Carroll 1988, Long 1995). As such, studying the feeding behavior of sharks has the potential to teach us a great deal about the function and evolution of vertebrate feeding systems (Wilga et al. 2000). In addition, extant sharks number about 467 (R. Martin personal communication) species inhabiting a wide range of marine and fresh water environments and possess great morphological diversity. Interestingly, no sharks are herbivorous, but their prey spans a variety of trophic levels ranging from zooplankton to the largest marine mammals. Although sharks feed opportunistically, they typically consume relatively small amounts of food (Pratt et al. 1982, Compagno 1984, Cortes 2000). Furthermore, compared to bony fishes, sharks retain a relatively simple feeding apparatus composed of a fused chondrocranium, jaws consisting of a palatoquadrate and Meckel's cartilage, and a hyoid arch composed of only three elements: a hyomandibula, ceratohyal, and basihyal. Despite this simple mechanism sharks utilize a variety of prey capture modes that include suction, ram, bite, bite and gouge, and filter feeding (Moss 1972, 1977, Frazzetta 1994). The rich fossil history that we have of shark teeth means that we have one of the most extensive indicators of vertebrate feeding mechanisms. Although much is known about tooth replacement and growth patterns, we still have an incomplete understanding of their function during feeding beyond that of grasping prey. A tremendous increase in our knowledge of feeding mechanics of bony fishes has taken place within the last three decades (Alexander 1967, Liem 1978, Lauder 1985), yet such studies on sharks are advancing at a much slower rate (Moss 1972, 1977, Nobiling 1977, Shirai & Nakaya 1992). A key finding in bony fishes is that most have the ability to modulate feeding behavior in response to differing prey sizes and types (Liem 1978, Sanderson 1988, Nemeth 1997). Early studies indicated that feeding behavior of sharks is a highly predictable sequence of events (Gilbert[1]). This has raised the question as to whether sharks have similar modulatory abilities as bony fishes, or is the ability to vary capture behavior in sharks related to different types of capture mechanisms (ram, suction, bite, filter). Furthermore, the size of sharks span a tremendous range, from the largest fishes in the oceans to diminutive deepwater species. How the scaling of feeding structures affects their feeding mechanics and behavior remains relatively unknown. And despite what every student of comparative anatomy learns about jaw suspension types and the relationship to jaw mobility and feeding in sharks, new evidence challenges some of these assumptions and we are beginning to understand the relationship between jaw suspension types and feeding mechanics.

The goal of this paper is to review past work on the feeding behaviors, mechanisms, and mechanics of sharks and to begin to integrate these topics in order to interpret relationships between feeding behavior and morphology. By posing a set of hypotheses and questions we hope to guide research into the next millennium. To clarify the use of various terms we offer the following definitions. The term mechanism refers to the anatomical parts of the feeding apparatus such as teeth, cartilages, muscles and connective tissue structures. Mechanics refers to the functioning of the feeding mechanism and deals with the motions of the various parts and the forces acting upon them. Kinematics simply deals with motion without reference to forces. Prey refers here to dietary items captured by natural feeding, and food refers to pieces of or whole food items offered under experimental conditions. The term feeding behavior is used in a more general ethological sense and also interchangeably with prey or food capture behavior. In this review we do not specifically cover batoids although we do occasionally discuss them. It is not our goal to review feeding ecology or diet per se, although we relate diet to feeding mechanisms and behavior. The reader is referred to dietary studies and reviews listed in the text. Lastly, the sensory system as it relates to feeding has been reviewed by Hodgson & Mathewson (1978), Hueter & Gilbert (1990), Bleckmann & Hofmann (1999) and elsewhere in this volume.

Discussion

Evolution of the feeding mechanism

The earliest fossil evidence of sharks consists of scales dating back to the Silurian Period and perhaps as early as the late Ordovician Period some 455 million years

[1] Gilbert, P.W. 1970. Studies on the anatomy, physiology, and behavior of sharks. Final report, Office of Naval Research, Contract Nonr-401(33): Project NR 104-471. 45 pp

ago (Schaeffer & Williams 1977, Carroll 1988, Long 1995, Monastersky 1996, Sansom et al. 1996). The earliest sharks, the Cladoselachida (extinct), had an amphistylic jaw suspension characterized by a long palatoquadrate cartilage with otic and orbital processes articulating with the neurocranium, a long mandible, a long and presumably non-suspensory hyomandibula, a large and mostly terminal mouth with multi-cuspid teeth, relatively small labial cartilages, and a body and caudal fin form similar to modern fast-swimming pelagic sharks (Figure 1). These predators had a dentition suited for seizing and tearing prey rather than shearing or sawing, and it is speculated that they captured prey by overtaking and engulfing it, although suction probably played a role in their feeding (Shaeffer 1967, Moy-Thomas & Miles 1971, Carroll 1988, Lund & Grogan 1997).

An early period of chondrichthyan radiation in the Paleozoic was followed by a period of elasmobranch radiation in the Mesozoic and Cenozoic. A ctenacanthid ancestor then gave rise to the neoselachians, which

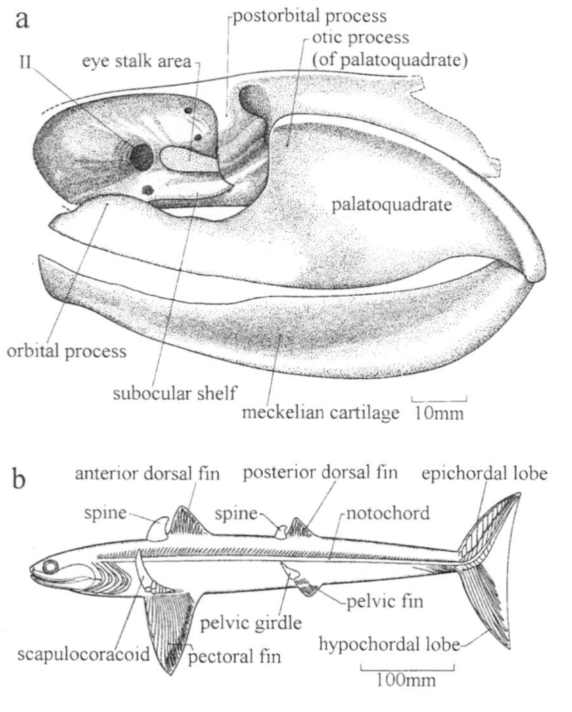

Figure 1. a – Restoration of the neurocranium and jaws of *Cladodus*, b – restoration of *Cladoselache fyleri*. The cladodont palatoquadrate shown in (a) had a large otic process which is not well represented in the whole animal reconstruction (b). The narrow suborbital ramus also extends anteriorly to the rostrum (Schaeffer 1967) (reproduced with permission from Moy-Thomas & Miles 1971).

includes all modern sharks, skates and rays. Three major groups comprise the neoselachians, the galeoids, squaloids, and batoids (but see Compagno 1973, Carroll 1988, Shirai 1996). The feeding apparatus in modern sharks has changed primarily in the proportions of the skeletal elements and in the mechanism of jaw suspension. Modern sharks have a sub-terminal mouth that opens ventrally, shorter jaws, more moveable hyomandibula that suspends the jaws, more protrusible palatoquadrate cartilage with a smaller otic process and a dentition better suited for sawing and shearing compared to ancestral sharks (Shaeffer 1967, Carroll 1988).

Diet and feeding activity

The diet of sharks is perhaps one of the most thoroughly investigated areas related to feeding, yet many studies are confined to relatively few species. Most of the work on shark diet is on carcharhiniform sharks. Sharks are generally asynchronous opportunistic feeders on the most abundant prey item, which are primarily other fishes. Feeding occurs in short bursts followed by longer periods of digestion when feeding is reduced (Wetherbee et al.[2]). Sharks in general eat only about 3–14% of their body weight per week (Budker 1971). Mean weight of recently ingested prey items as a percent of body weight is generally low: 2.25% for the Australian sharpnose shark, *Rhizoprionodon taylori* (Simpfendorfer 1998); 2.68% for lemon shark, *Negaprion brevirostris* (Cortes & Gruber 1990); and 4.23% for juvenile sandbar shark, *Carcharhinus plumbeus* (Medved et al. 1988). Examining 148 species representing 22 families, Cortes (2000) found that sharks are tertiary consumers, with significant differences in diet occurring among the six orders examined. Trophic level and body size is positively correlated, especially within the carcharhinid sharks, and within the Carcharhiniformes teleost fishes are the dominant prey item.

The diet of young lemon sharks, *N. brevirostris*, has been extensively studied, revealing that their diet is dominated by teleosts. Feeding is asynchronous, intermittent and stomach contents are evacuated in

[2] Wetherbee, B.M., S.H. Gruber & E. Cortes. 1990. Diet, feeding habits and consumption in sharks, with special reference to the lemon shark, *Negaprion brevirostris*. pp. 29–48. *In*: H.L. Pratt Jr., S.H. Gruber & T. Taniuchi (ed.) Elasmobranchs as Living Resources: Advances in the Biology, Ecology, Systematics, and the Status of the Fisheries, NOAA Tech. Rep. NMFS 90.

25–41 h. Daily ration is estimated at 1.5–2.1% of body weight per day (Cortes & Gruber 1990, Wetherbee et al.[2]). In southwest Florida, the bonnethead shark *Sphyrna tiburo* has a very homogeneous diet dominated by crustaceans, particularly blue crabs *Callinectes sapidus* (Cortes et al. 1996). On the contrary, the tiger shark *Galeocerdo cuvieri* is an opportunistic feeder in Hawaiian waters, undergoing ontogenetic dietary shifts from a diet dominated by teleosts and cephalopods, to a broader diet that includes elasmobranchs, turtles, land mammals, crustaceans, and indigestible items. The ontogenetic shifts in this species may be attributable to increased size, expanded range, and improved hunting skills (Lowe et al. 1996). Leopard sharks, *Triakis semifasciata*, undergo dietary shifts in Monterey Bay, California. Sharks less than 130–140 cm total length feed primarily on benthic invertebrates, whereas fishes were the most important items in larger sharks (see Lowe et al. 1996 for a summary of ontogenetic dietary shifts). However, the vast majority of dietary data in sharks is confined to stomach content analyses which vary greatly in the number of stomachs analyzed (Bigelow & Schroeder 1948, Budker 1971, Compagno 1984, Cortes 2000).

In general less is known about feeding activity patterns of sharks, and many of these studies are confined to easily observable or captive species, many of which are coastal. In addition to ontogenetic feeding shifts many sharks undergo seasonal or diel changes in diet and feeding activity patterns. Most white sharks, *Carcharodon carcharias*, at the South Farallon Islands, California are transient visitors, but a few sharks are short-term residents. The sharks apparently regulate the timing of their movements so they arrive at the Farallon Islands during the time of year that coincides with previously successful foraging on pinniped prey (Klimley & Anderson 1996). In contrast to young lemon sharks that show no diel feeding activity (Cortes & Gruber 1990), large tiger sharks *G. cuvieri*, which have large home ranges, feed near the bottom at night, but also feed at the surface during the day (Lowe et al. 1996, Holland et al.[3]). At French Frigate Shoal in the Hawaiian Islands, a tagged tiger shark showed a diurnal activity pattern. The shark spent most of its daytime activity on the outer reef where it swam close to the bottom. The shark occasionally swam to the surface to investigate prey and perhaps feed. At night, most of the shark's activity was centered above or beyond the drop-off (Tricas et al. 1981). Telemetric and digestive rate studies on blue shark *Prionace glauca* off Santa Catalina Island, California, indicate that these sharks forage in waters near the surface from around midnight through dawn, feeding on dispersed schools of anchovies and mackerel (Tricas 1979). Swell sharks *Cephaloscyllium ventriosum*, are nocturnal feeders that detect the bioelectric fields of their individual prey items (Tricas 1982). Horn sharks *Heterodontus francisci* off Santa Catalina Island, California, exhibit diel activity patterns of increased activity at night, particularly at 18:00 h. Few sharks are exposed in the day, instead seeking shelter among rocks, whereas sharks are frequently found in exposed locations at night, coinciding with their increased feeding on nocturnally active prey (Nelson & Johnson 1970, Finstad & Nelson 1975, Strong 1989). Large diurnal aggregations of scalloped hammerhead *Sphyrna lewini* occur at seamounts in the Gulf of California. These sites may serve as a refuge point at the center of foraging paths. During the day the sharks swim slowly back and forth over the seamount and do not forage. At night the sharks leave either as solitary animals or in small groups to forage in the pelagic environment (Klimley & Nelson 1981, 1984, Klimley et al. 1988, Klimley 1995). Small numbers of whitetip reef sharks *Triaenodon obesus* aggregate in caves during the day. At night they disperse and feed within and about the reefs, although some feeding during the day evidently occurs (Randall 1977).

Apparently many shark species segregate by sex, resulting in dietary differences that in turn result in different growth rates, with females growing to a larger size. This is attributed to differing reproductive pressures on sexes. The scalloped hammerhead, *S. lewini*, is a case in point. Females of this species move offshore at a smaller size than males, consequently females feed more on pelagic prey than males and with greater predatory success. Females therefore grow to reproductive size more rapidly than males (reviewed by Klimley 1987).

Feeding behavior – capture and approach

Sharks capture their prey in a variety of ways. During ram capture the shark swims over the relatively stationary prey, engulfs it whole or seizes it in its jaws

[3] Holland, K.N., S.M. Kajiura, A. Bush, B.M. Wetherbee, C.G. Lowe & C. Meyer. 1997. Movement patterns of tiger sharks in Hawaiian waters. Abstracts, American Society of Ichthyologists and Herpetologists 77th Annual Meeting, 26 June–2 July 1997, University of Washington, Seattle. p. 166.

and may use manipulation bites to reduce it in size. The food is then transported by hydraulic suction from the mouth through the pharyngeal cavity into the esophagus. Many species of carcharhinid and lamnid sharks primarily rely on ram feeding. Ram feeding sharks generally have a large gape in comparison to suction feeders (see below). Bonnethead sharks, *S. tiburo*, ram feed benthic food by depressing the mandible and scooping the food up as they swim over it (Wilga 1997, Wilga & Motta 2000). White sharks, *C. carcharias*, primarily ram capture their food sometimes approaching the food at great speeds (Tricas & McCosker 1984, Tricas 1985, Klimley 1994, Klimley et al. 1996). Oceanic whitetip sharks, *Carcharhinus longimanus*, employ a feeding behavior similar to ram capture. These sharks have been observed to swim slowly through schools of tuna with their mouth open at the surface, without snapping or chasing, simply engulfing prey that enters its mouth (Bullis 1961, Compagno 1984). During ram feeding, particularly at high swimming speeds, the predator might displace a wave of water anterior to the mouth in effect pushing the prey away from the open mouth. In this case they might incorporate what has been termed compensatory suction to overcome the inertia of water being displaced anteriorly to the moving head (Van Damme & Aerts 1997).

During ram feeding by white sharks, Tricas & McCosker (1984) and McCosker (1985) noted that the sharks often bite such prey as elephant seals, and then retreat until the prey lapses into shock or bleeds to death. The shark then returns to feed on the prey. Tricas & McCosker(op. cit.) referred to this as the 'bite and spit' strategy. However, Klimley (1994) and Klimley et al. (1996) proposed that white sharks hold the pinniped prey tightly in their mouth and drag it below the surface, often removing a bite from the prey in the process. The prey may be released underwater upon which it floats or swims to the surface and dies by exsanguination. Meanwhile, the shark follows the prey to the surface to begin feeding after it dies.

Inertial suction feeding involves a decrease in the pressure of the buccopharyngeal chamber such that the prey or food is pulled into the mouth. This is commonly referred to as suction feeding. There is a functional continuum from pure ram to pure inertial suction and sharks can use a combination of both (Norton & Brainerd 1993, Van Damme & Aerts 1997, Wilga & Motta 1998a). Although the vast majority of shark feeding behaviors have not been studied, morphological specialization for suction feeding is found in many shark lineages including the Squaliformes, Heterodontiformes, Carcharhiniformes, Orectolobiformes, Squatiniformes and the Batoidea, while suction feeding appears to be the predominant food capture behavior in the Orectolobiformes (Tanaka 1973, Moss 1977, Wu 1994, Fouts 1995, Clark & Nelson 1997, Wilga 1997, Ferry-Graham 1998b, Heupel & Bennett 1998, Wilga & Motta 1998a,b, Fouts & Nelson 1999, Pretlow-Edmonds 1999, Robinson 1999).

Specializations for suction feeding in sharks can include a relatively small mouth as compared to ram feeding sharks, small teeth, a laterally enclosed mouth such as by large labial cartilages, hypertrophied abductor muscles and rapid buccal expansion (Motta & Wilga 1999). The nurse shark, *Ginglymostoma cirratum*, is apparently an obligate suction feeder. When it approaches its food, the shark usually ceases swimming and initiates a very stereotyped food capture behavior. The mandible is rapidly depressed as the robust labial cartilages move anteriorly to laterally occlude the mouth and there is often no cranial elevation. It takes approximately 32 msec to open its mouth, one of the fastest food capture sequences of any shark recorded. Even in the absence of food when animals were conditioned to feed, the entire sequence occurred with little variation. This explosive suction action can generate sub-ambient pressures as low as −1 atm, and large nurse sharks can even dismember their food in this manner (Tanaka 1973, Motta & Wilga 1999, Robinson 1999, Motta personal observation). Specializations for suction prey capture in *G. cirratum* and other specialized suction feeding sharks are apparently functionally convergent with those of high speed inertial suction feeding bony fishes. Bony fishes that are suited for inertial suction feeding usually have a small, laterally enclosed mouth, reduced dentition, strongly developed abductor muscles, buccal expansion is large and rapid, the expansive phase is shorter in duration than the compressive phase, and peak hyoid depression occurs after peak gape. In contrast to *G. cirratum*, upper jaw protrusion in bony fishes may contribute significantly to buccal expansion (Lauder 1980, 1985, Muller et al. 1982, Liem 1993, Motta & Wilga 1999).

Biting, which may accompany ram feeding, may also occur when a shark approaches its prey or food, ceases swimming, and simply bites the prey or pieces off the prey. The negative buoyancy of sharks does not facilitate this behavior in mid-water, and the shark may try to swim in-place to maintain its position in the water column. The cookiecutter shark, *Isistius brasiliensis*, shows a unique biting behavior in which it employs its modified pharyngeal muscles, upper jaw, hyoid and branchial arches to suck onto its prey of pelagic fishes or marine mammals. Forming a seal with its fleshy lips, it then sinks its hook-like upper teeth and saw-like modified lower teeth into the prey, twists about its longitudinal axis to gouge out a plug of flesh, leaving a crater-like wound (Jones 1971, Compagno 1984, LeBoeuf et al. 1987, Shirai & Nakaya 1992).

Continuous ram filter feeding, such as in basking sharks, *Cetorhinus maximus*, occurs when the shark continuously swims forward with the mouth open. In this manner these sharks will actively seek and locate zooplankton patches on the surface (Sims et al. 1997, Sims & Merrett 1997). The megamouth shark, *Megachasma pelagios*, and perhaps the whale shark, *Rhincodon typus*, can employ intermittent suction filter feeding, generating suction with aperiodic pulses (Taylor et al. 1983, Diamond 1985, Compagno[4], Sanderson & Wassersug 1993, Clark & Nelson 1997, Martin & Naylor 1997). Whale sharks can also use continuous ram filter feeding, or hang vertically in the water column. In the latter case they will suck prey into the mouth, or rise vertically out of the water and sink back under water, creating an inflow of water and prey into their open mouths (Gudger 1941b, Springer 1957, Budker 1971, Colman 1997).

It often has been speculated that mouth position and body form in sharks is related to where they feed and how they approach the prey (Budker 1971). The subterminal mouth of modern sharks does not preclude them from feeding on prey in front of or above them. Sharks will approach surface or underwater food either with a direct head-on approach or roll on their side to bite at the food (Budker 1971). White sharks, *C. carcharias*, have been observed to approach and feed on underwater bait or a floating whale carcass either in their normal orientation, by rolling on their side, or rolling completely over so their ventral side is up (Pratt et al. 1982, Tricas & McCosker 1984). Surface feeding blue shark, *P. glauca*, approach schools of squid on the surface with an underwater approach or a surface charge. Small anchovies are captured from a normal swimming posture, but when capturing larger whole mackerel from behind blue sharks may roll on their side (Tricas 1979). Surface feeding blacknose, *Carcharhinus acronotus*, oceanic white tip sharks, *C. longimanus*, and white sharks, *C. carcharias*, have been observed to raise the head just prior to capture (Bullis 1961, Tricas & McCosker 1984, Frazzetta & Prange 1987). This might place the open mouth in line with food as the shark approaches (Frazzetta & Prange 1987).

Feeding on or in the benthos is not uncommon. Gray reef sharks, *Carcharhinus amblyrhynchos*, in Hawaii primarily feed near the bottom on reef associated teleosts and supplement their diet with invertebrates (Wetherbee et al. 1997). Rays are often taken by sharks, particularly hammerhead sharks (Gudger 1907, Budker 1971). Great hammerhead sharks, *Sphyrna mokarran*, have been observed to use their head to deliver powerful blows and to restrain rays on the substrate prior to biting pieces off the ray (Strong 1990). Leopard sharks, *T. semifasciata*, can apparently suck worms out of their burrows in addition to biting pieces off their benthic prey (Talent 1976, Compagno 1984). Small bonnethead sharks, *S. tiburo*, ram capture their food by depressing the mandible considerably as they swim over the food, catching the food either within the mouth or with the anterior mandibular teeth. Occasionally they swim in tight circles biting at benthic food (Wilga 1997, Wilga & Motta 2000). The epaulette shark, *Hemiscyllium ocellatum*, and the whitespotted bamboo shark, *Chiloscyllium plagiosum*, employ a feeding behavior uncommon among sharks. These sharks occasionally thrust their head into the sediment up to the level of the first gill slit, apparently using suction to capture their benthic prey of worms and crabs. They then winnow the prey from the sand in the buccopharyngeal cavity and eject the sand through the first gill slit (Heupel & Bennett 1998, Wilga personal observation).

Sharks will ambush, stalk, or lure their prey, perhaps forage cooperatively, and in some cases feed in large aggregations resulting in what is termed 'feeding frenzies'. By partially burying themselves in the soft substrate, Pacific angel sharks, *Squatina californica*, ambush demersal fishes. These sharks appear to actively select ambush sites within localized areas adjacent to reefs (Fouts 1995, Fouts & Nelson 1999). White sharks *C. carcharias*, will stalk prey downstream in oceanic or tidal currents (Pyle et al. 1996), while other sharks may lure prey to them. Luminescent tissue on the upper jaw of the megamouth

[4] Compagno, L.J.V. 1990. Relationships of the megamouth shark, *Megachasma pelagios* (Lamniformes: Megachasmidae), with comments on its feeding habits. pp. 357–379. *In*: H.L. Pratt, Jr., S.H. Gruber & T. Taniuchi (ed.) Elasmobranchs as Living Resources: Advances in the Biology, Ecology, Systematics, and the Status of the Fisheries, NOAA Tech. Rep. NMFS 90.

Figure 2. Illustration of sevengill shark, *Notorynchus cepedianus*, feeding by social facilitation: a – typical attack pattern begins when a group of sharks forms a loose circle around its intended victim, b – attempts to escape are prevented by the pack, which gradually tightens its circle around the prey, c – eventually, one or more sharks bite the prey, which stimulates the rest of the pack to converge on it, d – once the attack has been initiated, the rest of the pack quickly joins in to obtain its share of the food (reproduced with permission from Ebert 1991).

shark, *M. pelagios*, might attract euphausid shrimp and other prey into its mouth (Compagno[4]), the white tips on the pectoral fins of oceanic whitetip sharks, *C. longimanus*, might act as visual lures to aid in the capture of its rapid moving prey (Myrberg 1991), and bioluminescence in the cookiecutter shark, *I. brasiliensis*, might serve to lure pelagic predators from which it gouges chunks of flesh (Jones 1971, Widder 1998).

Although many species of sharks forage solitarily, in some cases aggregations of sharks will come together to feed. Cooperative foraging by sharks has been suggested repeatedly. Blacktip reef sharks, *Carcharhinus melanopterus*, and lemon sharks, *N. brevirostris*, were observed to apparently herd schools of fish against the shoreline and then feed on them (Eibl-Eibesfeldt & Hass 1959, Morrissey 1991) and oceanic whitetip sharks, *C. longimanus*, were observed to herd squid at night (Strasburg 1958). In one instance more than a hundred sandtiger sharks, *Carcharias taurus*, were observed to surround blue fish (Pomatomidae), force them into a school and then all dashed in to feed on the prey at the same instant (Coles 1915). Thresher sharks (*Alopias*) are reported to apparently work in groups to capture fish, using their long caudal fins to herd and stun fishes (Coles 1915, Budker 1971, Castro 1996). When hunting seals sevengill sharks, *Notorynchus cepedianus*, will circle the seal and prevent its escape (Figure 2). The circle is tightened and eventually one shark initiates the attack that stimulates the others to begin feeding (Ebert 1991). However, in many of these cases of supposed cooperative foraging it is not clear if the sharks are actually cooperating or the behavior (e.g. circling the prey) represents an aggregation of sharks to one prey item.

Feeding frenzies also resemble cooperative foraging behavior. Feeding frenzies involve aggregations of sharks attacking prey or food items indiscriminately. These sharks move at an accelerated speed, disregarding any injuries they may receive in the attack. Injured or hooked sharks are often attacked and consumed by the other sharks. These feeding bouts, which can involve as few as six sharks to hundreds of sharks, can end abruptly as they begin (Gilbert 1962, Vorenberg 1962, Springer 1967, Nelson 1969, Hobson 1963).

Feeding mechanisms and mechanics

The feeding apparatus and cranial anatomy of sharks have been extensively reviewed (e.g. Daniel 1934, Marinelli & Strenger 1959, Gohar & Mazhar 1964, Moss 1972, 1977, Nobiling 1977, Compagno 1988,

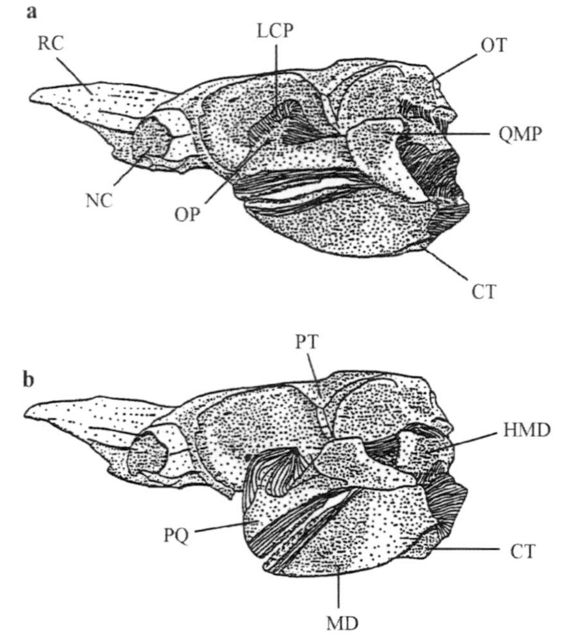

Figure 3. Left lateral view of the neurocranium, jaws and hyoid arch of a 74.5 cm TL spiny dogfish, *Squalus acanthias*, with the skin and muscles removed: a – at resting position, b – at peak upper jaw protrusion. QMP = quadratomandibularis process of palatoquadrate, CT = ceratohyal, HMD = hyomandibula, LCP = ethmopalatine ligament, MD = mandible or lower jaw, NC = nasal capsule, OP = orbital process of palatoquadrate, OT = otic capsule of cranium, PQ = palatoquadrate cartilage or upper jaw, PT = postorbital process, RC = rostral cartilage (reproduced with permission from Wilga & Motta 1998a).

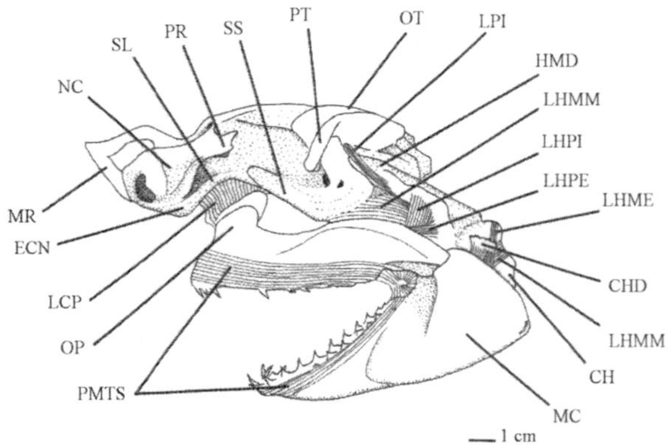

Figure 4. Left lateral view of the neurocranium, jaws and hyoid arch of a 122 cm TL lemon shark, *Negaprion brevirostris*, with the skin and muscles removed. Tendons and ligaments are indicated. CH = ceratohyal, CHD = constrictor hyoideus dorsalis tendon, ECN = ectethmoid condyle, HMD = hyomandibula, LCP = ethmopalatine ligament, LHME = external hyoid-mandibular ligament, LHMM = medial hyoid-mandibular ligament, LHPE = external hyomandibula-palatoquadrate ligament, LHPI = internal hyomandibula-palatoquadrate ligament, LPI = postspiracularis ligament, MC = Meckel's cartilage or lower jaw, MR = medial rostral cartilage, NC = nasal capsule, OP = orbital process of palatoquadrate, OT = otic capsule, PMTS = palatoquadrate-mandibular connective tissue sheath, PR = preorbital process, PT = postorbital process, SL = suborbital ledge, SS = suborbital shelf (reproduced with permission of Wiley-Liss, Inc., a subsidiary of John Wiley & Sons, Inc. from Motta & Wilga 1995).

Waller & Baranes 1991, Shirai & Okamura 1992, Frazzetta 1994, Wu 1994, Motta & Wilga 1995, 1999). The majority of the studies have been on the hard parts of the carcharhiniform cranial skeleton, although some have reviewed the joints, ligaments and muscles (Gadow 1888, Luther 1909, Goodey 1910, Daniel 1915, 1934, Allis 1923, Edgeworth 1935, Lightoller 1939, Marinelli & Strenger 1959, Moss 1972, Nobiling 1977, Motta & Wilga 1995, 1999).

The feeding apparatus is perhaps best known in the spiny dogfish, *Squalus acanthias*, and the lemon shark, *N. brevirostris*. The jaw suspension differs somewhat between the two. Squaloids have an orbitostylic type of jaw suspension in which the hyomandibula suspends the jaws from the cranium, and the palatoquadrate articulates with the orbital wall of the cranium by a relatively long orbital process (Marinelli & Strenger 1959, Maisey 1980, Wilga & Motta 1998a) (Figures 3, 4, 5, 6). The lemon shark has the jaws suspended from a more posteroventrally oriented hyomandibula, in contrast to the more laterally directed hyomandibula of the dogfish. The distal hyomandibula is braced against the mandibular knob of the mandible, and the ceratohyal is ligamentously bound to the distal hyomandibula and the mandible. The smaller orbital process of the lemon shark palatoquadrate is attached to the chondrocranial wall by the ethmopalatine ligament as it does in the dogfish (Moss 1965, 1972, 1977, Motta & Wilga 1995). In both species, the hyomandibula is ligamentously bound to the ceratohyal and in turn to the ventral basihyal which rests somewhat dorsal to the mandibular symphysis.

Electromyographic analyses reveal that during jaw opening a relatively conservative series of events occur in both species. Similar to the expansive phase described for teleost fishes (Liem 1978, Lauder 1985), the cranium is elevated by contraction of the epaxialis muscle, although cranial elevation need not occur (see pp. 134–137) (Motta et al. 1991, 1997, Wilga & Motta 1998a) (Figures 7, 8). Almost

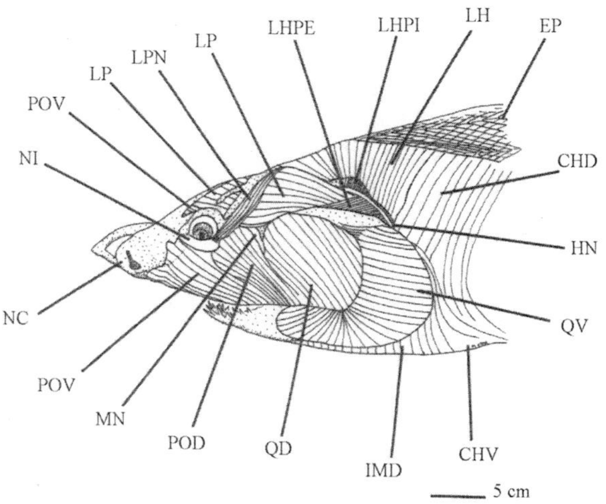

Figure 5. Left lateral view of the head and muscles of a 229 cm TL lemon shark, *Negaprion brevirostris*, with the skin removed and muscle fiber direction indicated. Myosepta of the epaxialis muscle (W-shape) are indicated in addition to the muscle fiber direction. The chondrocranial-palatoquadrate connective tissue sheath is removed. CHD = constrictor hyoideus dorsalis, CHV = constrictor hyoideus ventralis, EP = epaxialis, HN = hyomandibular nerve, IMD = intermandibularis, LH = levator hyomandibularis, LHPE = external hyomandibula-palatoquadrate ligament, LHPI = internal hyomandibula-palatoquadrate ligament, LP = levator palatoquadrati, LPN = levator palpebrae nictitantis, MN = mandibular branch of trigeminal nerve, NC = nasal capsule, NI = nictitating membrane, POD = dorsal preorbitalis, POV = ventral preorbitalis, QD = quadratomandibularis dorsal, QV = quadratomandibularis ventral (reproduced with permission of Wiley-Liss, Inc., a subsidiary of John Wiley & Sons, Inc. from Motta & Wilga 1995).

simultaneously, the mandible is depressed primarily by the action of the coracomandibularis muscle, and the basihyal-ceratohyal apparatus begins to depress due to contraction of the coracoarcualis and coracohyoideus muscles. The branchial apparatus is depressed by action of the coracobranchiales muscles. In the dogfish in particular, the labial cartilages are extended as the mandible is depressed and laterally occlude the mouth (Motta et al. 1991, 1997, Wilga & Motta 1998a). The compressive phase begins at peak gape as the mouth is maximally open, which is followed by the beginning of upper jaw protrusion and elevation of the mandible. Jaw adduction in both species is accomplished by contraction of the quadratomandibularis muscle. Various combinations of the preorbitalis and levator palatoquadrati muscles that are particular to each taxon protrude the upper jaw (see pp. 146–149). As the upper jaw protrudes, the folded and elastic ethmopalatine ligament

a

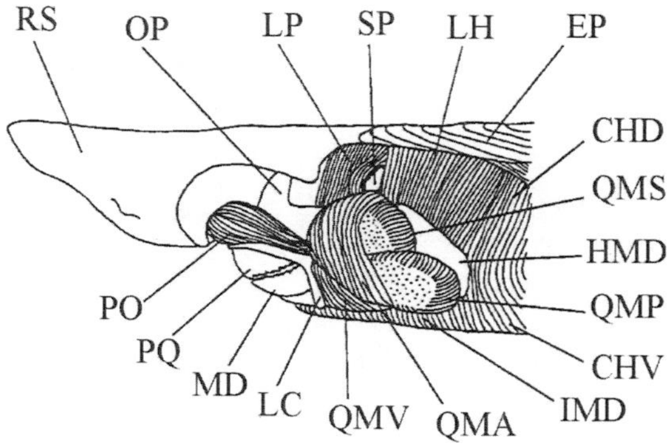

b

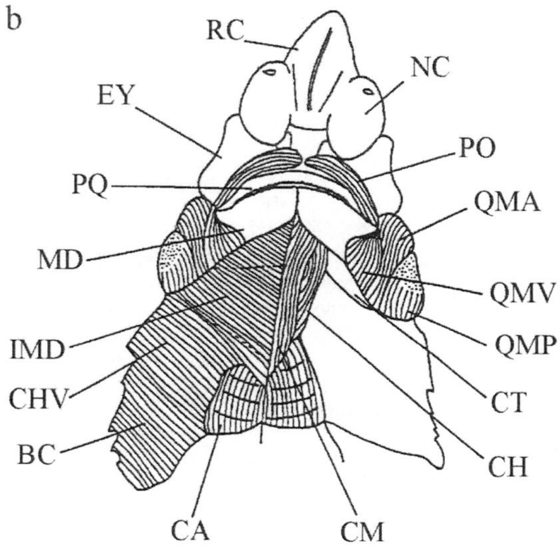

Figure 6. a – Left lateral view (74.5 cm TL), and b – ventral view (60 cm TL) of the head of the spiny dogfish, *Squalus acanthias*, with the skin and eye removed and muscle fiber direction indicated. Skin over the rostrum and cranium is left intact. Myosepta only of the epaxialis muscle are indicated. Raphe overlying quadratomandibularis is indicated by stippling. Anterior and posterior margins of the interhyoideus (deep to IMD) is indicated by dotted lines. BC = branchial constrictors, CA = coracoarcualis, CH = coracohyoideus, CHD = constrictor hyoideus dorsalis, CHV = constrictor hyoideus ventralis, CM = coracomandibularis, CT = ceratohyal, EP = epaxialis, EY = eye, HMD = hyomandibula, IMD = intermandibularis, LC = labial cartilages, LH = levator hyomandibularis, LP = levator palatoquadrati, MD = mandible or lower jaw, NC = nasal capsule, OP = orbital process of palatoquadrate, PO = preorbitalis, PQ = palatoquadrate or upper jaw, QMA = quadratomandibularis anterior, QMS = quadratomandibularis superficial, QMP = quadratomandibularis posterior, QMV = quadratomandibularis ventral, RC = rostral cartilage, RS = rostrum, SP = spiracularis (reproduced with permission from Wilga & Motta 1998a).

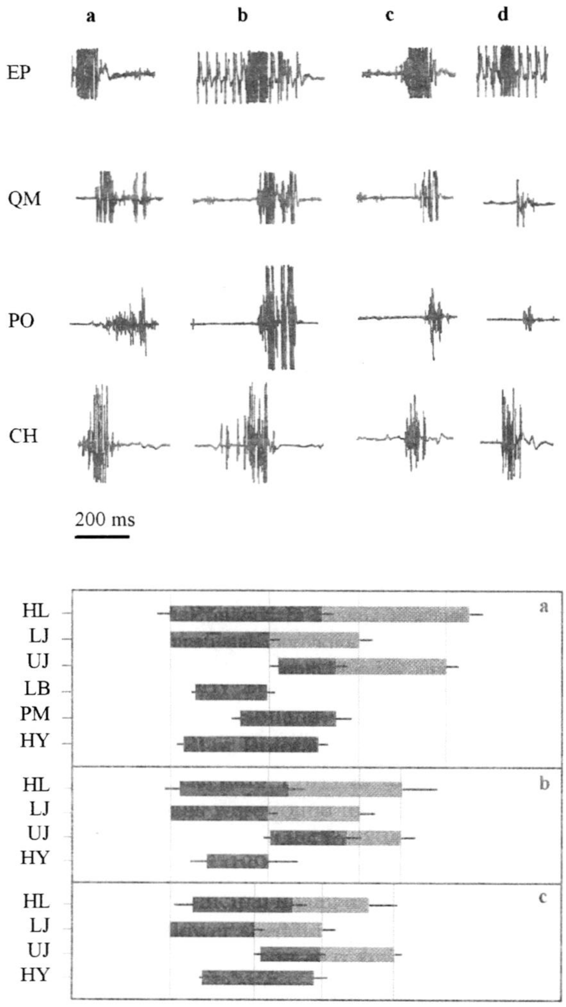

Figure 7. Motor activity and kinematic patterns for: a – suction capture, b – bite manipulation, c – suction transport, and motor activity for d – one head shake to one side in the spiny dogfish, *Squalus acanthias*. Motor activity is from four muscles: epaxialis, quadratomandibularis, preorbitalis, and coracohyoideus. Kinematic events are represented by black bars indicating start to maximum movement followed by gray bars indicating maximum to end of movement, with error bars indicating 1 standard error of the mean (SEM). Vertical lines indicate, from left to right, the start of the mouth opening, mouth closing, and jaw retraction phase, and end of the jaw retraction phase respectively. CH = coracohyoideus, EP = epaxialis, HL = head lift, HY = hyoid depression, LB = labial cartilage extension, LJ = lower jaw depression, PM = prey movement, PO = preorbitalis, QM = quadratomandibularis, UJ = upper jaw protrusion (reproduced with permission from Wilga & Motta 1998a).

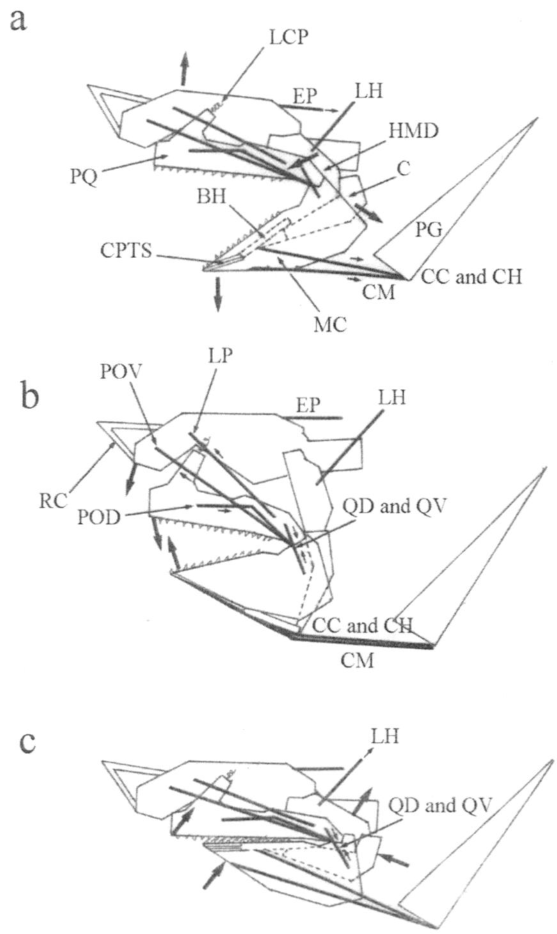

Figure 8. A model of chondrocranial, mandibular and hyoid arch kinetics during feeding in the lemon shark, *Negaprion brevirostris*: a – expansive phase, characterized by depression of the mandible and elevation of the cranium, b – compressive phase, characterized by elevation of the mandible, cranial depression, and palatoquadrate protrusion, c – recovery phase, characterized by hyomandibular and palatoquadrate retraction. Only the major components of the chondrocranium, mandibular, and hyoid arch are represented; the branchial arches are not included. Thick dark lines indicate muscles, large arrows indicate the movement of specific elements, and small arrows indicate direction of muscle contraction. BH = basihyal, C = ceratohyal, CC = coracoarcualis, CH = coracohyoideus, CM = coracomandibularis, CPTS = chondrocranial-palatoquadrate connective tissue sheath, EP = epaxialis, HMD = hyomandibula, LCP = ethmopalatine ligament, LH = levator hyomandibularis, LP = levator palatoquadrati, MC = Meckel's cartilage or lower jaw, OP = orbital process of palatoquadrate, PG = pectoral girdle, POD = dorsal preorbitalis, POV = ventral preorbitalis, PQ = palatoquadrate cartilage or upper jaw, QD = quadratomandibularis dorsal, QV = quadratomandibularis ventral, RC = rostral cartilage (modified and reproduced with permission from Motta et al. 1997).

between the orbital process of the upper jaw and the chondrocranium unfolds and stretches, and the distal end of the hyomandibula swings anteroventrally as the distal ceratohyal and basihyal complex continues to pivot posteroventrally. Consequently, peak hyoid depression occurs in the latter half of the compressive phase. In both species the mandible meets the maximally protruded upper jaw either with the food grasped between the teeth or after the food has been engulfed and passes through the buccal cavity. Finally, the recovery phase occurs as the palatoquadrate is retracted into its cranial seat. In the dogfish the dorsoventrally oriented levator palatoquadrati assists in its retraction, whereas in the carcharhinids the elastic ethmopalatine ligament assists. It is not known if the ethmopalatine ligament of squaloids is elastic. In both species however, the levator hyomandibularis retracts the hyomandibula helping to elevate the entire jaw apparatus (Motta et al. 1997, Wilga & Motta 1998a).

This kinematic sequence is similar to that reported for carcharhiniform sharks such as the blacknose *C. acronotus*, blacktip *Carcharhinus limbatus*, swell *C. ventriosum*, bonnethead *S. tiburo*, and Caribbean reef *Carcharhinus perezii* sharks although cranial elevation and upper jaw protrusion may be lacking in some bites (Frazzetta & Prange 1987, Ferry-Graham 1997a, 1998a, Wilga 1997, Wilga & Motta 2000, Motta personal observation) (Figure 9). This differs somewhat for the lamnid white shark, *C. carcharias*, in that peak upper jaw protrusion occurs well before the lower jaw is completely elevated, and cranial depression does not occur until the recovery phase rather than during the compressive phase (Tricas & McCosker 1984, Tricas 1985) (Figure 10). Prey capture, manipulation, and transport events in lemon *N. brevirostris*, spiny dogfish *S. acanthias*, and bonnethead *S. tiburo* sharks have a common kinematic and motor pattern sequence, but are distinguishable from each other by their duration and relative timing of individual kinematic events. Manipulation and transport events are typically shorter than capture events, although crushing manipulation events may be extensive in some species (Motta et al. 1997, Wilga 1997, Wilga & Motta 1998a,b, 2000).

In contrast to these ram-feeding sharks, suction feeding sharks from a variety of families such as the nurse shark, *G. cirratum* (Ginglymostomatidae), leopard shark, *T. semifasciata* (Triakidae), and horn shark, *H. francisci* (Heterodontidae), appear to exhibit an abbreviated kinematic sequence in which cranial elevation is reduced or lacking. In these sharks as well

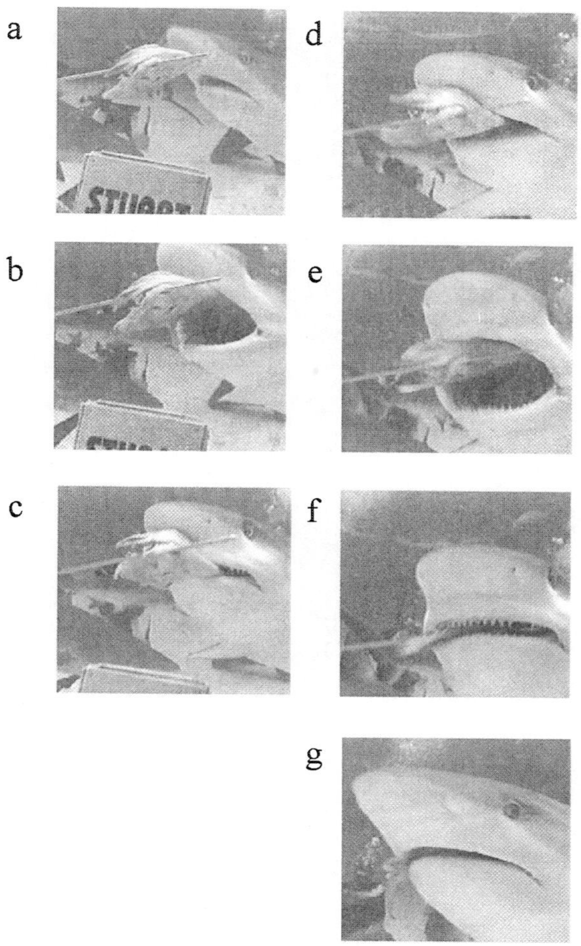

Figure 9. Food capture (a, b, c) and transport (d, e, f, g) sequence of a Caribbean reef shark, *Carcharhinus perezii*, feeding on a grouper (Serranidae) carcass offered on a pole near New Providence Island, Bahamas: a – beginning of the expansive phase marked by initiation of lower jaw depression, b – end of the expansive phase characterized by maximum gape with maximum mandible depression and cranium elevation (120 msec), c – end of the compressive phase when the cranium is depressed, the palatoquadrate protruded, and the mandible elevated as the food is seized (383 msec) (note exposed serrated upper teeth (also in f) during palatoquadrate protrusion), d – beginning of transport marked by mandible depression as the food is already grasped between the jaws, e – suction occurs as maximum gape is attained (note palatoquadrate is slightly protruded), f – the food has been sucked into the buccal cavity but is still attached to the feeding pole as the protruded upper jaw is closed against the elevated lower jaw, g – the palatoquadrate is finally retracted at the termination of the recovery phase. The horizontal width of the feeding box in (a) is 33 cm across for scale (from Pamela Christman personal communication).

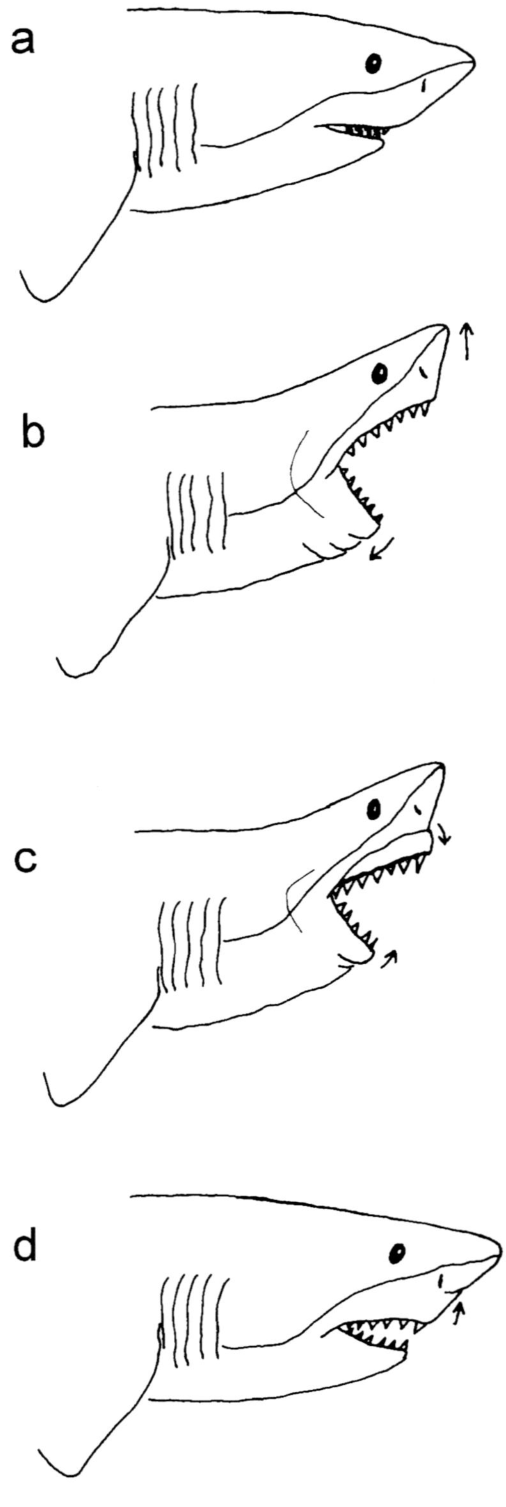

as the predominantly suction feeding spiny dogfish, *S. acanthias*, the labial cartilages protrude anteriorly as the lower jaw is depressed to effectively form a lateral enclosure of the mouth. This not only directs the suction anteriorly but may also prevent the food escaping from the sides of the mouth (Ferry-Graham 1997b, 1998b, Wilga & Motta 1998a, Pretlow-Edmonds 1999, Motta & Wilga 1999).

The duration of the entire feeding sequence in these sharks is relatively rapid considering their size. Bite duration, from the beginning of mandible depression to retraction of the jaws to their resting position is shorter for the suction feeding sharks (nurse 92 msec, horn 148 msec, leopard 170 msec). Bite duration is 200 msec for suction and 280 msec for ram feeding sequences in the dogfish (Wilga 1997, Wilga & Motta 1998a). Ram feeding sharks have much longer bite durations, such as the bonnethead 302 msec, lemon 309 msec, and swell shark 367 msec, with the white shark having the longest recorded duration of 405 msec (Tricas & McCosker 1984, Tricas 1985, Ferry-Graham 1997a, Motta et al. 1997, Wilga 1997, Wilga & Motta 1998a, 2000, Pretlow-Edmonds 1999).

Dentition and feeding

The teeth of sharks are arranged in rows on the palatoquadrate and Meckel's cartilage, and are functionally related to feeding behavior and diet. A tooth in the functional position at the edge of the jaw and its replacement teeth constitute a tooth family. Teeth in neighboring families form diagonal rows (James 1953, Reif 1976). The rate of replacement has not been demonstrated to be related to damage such as during feeding or mating, but is species specific and affected by age of the animal, diet, seasonal changes and water temperature.

Figure 10. Food capture kinematics of the white shark, *Carcharodon carcharias*, feeding on surface bait: a – shark just prior to initiation to feeding action (snout and lower jaw are at normal resting position), b – snout lift and lower jaw depression result in maximum gape, c – palatoquadrate protrusion rotates upper jaw forward and downward exposing upper teeth (lower jaw moves forward and upward; these two components comprise the actual bite), d – snout drop entails retraction of the palatoquadrate cartilage to its normal position beneath cranium. Snout drop at the end of the feeding bout during the recovery phase. Arrows indicate direction of jaw movements (reproduced with permission of the California Academy of Sciences, from Tricas & McCosker 1984).

For most species only a few teeth are replaced at a time, although some sharks have different replacement rates for upper and lower jaws (Moss 1967). The cookiecutter shark, *I. brasiliensis*, is unusual in that its relatively large lower triangular teeth are shed together as a complete set (Strasburg 1963). Replacement rates, as measured by the rate of movement of a tooth from the row lingual to the functional row to that of the functional row, varies from 9 to 12 days for the leopard shark, *T. semifasciata* (Reif et al. 1978), 9 to 28 days for the nurse shark, *G. cirratum* (Reif et al. 1978, Luer et al. 1990), 8 to 10 days for the lemon shark, *N. brevirostris* (Moss 1967), and about 4 weeks for *Heterodontus* (Reif 1976).

True homodonty, where the teeth in a jaw are all the same shape and show no abrupt change in size, is rare in recent and fossil sharks, but apparently exists in *Rhincodon* and *Cetorhinus*. Heterodonty, the term applied to significant change in size and shape of the teeth in different parts of the jaw, is common in recent and fossil sharks but the term should not be applied to the normal reduction in tooth size from the front to the rear of the mouth (Applegate 1965, Compagno 1988). Heterodonty within the mouth can include blade-like teeth more suited for cutting to long, pointed teeth suited for grasping and piercing. Horn sharks (Heterodontidae) and bonnethead sharks (Sphyrnidae), two distant taxa, both have anterior cuspidate teeth for grasping and posterior molariform crushing teeth (Figure 11) (Smith 1942, Peyer 1968, Budker 1971, Taylor 1972, Reif 1976, Nobiling 1977, Compagno 1984). Carcharhinid sharks have more cuspidate lower jaw teeth lacking serrations, with more blade-like serrated upper teeth (Figure 12) (Bigelow & Schroeder 1948, Compagno 1984, 1988). The lower teeth may be used to grasp the prey and facilitate rapid penetration into the tissue. The mandible is then elevated and the prey grasped between the upper and lower jaw teeth. The serrated upper teeth saw through the prey, often facilitated by rapid head shaking (Springer 1961, Moss 1972, 1977, Frazzetta & Prange 1987, Frazzetta 1988, 1994, Motta et al. 1997). Serrated teeth cut compliant material such as skin and muscle with a grip-and-rip mechanism where serrations trap and cut sections of material (Abler 1992). Squaloid sharks have blade-like teeth in both jaws with a large laterally pointed oblique smooth cusp that functions similarly during lateral head shaking to cut food (Compagno 1984, Wilga & Motta 1998a). We believe that protrusion of the upper jaw exposes the serrated or blade-like upper teeth facilitating their unobstructed lateral movement through the prey.

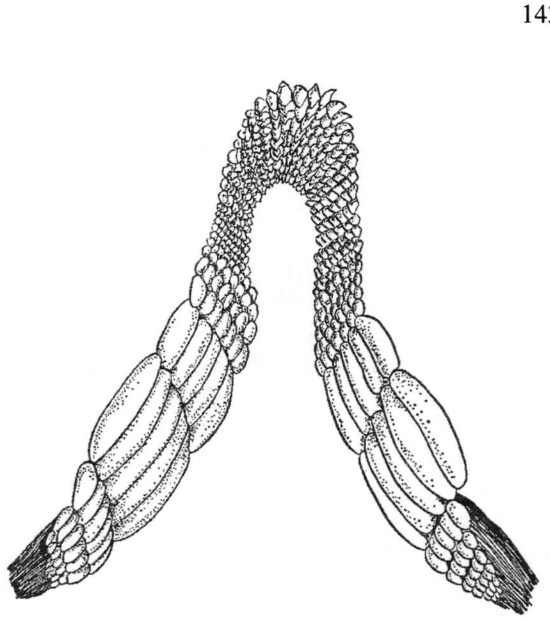

Figure 11. Dorsal view of the lower jaw and teeth of the Port Jackson shark, *Heterodontus* sp. showing the grasping teeth in the front of the jaw and the molariform or grinding teeth behind. Rostral tip of jaw at top of picture (reproduced with permission from Budker 1971, Columbia University Press).

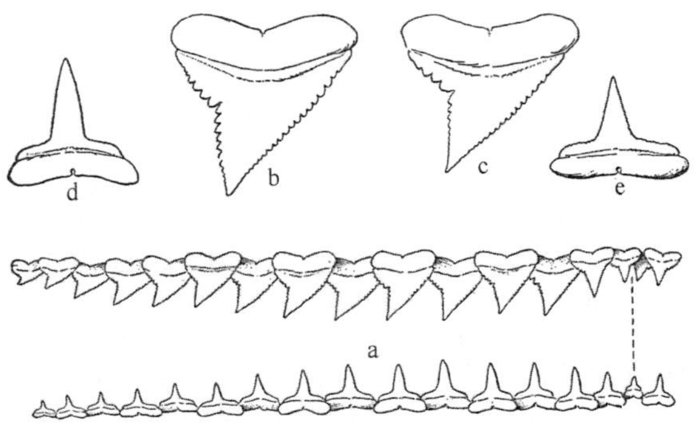

Figure 12. Teeth of silky shark, *Carcharhinus falciformis*: a – right-hand upper and lower teeth, b – fifth upper tooth, c – ninth upper tooth, d – second lower tooth, e - tenth lower tooth. Note serrations in upper teeth for cutting prey and smooth edges of lower teeth for impaling prey (Bigelow & Schroeder 1948, courtesy of the Sears Foundation for Marine Research, Yale University).

Frazzetta (1988, 1994) has proposed that the relatively loose fibrous connection of shark teeth to the jaw cartilage allows the teeth to conform to irregularities in soft tissue and guide around solid obstructions such as bone. Furthermore, he postulates that sharks can use muscle contraction to voluntarily control the strength of tooth fixation onto the jaw (Frazzetta 1994), a finding we have not been able to substantiate (Motta & Wilga 1995, Motta et al. 1997), but certainly deserves further investigation.

In contrast to many carcharhinid sharks which have out-turned tooth tips probably more suitable for piercing small fish (Frazzetta 1988), the white shark *C. carcharias* has the tips of the front teeth angled inward perhaps making them more effective at gouging chunks of flesh, grasping prey items or preventing prey escape from the mouth. During mouth closure the tooth crown angle formed between the jaw and the center-most teeth increases by 8.7 degrees as the jaw closes through an angle of 20–35 degrees, and then decreases by 15.7 degrees as the jaw is adducted through 35 degrees or more. Although the mechanism isn't clear, this is believed to facilitate a plucking action during feeding (Powlik 1995).

Ontogenetic changes in diet may be associated with ontogenetic changes in dentition. The shape of the teeth and number of tooth cusps in horn sharks (Heterodontidae) change with ontogeny. Rear replacement teeth gradually lose cusps, broaden at the base, and flatten along the crown. The more anterior recurved teeth have larger central cusps and fewer overall cusps with age. Juvenile *H. portusjacksoni* with more pointed teeth apparently take more soft-bodied prey than the adults (Smith 1942, Peyer 1968, McLaughlin & O'Gower 1971, Taylor 1972, Reif 1976, Nobiling 1977, Compagno 1984). Tricas & McCosker (1984) found that white sharks less than 3 m (TL) with relatively long and narrow teeth fed primarily on fish prey, while larger animals with broader teeth preferred marine mammals, a finding they attributed in part to differences in tooth morphology. This change in tooth morphology with growth is termed ontogenetic heterodonty and may include the appearance of or increase in tooth serrations with age as occurs in the night shark, *Carcharhinus signatus*, scalloped hammerhead, *S. lewini*, and some *Rhizoprionodon* species. However, the ecological implications of this are unclear (Raschi et al. 1982, Compagno 1988).

Sexual heterodonty occurs in many species of sharks, and in many cases the teeth of adult males are different in shape from those of females and immature males. The dimorphism is often confined to the anterior teeth, and in the carcharhinoids it is confined to smaller sharks that are less than one meter in length. Sexual heterodonty in sharks and rays appears to be related to courtship during which the males hold on to the females with their mouth, rather than feeding (Springer 1967, Compagno 1970, 1988, Ellis & Shackley 1995, Kajiura & Tricas 1996). Female sharks often have thicker skin than males due to a thicker stratum compactum layer of helically wound collagen fibers in the dermis. This might be related to the pre-copulatory biting behavior of male sharks (Motta 1977, Wainwright et al. 1978, Pratt 1979). The skin of female blue sharks, *P. glauca*, is more than twice the thickness of males, and exceeds the height of the male's dentition to reduce damage to the underlying muscle (Pratt 1979).

Seasonal change in feeding morphology occurs in basking sharks, *C. maximus*. Basking, and whale sharks, *R. typus*, are filter feeders that have specialized filamentous gill rakers that form a sieve used to filter feed. The rakers are formed from elongated and specialized mucous membrane denticles (Gudger 1941a, Peyer 1968). The sieve is apparently shed each year during winter, a period during which the sharks are believed not to feed (Parker & Boeseman 1954, Sims et al. 1997). Therefore, in many cases there appears to be a good ecomorphological relationship between diet, feeding behavior, and dental morphology in sharks. However, we still lack a good understanding of how teeth function during prey capture and manipulation.

Perhaps the least investigated aspect of teeth and their role in feeding involves the measurement of bite force. Bite force in large tiger sharks, *G. cuvieri*, can be so great as allow the shark to bite through the carapace and plastron of large chelonid sea turtles (Coles 1919, Springer 1961, Witzell 1987). Bite forces on the jaws and teeth have only been theoretically and experimentally verified for a few species (Nobiling 1977). Using an aluminum gnathodynamometer, Snodgrass & Gilbert (1967), Gilbert[1] and Evans & Gilbert (1971) estimated biting forces for a variety of sharks. Young silky sharks *Carcharhinus falciformis*, had biting forces ranging from 38 kg to 90.7 kg. In a series of experiments involving adult tiger sharks *G. cuvieri*, lemon sharks *N. brevirostris*, and dusky sharks *Carcharhinus obscurus*, they recorded a maximum force of 60 kg applied to a single tooth by a 2.5 m dusky shark. This was applied over an area of 2 square mm resulting in an estimated biting force of 30 kg per square mm. However, Frazzetta (1994) raises a cautionary note about such indirect measures of bite force. Clearly, direct measures of bite force remain an area of both theoretical and applied interest, as shark bites are also implicated in damage to underwater cables (Marra 1989).

Modulation of feeding behavior

Early studies on the feeding behavior of sharks assumed that prey capture behavior was stereotyped and consisted of a predictable series of events that show only limited variation (Gilbert[1], Tricas & McCosker 1984, Tricas 1985). However, several recent studies have noted that individual sharks may exhibit extensive modulation and variation in head and jaw movements, some in motor pattern as well, in response to different feeding situations (Moss 1972, Frazzetta & Prange 1987, Motta et al. 1997, Wilga 1997, Wilga & Motta 1998a,b, Ferry-Graham 1998b, Pretlow-Edmonds 1999). Modulation is defined as consistent repeatable differences in feeding trials in response to different feeding situations while variation is random differences among feeding trials (Chu 1989, Liem 1978, Nemeth 1997). Modulation may be important to individual survival because the ability to modulate feeding behavior may enable a predator to increase its dietary diversity or increase its capture success (Nemeth 1997).

Several species of sharks have been observed to modulate their feeding behavior in response to food size or type. According to the prey size hypothesis, sharks should modulate their capture behavior in response to prey size (Frazzetta & Prange 1987). The prey size hypothesis states that inertial suction or ram feeding should be used to capture small prey (much smaller than the diameter of the mouth), whereas biting should be used to capture medium (closer to the diameter of the mouth, but not larger) and large prey (equal to or larger than the diameter of the mouth). Based on the ram-suction index (RSI) (Norton & Brainerd 1993), the spiny dogfish, *S. acanthias*, uses suction to capture small food items (pieces of shrimp) while a combination of biting and suction is used to capture medium (pieces of herring) and large food items (whole herring) (Wilga 1997). In addition, feeding trials were shorter in duration when capturing small food items compared to large food items. In contrast, RSI values for the leopard shark, *T. semifasciata*, show that relatively greater suction is used to capture medium sized food items than when feeding on small food items. Little modulation in capture kinematics was found in response to elusivity of simulated prey in the leopard shark during suction feeding, however, the chosen food items may not have been perceived as elusive to the shark (Ferry-Graham 1998b). Food capture kinematics in bonnethead, *S. tiburo*, and juvenile swellsharks, *C. ventriosum*, generally consists of a ram-dominated capture bite that apparently is not modulated by food type or size (Wilga 1997, Wilga & Motta 2000, Ferry-Graham 1997a). Therefore, lemon, blacktip, blacknose, spiny dogfish and leopard sharks appear to be capable of modulating the contribution of suction and ram to capture different sized food items, although not always in accordance with the prey-size hypothesis, while bonnethead and swellsharks do not show modulation.

Kinematic studies using different food types or presentation techniques also provide differing results on the modulatory capabilities of shark feeding behavior. Many sharks have been observed to vary their attack approach in response to the position of the food in the water column (Moss 1972, Tricas & McCosker 1984, Fouts & Nelson 1999, Pretlow-Edmonds 1999). Several different approaches have been observed for the tiger, *G. cuvieri*, lemon, *N. brevirostris*, and dusky, *C. obscurus*, sharks: small food items are picked up from the substrate by impaling them on the lower jaw teeth, small food items in mid-water are taken by simply opening the mouth and swimming at the food, food items at the surface are grasped and bitten as the head and back are quickly elevated, and large mouthfuls are gouged from food items too large to be swallowed whole and are usually accompanied by vigorous shakes of the head and body. White sharks, *C. carcharias*, usually engulf surface bait at normal swimming speeds but may also take the bait in an accelerated rush, while underwater food may be engulfed from an upright posture or after the shark has rolled onto its side. In contrast, the horn shark, *H. francisci*, shows no modulation of suction capture kinematics when presented with food at rest on the substrate, attached to the substrate, or pressed into a feeding tube (Pretlow-Edmonds 1999). However, food capture behavior is altered by the use of additional strikes and vigorous body movements until attached or tube-hidden food is acquired (Strong 1989, Pretlow-Edmonds 1999). Similarly, food capture kinematics of several specialized suction feeding orectolobid and squatinid sharks (nurse, *G. cirratum*, spotted wobbegong, *Orectolobus maculatus*, epaulette, *Hemiscyllium ocellatum*, and angel, *Squatina californica*) are also stereotyped and for the nurse shark the kinematics change little despite the presence or absence of food (Wu 1994, Fouts & Nelson 1999, Robinson

1999). Therefore, some sharks show modulation in pre- or post-capture behavior in response to food items rather than in capture mechanism. In addition, morphologically specialized suction and ram feeders appear to have less modulatory ability in the capture mechanism than more generalized feeders that use a combination of these behaviors.

The specific behavior of head and body shaking during feeding varies among shark species (Springer 1961, Moss 1972, Motta et al. 1997, Wilga 1997, Wilga & Motta 1998a). Lemon and bonnethead sharks shake their head rapidly around the pectoral girdle axis. In contrast, tiger, spiny dogfish, and leopard sharks shake the head and entire body back and forth. Still other sharks will roll and spin the body around the prey. This head and body shaking behavior appears to be an efficient means to cut and remove mouth-sized pieces from large prey items.

Differences in the pattern of muscle activity have also been measured during feeding in sharks. Differential activation occurs in the retractor muscles of the palatoquadrate (levator palatoquadrati) and the hyomandibula (levator hyomandibularis) in the spiny dogfish, *S. acanthias*, and the Atlantic guitarfish, *Rhinobatos lentiginosus*, in capture, manipulation, and transport events during feeding (Wilga & Motta 1998a,b). These muscles elevate the upper jaw and hyoid arch respectively retracting the jaws in the recovery phase of capture and transport events. In contrast, these muscles are active during the mouth opening phase in manipulation events. It appears that modulation occurs during manipulatory events when the food is being repositioned within the buccal cavity rather than in the more rapid suction events such as capture and transport. Carcharhinid sharks, such as the lemon shark, *N. brevirostris*, and bonnethead shark, *S. tiburo*, show similar overall motor patterns among the three feeding events (Motta et al. 1997, Wilga 1997, Wilga & Motta 2000). Additional research is needed to determine the extent of this differential muscle activity among feeding behaviors.

Scaling of feeding kinematics

Ontogenetic changes in diet (Wetherbee et al.[2], Simpfendorfer 1992, Lowe et al. 1996) feeding behavior (McCosker 1985), suction forces (Tanaka 1973), and feeding mechanisms (Beebe 1941, Raschi et al. 1982, Tricas & McCosker 1984, Ellis & Shackley 1995) have been found in some shark species. However, the effect of scale on the feeding mechanics of sharks has only recently been addressed. The magnitude of head and jaw kinematics during feeding of hatchling and one-year-old swellsharks, *C. ventriosum*, doubled with a doubling of shark length (Ferry-Graham 1998a). Subsequent scaling analysis found that growth in the swellshark scaled isometrically. However, hatchling food captures were predominately ram while one-year-old captures were predominately suction. Hatchlings were more active during capture and may be overtaking the food before it can be drawn into the mouth by suction inflow. Alternatively, hatchling sharks may be too small to use suction effectively, suggesting that there may be an absolute size at which a particular food capture behavior may be effective (Ferry-Graham 1998a). In addition to this assessment of predator-size scaling with prey size, data on prey-size scaling with predator size will provide information on modulatory abilities of sharks.

In a broad ontogenetic study of the feeding mechanics (12 individuals ranging from 33 cm to 254 cm total length), it was found that the feeding apparatus and kinematics of the nurse shark, *G. cirratum*, also scaled isometrically (Robinson 1999). However, the duration of the feeding events had a tendency to increase with negative allometry, i.e., larger sharks tended to take longer to perform the same kinematic feeding sequence than smaller nurse sharks. Physiological constraint on muscular contraction dynamics was suggested as the underlying factor causing the increase in the timing variables (Robinson 1999), supporting a model proposed by Wainwright & Richards (1995) for teleost fishes. Similar results were found in the bonnethead shark, *S. tiburo*, where peak kinematic events were attained later in larger sharks than smaller sharks (Wilga 1997, Wilga & Motta 2000). The ecological implications of scaling of prey capture kinematics, if any, are not clear.

Mechanics and function of jaw protrusion

Protrusion of the upper jaw away from the cranium during the jaw closing phase is an integral part of feeding behavior in most sharks and likely serves numerous functions. Protrusion may shift the entire jaw apparatus away from the cranium and expose the teeth to allow more efficient bites and manipulation of prey (Frazzetta 1994). Protrusible jaws might provide the shark with a versatile yet hydrodynamic subterminal mouth (Tricas & McCosker 1984). Protrusion of the upper jaw is also believed to facilitate the cutting action of the teeth and allow deep gouging bites to be made into oversized prey (Moss 1977, Tricas & McCosker 1984). Upper jaw protrusion may enable the shark to grasp items from the substrate with more precision (Moss 1972). The teeth may be reoriented during protrusion of the upper jaw for increased grasping ability (Frazzetta & Prange 1987). In addition, nearly simultaneous protrusion of the upper jaw while the lower jaw is elevating may also provide the shark with a better grasp of struggling or elusive prey (Frazzetta & Prange 1987). However, few of these ideas have been experimentally tested.

There are quantitative data to suggest that one function of upper jaw protrusion in some sharks may be to assist in jaw closure as Frazzetta & Prange (1987) hypothesize. In the spiny dogfish *S. acanthias*, the lemon shark *N. brevirostris* and the Atlantic guitarfish *R. lentiginosus* protrusion of the upper jaw significantly decreases the jaw closing distance necessary for the lower jaw to travel before meeting the upper jaw by 27–64% (Motta et al. 1997, Wilga & Motta 1998a,b). However, in sevengill *N. cepedianus* (16%) bonnethead *S. tiburo* (14%) and swell *C. ventriosum* sharks upper jaw protrusion is slight and does not contribute significantly to decreasing this distance (Ferry-Graham 1997a, Wilga 1997, Wilga & Motta 2000). Note that protrusion of the upper jaw occurs independently of cranial depression (Motta et al. 1997, Ferry-Graham 1998b, Wilga & Motta 1998a,b).

Another function of protruding the upper jaw may be to disable prey. Capture of large prey often elicits vigorous lateral head shaking in the tiger *G. cuvieri*, white *C. carcharias*, blacknose *C. acronotus*, blacktip *C. limbatus*, lemon *N. brevirostris*, bonnethead *S. tiburo*, spiny dogfish *S. acanthias*, leopard *T. semifasciata* sharks as well as other sharks (Springer 1961, Moss 1972, Frazzetta & Prange 1987, Frazzetta 1994, Motta et al. 1997, Wilga 1997, Wilga & Motta 1998a, Wilga & Motta 2000). This head shaking behavior is thought to be used for cutting or gouging smaller pieces from large prey. Cutting is facilitated by protrusion of the upper jaw into the prey (Springer 1961, Hobson 1963, Gilbert[1], Moss 1972, 1977, Tricas & McCosker 1984, Frazzetta & Prange 1987, Frazzetta 1988, 1994, Powlik 1995). Vigorous lateral head shaking with the prey grasped between the lower jaw and the protruded upper jaw results in the prey being cut in two pieces in the spiny dogfish, *S. acanthias*, and the lemon shark, *N. brevirostris* (see pp. 142–145) (Motta et al. 1997, Wilga & Motta 1998a). Rapid upper jaw protrusion in the horn shark, *H. francisci*, may be used like a striking chisel to remove attached prey, such as sea urchins, from the substrate (Pretlow-Edmonds 1999). Furthermore, the white shark, *C. carcharias*, may prolong head elevation when feeding on whales in order to deliver multiple bites by repeated protrusion and retraction of the upper jaw until a consumable chunk is removed (Pratt et al. 1982).

Several mechanisms of jaw protrusion have been described from observational and manipulative studies of feeding in sharks (Luther 1909, Haller 1926, Zlabek 1931, Alexander 1967, Moss 1972, Nobiling 1977, Tricas & McCosker 1984, Frazzetta & Prange 1987, Frazzetta 1994, Shirai & Okamura 1992, Wu 1994). Some of these hypotheses have been supported by a few functional studies that have correlated head and jaw movements during feeding with simultaneous muscle activity. The feeding mechanism of the spiny dogfish, *S. acanthias*, may serve as a general mechanism for squaliform sharks (Wilga & Motta 1998a). The preorbitalis muscle [homologous to the ventral preorbitalis in carcharhiniform sharks (Moss 1972, Compagno 1988)] produces an anteriorly directed force near the posterior region of the jaw (Figure 13). This forces the orbital process of the upper jaw to slide ventrally along the orbital wall and ethmopalatine groove to protrude the upper jaw. As the upper jaw is protruding, the orbital process slides ventrally within the sleeve-like ethmopalatine ligament until the ligament becomes taught at which time upper jaw protrusion is complete. As the upper jaw protrudes, the entire jaw moves anteroventrally while the hyomandibula passively follows. The distal end of the hyomandibula is pulled ventrally and only slightly anteriorly. Since the action of an adductor muscle is to bring two elements closer together, contraction of the quadratomandibularis not only elevates the lower jaw, but may also pull the upper jaw away from the cranium toward the lower jaw. In this way, the quadratomandibularis may assist the preorbitalis in protruding the upper jaw, which can be protruded a considerable distance.

The mechanism of upper jaw protrusion in carcharhiniform sharks differs slightly from that in squaliform sharks. The carcharhiniform mechanism has been proposed in several studies (Luther 1909, Moss 1972, Frazzetta & Prange 1987, Frazzetta 1994) and has largely been supported in functional studies of feeding in the lemon shark, *N. brevirostris*, and the bonnethead shark, *S. tiburo* (Motta et al. 1997, Wilga 1997,

Wilga & Motta 2000). Carcharhiniform sharks have a derived condition in which the levator palatoquadrati muscle is oriented more anteroposteriorly instead of dorsoventrally as in dogfish (Nakaya 1975, Moss 1972, Compagno 1988). In this orientation, the levator palatoquadrati muscle can assist the dorsal and ventral preorbitalis muscle (carcharhiniform sharks have two divisions of the preorbitalis muscle) in protruding the upper jaw (Figures 5, 8, 13). The dorsal division of the preorbitalis pulls the palatoquadrate ventrally as the ventral division of the preorbitalis and the levator palatoquadrati muscles pull it anterodorsally. Similar to the dogfish, the orbital process of the palatoquadrate is forced to glide on the ethmopalatine groove, and the resultant reaction force drives the upper jaw anteriorly and ventrally to protrude it. As the upper jaw is protruded, the rope-like ethmopalatine ligament unfolds (folded in the resting position) until it becomes taught, halting upper jaw protrusion. As the upper jaw protrudes, the jaws and the distal end of the hyomandibula also swing anteroventrally but to a greater extent than the spiny dogfish. Contraction of the quadratomandibularis muscle might also assist upper jaw protrusion as described above (Moss 1965).

Still another mechanism for protruding the jaws was proposed for orectolobiform sharks using cineradiography to visualize cartilage movements of the head and jaws in the spotted wobbegong, *O. maculatus* (Wu 1994). Compression of the two sides of the hyomandibula and lower jaw by the interhyoideus and intermandibularis muscles, respectively, is thought to protrude the jaws anteriorly, resulting in a more acute symphyseal angle. Concomitant with hyomandibular compression is rotation of the ceratohyal around a posterior process on the lower jaw that may push the hyomandibulae and jaws anteroventrally to protrude the jaws. However, the orectolobiform mechanism, as well as heterodontiform and lamniform mechanisms, await confirmation by detailed functional studies. The mechanics of upper jaw protrusion differ among taxa,

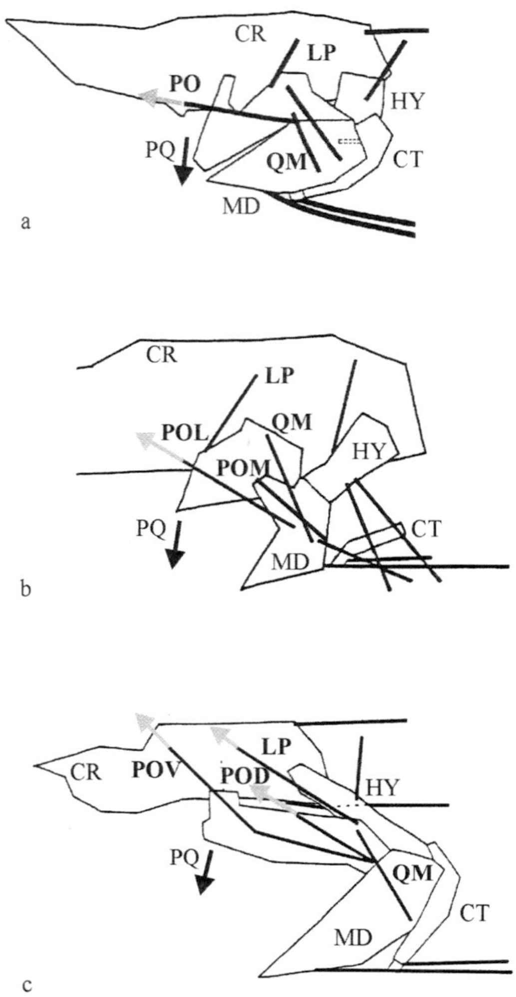

Figure 13. Mechanisms of upper jaw protrusion in elasmobranchs: a – spiny dogfish, *Squalus acanthias*. The PO muscle pulls the upper jaw anteriorly, forcing the orbital process of the upper jaw to slide ventrally along the orbital wall to protrude the upper jaw. As the upper jaw protrudes, the entire jaw moves anteroventrally while the distal hyomandibula is pulled passively ventrally and slightly anteriorly. b – Atlantic guitarfish, *Rhinobatos lentiginosus*. The POL muscle pulls the upper jaw anteriorly while the POM pulls the upper jaw ventrally. As the upper jaw protrudes, the entire jaw moves ventrally while the distal hyomandibula is pulled passively ventrally and slightly posteriorly. c – bonnethead shark, *Sphyrna tiburo*. The POD pulls the upper jaw ventrally while the POV and LP muscles pull it anterodorsally, forcing the orbital process of the palatoquadrate to glide on the ethmopalatine groove driving the upper jaw anteriorly and ventrally. During upper jaw protrusion, the jaws and the distal hyomandibula also swing anteroventrally to a greater extent than the spiny dogfish. Note that the QM can assist in protruding the upper jaw in all three species by pulling the upper jaw towards the lower jaw. Solid black lines indicate muscles with gray arrows indicating direction of action. Back arrows indicate direction of cartilage movement. Muscle names are in bold font, skeletal elements are in normal font. CR = cranium, CT = ceratohyal, HY = hyomandibula, LP = levator palatoquadrati, MD = mandible, PO = preorbitalis, POD = preorbitalis dorsal division, POL = preorbitalis lateral division, POM = preorbitalis medial division, POV = preorbitalis ventral division, PQ = palatoquadrate, QM = quadratomandibularis (reproduced with permission from Wilga 1997).

presumably due to the evolutionary differences in the suspensory apparatus and myology.

Relationship of jaw suspension to feeding

As noted above, upper jaw protrusion is a key feature of feeding behavior in most sharks. In turn, jaw suspension has been presumed to be a major factor in determining the extent of upper jaw protrusion, as well as the magnitude of the gape (Schaeffer 1967, Moss 1972, Maisey 1980). The types of jaw suspension present in sharks are defined primarily by the amount of support that the hyomandibular element of the hyoid arch gives to the mandibular arch and secondarily by the number of connections and ligaments between the palatoquadrate and the cranium (Figure 14) (Huxley 1876, Gregory 1904, Maisey 1980, Lund & Grogan 1997, Grogan et al. 1999). In amphistylic jaw suspension (i.e., *Paleospinax*), the palatoquadrate or upper jaw is suspended from the cranium by an ethmoid articulation with the ethmoid region of the cranium anteriorly, by a relatively small hyomandibula that is presumed to contribute little to jaw support posteriorly, and by a postorbital articulation between the cranium and palatoquadrate (Figure 14a). Some researchers propose that the tight postorbital articulation severely limits or prevents protrusion of the upper jaw in amphistylic sharks (Compagno 1977) while others suggest that limited upper jaw protrusion is present (Moss 1972, 1977, Zangerl & Williams 1975). In contrast, the relatively large hyomandibula is thought to contribute more to supporting the jaws than the anterior ethmoid articulation in hyostylic jaw suspension (i.e., carcharhiniform, lamniform, orectolobiform, and heterodontiform sharks) (Figure 14b). In hyostyly, the ethmoid articulation is relatively loose and all researchers agree that these sharks have freely protractile jaws (Huxley 1876, Maisey 1980, Compagno 1988). Orbitostylic jaw suspension is similar to hyostyly except instead of an ethmoid articulation, there is an orbital articulation between the palatoquadrate and orbital region of the cranium (i.e., squaliform sharks) (Figure 14c) (Maisey 1980). Some presume that the orbital articulation is so tight as to limit protrusion of the upper

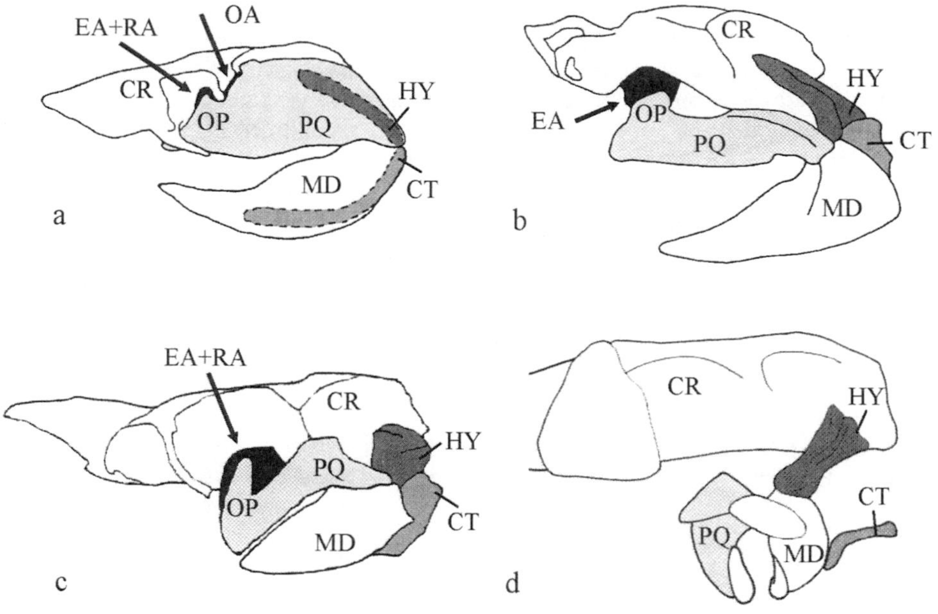

Figure 14. Jaw suspension type in: a – Hexanchiformes (amphistylic), b – Carcharhiniformes (hyostylic), c – Squaliformes (orbitostylic), d – Batoidea (euhyostylic). Black = ethmopalatine ligament, light gray = palatoquadrate or upper jaw, medium gray = ceratohyal, dark gray = hyomandibula, white = cranium and mandible. CR = cranium, CT = ceratohyal, EA = ethmopalatine articulation, HY = hyomandibula, MD = mandible or lower jaw, OA = otic articulation, OB = orbital process, PQ = palatoquadrate or upper jaw, RA = orbital articulation (reproduced with permission from Wilga 1997).

jaw (Schaeffer 1967, Compagno 1977) while others report that the jaws are freely protractile (Haller 1926, Moss 1977). Note that there is an ethmopalatine ligament interconnecting the cranium and the palatoquadrate in ethmoid and orbital articulations. Batoids (skates, rays and guitarfishes) have an euhyostylic type of jaw suspension in which the jaws are suspended solely by the hyomandibulae and lack any ligamentous connections or articulations between the palatoquadrate and the cranium (Figure 14d) (Gregory 1904, Luther 1909, Schaeffer 1967). Accordingly, most agree that euhyostylic batoids are capable of extreme jaw protrusion.

Contrary to popular belief, jaw suspension type is not a good predictor of mobility of the jaws (as measured by upper jaw protrusion distance during feeding in sharks) as concluded by a quantitative study of kinematic and morphological variables of five elasmobranchs (Wilga unpublished). The bonnethead *S. tiburo* is a hyostylic shark and has nearly the same weak jaw protrusion (10% head length, HL) as the sevengill *N. cepedianus* (9% HL), which is an orbitostylic shark (some categorize it as amphistylic). In addition, the spiny dogfish *S. acanthias* (29% HL), which is an orbitostylic shark (some categorize it as hyostylic) and the Atlantic guitarfish *R. lentiginosus* (26% HL), which is euhyostylic, both have extensive upper jaw protrusion. Furthermore, the lemon shark *N. brevirostris*, another hyostylic shark, has only intermediate protrusion (18% HL). Anatomical dissection reveals that the spiny dogfish has a relatively longer ethmopalatine ligament than the lemon shark, which has a longer ligament than the bonnethead and sevengill, with the ligament being absent in the guitarfish. Thus, in this study, the best predictor of jaw mobility appears to be either the presence of a long ethmopalatine ligament or its complete absence, both of which permits the upper jaw to move further from the cranium.

Studies based on anatomy of heterodontiform and orectolobiform sharks, both of which are classified as hyostylic, suggest that their extensive ethmoid articulation greatly restricts upper jaw protrusion and makes the jaw suspension functionally amphistylic (Holmgren 1940, 1942, Smith 1942, Schaeffer 1967, Maisey 1980). As predicted, several of these sharks, horn *H. francisci* (Heterodontiformes, 7% HL), nurse *G. cirratum* (Orectolobiformes, 12% HL) and the epaulette *H. ocellatum* (Orectolobiformes, 9% HL) have relatively little upper jaw protrusion (Wu 1994, Pretlow-Edmonds 1999). However, the spotted wobbegong *O. maculatus* (Orectolobiformes) can protrude its upper jaw up to 33% of its head length (Wu 1994). It appears that there is a wide range of variation in the extent of upper jaw protrusion within any particular jaw suspension type and generalizations regarding jaw suspension and upper jaw protrusion ability are not supported.

Long posteriorly directed hyomandibulae has been presumed to allow the jaws to swing forward and protrude the upper jaw a greater distance compared to species with shorter more laterally directed hyomandibulae, and secondly to swing laterally when the mouth opens increasing the lateral and vertical gape distance (Moss 1972, 1977). A recent study on the morphology of the jaw suspension shows that long posteriorly directed hyomandibulae do not promote greater jaw protrusion during feeding in the five species examined (Wilga unpublished), but they do correlate directly with a larger vertical gape. This is because having a longer hyomandibula is simply a function of having a longer jaw. The hyomandibula connects the jaws to the cranium and therefore a longer jaw at the same angle of mouth opening will have a larger vertical gape. The sevengill, bonnethead, and lemon sharks all have long posteriorly directed hyomandibulae, long jaws and larger vertical gapes. In contrast the spiny dogfish has short laterally directed hyomandibulae and a smaller vertical gape, while the guitarfish has anteriorly directed intermediate length hyomandibulae but also has a small vertical gape (Wilga 1997). According to this study, longer hyomandibulae do appear to confer a larger vertical gape but have no effect on jaw protrusion.

Interestingly, those sharks studied with relatively longer jaws also capture prey using bite or ram feeding, while those sharks with shorter jaws use inertial suction to capture food. In bony fishes, characteristics such as relatively long jaws with restricted jaw mobility have been predicted to be specializations for ram feeding, while fishes with small gapes and kinetic jaws are proposed to be specialized for suction feeding (Norton 1995). Larger gapes are proposed to improve the capture success of ram or biting predators by increasing the volume of the capture apparatus (Norton 1991, 1995). In contrast, smaller gapes improve the capture success of suction feeding predators by increasing the velocity of water flow into the mouth (see pp. 134–137). More research is needed on the morphology of the feeding apparatus and feeding mechanics before such generalizations can be accepted.

Thus, upper jaw mobility varies widely by jaw suspension type and predictions of jaw mobility based on jaw suspension type are usually not supported. More detailed morphological and functional analyses are needed to further investigate the relationship of the jaw suspension to the feeding mechanism, in particular, studies are lacking on lamniform sharks.

Future directions

Our knowledge of the feeding behaviors, mechanisms, and mechanics of sharks has increased immensely since the early 1900s. There now exists a relatively good understanding of the diet of a subset of species, we have begun to model the mechanics of the feeding apparatus of a handful of species, and understand dental structure, tooth replacement and growth. An understanding of feeding behavior still lags behind many other aspects of their feeding biology, primarily because of the difficulty of recording natural feeding behavior of these predators. Consequently, a great deal of the data collected on their feeding behavior and kinematics is gleaned from their diets, fortuitous observations, and a handful of controlled feeding experiments either in the laboratory or field. As much as we have advanced in our understanding of these predators we have just begun in other areas of research. For example, we have little understanding of the evolution of feeding behaviors such as suction, ram, and biting in elasmobranchs. We therefore pose a variety of hypotheses and questions that remain to be addressed.

The functional specialization for suction-feeding hypothesis
Paleontological evidence and preliminary data on prey capture by more basal sharks including sixgill and sevengill sharks indicates that early sharks primarily grasped their prey with negligible suction. We hypothesize that from an early shark ancestor that was morphologically suited for biting and tearing its prey yet could undoubtedly employ some suction, morphological and functional specialization for suction feeding has arisen repeatedly in numerous elasmobranch clades.

The aquatic suction-feeding functional convergence hypothesis
Sharks have a relatively simple feeding apparatus composed of fewer anatomical elements compared to that of teleost fishes. Despite this, we hypothesize that due to hydrodynamic constraints placed upon aquatic feeding vertebrates, many of the morphological and functional specializations for suction feeding that have arisen in sharks are convergent with those of inertial suction feeding bony fishes.

The feeding modulation hypothesis
Ram and biting carcharhinid and lamnid sharks appear to have a versatile and modulated prey capture kinematic sequence that varies with prey or food type, size and location. Conversely, suction feeding sharks appear to have capture events of shorter duration that are more stereotyped. We hypothesize that sharks which are morphologically specialized for inertial suction will have capture events that are more stereotyped with less modulatory ability than capture events from sharks that ram or bite. Furthermore, rapid suction captures will be preprogrammed stereotyped bites that go to completion once initiated, regardless of sensory input.

The independence of jaw suspension and feeding behavior hypothesis
Traditional studies have categorized different shark taxa based upon the anatomical arrangement of the jaws and cranium, and the suspension of the jaws. Preliminary work indicates that the extent of jaw protrusion, which has been proposed to be related to different prey capture behaviors, is not simply related to traditional jaw suspension types. We hypothesize that the overly simplistic categorization of traditional jaw suspension types (amphistylic, hyostylic) is not a good predictor of jaw mobility, and has a complex relationship with prey capture behavior (suction, ram biting, biting and gouging, filter feeding) and feeding performance.

In addition to these hypotheses we also pose questions that remain to be addressed: (1) How does prey capture differ among sharks, and are there still unexplored methods of prey capture yet to be revealed? (2) How do teeth function during prey capture and manipulation in the different feeding types? (3) Do most sharks have modulatory abilities similar to bony fishes? and (4) What are the functions of upper jaw protrusion in sharks? These hypotheses, questions, and their answers will help to guide research on the feeding biology of sharks into the next millennium.

Acknowledgements

The authors wish to thank the organizers of this symposium for the invitation to include this review in the proceedings. This paper represents, in part, the plenary address 'Feeding behavior and mechanics of sharks: past, present and future' by P.J.M. to the 78th annual joint meeting of American Society of Ichthyologists and Herpetologists, Herpetologist League, Society for the Study of Amphibians and Reptiles, American Elasmobranch Society, and the Canadian Association of Herpetologists hosted by the University of Guelph, Guelph, Ontario, Canada on 16–22 July 1998. The organizers of that meeting are gratefully acknowledged. Research by both authors was in part funded by NSF grants DEB 9117371 and IBN 9807863 to P.J.M and by NSF grant 93-02844 to C.D.W. We thank Pamela Christman and Stuart Cove's Dive South Ocean for video footage, and L. Ferry-Graham, T. Frazzetta and T.C. Tricas for their comments on this manuscript.

References cited

Abler, W.L. 1992. The serrated teeth of tyrannosaurid dinosaurs, and biting structures in other animals. Paleobiology 18: 161–183.

Alexander, R.McN. 1967. Functional design in fishes. Hutchinson University Library, London. 160 pp.

Allis, E.P. Jr. 1923. The cranial anatomy of *Chlamydoselachus anguineus*. Acta. Zool. 4: 123–221.

Applegate, S.P. 1965. Tooth terminology and variation in sharks with special reference to the sand shark, *Carcharias taurus* Rafinesque. Los Angeles County Museum Contributions in Science 86: 1–18.

Beebe, W. 1941. External characters of six embryo nurse sharks, *Ginglymostoma cirratum* (Gmelin). Zoologica 26: 9–12.

Bleckmann, H. & M.H. Hofmann. 1999. Special senses. pp. 300–328. *In:* W.C. Hamlett (ed.) Sharks, Skates, and Rays, The Biology of Elasmobranch Fishes, The Johns Hopkins University Press, Baltimore.

Bigelow, H.B. & W.C. Schroeder. 1948. Sharks. pp. 59–546. *In:* J. Tee-Van (ed.) Fishes of the Western North Atlantic, Memoir Sears Foundation for Marine Research, New Haven.

Budker, P. 1971. The life of sharks. Columbia University Press, New York. 222 pp.

Bullis, H.R. 1961. Observations on the feeding behavior of white-tip sharks on schooling fishes. Ecology 42: 194–195.

Carroll, R.L. 1988. Vertebrate paleontology and evolution. W.H. Freeman and Company, New York. 698 pp.

Castro, J.I. 1996. The sharks of North American waters. Texas A & M University Press, College Station. 180 pp.

Chu, C.T. 1989. Functional design and prey capture dynamics in an ecologically generalized surfperch (Embiotocidae). J. Zool., Lond. 217: 417–440.

Clark, E. & D.R. Nelson. 1997. Young whale sharks, *Rhincodon typus*, feeding on a copepod bloom near La Paz, Mexico. Env. Biol. Fish. 50: 63–73.

Coles, R.J. 1915. Notes on the sharks and rays of Cape Lookout, N.C. Proc. Biol. Soc. Washington 28: 89–94.

Coles, R.J. 1919. The large sharks of cape lookout, North Carolina. The white shark or maneater, tiger shark and hammerhead. Copeia 1919: 34–43.

Colman, J.G. 1997. A review of the biology and ecology of the whale shark. J. Fish Biol. 51: 1219–1234.

Compagno, L.J.V. 1970. Systematics of the genus *Hemitriakis* (Selachii: Carcharhinidae), and related genera. Proc. Calif. Acad. Sci. ser. 4, 38: 63–98.

Compagno, L.J.V. 1973. Interrelationships of living elasmobranchs. pp. 15–61. *In:* P.H. Greenwood, R.S. Miles & C. Patterson (ed.) Interrelationships of Fishes, Zool. J. Linn. Soc. Supplement 1(53).

Compagno, L.J.V. 1977. Phyletic relationships of living sharks and rays. Amer. Zool. 17: 303–322.

Compagno, L.J.V. 1984. FAO species catalogue, Vol. 4. Sharks of the World. FAO Fisheries Synopsis No. 125, 4: 1–655.

Compagno, L.J.V. 1988. Sharks of the order Carcharhiniformes. Princeton University Press, Princeton. 486 pp.

Cortes, E. 2000. Standardized diet compositions and trophic levels of sharks. ICES J. Mar. Sci. (in press).

Cortes, E. & S.H. Gruber. 1990. Diet, feeding habits and estimates of daily ration of young lemon sharks, *Negaprion brevirostris* (Poey). Copeia 1990: 204–218.

Cortes, E., C.A. Manire & R.E. Hueter. 1996. Diet, feeding habits, and diel feeding chronology of the bonnethead shark, *Sphyrna tiburo*, in southwest Florida. Bull. Mar. Sci. 58: 353–367.

Daniel, J.F. 1915. The anatomy of *Heterodontus francisci* II: the endoskeleton. J. Morphol. 26: 447–493.

Daniel, J.F. 1934. The elasmobranch fishes. University of California Press, Berkeley. 332 pp.

Diamond, J.M. 1985. Filter-feeding on a grand scale. Nature 316: 679–680.

Ebert, D.A. 1991. Observations on the predatory behaviour of the sevengill shark *Notorynchus cepedianus*. S. Afr. J. Mar. Sci. 11: 455–465.

Edgeworth, F.H. 1935. Cranial muscles of vertebrates. Cambridge University Press, Cambridge. 493 pp.

Eibl-Eibesfeldt, I. & H. Hass. 1959. Erfahrungen mit Haien. Zeit. Tierpsychol. 16: 733–746.

Ellis, J.R. & S.E. Shackley. 1995. Ontogenetic changes and sexual dimorphism in the head, mouth and teeth of the lesser spotted dogfish. J. Fish Biol. 47: 155–164.

Evans, W.E. & P.W. Gilbert. 1971. The force of bites by the silky shark (*Carcharhinus falciformis*) measured under field conditions. Naval Undersea Research and Development Center, San Diego. 8 pp.

Ferry-Graham, L.A. 1997a. Feeding kinematics of juvenile swellsharks, *Cephaloscyllium ventriosum*. J. Exp. Biol. 200: 1255–1269.

Ferry-Graham, L.A. 1997b. Effects of prey size and elusivity on prey capture kinematics in leopard sharks, *Triakis semifasciata*. Amer. Zool. 37: 82A.

Ferry-Graham, L.A. 1998a. Feeding kinematics of hatchling swellsharks, *Cephaloscyllium ventriosum* (Scyliorhinidae): the importance of predator size. Mar. Biol. 131: 703–718.

Ferry-Graham, L.A. 1998b. Effects of prey size and mobility of prey-capture kinematics in leopard sharks, *Triakis semifasciata*. J. Exp. Biol. 201: 2433–2444.

Finstad, W.O. & D.R. Nelson. 1975. Circadian activity rhythm in the horn shark, *Heterodontus francisci*: effect of light intensity. Bull. S. Calif. Acad. Sci. 74: 20–26.

Fouts, W.R. 1995. The feeding behavior and associated ambush-site characteristics of the Pacific angel shark, *Squatina californica*, at Santa Catalina Island, California. M.S. Thesis, California State University, Long Beach. 147 pp.

Fouts, W.R. & D.R. Nelson. 1999. Prey capture by the Pacific angel shark, *Squatina californica*: visually mediated strikes and ambush-site characteristics. Copeia 1999: 304–312.

Frazzetta, T.H. 1988. The mechanics of cutting and the form of shark teeth (Chondrichthyes, Elasmobranchii). Zoomorphology 108: 93–107.

Frazzetta, T.H. 1994. Feeding mechanisms in sharks and other elasmobranchs. Adv. Comp. Environ. Physiol. 18: 31–57.

Frazzetta, T.H. & C.D. Prange. 1987. Movements of cephalic components during feeding in some requiem sharks (Carcharhiniformes: Carcharhinidae). Copeia 1987: 979–993.

Gadow, H. 1888. On the modifications of the first and second visceral arches, with special reference to the homologies of the auditory ossicles. Philos. Trans. Royal Soc. Lond. 179B: 451–485.

Gilbert, P.W. 1962. The behavior of sharks. Sci. Amer. 207: 60–68.

Gohar, H.A.F. & F.M. Mazhar. 1964. The internal anatomy of Selachii from the North Western Red Sea. Publ. Mar. Biol. Stat. Al-Ghardaqa, Egypt 13: 145–240.

Goodey, T. 1910. A contribution to the skeletal anatomy of the frilled shark, *Chlamydoselachus anguineus* (Gar.) Proc. Zool. Soc. London 2: 540–571.

Gregory, W.K. 1904. The relations of the visceral arches to the chondrocranium. Biol. Bull. 7: 55–69.

Grogan E.D., R. Lund & D. Didier. 1999. Description of the chimaerid jaw and its phylogenetic origins. J. Morphol. 239: 45–59.

Gudger, E.W. 1907. A note on the hammerhead shark (*Sphyrna zygaena*) and its food. Science 25: 1005.

Gudger, E.W. 1941a. The feeding organs of the whale shark, *Rhineodon typus*. J. Morphol. 68: 81–99.

Gudger, E.W. 1941b. The food and feeding habits of the whale shark, *Rhineodon typus*. Journal of the Elisha Mitchell Scientific Society 57: 57–72.

Haller, G. 1926. Über die Entwicklung, den Bau und die Mechanik des Kieferapparates des Dornhais (*Acanthias vulgaris*). Zeitschr. Mikrosk. Anat. Forsch. 5: 749–793.

Heupel, M.R. & M.B. Bennett. 1998. Observations of the diet and feeding habits of the epaulette shark, *Hemiscyllium ocellatum* (Bonnaterre), on Heron Island Reef, Great Barrier Reef, Australia. Mar. Freshwater Res. 49: 753–756.

Hobson, E.S. 1963. Feeding behavior in three species of sharks. Pac. Sci. 17: 171–194.

Hodgson, E.S. & R.F. Mathewson. (ed.) 1978. Sensory biology of sharks, skates, and rays. Office of Naval Research, Arlington. 666 pp.

Holmgren, N. 1940. Studies on the head of fishes. Embryological, morphological, and phylogenetical researches part I. Development of the skull in sharks and rays. Acta. Zool. 21: 51–267.

Holmgren, N. 1942. Studies on the head of fishes. An embryological, morphological, and phylogenetical study part III. The phylogeny of elasmobranch fishes. Acta. Zool. 23: 129–261.

Hueter, R.E. & P.W. Gilbert. 1990. The sensory world of sharks. pp. 48–55. *In:* S.H. Gruber (ed.) Discovering Sharks, American Littoral Society, Highlands.

Huxley, T. H. 1876. On *Ceratodus fosteri*, observation on the classification of fishes. Proc. Zool. Soc. London 1876: 24-59.

James, W.W. 1953. The succession of teeth in elasmobranchs. Proc. Zool. Soc. Lond. 123: 419–475.

Jones, E.C. 1971. *Isistius brasiliensis*, a squaloid shark, the probable cause of crater wounds on fishes and cetaceans. U.S. Fish. Bull., 69: 791–798.

Kajiura, S.M. & T.C. Tricas. 1996. Seasonal dynamics of dental sexual dimorphism in the Atlantic stingray, *Dasyatis sabina*. J. Exp. Biol. 199: 2297–2306.

Klimley, A.P. 1987. The determinants of sexual segregation in the scalloped hammerhead shark, *Sphyrna lewini*. Env. Biol. Fish. 18: 27–40.

Klimley, A.P. 1994. The predatory behavior of the white shark. Amer. Sci. 82: 122–133.

Klimley, A.P. 1995. Hammerhead city. Nat. Hist. 104: 32–39.

Klimley, A.P. & S.D. Anderson. 1996. Residency patterns of white sharks at the South Farallon Islands, California. pp. 365–373. *In:* A.P. Klimley & D.G. Ainley (ed.) Great White Sharks, The Biology of *Carcharodon carcharias*, Academic Press, New York.

Klimley, A.P., S.B. Butler, D.R. Nelson & A.T. Stull. 1988. Diel movements of scalloped hammerhead sharks, *Sphyrna lewini* Griffith and Smith, to and from a seamount in the Gulf of California. J. Fish Biol. 33: 751–761.

Klimley, A.P. & D.R. Nelson. 1981. Schooling of the scalloped hammerhead shark, *Sphyrna lewini*, in the Gulf of California. U.S. Fish. Bull. 79: 356–360.

Klimley, A.P. & D.R. Nelson. 1984. Diel movement patterns of the scalloped hammerhead (*Sphyrna lewini*) in relation to El Bajo Espiritu Santo: a refuging central-position social system. Behav. Ecol. Sociobiol. 15: 45–54.

Klimley, A.P., P. Pyle & S.D. Anderson. 1996. The behavior of white sharks and their pinniped prey during predatory attacks. pp. 175–191. *In:* A.P. Klimley & D.G. Ainley (ed.) Great White Sharks, The Biology of *Carcharodon carcharias*, Academic Press, New York.

Lauder, G.V. 1980. Evolution of the feeding mechanism in primitive actinopterygian fishes: a functional anatomical analysis of *Polypterus*, *Lepisosteus*, and *Amia*. J. Morphol. 163: 283–317.

Lauder, G.V. 1985. Aquatic feeding in lower vertebrates. pp. 210–229. *In:* M. Hildebrand, D.M. Bramble, K.F. Liem & D.B. Wake (ed.) Functional Vertebrate Morphology, Belknap Press, Cambridge.

LeBoeuf, B.J., J.E. McCosker & J. Hewitt. 1987. Crater wounds on northern elephant seals: the cookiecutter shark strikes again. U.S. Fish. Bull. 85: 387–392.

Liem, K.F. 1978. Modulatory multiplicity in the functional repertoire of the feeding mechanism in cichlids. I. Piscivores. J. Morphol. 158: 323–360.

Liem, K.F. 1993. Ecomorphology of the teleostean skull. pp. 422–452. *In:* J. Hanken & B.K. Hall (ed.) The Skull, Functional and Evolutionary Mechanisms, Vol. 3, University of Chicago Press, Chicago.

Lightoller, G.H.S. 1939. Probable homologues. A study of the comparative anatomy of the mandibular and hyoid arches and their musculature. Part I: Comparative Morphology. Trans. Zool. Soc. Lond. 24: 349–444.

Long, J.A. 1995. The rise of fishes. Johns Hopkins University Press, Baltimore. 220 pp.

Lowe, C.G., B.M. Wetherbee, G.L. Crow & A.L. Tester. 1996. Ontogenetic dietary shifts and feeding behavior of the tiger shark, *Galeocerdo cuvier*, in Hawaiian waters. Env. Biol. Fish. 47: 203–211.

Luer, C.A., P.C. Blum & P.W. Gilbert. 1990. Rate of tooth replacement in the nurse shark, *Ginglymostoma cirratum*. Copeia 1990: 182–191.

Lund, R. & E.D. Grogan. 1997. Relationships of the Chimaeriformes and the basal radiation of the Chondrichthyes. Rev. Fish Biol. Fisheries 7: 65–123.

Luther, A. 1909. Untersuchungen über die vom n. trigeminus innervierte Muskulatur der Selachier (Haie und Rochen) unter Berucksichtigung ihrer Beziehungen zu benachbarten Organen. Acta. Soc. Sci. Fenn. 36: 1–176.

Maisey, J.G. 1980. An evaluation of jaw suspension in sharks. Amer. Mus. Novit. 2706: 1–17.

Marinelli, W. & A. Strenger. 1959. Vergleichende Anatomie und Morphologie der Wirbeltiere. III. Lieferung (*Squalus acanthias*). Franz Deuticke, Vienna. 305 pp.

Marra, L.J. 1989. Sharkbite on the SL submarine lightwave cable system: history, causes, and resolution. J. Ocean Engineer 14: 230–237.

Martin, A.P. & G.J.P. Naylor. 1997. Independent origins of filter-feeding in megamouth and basking sharks (order Lamniformes) inferred from phylogenetic analysis of cytochrome b gene sequences. pp. 39–50. *In:* K. Yano, J.F. Morrissey, Y. Yabumoto & K. Nakaya (ed.) Biology of Megamouth Shark, Tokai University Press, Tokyo.

McCosker, J.E. 1985. White shark attack behavior: observations of and speculations about predator and prey strategies. Mem. South. Calif. Acad. Sci. 9: 123–135.

McLaughlin, R.H. & A.K. O'Grower. 1971. Life history and underwater studies of a heterodont shark. Ecol. Monog. 41: 271–289.

Medved, R.J., C.E. Stillwell & J.G. Casey. 1988. The rate of food consumption of young sandbar sharks (*Carcharhinus plumbeus*) in Chincoteague Bay, Virginia. Copeia 1988: 956–963.

Monastersky, R. 1996. The first shark: to bite or not to bite? Science News 149: 101.

Morrissey, J.F. 1991. Home range of juvenile lemon sharks. pp. 85–86. *In:* S.H. Gruber (ed.) Discovering Sharks, American Littoral Society, Highlands.

Moss, S.A. 1965. The feeding mechanisms of three sharks: *Galeocerdo cuvieri* (Peron & Le Sueur), *Negaprion brevirostris* (Poey), and *Ginglymostoma cirratum* (Bonnaterre). Doctoral Dissertation, Cornell University, Ithaca. 139 pp.

Moss, S.A. 1967. Tooth replacement in the lemon shark, *Negaprion brevirostris*. pp. 319–329. *In:* P.W. Gilbert, R.F. Mathewson & D.P. Rall (ed.) Sharks, Skates, and Rays, John Hopkins Press, Baltimore.

Moss, S.A. 1972. The feeding mechanisms of sharks on the family Carcharhinidae. J. Zool. Lond. 167: 423–436.

Moss, S.A. 1977. Feeding mechanisms in sharks. Amer. Zool. 17: 355–364.

Motta, P.J. 1977. Anatomy and functional morphology of dermal collagen fibers in sharks. Copeia 1977: 454–464.

Motta, P.J., R.E. Hueter & T.C. Tricas. 1991. An electromyographic analysis of the biting mechanisms of the lemon shark, *Negaprion brevirostris*: functional and evolutionary implications. J. Morphol. 210: 55–69.

Motta, P.J., T.C. Tricas, R.E. Hueter & A.P. Summers. 1997. Feeding mechanism and functional morphology of the jaws of the lemon shark, *Negaprion brevirostris* (Chondrichthyes, Carcharhinidae). J. Exp. Biol. 200: 2765–2780.

Motta, P.J. & C.D. Wilga. 1995. Anatomy of the feeding apparatus of the lemon shark, *Negaprion brevirostris*. J. Morphol. 226: 309–329.

Motta, P.J. & C.D. Wilga. 1999. Anatomy of the feeding apparatus of the nurse shark, *Ginglymostoma cirratum*. J. Morphol. 241: 33–60.

Moy-Thomas, J.A. & R.S. Miles. 1971. Paleozoic fishes. Chapman and Hall, London. 259 pp.

Muller, M., J. Osse & J.H.G. Verhagen. 1982. A quantitative hydrodynamic model of suction feeding in fish. J. Theor. Biol. 95: 49–79.

Myrberg, Jr., A.A. 1991. Distinctive markings of sharks: ethological considerations of visual function. J. Exp. Zool. (Suppl.) 5: 156–166.

Nakaya, K. 1975. Taxonomy, comparative anatomy and phylogeny of Japanese catsharks, Scyliorhinidae. Mem. Fac. Fish. Hokkaido Univ. 23: 1–94.

Nelson, D.R. 1969. The silent savages. Oceans 1: 8–22.

Nelson, D.R. & R.H. Johnson. 1970. Diel activity rhythms in the nocturnal, bottom-dwelling sharks, *Heterodontus francisci* and *Cephaloscyllium ventriosum*. Copeia 1970: 732–739.

Nemeth, D.H. 1997. Modulation of attack behavior and its effect on feeding performance in a trophic generalist fish, *Hexagrammos decagrammos*. J. Exp. Biol. 200: 2155–2164.

Nobiling, G. 1977. Die Biomechanik des Kiefferapparates beim Stierkopfhai (*Heterodontus portusjacksoni* = *Heterodontus philippi*). Adv. Anat., Embr. & Cell Biol. 52: 1–52.

Norton, S.F. 1991. Capture success and diet of cottid fishes: the role of predator morphology and attack kinematics. Ecology 72: 1807–1819.

Norton, S.F. 1995. A functional approach to ecomorphological patterns of feeding in cottid fishes. Env. Biol. Fish. 44: 61–78.

Norton, S.F. & E.L. Brainerd. 1993. Convergence in the feeding mechanics of ecomorphologically similar species in the Centrarchidae and Cichlidae. J. Exp. Biol. 176: 11–29.

Parker, H.W. & M. Boeseman. 1954. The basking shark (*Cetorhinus maximus*) in winter. Proc. Zool. Soc. Lond. 124: 185–194.

Peyer, B. 1968. Comparative odontology. University of Chicago Press, Chicago. 347 pp.

Powlik, J.J. 1995. On the geometry and mechanics of tooth position in the white shark, *Carcharodon carcharias*. J. Morphol. 226: 277–288.

Pratt, Jr., H.L. 1979. Reproduction in the blue shark, *Prionace glauca*. U.S. Fish. Bull. 77: 445–470.

Pratt, Jr., H.L., J.G. Casey & R.B. Conklin. 1982. Observations on large white sharks, *Carcharodon carcharias*, of Long Island, New York. U.S. Fish. Bull. 80: 153–156.

Pretlow-Edmonds, M.A. 1999. Prey capture kinematics of the horn shark, *Heterodontus francisci*. M.S. Thesis, University of South Florida, Tampa. 64 pp.

Pyle, P., A.P. Klimley, S.D. Anderson & R.P. Henderson. 1996. Environmental factors affecting the occurrence and behavior of white sharks at the Farallon Islands, California. pp. 281–291. *In:* A.P. Klimley & D.G. Ainley (ed.) Great White Sharks, The Biology of *Carcharodon carcharias*, Academic Press, New York.

Randall, J.E. 1977. Contribution to the biology of the whitetip reef shark (*Triaenodon obesus*). Pac. Sci. 31: 143–164.

Raschi, W., J.A. Musick & L.J.V. Compagno. 1982. *Hypoprion bigelowi*, a synonym of *Carcharhinus signatus* (Pisces: Carcharhinidae), with a description of ontogenetic heterodonty in this species and notes on its natural history. Copeia 1982: 102–109.

Reif, W. 1976. Morphogenesis, pattern formation and function of the dentition of *Heterodontus* (Selachii). Zoomorphologie 83: 1–47.

Reif, W.-E., D. McGill & P. Motta. 1978. Tooth replacement rates of the sharks *Triakis semifasciata* and *Ginglymostoma cirratum*. Zoll. Jb. Anat. Bd. 99: 151–156.

Robinson, M.P. 1999. Patterns of growth and the effects of scale on the feeding kinematics of the nurse shark, *Ginglymostoma cirratum*. M.S. Thesis, University of South Florida, Tampa. 78 pp.

Sanderson, S.L. 1988. Variation in neuromuscular activity during prey capture by trophic specialists and generalists (Pisces: Labridae). Brain Behav. Evol. 32: 257–268.

Sanderson, S.L. & R. Wassersug. 1993. Convergent and alternative designs for vertebrate suspension feeding. pp. 37–112. *In:* J. Hanken & B.K. Hall (ed.) The Skull, Volume 3, University of Chicago Press, Chicago.

Sansom, I.J., M.M. Smith & M.P. Smith. 1996. Scales of thelodont and shark-like fishes from the Ordovician of Colorado. Nature 379: 628–630.

Schaeffer, B. 1967. Comments on elasmobranch evolution. pp. 3–35 *In:* P.W. Gilbert, R.F. Mathewson & D.P. Rall (ed.) Sharks, Skates and Rays, Johns Hopkins Press, Baltimore.

Schaeffer, B. & M. Williams. 1977. Relationship of fossil and living elasmobranchs. Amer. Zool. 17: 293–302.

Shirai, S. 1996. Phylogenetic interrelationships of neoselachians (Chondrichthyes: Euselachii). pp. 9–34. *In:* M.L.J. Stiassny, L.R. Parenti & G.D. Johnson (ed.) Interrelationships of Fishes, Academic Press, San Diego.

Shirai, S. & K. Nakaya. 1992. Functional morphology of feeding apparatus of the cookie-cutter shark, *Isistius brasiliensis* (Elasmobranchii, Dalatiinae). Zool. Sci. 9: 811–821.

Shirai, S. & O. Okamura. 1992. Anatomy of *Trigonognathus kabeyai*, with comments on feeding mechanism and phylogenetic relationships (Elasmobranchii, Squalidae). Japan. J. Ichthyol. 39: 139–150.

Simfendorfer, C.A. 1992. Biology of tiger sharks (*Galeocerdo cuvier*) caught by the Queensland shark meshing program off Townsville, Australia. Aust. J. Mar. Freshwater Res. 43: 33–43.

Simpfendorfer, C.A. 1998. Diet of the Australian sharpnose shark, *Rhizoprionodon taylori*, from northern Queensland. Mar. Freshwater Res. 49: 757–761.

Sims, D.W., A.M. Fox & D.A. Merrett. 1997. Basking shark occurrence off south-west England in relation to zooplankton abundance. J. Fish Biol. 51: 436–440.

Sims, D.W. & D.A. Merrett. 1997. Determination of zooplankton characteristics in the presence of surface feeding basking sharks, *Cetorhinus maximus*. Mar. Ecol. Prog. Ser. 158: 297–302.

Smith, H.M. 1942. The heterodontid sharks: their natural history and the external development of *Heterodontis (Cestracion) japonicus* based on notes and drawings by Bashford Dean. pp. 651–770. *In:* The Bashford Dean memorial volume-Archaic fishes, Art. 8, American Museum of Natural History, New York.

Snodgrass, J.M. & P.W. Gilbert. 1967. A shark-bite meter. pp. 331–337. *In:* P.W. Gilbert, R.F. Mathewson & D.P. Rall (ed.) Sharks, Skates, and Rays, Johns Hopkins Press, Baltimore.

Springer, S. 1957. Some observations of the behavior of schools of fishes in the Gulf of Mexico and adjacent waters. Ecology 38: 166–171.

Springer, S. 1961. Dynamics of the feeding mechanism of large galeoid sharks. Amer. Zool. 1: 183–185.

Springer, S. 1967. Social organization of shark populations. pp. 149–174. *In:* P.W. Gilbert, R.F. Mathewson & D.P. Rall (ed.) Sharks, Skates, and Rays, Johns Hopkins Press, Baltimore.

Strasburg, D.W. 1958. Distribution, abundance, and habits of pelagic sharks in the central Pacific Ocean. Fish. Bull. U.S. Fish Wildlife Serv. 58: 335–361.

Strasburg, D.W. 1963. The diet and dentition of *Isistius brasiliensis*, with remarks on tooth replacement in other sharks. Copeia 1963: 33–40.

Strong, W.R. Jr. 1989. Behavioral ecology of horn sharks, *Heterodontus francisci*, at Santa Catalina Island, California, with emphasis on patterns of space utilization. M.S. Thesis, California State University, Long Beach. 252 pp.

Strong, W.R. Jr. 1990. Hammerhead shark predation on stingrays: an observation of prey handling by *Sphyrna mokarran*. Copeia 1990: 836–840.

Talent, L.G. 1976. Food habits of the leopard shark, *Triakis semifasciata*, in Elkhorn Slough, Monterey Bay, California. Calif. Fish Game 62: 286–298.

Tanaka, S.K. 1973. Suction feeding by the nurse shark. Copeia 1973: 606–608.

Taylor, L.R. Jr. 1972. A revision of the shark family Heterodontidae (Heterodontiformes, Selachii). Doctoral Dissertation, University of California, San Diego. 176 pp.

Taylor, L.R., L.J.V. Compagno & P.J. Struhsaker. 1983. Megamouth – a new species, genus, and family of lamnoid shark (*Megachasma pelagios*, family Megachasmidae) from the Hawaiian Islands. Proc. Calif. Acad. Sci. 43: 87–110.

Tricas, T.C. 1979. Relationships of the blue shark, *Prionace glauca* and its prey species near Santa Catalina Island, California. U.S. Fish. Bull. 77: 175–182.

Tricas, T.C. 1982. Bioelectric-mediated predation by swell sharks, *Cephaloscyllium ventriosum*. Copeia 1982: 948–952.

Tricas, T.C. 1985. Feeding ethology of the white shark, *Carcharodon carcharias*. Mem. South. Calif. Acad. Sci. 9: 81–91.

Tricas, T.C. & J.E. McCosker. 1984. Predatory behavior of the white shark (*Carcharodon carcharias*), with notes on its biology. Proc. Calif. Acad. Sci. 43: 221–238.

Tricas, T.C., L.R. Taylor & G. Naftel. 1981. Diel activity of the tiger shark, *Galeocerdo cuvier*, at French Frigate Shoals. Copeia 1981: 904–908.

Van Damme, J. & P. Aerts. 1997. Kinematics and functional morphology of aquatic feeding in Australian side necked turtles (Pleurodira: *Chelodina*). J. Morphol. 233: 113–125.

Vorenberg, M.M. 1962. Cannibalistic tendencies of lemon and bull sharks. Copeia 1962: 455–456.

Wainwright, P.C. & B.A. Richard. 1995. Scaling the feeding mechanism of the largemouth bass (*Micropterus salmoides*): motor pattern. J. Exp. Biol. 198: 1161–1171.

Wainwright, S.A., F. Vosburgh & J.H. Hebrank. 1978. Shark skin: function in locomotion. Science 202: 747–749.

Waller, G.N.H. & A. Baranes. 1991. Chondrocranium morphology of northern Red Sea triakid sharks and relationships to feeding habits. J. Fish Biol. 38: 715–730.

Wetherbee, B.M., G.L. Crow & C.G. Lowe. 1997. Distribution, reproduction and diet of the gray reef shark, *Carcharhinus amblyrhynchos*, in Hawaii. Mar. Ecol. Prog. Ser. 151: 181–189.

Widder, E.A. 1998. A predatory use of counterillumination by the squaloid shark, *Isistius brasiliensis*. Env. Biol. Fish. 53: 267–273.

Wilga, C.D. 1997. Evolution of feeding mechanisms in elasmobranchs: a functional morphological approach. Doctoral Dissertation, University of South Florida, Tampa. 195 pp.

Wilga, C.D. & P.J. Motta. 1998a. Conservation and variation in the feeding mechanism of the spiny dogfish, *Squalus acanthias*. J. Exp. Biol. 201: 1345–1358.

Wilga, C.D. & P.J. Motta. 1998b. Feeding mechanism of the Atlantic guitarfish, *Rhinobatos lentiginosus*: modulation of kinematic and motor activity. J. Exp. Biol. 201: 3167–3184.

Wilga, C.D. & P.J. Motta. 2000. Kinematics and motor pattern of prey crushing in the bonnethead shark, *Sphyrna tiburo*. J. Exp. Zoology (in press).

Wilga, C.D., P.C. Wainwright & P.J. Motta. 2000. Evolution of jaw mechanics in vertebrates: insights from Chondrichthyes. Biol. J. Linn. Soc. (in press).

Witzell, W.N. 1987. Selective predation on large cheloniid sea turtles by tiger sharks (*Galeocerdo cuvier*). Japan. J. Herp. 12: 22–29.

Wu, E.H. 1994. A kinematic analysis of jaw protrusion in orectolobiform sharks: a new mechanism for jaw protrusion in Elasmobranchs. J. Morphol. 222: 175–190.

Zangerl, R. & M.E. Williams 1975. New evidence on the nature of the jaw suspension in paleozoic anacanthous sharks. Paleontology 18: 333–341.

Zlabek, K. 1931. Sur la constitution et le mecanisme de l'appereil maxillaire de la roussette. Archs. Anat. Histol. Embryol. 14: 83–121.

Environmental Biology of Fishes **60:** 157–188, 2001.

A review of elasmobranch reproductive behavior with a case study on the nurse shark, *Ginglymostoma cirratum*

Harold L. Pratt, Jr.[a] & Jeffrey C. Carrier[b]
[a]*Apex Predators Program, US DOC, NOAA, NMFS, NEFSC, Narragansett Laboratory, 28 Tarzwell Drive, Narragansett, RI 02882, U.S.A. (e-mail: Wes.Pratt@NOAA.gov)*
[b]*Department of Biology, Albion College, Albion, MI 49224, U.S.A.*

Received 14 January 1999 Accepted 3 May 2000

Key words: mating, courtship, copulation, sexual segregation, tagging, habitat, fishery management, nursery grounds, pheromones, sharks, skates, rays

Synopsis

Elasmobranch reproductive behavior has been inferred from freshly caught specimens, laboratory examinations of reproductive structures and function, or determined from direct observations of captive or free swimming wild animals. Several general behaviors have been described including seasonal sexual segregation, courtship and copulation. Courtship behavior was inferred for many species from the presence of scars and tooth cuts on the female's body, and noted in more detail from underwater observations. Copulation has been directly observed in captive settings for several species of elasmobranchs in large aquaria, and in the wild for three species of urolophids and for *Triaenodon obesus* and *Ginglymostoma cirratum*. A detailed 'case history' of nurse shark reproductive behavior is presented that may be used as a template for future work on shark reproductive behavior of other species. Our studies, using diver identifiable tags and *in situ* behavioral observations, provide unprecedented information on social structure and mating behavior in this species. Since 1993, 115 *G. cirratum*, 45 adults and 70 juveniles have been tagged in the Dry Tortugas, Florida. Observations show that adult males visit the study site every year with three males dominant. Individual adult females visit the study area to mate in alternate years. Polygyny and polyandry are common. Future research on reproductive behavior of elasmobranchs should address questions on male access to females, sexual selection and dominance hierarchies.

Introduction

Sharks are the focus of much attention because of their real and perceived relationships with man, both as consumers and as the consumed (Lent 1999) and increasingly, for their role in marine ecosystems. Skates and rays are important because their fisheries are significant. Elasmobranchs have complex reproductive life histories, specialized organ systems for reproduction and complex precopulatory and copulatory behavior. Many have large brains overlapping some mammals and birds in brain-body ratios (Northcutt 1977, 1978). Like mammals, elasmobranchs have internal fertilization and most have relatively long gestation periods, up to 22 months in the spiny dogfish, *Squalus acanthias*. In both groups, extended development may be necessary to produce central nervous systems capable of mediating complex social and sexual behavior (Demski 1990a). The discovery of large groupings of scalloped hammerheads, *Sphyrna lewini* (Klimley 1987), dominance hierarchies in sand tigers, *Carcharias taurus* (Gordon 1993) and bonnetheads, *S. tiburo* (Myrberg & Gruber 1974) and complex pair and group mating behaviors in nurse sharks, *Ginglymostoma cirratum* (Carrier et al. 1994), indicate a high degree of sophistication in their reproductive behavior.

Much is known about elasmobranch reproduction and development (Wourms 1977, 1981, Dodd 1983, Gilmore 1983) but the quantitative study of elasmobranch reproductive behavior is a relatively new endeavor. Studies of reproductive behavior have lagged behind other life history and fishery studies largely because of the paucity of direct observations (Bres 1993). Demski (1990a) briefly reviewed elasmobranch reproductive behavior as part of his discussion of a neuroendocrine mechanism controlling the sexual development and behavior of sharks and rays. Bres (1993) reviewed shark behavior, including reproduction and noted that few species of sharks have been directly observed in courtship and mating activities. Elasmobranch reproductive behavior has been implied from freshly caught specimens, laboratory studies of reproductive structures and function, or determined from direct observations of captive or free swimming wild animals. The study of elasmobranch reproductive behavior has become increasingly necessary because of their value in the commercial fishery. The reproductive behavior, habits, breeding areas, sexual segregation of females and nursery grounds of sharks must be thoroughly understood for successful conservation and management. Our ongoing work on nurse shark reproductive behavior (Carrier et al. 1994, Carrier & Pratt 1998) permits us to add this unique project as a case study. Our approach can be applied directly to other nurse shark populations as well as other elasmobranchs, and with technique modification, to studies of the reproductive behavior of other species of deeper water sharks, skates and rays. It is time for a complete review and synthesis of this topic. Further, we have updated the terminology used in studies of elasmobranch reproductive behavior in the Appendix.

Overview of reproductive structures and functions

Elasmobranchs are a diverse taxonomic group but have, to varying degrees, similar, often complex life histories and unique organ systems (claspers, siphon sacs, sperm storage) for reproduction. Successful mating requires the male to hold the female and to introduce one, or in a few cases both claspers into the cloaca and common vagina. Repeated copulations may be necessary in some species for successful fertilization. Sperm viability and the extent to which different species, both male and female, may store sperm has yet to be detailed for most species. Recent research has also revealed the role of the electrosensory system and the neuroendocrine system in reproductive behavior. A brief review of important adaptations in elasmobranch reproductive biology is provided in the following sections to add background for the review.

Teeth: biting and holding

Among male elasmobranchs, 'biting' (see Glossary, Appendix) or holding by some means seems universal. Springer (1967, p. 157) stated: 'Among the larger species, some cooperation on the part of the female seems necessary'. Indeed in most species in which mating behavior has been observed, the male bites and often holds the female, apparently as a precopulatory releasing mechanism to invoke female acquiescence during courtship, to facilitate insertion of the clasper and to maintain the proper position and proximity until sperm transfer is complete (Figure 1). During the mating season, female elasmobranchs will usually bear mating marks on their fins, flanks and elsewhere on their bodies depending on male biting, 'grasping' and courtship behaviors and the presence and employment of dentition, horns, spines and denticles. Most commonly found are tooth cuts and abrasions on the pectoral fins. Gordon (1993) found several types of biting behavior in captive sand tiger sharks. Males bit females to mate, but also snapped at other smaller species in the tank, but never at each other. The mature females

Figure 1. Male nurse shark (left) of the Dry Tortugas study stock bites and holds the entire left pectoral fin of a tagged female while copulating with the left clasper in crossed position (near surface). The female with cupped pelvic fins remains motionless as her head is pressed into the algae covered coral rubble bottom of the shallow study area. Sharks are 2.3–2.7 m in total length.

also bit the mating males as noted in wild individuals by Gilmore et al. 1983. These observations indicate that there may be sexual hierarchies which are maintained through biting behavior (Gordon 1993). These hierarchies may serve to synchronize the mating sequence in some sharks.

Social hierarchies and interactions may precede or follow active mating, and carry over into non-reproductive periods particularly in species which remain together or repeatedly use particular mating sites over several annual cycles like the sand tiger (Gilmore et al. 1983) and the nurse shark (Carrier et al. 1994). However not all bite marks are mating marks. Bites around the head and gills of large female shortfin makos, *Isurus oxyrinchus*, and white sharks, *Carcharodon carcharias* (Pratt et al. 1982), are not usually accompanied by signs of recent activity around the cloaca or recently introduced sperm. All of the makos surveyed with these marks have been in a reproductively resting condition and we conclude that some of these head and gill area bites, slashes and resultant scars, while possibly involving hierarchies, usually result from agonistic interactions over food rather than mating activity.

Sexual dimorphism

Sexual dimorphism occurs in skin thickness of maturing and adult females of some shark species. This is most evident in the blue shark, *Prionace glauca*, where dermal layers of the body surface are significantly thicker than that of same size males in a corresponding region (Figure 2). The dermis of large female sandbar sharks, *Carcharhinus plumbeus*, is also measurably thicker than same sized males (Pratt unpublished data). This is probably a characteristic of carcharhinids and many other species. Female blue sharks receive so much precopulatory biting that they often appear to be severely wounded while on the seasonal mating grounds (Figure 3) (Stevens 1974, Pratt 1979). Male bites on females may serve as a stimulus to induce mating 'acceptance' by the females (Springer 1967). Deep cuts could have a high potential to damage females due to blood loss or infection. Both male and female nurse sharks have a thick, dense integument with tile-like denticles (Compagno 1984) to resist damage during mating and in their lives around rock and coral.

Special dimorphism of the teeth of males in a few species of small sharks and rays has developed to make

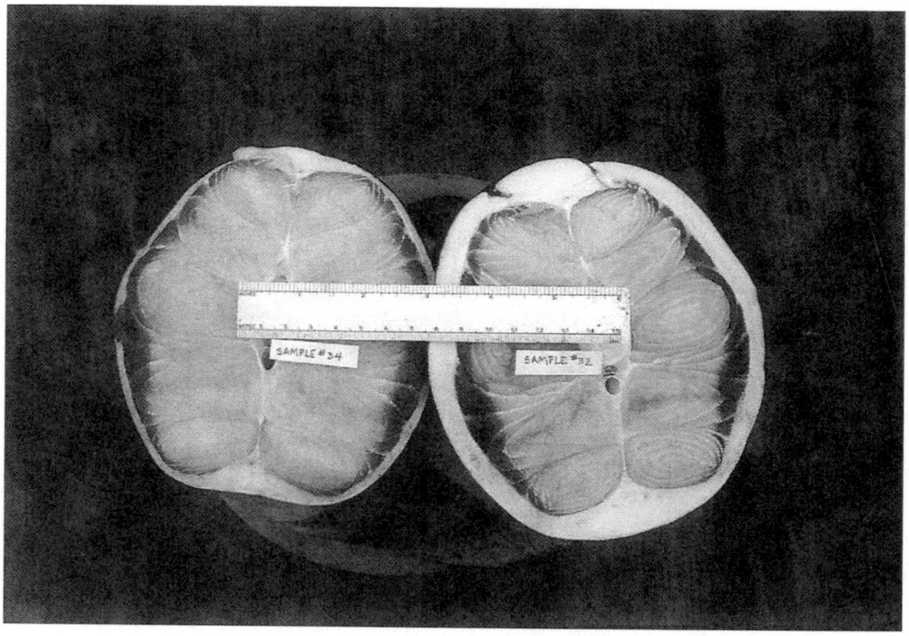

Figure 2. Sexual dimorphism in the skin of the blue shark. Cross sections immediately behind the pelvic fins of similar sized (~180 cm total length) male (left) and female (right) sharks.

Figure 3. Mating bites on a free swimming female blue shark in New York Bight mating grounds. Inset photo is right side of the same female.

'courtship biting' more effective. The teeth of some male squaloids, notably in the genus *Deania* and three scyliorhinids, *Halaelurus chilensis*, *H. bivius* and *Apristurus riveri*, have longer teeth with fewer cusps than females (Springer 1967). Dental sexual dimorphism is widespread in batoids (Bigelow & Schroeder 1953). McCourt & Kerstitch (1980) reported on mating behavior and sexual dimorphism of dentition in the stingray, *Urolophus concentricus* from the Gulf of California. Kajiura & Tricas (1996) quantified seasonal changes in male tooth morphology in the male Atlantic stingray, *Dasyatis sabina*, from female-like molarform to a recurved cuspidate form during the reproductive season (Figure 4). In contrast, female dentition remained unchanged throughout the year.

Pectoral fin biting and sexual dimorphism in teeth (seasonal or permanent) and other structures for reproductive purposes may be more common among sharks and rays than their absence, but more detailed analyses are needed. Even the relatively small teeth of the basking shark, *Cetorhinus maximus*, and those found in the genus *Manta* may persist mainly for use in courtship biting. Other yet undescribed sexually dimorphic structures probably exist in elasmobranch fishes to facilitate holding and to thicken skin, pectoral fins, skeletal supports and other somatic parts against the assault of teeth, jaws and claspers.

Claspers

Internal fertilization utilizing paired intromittent organs, the claspers, is a characteristic of all elasmobranch fishes. In copulation, at least one of the two mature claspers rotates forward (Leigh-Sharpe 1920). As the clasper rotates medially under the male cloaca, the single urogenital papilla engages the opened apopyle on the dorsal surface of the clasper to permit sperm transfer into the clasper tube. The clasper's terminal cartilages (Figure 5), in some species bearing sharpened ridges, hooks or barbs, expand out when flexed to lock the clasper in the vagina, often leaving abrasions and lesions in the vaginal wall. In sharks, it is generally agreed (Gilbert & Heath 1972) that the propulsive power for sperm transfer to the female is provided by contractions of the subdermal siphon sacs that connect with each clasper base. If the clasper is used in a splayed position (that is, not under the urogenital papilla) then sperm must be transferred to the clasper groove prior to copulation. Gilbert & Heath (1972) give an excellent account of clasper and siphon sac function in *Squalus acanthias* and *Mustelus canis*. It is possible that in some or all species of sharks, the role of the siphon sacs is not sperm propulsion as suggested by Gilbert & Heath (op.cit.), but to assist sperm competition and wash any rival sperm from the female's

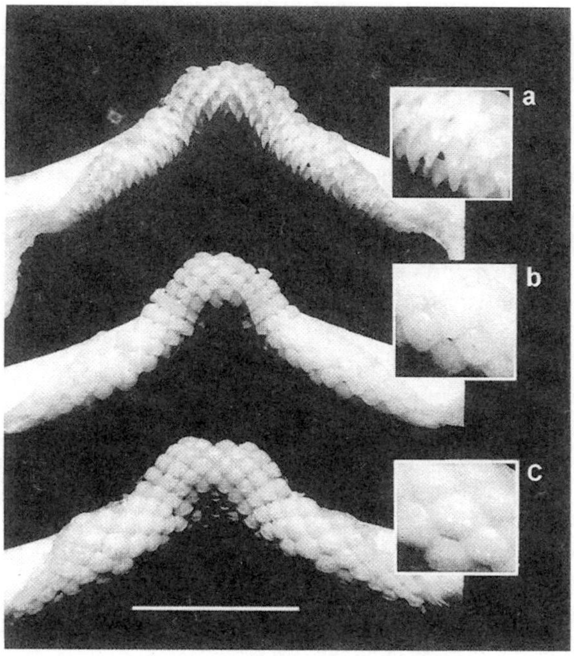

Figure 4. Upper jaw dentition of a male stingray, *Dasyatis sabina*, sampled during the mating season: a – The cuspidate dentition is evident not only in the first 3–4 rows of functional teeth but also in the developing teeth. In contrast, the upper jaw dentition of a male sampled during the non-mating season. b – is molarform and similar to the female dentition. c – Insets show enlarged views of the teeth. Scale bar = 1.0 cm (from Kajiura & Tricas 1996).

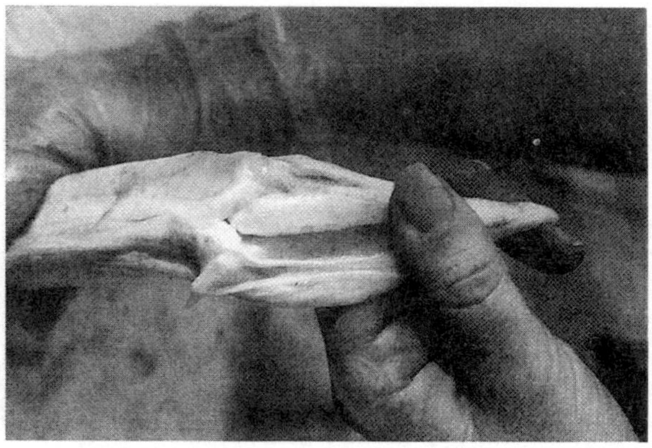

Figure 5. Terminal cartilages of the left clasper in the shortfin mako, *Isurus oxyrinchus*. When the tip is deflected downward, the end pieces splay open to reveal the clasper groove and deflects the spur laterally to engage the vaginal wall.

vagina before copulation (Eberhard 1985). Most skates and rays have a clasper gland that may aid copulation by providing a lubricant, serve as an energy source, or may facilitate sperm capacitation or reception by relaxing the vagina (LaMarca 1964).

During nurse shark mating, only one clasper has thus far been observed to be used in copulation and copulation attempts. For example, the left clasper is used to copulate with a female located on the male's left side (Carrier et al. 1994). It is not clear for other species whether one or both claspers are used. Leigh-Sharpe (1920) killed two *Scyliorhinus canicula* in copula and found both claspers inserted and Uchida et al. (1990) observed *Rhinoptera javanica* insert two claspers while mating in the Okinawa Expo Aquarium. However, most detailed accounts of copulation report the use of only a single clasper (Klimley 1980, Tricas & LeFeuvre 1985, Carrier et al. 1994). The claspers of sawfishes and some rays are positioned lateral of the midline and are too far apart for simultaneous use. The use of single claspers may be the rule in sharks, but multiple copulations may be required to inseminate all of the ova in both oviducts. Gordon (1993) provided evidence that the claspers of the sand tiger shark may be 'splayed' laterally at ninety degrees or crossed at a strong angle, seemingly at will while swimming (see below). Dral (1980) reported that *Chiloscyllium griseum*, can also cross and splay its claspers. It is not known which other species have this ability and its reproductive significance.

All known species of sharks have two functional oviducts, but gravid females often have developing embryos in one oviduct and infertile ova in the other. This is rare in smaller species but frequent among larger ones (Springer 1967) and may be the result of limited matings where only one clasper is used and consequently, only one oviduct is filled (unilateral fertilization). In sand tiger sharks, Gilmore and his colleagues have shown that both oviducts are always in complete synchrony with regard to physical condition, oviducal gland activity, fertilization and capsulation of ova and embryos as well as uterine condition (Gilmore et al. 1983, Gilmore 1993). Some rays (for example, stingrays) have only one functional oviduct and one copulation by one male may suffice. Repeated symmetrical insemination may especially be needed in species with a minimum of body flexibility.

The mating pair separates without forming a pair bond and each may continue to mate polygamously. Little is known about copulation frequency and polygamy in elasmobranchs. Snelson et al. (1988) found protracted mating in Atlantic stingrays, *Dasyatis sabina*, seven months prior to the one month of synchronous ovulation and fertilization.

Maruska et al. (1996) speculated that such promiscuous behavior by the male might serve a preparatory function for the female such as inducing steroidogenesis, oocyte growth, or ovulation.

Indicators of sexual activity: vagina, sperm and ova

In the mature female, the common vagina is found dorsal to the vent in the anterior portion of the cloaca. In a large shark the vagina may be 20–30 cm in length and bifurcates to join the paired uteri (Castro 1983). The vagina of the blue shark is thick-walled to receive the sharp terminal edges of the clasper (Pratt 1979). Abrasions, lesions and dark bruises are common on vaginal walls during and after the mating season and are a good index of sexual activity. The vagina will sometimes appear red and hematose during and after parturition as well. In nurse sharks, after much mating, the walls of the cloaca, the skin of the pelvic fins around the cloaca and even the abdomen of the female will be hematose.

Sperm may be stored for some period as sperm packets (spermatophores or spermatozeugmata) in the mature male's ampullae epididymides before transfer to the female. Pratt & Tanaka (1994) reported that 14 species of six families (Carcharhinidae, Scyliorhinidae, Sphynidae, Rajidae, Torpedinidae, and Chimaeridae) possessed spermatozeugmata and six species in three families (Allopiidae, Odontaspidae, and Lamnidae) possessed spermatophores. During or immediately before copulation, sperm exit the paired ampullae to the urogenital sinus and then are injected into the apopyle and clasper groove of one forward-rotated and copulating clasper. From the clasper groove, the sperm are propelled into the female's vagina and uterus either through the hydraulic action of the corresponding muscular subdermal siphon sac or by the action of smooth muscles in the walls of the ampullae epididymides. This transfer may be further augmented by changes in hydrostatic pressure within the peritoneal cavity generated by the male's thrusting behavior during copulation (Jones & Jones 1982).

Female elasmobranchs may also store sperm (Metten 1941, Prasad 1944, 1945, Pratt 1979, 1993). For example, histological analysis of bitten female blue sharks caught and dissected a short time after courtship and copulation reveals large volumes of fresh sperm in their reproductive tracts, especially in the lumen of the oviducal glands (Pratt 1979). Similarly, sperm packets were found in the uteri of the finetooth, *Carcharhinus isodon* (Castro 1993). Carrier et al. (1994) sampled the copious quantities of single (unpackaged) sperm suspended in voluminous sheets of a mucus-like material found floating in the water and tangled in benthic algae during and after nurse shark copulations. Because sperm competition is probable in this species, superfluous sperm production probably occurs during nurse shark courtship and mating.

Some females may delay fertilization and store sperm for many months in the shell gland (Pratt 1993). Other species immediately fertilize ova following copulation (Gilmore et al. 1983). Repeated fertilizations may be necessary to fertilize all ova in both oviducts if sperm is not stored. Ova are apparently released sequentially, usually alternating between right and left uterine tracts. Fertilization occurs in the shell gland for elasmobranchs that store sperm and in the upper oviduct above it or in its vicinity for those that do not (Gilmore et al. 1983).

Freshly caught mature males are frequently reported with sperm, spermatophores and spermatozeugmata oozing from claspers, cloaca or the urogenital pore and judged by these criteria to be sexually active (Clark & von Schmidt 1965, Pratt 1979, Francis 1996). Caution must be exercised. The stress of capture occasionally causes spontaneous abortions and/or loss of stomach contents in sharks. The loss of sperm during the stress of capture does not necessarily indicate mating activity unless accompanied by hematose claspers, possibly with lesions and abrasions such as has been noted for lemon sharks, *Negaprion brevirostris*, nurse sharks (Clark & von Schmidt 1965) and blue sharks (Pratt 1979).

Physiological control mechanisms

Electrosensory system

Tricas et al. (1995) found that the electrosensory system serves an important function in the social and reproductive behavior of round stingrays, *Urolophus halleri*, and probably other elasmobranchs. The rays produce a weak electric field that is partially modulated rhythmically by the movements of the spiracles and gills during ventilation. This signal is detected by other rays via the ampullae of Lorenzini, which are found in pores located on the dorsal and ventral cranial surfaces of elasmobranchs. This sensory system is used by reproductively active males to locate mates and by females to locate buried consexuals (Figure 6). Other senses such as olfaction have been implicated in shark courtship behavior (Johnson & Nelson 1978) but remain to be experimentally confirmed.

Neuroendocrine control of courtship and copulation

Although there are exceptions, especially in tropical areas, mating behavior usually occurs seasonally. Temperature and photoperiod may be the major environmental regulators of the neuroendocrine mechanisms that drive sexual development and behavior. Lunar and solar influences may play a role in some cases. Dodd (1983) provides a comprehensive overview of reproduction and endocrine control in cartilaginous fishes. Temperature and photoperiod control the endocrine regulation of spermatogenesis in the small spotted catshark, *Scyliorhinus canicula* (Dobson & Dodd 1977). Hormonal control of sexual development and behavior is well documented in several elasmobranch species (Sumpter & Dodd 1979, Rasmussen et al. 1992, Wright & Demski 1993). Demski (1990a) found that gonadotropin-releasing hormones mediate brain control of testicular and ovarian development via a brain-pituitary axis and speculates that the 'following' behavior noted by Myrberg & Gruber (1974) and Klimley (1980) may be triggered by female pheromones which trigger sexual behavior in males. Rasmussen & Gruber (1990) found elevated levels of the steroid hormones estradiol and testosterone in the serum of free-ranging, actively courting female lemon sharks.

Feeding and liver observations in mating sharks

Springer (1967) found that immature and female sharks of the coastal Florida-Gulf and Mexico-Caribbean regions have large livers containing a high proportion of oil. In the same season, sexually mature males have thin livers with low oil content. He postulated that this might be due to prolonged fasting during courtship. Alternatively, thin livers may also result from male activity during courtship such as increased 'patrolling' and aggressive behavior required to acquire and mate with females. Other factors may include the reproductive costs associated with production of vast quantities of sperm in those species in which males engage in sperm competition. Future studies may reveal male reproductive costs to be as high for males of some species as for females.

Springer (1960) noted a correlation between fresh tooth cuts and the appearance of early embryos in the sandbar shark. In addition, he recorded 'mating scars' on female tiger sharks, dusky sharks and bull sharks, but scars on males were scarce. He proposed that in some species, the larger females might occasionally respond to male biting during mating attempts

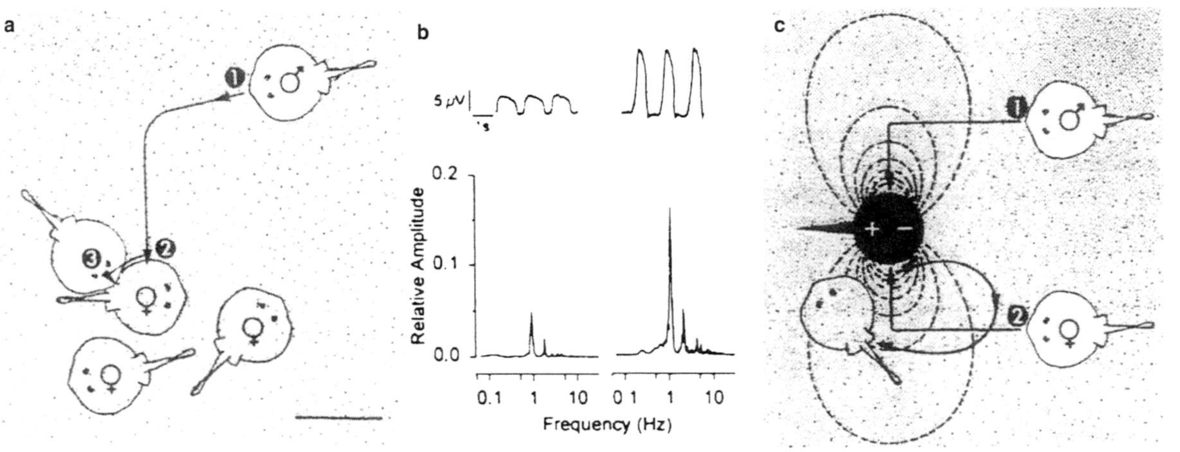

Figure 6. Orientation responses by round stingrays, *Urolophus halleri*, to buried females and bioelectric stimuli produced by conspecifics during the mating season: a – males localize, orient towards, and inspect buried females in the wild population, b – female bioelectric potentials, c – diagram of orientation response by round stingrays to a synthesized phasic bioelectric field (from Tricas et al. 1995).

by killing the male. This is supported to some extent by Gordon's (1993) captive observations of female sand tiger shark's occasionally biting mating males.

Reproductive behavior

Louis Agassiz (1871, p. 340) recounts Aristotle as reporting what must be the first record of elasmobranch mating behavior: 'The cartilaginous fishes in copulation hang together after the fashion of dogs, the long tailed ones mounting the others, unless the latter have a thick tail preventing this, when they will come together belly to belly.' Since those initial observations, a more complete understanding of elasmobranch reproductive behavior has emerged. The following account of key observations and conclusions is presented by category and then summarized into an overview of apparent patterns of behavior commonly found in most elasmobranchs.

Sexual segregation

There is a general tendency for some species of elasmobranchs of the same size and sex to associate and travel in groups. The migration and distribution of sharks is influenced both directly and indirectly by water temperatures (Casey & Kohler 1992). Consequently some sharks migrate seasonally, apparently following temperature isotherms and probably seasonal movements of prey species. Sharks that have been reported to segregate by sex include school sharks, *Galeorhinus galeus* (Olsen 1954), sandbar sharks (Springer 1960), bonnetheads, *Sphyrna tiburo* (Myrberg & Gruber 1974), blue sharks (Pratt 1979), sand tiger sharks (Gilmore et al. 1983), grey reef sharks, *Carcharhinus amblyrhynchos* (McKibben & Nelson 1986), scalloped hammerheads (Klimley 1985, 1987) and spiny dogfish (Callard et al. 1988).

Migrations often take sharks to specific places where reproductive events may occur (Springer 1967). Economakis & Lobel (1998) observed aggregations of female grey reef sharks in warm shallow waters at Johnston Atoll and suggest that this behavior may hasten embryonic development. They did not notice scarring of females or males in the area, so aggregation in this instance is probably not for mating activity.

Segregation may also increase individual fitness. Demski (1990b) noted that in those elasmobranchs that form schools, population density may play an important role in reproduction and suggested that in some birds and mammals, the sight and sounds of a critical mass of individuals is necessary to trigger mating behavior. Klimley (1985) suggested that scalloped hammerhead females compete for position near the center of the school. The location of larger females within the school might have implications for mating success in a behavior pattern similar to Hamilton's (1971) 'selfish herd' premise, a behavior used by schooling species to escape predation. Male hammerheads were rare in Klimley's (1985) observations and entered the school performing 'torso thrust', usually accompanied by 'clasper flexion'. Female segregation in scalloped hammerhead and other species of sharks may spread out male selective pressure and could aid in female choice in mating with or 'avoiding' a particular male. Frequent mating may have a high cost for females due to blood loss or possible infection from bites or manipulations and also to males resulting from possible female defensive maneuvers (Springer 1967, Gordon 1993). On the southern New England continental shelf in summer, nearly every female blue shark bears fresh, deep dermal lacerations from male tooth cuts (Figure 3). Dissection of these females typically shows vaginal cuts and edema and reproductive tracts that have large volumes of fresh sperm in the lumen of the uteri and oviducal gland (Pratt 1979). If sexual selection is regarded as competition between the sexes (Krebs & Davies 1993), then segregating away from the opposite sex when not engaged in mating activity may be the best behavior for both breeding success and survival from repeated and prolonged mating attempts.

Observations of mating in captive elasmobranchs

The earliest documentation of mating behaviors and copulation in the reclusive elasmobranchs has come from the captive environment (Table 1). Observations of aquarium specimens are often serendipitous or anecdotal. However, many modern aquaria have veterinary and research staff interested not in just the health, but in the research potential of their captive animals. Additionally, constant video monitoring within large aquaria has become increasingly more prevalent and offers the promise for continuous observation. The other source of significant contributions has come from the research community either working in the laboratory or in collaboration with aquaria.

Table 1. Summary of observed courtship and mating behaviors in elasmobranch fishes.

General behavior and species	Descriptions/Notes	References
Sharks		
Precopulatory and courtship		
Following		
Carcharhinus melanopterus	'Close follow' near female's vent possibly olfactory-mediated.	2
Ginglymostoma cirratum	Male and female swim parallel and synchronously side by side.	3
Negaprion brevirostris	Swimming with body axes in parallel.	4
Female avoidance		
Carcharias taurus	Female 'shields' with pelvics close to substrate.	
Ginglymostoma cirratum	'Lying on back' the female rests motionless and rigid.	1
	Female 'pivots and rolls' on her back when a male bites her pectoral fin.	3
Female acceptance		
Carcharias taurus	'Submissive' body, 'cupping' and 'flaring' of pelvic fins.	1
Ginglymostoma cirratum	Female arches body toward male, 'cups' pelvic fins.	5
Biting		
Heterodontus francisci	Male bites and wraps female pectoral fin body, tail, gills.	7
Scyliorhinus retifer	Male bites and wraps female pectoral fin body, tail, gills.	8
Scyliorhinus torazame	Male bites and wraps female pectoral fin body, tail, gills.	9
Ginglymostoma cirratum	Male bites and holds female's pectoral fin.	3, 5
Carcharhinus sp.	Male bites and holds female's pectoral fin.	10
Triaenodon obesus	Male bites and holds female's pectoral fin.	9, 13
Positioning and alignment		
Ginglymostoma cirratum	'Nudging' female into position with head.	3
Ginglymostoma cirratum	After 'pectoral bite' male rolls female, then aligns for insertion.	5
Ginglymostoma cirratum		5
Sphyrna lewini	'Torso thrust' with 'clasper flexion' possibly filling siphon sacs.	6
Carcharias taurus	'Crossing' or 'splaying' claspers as position requires.	1
Group		
Ginglymostoma cirratum	Multiple males compete or cooperate for a mate. A cooperative behavior or a single male 'blocking' a mating pair.	5
Insertion and copulation	Insertion of one or more claspers into the cloaca leading to ejaculation.	5
Copulatory		
Male bites female while at rest		
Heterodontus francisci	Male wraps around females body.	7
Scyliorhinus retifer	Male wraps around females body.	8
smaller shark sp.	Male wraps around females body.	7, 8, 11, 12
Triaenodon obesus	Heads to substrate, sharks undulate to keep tails elevated.	13
Ginglymostoma cirratum	'Lying parallel on substrate' less than two pectoral widths apart during bouts of 'parallel swimming'.	3
Ginglymostoma cirratum	Heads to substrate, tails elevated or lying parallel. 'Copulation' sometimes in groups of many males.	5
Heterodontus francisci	Male crosses female's body, rhythmic motion for up to 35 m.	7
Parallel swimming 'in copula'		
Negaprion brevirostris	Coordinated pair swimming while copulating.	4
Carcharodon carcharias	Possible coordinated pair swimming while copulating.	14
Polygyny		
Ginglymostoma cirratum	Males will mate with many females over several weeks.	This report
Polyandry		
Ginglymostoma cirratum	Females will mate with many males over several weeks.	This report
Post copulatory	Pair remains together or departs rapidly.	5
Stalking		
Carcharias taurus	Male aggression toward other species in a captive environment.	1
Batoids		
Precopulatory and courtship		
Following		

Table 1. (*Continued*).

General behavior and species	Descriptions/Notes	References
Aetobatus narinari	Rapid 'chase', close to tail of female.	9
Manta birostris	Rapid 'chase', close to tail of female.	15
Myliobatis californica	Male ventral to female with wingbeats synchronized.	16
Myliobatis californica	Males 'follow' females.	17
Female avoidance		
Urolophus halleri	Females bury in sand to 'avoid' males.	18
Aetobatus narinari	Females raise back out of water and slap wings on surface in response to male nipping.	16
Urolophus halleri	Females spine males with caudal spine.	19
Female acceptance		
Raja eglanteria	'Back arching', 'pectoral fin undulations' to attract males.	20
Biting		
Aetobatus narinari	'Gouging', bites on female dorsal surface.	9
Rhinoptera bonasus	'Gouging', bites on female dorsal surface.	16
Rhinoptera javanica	'Gouging', bites on female dorsal surface.	9
Manta birostris	Male grasps pectoral fin tips (nipping).	15
Group		
Dasyatid and myliobatid rays	Common for multiple males to 'follow' single females.	9, 16, 17
Rhinoptera javanica	Many captive males overwhelmed a female for multiple matings. Mortality sometimes resulted from wounds and exhaustion.	9
Other behaviors		
Aetobatus narinari	Males 'bob' and 'sway' while 'following' 'avoiding' females.	16
Copulatory		
While reposed on bottom		
Raja eglanteria	Copulate for one to four hours while at rest on bottom. Male holds trailing edge of females pectoral fin, swings tail beneath hers and inserts one clasper.	20
While swimming		
Manta birostris	Copulation near the surface, abdomen to abdomen.	15
Aetobatus narinari	Mating abdomen to abdomen in the mid-depths of the tank. Insertion time was 0.5–1.5 min.	9
Rhinoptera javanica	Starts at the surface or mid-depth, abdomen to abdomen, continues on the bottom.	9
Rhinoptera bonasus	Starts at the surface or mid-depth, abdomen to abdomen, continues on the bottom.	9
Polyandry		
Aetobatus narinari	A captive female mated many times in succession with 3 to 4 males in 1 h.	9
Rhinoptera javanica	Multiple matings common.	9
Post copulatory		
Manta birostris	Male remains attached to pectoral fin tip briefly.	15

Reference key: 1 – Gordon 1993, 2 – Johnson & Nelson 1978, 3 – Klimley 1980, 4 – Clark 1963, 5 – Carrier et al. 1994, 6 – Klimley 1985, 7 – Dempster & Herald 1961, 8 – Castro et al. 1988, 9 – Uchida et al. 1990, 10 – Clark 1975, 11 – Gilbert & Heath 1972, 12 – Dral 1980, 13 – Tricas & LeFeuvre 1985, 14 – Francis 1996, 15 – Yano et al. 1999, 16 – Tricas 1980, 17 – Feder et al. 1974, 18 – Tricas et al. 1995, 19 – Michael 1993, 20 – Luer & Gilbert 1985.

The first published information about shark copulation came from studies of the small spotted catshark, held in relatively small aquaria. Males of small flexible sharks may initially bite the female as part of precopulatory behavior and usually 'wrap' their bodies around the female while inserting one clasper for 20 min or longer (Bolau 1881, Schensky 1914, Hardy 1959, Gilbert 1981).

The observations of Castro et al. (1988) of the chain dogfish, *Scyliorhinus retifer*, are typical. Mating by the chain dogfish had only been observed once in a 10-year study. One female swam very actively with a male in tight circles near the bottom of the tank. The male bit the female four times in the gill region, flanks or tail within 30 min. The female sometimes avoided the male and at other times they swam together or rested apart motionless. Eventually the male bit the female's tail and would not release it. The female struggled, violently at first, then became listless. The male moved its bite up the left ventral flank until it reached the left pectoral axilla, wrapped its body around the female and flexed its left clasper to the right, toward the midline, for copulation. The pair remained nearly motionless during 30 sec of copulation. The male released its bite but remained coiled around the female. The female then broke away and both swam to separate areas of the tank. Biting behavior was observed on the following days indicating that mating may occur repeatedly in this species.

Similar observations exist for other small captive sharks. After a rapid and violent chase in the Steinhart Aquarium, horn shark males, *Heterodontus fancisci*, caught the female by biting the body and even the gills until a good grasp of the pectoral fin was obtained (Dempster & Herald 1961). If the male was positioned laterally to the female's right side, the male then held the left pectoral fin in his mouth and rotated the right clasper toward the female. The male then laid across the female's body, inserted the clasper and copulated up to 35 min. The male epaulette shark, *Hemiscyllium ocellatum*, bites and holds the female for copulation in a similar manner (West & Carter 1990, Michael 1993). Dral (1980) described captive mating of the grey bamboo shark, *Chiloscyllium grisium*, as mating side-to-side with the male biting the female's pectoral fin. One clasper is used in *Chiloscyllium*; it may be crossed or uncrossed.

Clark (1963, p. 145) reported an observed copulation in the 12 × 18 m Cape Haze Marine Laboratories seaside stockaded shark pen: 'At midnight on 1 May 1959, Dr. and Mrs. Dugald Brown, who were working in the laboratory, walked out on the dock and saw two lemon sharks in copula. The sharks were side by side, heads slightly apart but the posterior half of their bodies in such close contact and the swimming movements so perfectly synchronized that they gave the appearance of a single individual with two heads, as they swam in slow counterclockwise circles around the pen. When the observers left the dock a half-hour later, the sharks were still in copula.' Although no other copulations were seen, Clark noted that the claspers of the male looked swollen and pink. On 10 Dec 1959 the female died and was found to contain six embryos.

Myrberg & Gruber (1974) studied adult male bonnetheads in the Miami Seaquarium's outdoor shark channel exhibit. They found that males sometimes flex one clasper while swimming to perform a clasper thrust which may be related to 'torso thrust', a swimming behavior with clasper flexion recognized by Klimley (1985) in scalloped hammerheads. Demski (1990a) suggests that these thrusts may also be a visual signal in a ritualized courtship display. They may also be part of a necessary action pattern by males to fill the siphon sacs prior to copulation (Gilbert 1981).

Klimley (1980) was the first to detail courtship and copulation of the nurse shark from his observations in the shark channel exhibit at the Miami Seaquarium. He based his findings on 15 courtship attempts and one copulation in the channel and from interpretations of photographs taken from land and boats of wild matings at the Dry Tortugas, FL. He observed: (1) 'parallel swimming' less than two pectoral fin widths apart. This is similar to Clark's (1963) lemon shark observation. Klimley states that this may not be related to the olfactory mediated event mentioned by Johnson & Nelson (1978) because the sharks' heads are too close together. This behavior was brief and uncommon in the field observations of nurse sharks by Carrier et al. (1994). Klimley also reported, (2) 'lying parallel on substrate', and (3) 'pectoral biting', which did not accompany copulation in Klimley's (1980) study, (4) 'pivot and roll' (Carrier et al. 1994, classified this as an 'avoidance' maneuver), (5) 'lying on back' (female), (6) 'nudging', (7) 'male on top of female', and (8) 'lying on back' (male). Copulation lasted for two min. Klimley also noticed chromatic patterns, where the courting males were dark brown, the females a lighter brown. In the four photographic sequences he obtained from wild matings, the males were always darker than the females.

Uchida et al. (1990) showed how remarkable captive work can be with the advent of large aquaria. They summarized the worldwide observations of elasmobranchs bred in captivity from the International Zoo Yearbooks (1963-1986). A total of 25 species in six families of sharks and 18 species in eight families of rays were bred in captivity (excluding animals taken from the sea in a fertilized or gravid condition). Mating was observed in two shark species at the aquarium. The cloudy catshark, *Scyliorhinus torazame*, was observed mating twice and whitetip reef sharks, *Triaenodon obesus*, were also observed mating twice. Copulation duration was from 15 sec to 4 min in both species of sharks. The mating whitetip reef sharks were born in the tank and were from the same litter. The male bit the female on her right pectoral fin and used his medially flexed right clasper in copulation (Figure 7), a pattern that is frequent among the larger sharks.

Gordon (1993) reported three instances of pre-copulatory and copulatory behavior of two adult males and one adult female captive sand tiger sharks held at Oceanworld Manly, Sydney, Australia. Events were evidently precipitated by temperature as lighting conditions were kept at a constant 15 : 9 light-dark ratio and water temperatures followed coastal conditions of 15 °C in winter and 24 °C in summer. Mating behavior took place in the warmer months between November and January. Several distinctive behaviors related to mating and dominance were identified, some of which help clarify field observations including 'stalling', an individual shark stopping all forward movement and hovering above the bottom. This behavior may be peculiar to those few sharks like sand tiger sharks whose body densities are very close to sea water, and permits them to hover without sinking when they stop swimming. Both sexes were observed to perform 'courtship bites' to facilitate mating or as a reaction by females to a persistent male. Behaviors noted for males include 'tailing', 'nosing', (Figure 8a,b), 'stalking', 'snapping', 'clasper flexion', 'splaying' claspers, and 'crossing' claspers (Figure 8f,g). Behaviors observed in females were 'submissive', 'cupping' (see also Carrier et al. 1994, figure 6a), 'flaring' (Figure 8d,e); and 'shielding'.

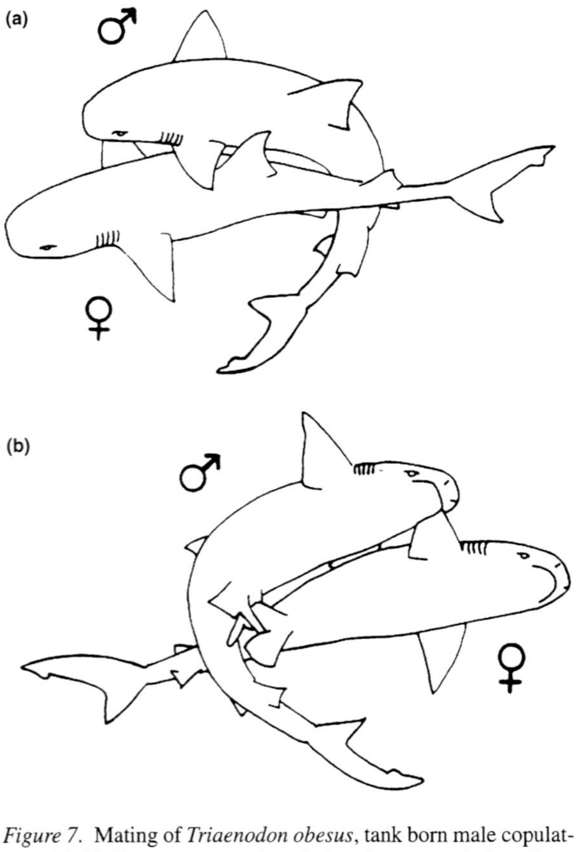

Figure 7. Mating of *Triaenodon obesus*, tank born male copulating with his litter mate, a = dorsal view, b = ventral view (from Uchida et al. 1990).

Gordon (1993) provided an excellent sequence of mating activities which began when the two male *C. taurus*, having become reluctant to feed, increased their swimming speed, accompanied by clasper flexion and eventual interest in their female tank mates. This increased activity supports Springer's (1967) contention that increased swimming associated with increased reproductive activity resulted in poor liver condition in mating males. Gordon (1993) was first to document a reproductive dominance hierarchy. Aggressive displays such as snapping and stalking became frequent, mostly toward species of smaller *Carcharhinus* in the tank. The males did not become interested in the female until she slowed, moved to the sand area and started cupping her pelvic fins. The two males then became competitive toward each other, circling and tailing, until the alpha male forced the beta male out of the sand area. The female bit the male prior to copulation. She exhibited shielding behavior for several days and then resumed cupping and flaring. The female gradually changed her swimming position and began displaying the submissive behavior. The alpha male swam in increasingly larger circles and began splaying its claspers, then approached the female and exhibited tailing and nosing. Copulation occurred as the male bit into the right flank and trailing edge of the pectoral fin of the female. The male swam side by side with the female, copulating with the right clasper for one to two min. After copulation, the male showed little interest in the female. Gordon (1993) speculates that a chemical stimulant (pheromone) attracted the males in his study. In contrast to Klimley's (1980) findings for the nurse sharks, he reports lighter colored males and darker females in precopulatory sand tiger sharks.

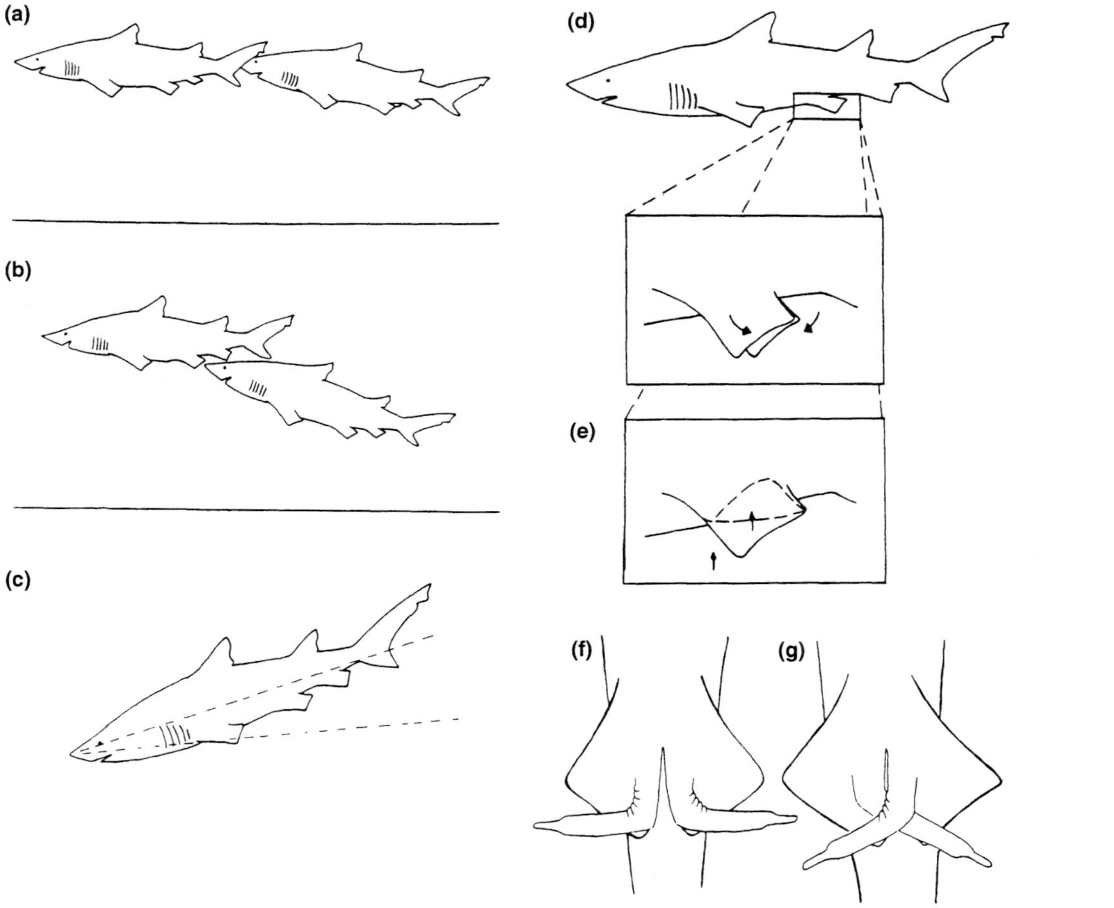

Figure 8. Mating displays and behaviors: a – 'tailing' performed by two males, b – 'nosing' female above, male below, c – 'submitting', d – 'cupping', e – 'flaring' of female pelvic fins, f – 'splaying', g – 'crossing' of claspers (from Gordon 1993).

Mating in skates and rays is complex and can take several forms (Table 1). Some small skates mate with ventral surfaces opposed (Wourms 1977). Some larger rays can apparently mate with the female above or below the male (Tricas 1980, Uchida et al. 1990, Yano et al. 1999). Demski (1990a) remarks that the variety of mating positions observed in elasmobranchs is probably related to the great differences in body form and swimming habits found in these fishes. Working in a captive setting, Libby & Gilbert (1960) and Luer & Gilbert (1985) discovered that male and female clearnose skates, *Raja eglanteria*, mate side by side with the pair in an upright position. The male bites the caudal margin of the female's pectoral, bends his tail 75 degrees beneath hers and inserts one clasper, flexed ('splayed') medially 90 degrees, into her oviduct. The female skate often attracts the attention of males with a precopulatory behavior that involves 'back arching' and 'pectoral fin undulations'. They found that the head and alar spines of the male may pierce the skin of the female to obtain purchase during the one to four hour copulation. They never observed both claspers inserted simultaneously and doubted if it was possible, though such an observation was made by Fowler (1906b) for clearnose skates. Luer & Gilbert (1985) found that if the male grasps the female's left pectoral and swings his tail beneath hers, the left clasper is flexed and inserted. The opposite clasper is used if the male bites the right pectoral fin. These are also the conclusions of Carrier et al. (1994) for nurse shark pectoral grasping and clasper use. One or more male round stingrays of the species *Urolophus halleri* and *U. aurantiacus* have also been observed to follow females to bite their pectoral fin edges (Michael 1993, Nordell 1994). Dugger (1987) illustrates *U. jamaicensis* mating with both male and female biting the pectoral fin of its mate. Uchida et al. (1990) observed mating and copulation once in the spotted eagle ray, *Aetobatus narinari*, and six times in the cownose ray, *Rhinoptera javanica*. Copulation duration was from 20 sec to 1 min in both species of rays. At the Okinawa Expo Aquarium, the spotted eagle rays have enough aquarium space for what Uchida et al. (1990, p. 223) call a 'typical pattern of mating behavior'. The male chases the female in mid water, then nibbles ('gouges' Tricas 1980) on her dorsal surface. The female stops swimming to begin copulation. The male bites the female on a pectoral fin and bends one clasper forward, then attempts an abdomen to abdomen copulation with either clasper, usually in mid-water (Figure 9). Uchida et al. (1990) noted that the cownose ray, *Rhinoptera bonasus*, has also been documented to receive mating bites.

Remarkable as they are, captive mating and subsequent behavioral observations must be interpreted with caution. Confined animals, even in the largest aquaria, may still be restricted in ways we cannot understand or appreciate. For example, the mated female nurse shark studied by Klimley (1980) at the Miami Seaquarium could not utilize extensive shallow waters to limit the male's access and was probably denied the whole range of avoidance behaviors available to females observed by Carrier et al. (1994). These limitations will also modify the response's of males. Further, Rasmussen & Gruber (1990) reported a precipitous drop in serum hormone levels for a captive reproductively active female lemon shark and a male nurse shark. They related these hormonal changes to the stress of recent capture and subsequent short-term captivity and indicate physiological changes had occurred that might manifest themselves through modifications in behavior.

Field observations of courtship and copulation

Although elasmobranch copulation must be a common occurrence in the sea, it is poorly documented. Necropsies of fishery landings have revealed evidence of mating activities through internal and external examination and provide much detail. Field observations of mating activities are uncommon and usually serendipitous due to the concealed nature of most pairings and the difficulty of quantifying events. Because the observer's usual viewpoint of such events is generally from above, much information is missed.

Courtship behavior was inferred from the dissection of the basking shark (Matthews 1950), sandbar shark (Springer 1960), blue shark (Stevens 1974, Pratt 1979), sand tiger shark (Gilmore et al. 1983), blacktip shark, *Carcharhinus limbatus* (Castro 1996), and others. External and internal examinations of these species have revealed tooth cuts, abrasions and punctures on the female's body and, in some cases, wounds to the cloaca and vaginal walls, presumably caused by mating activities and clasper terminal edges after insertion. Springer (1967, p. 157) wrote, 'Among the larger carcharhinid sharks . . . males harass the females persistently and violently to induce cooperation in mating . . . because the jaw armature of the males is such that the male could probably not bite and hold the female without producing a very severe injury'. He noted that the relatively inflexible male carcharhinid trunk region cannot by itself hold the female for mating as do the more supple catsharks and dogfish. Springer (1960) speculated that this harassment in sharks represents a form of courtship. Direct observation is needed to clarify and further refine such records. In other studies using necropsy, the presence of fresh sperm in the uteri or shell gland in blue sharks with fresh bite marks was reported by Pratt (1979) and for the finetooth and blacktip by Castro (1993, 1996).

Francis (1996) includes evidence of suspected mating by white sharks in Australia and New Zealand waters. Most records are of bite marks on females or observations of seminal fluid oozing from claspers or the genital papillae of adult males, though, as mentioned above, the stress of capture often causes mature sharks to leak semen. The three field reports of probable copulation in free living sharks, swimming side-by-side were, like Clark's (1963) record, seen from above. In the first, Francis (1996, p. 171) reports a personal communication from A. Strachan, a fur seal observer with the New Zealand Department of Conservation. Sometime before 21 Dec 1991 (probably in November), she made the following observation about white sharks: 'I have unwittingly been fortunate to witness a mating (between two white sharks). I had thought at the beginning they were fighting as one animal appeared to be attempting to grasp the other with its great mouth, making great gouges in its side. However, they had eventually become motionless, one under the other, turning over from time to time, belly to belly. This

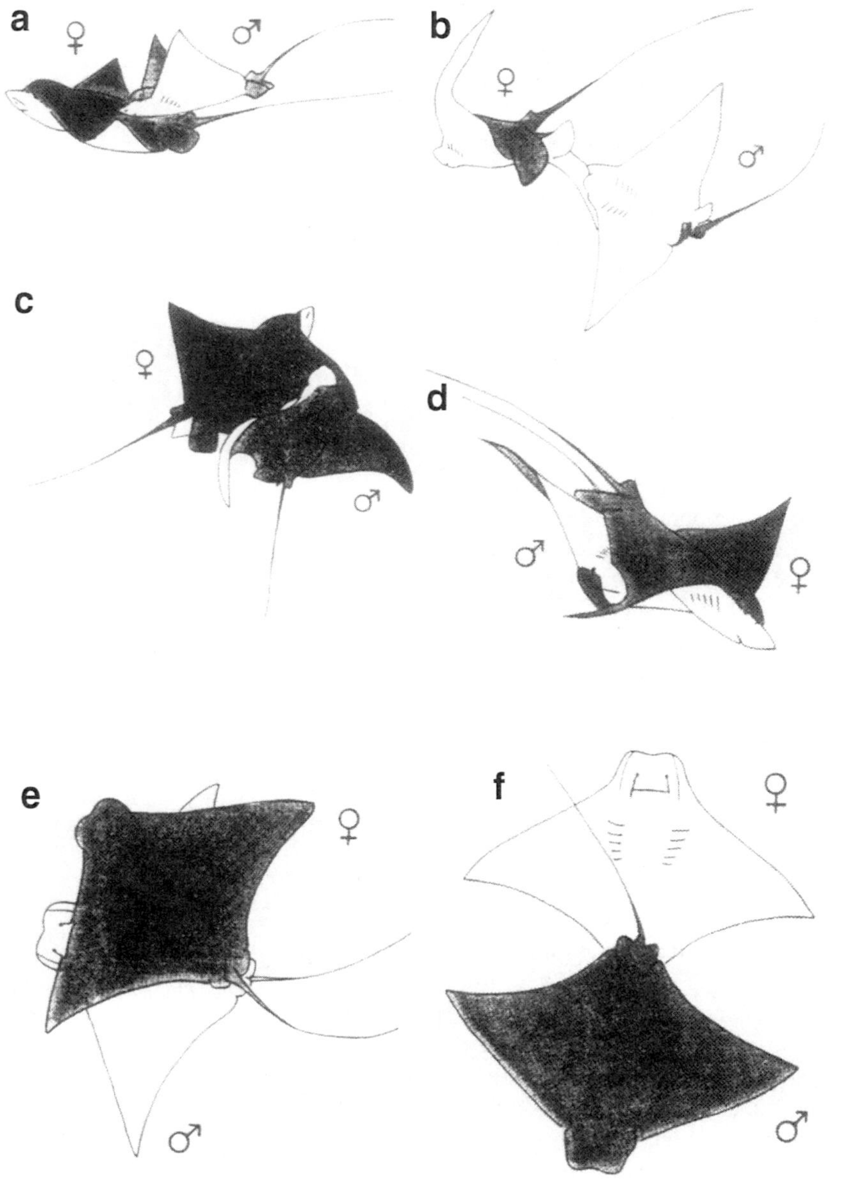

Figure 9. Typical mating behavior of *Aetobatus narinari*: a – 'chasing' male nibbles female's dorsal surface; female stops swimming to allow copulation, b – male bites female on pectoral fin (left or right, front or rear margin). Either clasper may be bent, c – male moves to abdomen-to-abdomen posture, d – male inserts one clasper and copulates. Copulation of *Rhinoptera javanica*: e – male inserting clasper in abdomen-to-abdomen position from beneath the female, f – pair swiveling horizontally, 180° on axis of inserted clasper (from Uchida et al. 1990).

obvious copulation lasted some forty minutes before the animals finally parted and glided off in opposite directions.'

Ebert (1996) includes a report of the copulation of two sevengill sharks, *Notorynchus cepedianus*, in Luderitz Bay, South Africa. The sharks were swimming slowly at the surface in water 3 m deep, 200 m from shore. The male's jaws were clamped on the female's flank, just behind the left pectoral fin. The bodies were side to side with the male's sagging below the female's. The number of claspers inserted was not discernible. Harvey-Clark et al. (1999) interpreted five min of aerial video footage and still photographs taken of 13 basking sharks in close circling formation and overlapping interactions as being of mating behavior. Copulation cold not be confirmed.

To elucidate behavior based on such brief, unplanned and suboptimal conditions is difficult and can lead to errors. Evidence of recent mating is valuable to establish the time and place and other particulars of mating. However, underwater field observations, though at times troublesome, are necessary to obtain details of mating behavior. Larger species are often reclusive, difficult to find and sometimes dangerous to approach when reproductively active. Diving methodology and underwater photographic equipment allow the observation and recording of elasmobranch reproductive behaviors in the field. Videography in particular, provides a lasting, detailed record for analysis and quantification of underwater behaviors not previously possible. The few planned studies will continue to be augmented by opportunistic observations and images made by a growing number of sport and ecotourist divers. Successful observation requires that the presence and actions of the observers does not modify the natural behaviors under study.

In situ reproductive behavioral observations began when Eugenie Clark (1975) became the first to document courtship biting by sharks in the open sea. In the early 1970's, Donald Nelson, working with Richard Johnson, pioneered underwater shark behavioral science by using SCUBA off Eniwetok Atoll to describe 'stereotyped displays' in the grey reef shark, *C. menisorrah* (= *amblyrhynchos*) (Johnson & Nelson 1973). Johnson & Nelson (1978) reported pre-copulatory behavior in the blacktip reef shark, *C. melanopterus*, and the whitetip reef shark, *T. obesus*. In both species the male followed very closely behind the female, with its snout less than 30 cm behind the lead shark's tail. The female held its tail up in an unnaturally erect posture that permitted the male to orient its snout to her vent. Johnson & Nelson (1978) hypothesized olfaction-mediated recognition and pairing after observing males intercepting females when visual and sound cues were absent. They observed one animal making a sudden turn to follow the path of another that was initially out of sight. The two sharks then swam together in a coordinated manner suggesting pair formation (Figure 10). They were later able to identify that the leader was a female and the follower a male. This behavior is termed by them a 'close follow' and is similar to Klimley's 'parallel swimming' and 'following' by Carrier et al. (1994).

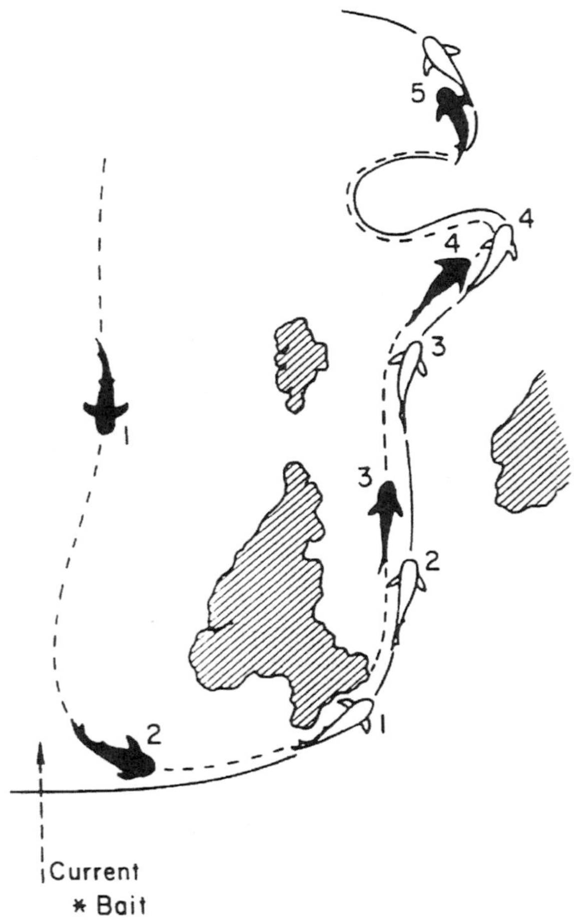

Figure 10. Field observations of possible pair-formation and initiation of close-following behavior in blacktip reef sharks, *Carcharhinus melanopterus*. Solid line indicates path of lead shark (in white) presumed to be female, dashed line indicates path of following shark (in black) presumed to be male. Like numbers indicate relative positions at a given time. At positions 1 and 2 visual contact between sharks was obstructed, indicating olfactory mediation. At position 4, close-following behavior initiated. Positions 4 through 5 characterized by increasingly sinuous swimming. The arrow indicates the direction of the current and the asterisk, the position of bait used to attract the sharks. The observations indicate that pair-formation may be initiated by olfactory cues. Not drawn to scale (after Demski 1990a as redrawn from Johnson & Nelson 1978).

Two mating whitetip reef sharks were observed and photographed copulating in 7 m of water off Molokini in the Hawaiian Islands (Figure 11) (Tricas & LeFeuvre 1985). The sharks copulated in a parallel orientation with the female's head on the bottom and their

Figure 11. Reef white tip sharks, *Triaenodon obesus*, in copulation: a – sharks in stationary parallel orientation on the bottom. The male has a left pectoral grasp on the female whose head is against the rocky bottom, b – ventral view showing intromission. Clasper of the male (at left) is bent medially and inserted in the cloaca of the female. Siphon sac of the male is distended (from Tricas & LeFeuvre 1985).

bodies angled upward 45 degrees into the water column. Pair contact was maintained by the male's tight oral grasp of the female's entire left pectoral fin and by his left clasper anchored one-third of its length into the female's cloaca. The pair remained upright by subtle, often synchronized undulations of their bodies and tails. They were the first to formally suggest that male biting served to maintain the pair bond during copulation.

Work on skates and rays shows similar patterns of courtship with chasing, grasping and biting behavior. Price (1967) observed courtship behavior in longline-caught clearnosed skates. He found one to two males firmly attached to the trailing edge of a female's pectoral fins in October in Chesapeake Bay. Brockmann (1975) observed two stingrays swimming together at the surface. The male was on the back of the female and overlapped her for about one-third of her length. The ventral surface of the male curved forward and covered that of the female. Copulation was not confirmed.

Working off Eniwetok Atoll, Tricas (1980) observed what appears to be complex courtship activity in spotted eagle rays. Male eagle rays dive upon and 'gouge' the female's back with their lower toothplate and bite the caudal margins of the female's pectoral fins. The females would then surface, apparently to make the male's dorsal approach impossible. The male swam behind the female with a sinuous 'bobbing' motion, occasionally circling the female. Feder et al. (1974) observed a pair of male bat rays, *Myliobatis californica*, following a female with one ray pushing the female in her 'genital region'. Tricas (1980) also noted a brief observation of dorsal clasper flexion in a bat ray *Myliobatis californica*, and a possible insertion attempt as the male swam in position just under and slightly behind a larger female. Reed & Gilmore (1981) found that male roughtail stingrays, *Dasyatis centroura*, also bite the pelvic fins of the females, inflicting scars during mating behavior.

Underwater observations of yellow stingrays, *Urolophus jamaicensis*, reveal that the males and females simultaneously grasp the leading edge of each other's pectoral fins. One or more of the usually smaller males may pursue one female. When mating begins, the ray grasps its partners fin edge in its mouth and flips over at a slight angle to its mate. The coupling is brief and the partners go their own way (Dugger 1987). Male round stingrays also bite female pectoral fins (Nordell 1994). Tricas et al. (1995) discovered that reproductively active male round stingrays use their electrosensory system to locate females during courtship. Females also used their electrosense to locate buried consexuals (Tricas et al. 1995, figure 6).

The mating behavior of wild Manta rays, *Manta birostris*, was observed by Yano et al. (1999) on 11 July 1997 while diving in the waters off Ogasawara Islands, Japan. They recorded the abdomen-to-abdomen courtship behavior of two males ~4.5 m disc width (DW) and one female ~5 to 5.5 m DW with underwater photography and videography. The males chased ('followed') behind the tail of the fast swimming female for 20–30 min. The males nipped and held the tip of the female's left pectoral fin during each copulation event, both of which occurred within one meter of the surface. Based on observations, Yano et al. (1999) divided the mating behavior of the

Manta rays into the following sequence: (1) 'chasing', the male rapidly follows behind the tail of the female and attacks her several times, (2) 'nipping', the male nips the tip of the pectoral fin of the female and then moves to the ventral surface of the female, (3) 'copulating', the male inserts a clasper into the cloaca of the female and copulates abdomen-to-abdomen, up to 123 sec (Figure 12), (4) 'post-copulating', and (5) 'separating'.

Fowler (1906a) was the first to describe mating nurse sharks around the Marquesas Islands, Florida.

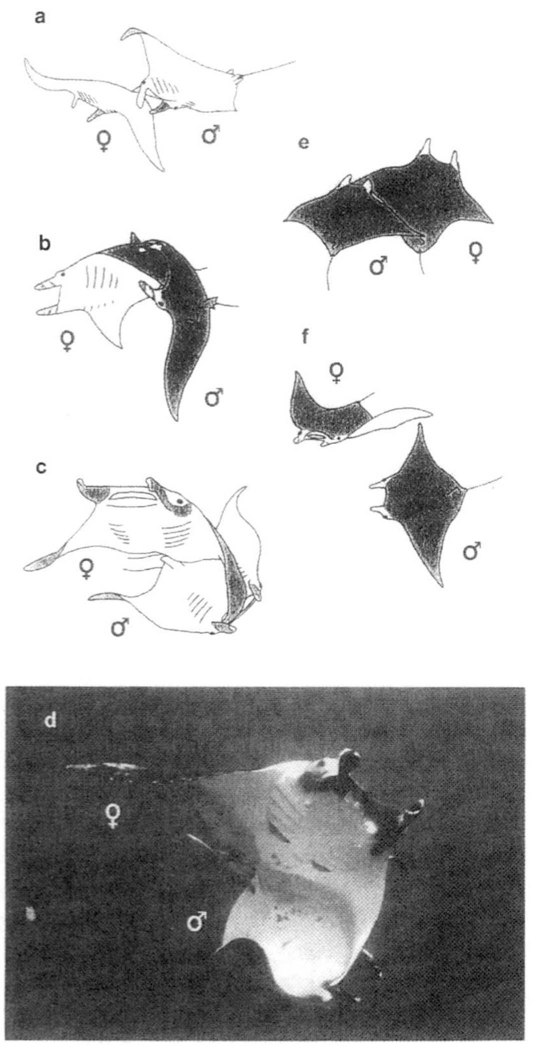

Figure 12. Mating *Manta birostris*: a – 'chasing' ('following'), b – 'biting', c – 'copulating', d – 'copulating' (photographic detail), e – 'post copulating', f – 'separating' (from Yano et al. 1999).

He described the exposed dorsal fins of females in shallow water and noted four males swimming around a female. Gudger (1912) made early observations of mating of free-living nurse sharks in the Dry Tortugas, Florida. Rouse (1985) provided fortuitous underwater photographs of nurse sharks mating on the bottom in 35 m of water off the East Coast of Florida. Carrier, Pratt and collaborators from 1992 to the present, added detailed underwater field observations of free-living nurse shark reproduction to those of Fowler (1906a), Gudger (1912), Klimley & Rouse (1985) (Carrier et al. 1994, Pratt & Carrier 1995). Using underwater videography, we conducted the first detailed study of mating in the same shallow lagoon described by Gudger (1912). Observations began during a nine day period in 1992, when fifty 'mating events' were documented. At least ten different adult males and females were discovered mating by sightings of their light colored ventral surfaces, pectoral fins and tails as they thrashed and rolled in shallow water (0.5–2 m). Most mating attempts (92%) ended without copulation after one to two min. Four steps of mating were identified: 'precoupling' which included the 'following' behavior similar to Klimley's 'parallel swimming'; 'coupling', in which at least one female pectoral fin was grasped; 'positioning and alignment' where rolling and body position were similar to Klimley's 'nudging' and a prelude to 'insertion and copulation'. Copulation was observed and photographed in four of the 1992 events. Sharks copulated in right and left, side-to-side and ventral-to-ventral positions. Female avoidance was observed in which a female sought shallow water, presumably to avoid copulation and possibly to engage in sexual selection by controlling male access and mating. Ten of the sixteen events observed in 1992 were accompanied by multiple males. Several unique male 'group behaviors' were described in which males would compete to grasp a female pectoral fin or possibly cooperate to aid other male's attempts to copulate.

A case study of reproduction and mating behavior in the nurse shark, *Ginglymostoma cirratum*

Although the nurse shark is abundant and widely distributed throughout the Florida Keys, Bahamas and Caribbean Sea, little is known of the details of its reproductive life history. Since 1991, we have conducted cooperative studies with biologists from Albion College, SeaWorld Adventure Parks, Orlando,

Florida, Mote Marine Laboratory, NMFS, SEFSC and other institutions in the Dry Tortugas National Park, Florida. This is currently the only directed ongoing in situ study of shark reproductive behavior. Our recent investigations include: (1) determining nurse shark distribution in the local reefs and entire archipelago using tagging, (2) defining elements of social structure with DNA fingerprinting and (3) elucidating behaviors, such as dominance hierarchies, sperm competition, male rivalry, possible male cooperation and kin interactions using direct field observations and sampling of tagged adult, neonate, and juvenile nurse sharks, (4) characterization of the study area as a critical habitat for mating and management of the site to ensure reproductive success (Carrier & Pratt 1998).

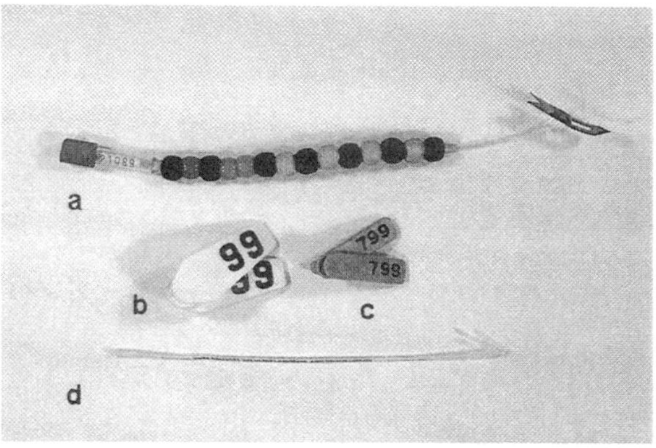

Figure 13. Four external passive tags used on nurse sharks in the Dry Tortugas study: a – NMFS 'M' type dart tag with strings of coded beads for adults, b – jumbotag for adult dorsal fins, c – roto-tag for juvenile dorsal fins, d – nylon dart tag for juveniles.

Adult tagging

Our earliest observations in 1992 and 1993 were hindered by lack of shark identification. Only six (10%) of the stock of mating nurse sharks have distinctive natural markings. Some natural marks heal rapidly, grow over or change in time. Fin notches look alike. Thus we chose tagging, though temporarily stressful to the shark, as the best method to identify individuals.

Tagging began in 1993 to help understand the dynamics of mating events. Six different types of tags were used in this study. Four external passive tags were placed on dorsal fins and in the shark's body at the dorsal fin base (Figure 13). Identification of sexually active individuals was accomplished by tagging mature males and females with passive, diver identifiable body tags using a modified speargun. NMFS 'M' type dart tags were chosen because they have good return rates on nurse sharks (Carrier 1985, Kohler et al. 1998). They were easily modified for diver recognition by adding strings of coded beads made of 6 × 6 mm styrene plastic. After the second year, the 'M' tags, particularly on the males, became so overgrown with biofouling that they could not be 'read' by diving observers. After several trials, we switched to beads and alternate short sections of copper tubing painted with different colors of copper based boat bottom paint, that, in conjunction with the copper beads, resisted biofouling. Some nurse sharks in our study have probably lost their tags to coral bottoms, predators and other marine life. Lost tags are an inevitable long-term problem in tag studies.

Photographs of external dart tags with coded beads provide a permanent record of 'visual resightings' because natural markings were seldom encountered in this study. Both still and video photography were utilized for recording mating behavior. Videography was useful in validating identification of animals but primarily in enumerating and quantifying the complex behaviors associated with mating. We equipped the video housings with external hydrophones to permit the addition of field observer voice records directly to videotape for later transcription, and with underwater still cameras to combine videography with still photography.

To provide a durable marker, net captured adult nurse sharks were also tagged internally with passive integrated transponder (P. I. T.) tags approximately 10 mm under the dermis, beneath the right base of the first dorsal fin. Nylon cattle ear 'Jumbotags' are placed in the first or second dorsal fin. Remote observations of tagged and untagged sharks were made from a temporary 7 m high tower constructed of scaffolding and from small boats and kayaks. In situ observations were made by free diving and photography. The size of the beads and their unique color combinations and the 'Jumbotags' in the dorsal fins facilitated identification of individuals by divers and by surface observers, even when using binoculars from relatively long distances. Blood and tissue samples were taken from all sharks which were then injected with oxytetracycline at 30 mg per 1 kg of body weight, tagged and released.

Between 1993 and 1998, we have tagged 45 adults and 70 juveniles in the Dry Tortugas study stock (Figure 14a). Of the 51 adults identified, 46 were subsequently resighted (Figure 14b). Observations from tagging and natural markings indicate that most adult

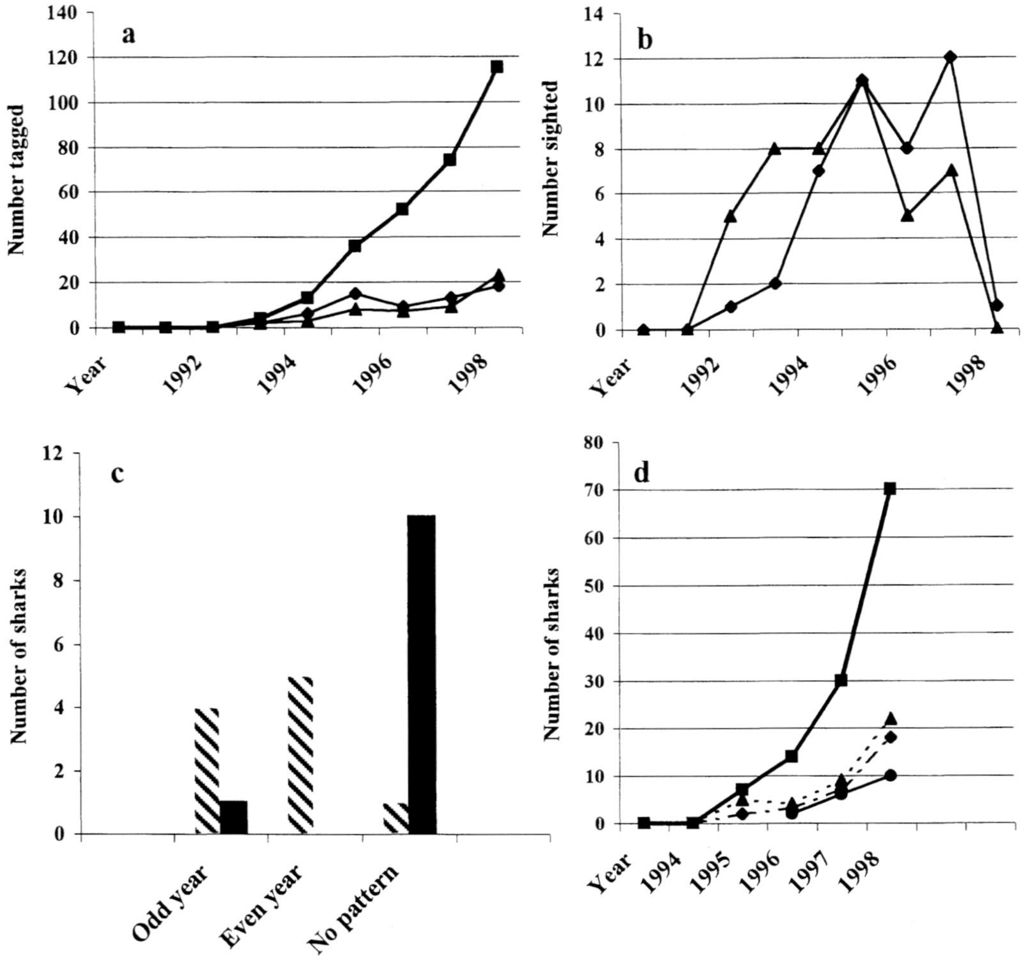

Figure 14. Tagging study of nurse sharks: a – adult (n = 45) and juvenile (n = 70) nurse sharks tagged by year, ▲ = males, ♦ = females, ■ = cumulative sum, b – tagged adults sighted each year (n = 46), c – numbers of tagged males and females that returned to study site more than one year by year of return, odd years 1993, 1995, 1997, even, 1994, 1996, 1998, or no pattern of return (numbered, cross-hatched bars = females, solid bars = males), d – juveniles tagged and recaptured by year. Upper solid bar represents cumulative tagged, lower solid bar represents juvenile recaptures and re-releases by year.

males visit the study site every year, sometimes in pairs or larger groups, with three males dominant in our observations since 1992. Typically the males enter the lagoon and rapidly swim a search pattern apparently seeking females. After five to 60 min they depart the lagoon. Males may repeat these sorties alone or in company with three or more males several times a day. Recaptures from sightings of our modified dart tags show that adult females visit the study area to mate in alternate years (Figure 14c), roughly half of the females being present in any one year. This biennial mating pattern is common in elasmobranch females and probably allows the post-partum females the time to rebuild reproductive reserves before mating again.

Telemetry

Ultrasonic telemetry is a powerful tool to determine the fine scale daily activities of a few sharks and is used to track male and female adult sharks when they leave the lagoon to determine their home ranges and activity cycles. Selected individuals were tagged with Vemco (Shad Bay, Nova Scotia) or Sonotronics

(Tucson, AR) ultrasonic transmitters for telemetry. Receivers and hydrophones were deployed from small boats and kayaks to track tagged sharks. Transmitters were always accompanied by a NMFS 'M' tag designed to remain after the transmitter breaks away.

Males tagged with ultrasonic transmitters in 1996 and 1997 showed a great deal of active swimming in their local patrolling behaviors supporting Springer's (1967) remarks of increased activity of males during the mating season. Males repeatedly patrolled the coast of the bracketing island in 2–12 m of water, resting at intervals near reef structure.

Six tagged adult females were released back into the study area after nearly 12 months of confinement at SeaWorld. All females rapidly departed the area at release. A transmitter placed on one of these female on 3 Jun 1998 at 10:27 h showed movement south for 8 h after tagging and release. The shark traveled along the 10 m depth contour, then southeast into open, deeper waters until lost at 18:39 h. These observations indicate a different behavior than seen in the females seeking to stay in the lagoon and adds support to the 'male access' and 'sexual selection' hypotheses (see below). We believe that females released from a year of confinement in SeaWorld are physiologically unprepared to be in a mating area (wrong alternate year for these individuals) and quickly leave the mating grounds possibly to return to their home range. Post confinement shock syndrome may also play a role. More information is necessary to elucidate these ideas.

Mating behaviors

Between 1992 and 1998, we have witnessed over 380 nurse shark mating events at the Dry Tortugas, 30 (8%) ending with copulation (see 'avoidance', below). Polygyny and polyandry have been found to be common, often with daily multiple matings. Typically, in the first and second weeks of June, females arrive in the shallow, 0.8 hectare, study lagoon. The lagoon is enclosed by a spit of land that is submerged at high tide, its western side is a harbor of refuge 4–6 m deep, the bottom is mixed turtle grass, *Thalassia testudinum*, and manatee grass, *Syringodium filiforme*, grading up to algae covered coral rubble with occasional small coral patch reefs. Water depths varied from 0–1.8 m.

Some females are chased into the shallow waters, swimming rapidly before the pursuing males. In some instances females were observed working their way slowly down the island's outer coast toward the lagoon. They were repeatedly driven to shoal against the rocky, algae covered, wave-beaten shore by two to four following males, before their arrival in the lagoon. With male pursuit a female movement of 300 m may take 5–6 h. Females will often be in the company of 1 to 4 males that will circle in the shallow waters with them. We have noticed that males sharks will also closely follow each other during the mating season while in pursuit of females and while patrolling. We have conducted discreet in situ inspections of all sharks when they first arrive to look for past tags and body condition. All newly arriving females exhibited fresh bite marks on their pectoral fins that varied in severity from mild scratches to bleeding, open wounds. Except for occasional healed scars and fin notches, males are generally unmarked.

Avoidance

When approached by patrolling males, females apparently reluctant to mate, 'avoid' the males by swimming to the shallowest limits of the lagoon that they can reach (< 40 cm depth) and stop nearly aground. If a non-receptive female was caught in a pectoral grasp by a male, she attempted to escape by 'arching' against the bottom and away from the male, twisting her body and cloaca away and, at times, out of the water. If the male did not lose interest within a few moments, she often 'pivoted and rolled' towards the male, turning over her trapped pectoral fin and his head, leveraging against the male's body and substrate, then suddenly twisted out of his grasp and rapidly swam away (Figure 15a-i). Variations of this behavior were observed hundreds of times in the course of our study.

Refuging

What brings nurse sharks back to this specific shallow lagoon is still unclear. We believe that nurse sharks seasonally migrate to the 0.8 hectare mating site from the entire 120 nautical square mile island group to reproduce. Although mid-sized queen conchs, *Strombus gigas*, spiny lobsters, *Panulirus argus*, crabs and fish are abundant, no adult feeding, or evidence thereof, has been observed in the lagoon; thus the movement inshore does not include feeding.

The presence of fresh pectoral bite marks on all females when they first arrive in the study lagoon indicates recent, sometimes vigorous, mating activity. Their health or reproductive success may be jeopardized by excessive male attention and competition out among the reefs. Furthermore, a female may use this

a
b
c
d
e
f
g
h
i
j

warm, quiet refuge as her large (~100 mm) ova descend into the uterus in cycles over a week's time. When she is physiologically ready for fertilization, she will once again seek and permit copulation. Several copulations may be needed before a clutch of ova is fertilized. We conclude that the study site apparently serves as a mating refuge or retreat for reproductively active females who are probably exuding pheromones or other similar reproductive olfactory cues.

Sexual selection

The vantage point of shallow water may also allow females to engage in mate selection. The refuging female is observed to retreat repeatedly from smaller, lighter and perhaps younger, weaker or inexperienced males. Some of these smaller males are very persistent and will approach females hourly in the lagoon. In contrast, the arrival of one of the dominant males often elicits a response in the female to remain in the lagoon's deeper water, or at least not to retreat from this more desirable, or perhaps more inescapable, mate. In his study of schooling aggregations, Klimley (1985) suggested that in female *S. lewini*, segregation and a preferred location in the school could facilitate female choice and selection in mating with or avoiding a particular male.

Nurse shark copulation

Occasionally a resting female nurse shark will copulate with a male in the deeper parts of the study area. In these cases, video sequences show that avoidance behavior terminates, or rarely, never starts. The submissive postures noted by Gordon (1993) have not yet been noted for the nurse shark. At this time, the male's 'pectoral bite' doesn't cause an 'accepting' female to 'pivot and roll' (Klimley 1980). If a mating event with an accepting female starts in shallow water (< 0.5 m), she will not struggle as the accepted male attempts to 'carry' her by her pectoral fin to water deep enough to permit 'positioning and alignment' (Figure 15j) and copulation. If successful, the female's rostrum usually becomes positioned, perhaps intentionally by the male, nose downward, firmly against the substrate (Figure 1). In our observations, the male will cross the distal clasper toward the female, catch her tail between his tail and clasper ('tail tuck'), and arch his body toward her, bringing his single clasper between her 'cupped' pelvic fins and into her cloaca and vagina (see Behavioral summary below). The female remains apparently relaxed and arches toward the male while cupping her pelvic fins probably to guide and aid insertion. As in other elasmobranch species observed, it is the male that is most active, thrashing the water to position the female and presumably to aid sperm transfer. Copulation lasts for about 2 min (Carrier et al. 1994). Whether mating is by submission or choice is not completely understood at this time. Because of the violent nature of the coupling, it often appears that the females are taken by force. However, close examination of the copulatory events on videotape show that some females will aid certain males some of the time.

Group mating activities where one female is approached by more than one male was witnessed in ten of the sixteen events observed in 1992. Males were observed to surround a female, competing for an effective bite on either pectoral fin. On several occasions, one male was observed to 'block' the heads of the mating pair with its body, a possible cooperative action to aid other male's attempts to copulate (Carrier et al. 1994, figure 7).

Behavioral summary: the emerging pattern

The literature described in the previous sections and the case study reveal behaviors that seem common to most elasmobranchs investigated to date. In this summary references to specific authors are omitted for clarity.

Elasmobranch courtship begins when one animal signals a potential mate that it is reproductively receptive to 'copulation'. In most instances, the female probably initiates mating but work on stingrays and captive sand tigers show that the male may sometimes initiate mating. The precopulatory sensory cues are probably principally olfactory, with females employing chemical stimulants (pheromones) that may be combined with motor displays. These include female 'back arching',

Figure 15. Female 'pivot and roll', an avoidance behavior: a – a 2.5 m female nurse shark (left) held in a left pectoral grasp by the dominant male (right), b,c – female rolls over the male's head, then, d – upside down and still held, e,f – twists to find leverage against the substrate, and g,i – pulls her pectoral free of his grasp to escape, j – female nurse shark (lower) relaxed in a right pectoral grasp, being carried by a male to deeper water. Sperm from previous matings in a mucoid 'rope' oozes from the female's cloaca (bottom).

'submissive' body tilting, 'pectoral fin undulations', pelvic fin 'cupping' and 'flaring' and perhaps other, as yet undescribed physical cues. Males may cease feeding, become aggressive toward other species and may compete for dominance with other males of the same species. One male may approach the female either alone or in a 'group' and attempt to 'follow', perhaps 'nosing' her cloaca. The female may raise her tail, or incline her body. After preliminary interactions, the males mouth is used for first contact in most elasmobranchs. Typically the male moves up the female's body to 'bite' her pectoral fin (Figure 16a). Biting may act as a releaser for female mating behaviors and precopulatory biting may stimulate ovulation and other physiological changes. Batoids may also bite and 'gouge' and hold females with spines and denticles. Female round stingrays can bite the males in return. Males of some larger shark species (nurse sharks, carcharhinids) concentrate on one of the pectoral fins. More flexible species (blue sharks, catsharks and horn sharks) may initially 'grasp' anywhere on the female's body, usually behind the eyes, including the pectoral fins. Smaller and more limber sharks are sometimes able to copulate from a body bite or even a body 'wrap' without using a 'courtship bite', but the rule in larger, less flexible sharks seems to necessitate an unshakable oral grasp of the pectoral fin.

After attaining a hold on the female, actions vary with species and female cooperation. The male nurse shark, if in shallow water (< 1 m), will lift and 'carry' the female to deeper water (1.5–2 m) or may attempt to do so. In nurse sharks, whitetip reef sharks and possibly hammerheads, the paired sharks may seek (or fall) to the sea floor where the female's rostrum becomes positioned, perhaps intentionally by the male, against the substrate to balance his thrusts. The male then catches the female's tail between his tail and more the distant (crossed) clasper and begins copulation. Some species, such as the hornshark, reportedly can use either clasper. The whitetip reef shark male uses the clasper on the same side as the grasped female (Figure 11). Some elasmobranchs may have the option of using either clasper in either direction, 'crossed' or 'splayed' (Figure 8g,f). Skates and some rays copulate on the sea floor; other pelagic or more powerful animals, such as the spotted eagle rays and mantas, almost perform a two-part ballet in mid-water.

In response to the male's maneuvers, an 'accepting' female nurse shark 'arches' her body toward the male, 'cups' her pelvic fins upward, becomes immobile and

Figure 16. Mating behavior and copulation in the nurse shark: a – a male nurse shark (left) attempts a right pectoral grasp after following a refuge seeking female in shallow water, b – the nurse shark clasper partially inserted on the cloaca of a female during copulation.

appears to become docile and receptive. The female spotted eagle ray stops in flight to permit a grasp. The clasper of nurse sharks and other species may be partly or totally inserted into the cloaca and presumably into the common vagina and perhaps one uterus (Figure 16b). It is locked in place as the terminal clasper structures take the form of an expanded reverse funnel. The clasper rhipidion may also bear sharpened ridges, hooks and spurs that find purchase in the female's thickened vaginal wall.

An 'avoiding', presumably unaccepting female will retreat from the approaching aggressive male. A female sand tiger may 'shield' her cloaca by swimming near the substrate. If a female nurse shark is caught in an unwanted 'pectoral bite', she will first 'arch' her body away from the male, bend towards him, then twist out of his grasp with a 'pivot and roll' to escape. Some males will let go of the female pectoral fin spontaneously after 4–5 min of waiting, perhaps in exhaustion or as a consequence of oxygen deprivation due to the blockage of the buccal cavity by the large pectoral fin of the female.

If the male is accepted or endured, copulation ensues and the female shark becomes relatively immobile in those species so far observed (Figure 1) (nurse shark, chain dogfish, spotted dogfish, whitetip reef sharks, spotted eagle rays, manta rays and anecdotally, lemon sharks, white sharks and sevengill sharks). The male thrusts and rotates his tail, probably accompanied by internal contractions of the ampullae epididymides and perhaps the siphon sacs, for several seconds to a few minutes in larger species of sharks to many minutes and even hours in certain skates. Species may copulate on the sea floor (whitetip reef shark, nurse shark, chain dogfish), in mid water (the larger rays) or at the sea surface (lemon, white and sevengill).

At the conclusion of insemination in the nurse shark, thrusting stops, both are suddenly still and then the male very slowly withdraws his clasper. A mucus mass, apparently containing some sperm, stretches out between clasper and female cloaca, then breaks. The female remains motionless, still cupping her pelvic fins. The pair then becomes active again and either breaks apart rapidly to swim away separately (cat shark, nurse shark), or the male may release his pectoral grasp and remain motionless on the sea floor as if in recovery (nurse shark). Male nurse sharks have been observed to leave the mating area and mate again the next day. Females may rest in the area of mating, may seek shallow water immediately or may mate with another male within minutes.

Future directions for studies of elasmobranch reproductive behavior

The previously described studies, including our work on *G. cirratum*, identify a number of hypotheses and research considerations that may serve to guide future studies of elasmobranch reproductive behavior.

The male access hypothesis

Females of some species of elasmobranchs may have the ability to limit male access when they are reproductively attractive. Unlimited male contact may jeopardize their health or reproductive success. Klimley (1985) suggested that female *S. lewini* maintained a preferred location in the school to control male access. Tricas et al. (1995) found that the electrosensory system of *U. halleri* is used by reproductively active males to locate buried otherwise inaccessible mates (Figure 6). As seen above, nurse sharks retreat to shallow water when approached by smaller or apparently undesirable males and will remain for days in this shallow lagoon, an apparent refugia. The access hypothesis may be tested in a sexually active species by observing the interactions of tagged or uniquely marked males and females.

The sexual selection hypothesis

Females of many vertebrate species are discriminate in mate selection (Krebs & Davies 1993). In female *S. lewini* segregation and a preferred location in the school could facilitate female choice and selection in mating with or avoiding a particular male (Klimley 1985). The vantage point of shallow water may allow female nurse sharks to engage in mate selection. The 'refuging' female is observed to retreat repeatedly from smaller, lighter colored males which are perhaps younger, weaker or inexperienced. The arrival of one of the larger, darker, dominant males often elicits a response in the female to remain in the lagoon's deeper water, or at least not to retreat when approached. The sexual selection hypothesis may be tested by observing and quantifying the interactions of tagged or uniquely marked males and females.

The dominance hierarchy hypothesis

Elasmobranchs may exhibit dominance hierarchies within wild reproductively active groups. The observations of Gordon (1993) revealed a distinct dominance hierarchy in captive mating male sand tigers. Gilmore et al. (1983) noted that in the field, mature females displayed mating bites, but also apparently bit the mating males. In our study of nurse sharks, we have seen more successful copulations by several larger males, which we have identified as dominant, than by any other males. It is possible that intrasexual hierarchies are expressed through biting behavior and may serve to synchronize mating. Social hierarchies and related behaviors may carry over from nonreproductive periods particularly in species like the nurse shark and sand tiger which remain together or repeatedly use particular mating sites over several annual cycles. This hypothesis may be tested using marked animals through both conventional tagging and with telemetry to follow and quantify individual mating success.

Other research

Captive studies
Captive studies can provide valuable information on reproductive and physiological processes. They provide valuable observations on mating related behavior if animal size is appropriate to aquarium size and habitat. For example, blood work on temporal changes in reproductive hormone levels, sonography, endoscopy and environmental manipulation are not easily conducted on sharks in the wild, but are relatively easy when conducted in aquarium facilities. Furthermore, captive studies are useful to detect pregnancy in live animals, characterize embryonic development and develop DNA assays for genetic relatedness. The authors acknowledge the limitations of captive studies.

Genetic studies
Various mating behaviors can enhance reproductive fitness and genetic diversity, even in non-migratory populations. Genetic identities of local species groups can reveal the extent to which immigration and emigration influences gene flow. While multiple copulations by different males are known for many species of female elasmobranchs, it is unknown whether a single male fathers an entire litter. The kinship relationships and family trees revealed by DNA microsatellite and standard field techniques, can greatly increase the understanding of mating social dynamics and mating behaviors within populations. For example, proposals to harvest only male sharks may be inappropriate if female submission, sexual selection, copulation or fertilization required participation by multiple males.

Anthropogenic influences
Anthropogenic influences from increased coastal urban development and human recreational activity can impair elasmobranch reproduction. In our observations of nurse shark mating systems, we have noted the disruptive effects of recreational boaters and scientists to mating success. As a result, we have worked in cooperation with the National Park Service and NOAA's Florida Key's Marine Sanctuary to close our study area and the associated identified mating and nursery ground to boat traffic by using navigation control buoys during the peak of the mating season (Carrier & Pratt 1998). Further work should identify essential habitats used by other elasmobranch species for the management of critical habitats.

Remote sensing
Remote sensing equipment and video tags are new techniques that can be applied to the study of elasmobranch behavior. Most previous work on elasmobranch reproductive behavior were point-of-time studies which lacked the technology to continuously monitor reproductively active animals. The use of telemetry can produce valuable movement records if the subjects are captured and tagged during reproductively important times. The use of hydrophone data loggers on the sea floor can provide long term data sets for species with small home ranges or activities during the mating season. Video cameras, that are placed on free-swimming sharks may provide valuable behavioral information but currently these devices have a restricted field of view and a short recording time. The behaviors associated with mating occur unpredictably and may last substantially longer than can be recorded. If such devices incorporated global positioning capabilities and could be activated by some mating behavior (e.g. clasper flexion) their utility would be greatly increased.

Concluding remarks

While tagging studies and photography may provide many details on courtship and copulation, studies on elasmobranch reproductive behavior cannot be conducted in isolation. An understanding of the ultimate and proximate causes of specific behaviors will arise from integration of multiple techniques such as physiology, biochemistry, molecular genetics and population ecology.

The limited studies of life history styles of elasmobranchs (Pratt & Casey 1990, Hoenig & Gruber 1990) reveal that these animals have slow growth, mature relatively late in life and have low fecundity. The extent to which a population is able to replenish its numbers has a profound impact on the rate at which animals can be removed from the breeding stock while not jeopardizing the natural ability of the population to recover. Most studies concentrate on one or two aspects of life history rather than on a comprehensive examination of all factors that ultimately influence reproductive success. Nevertheless, studies of reproductive behavior must be conducted in concert with studies of anatomy, age and growth, food habits, segregation, movement patterns and distribution in order to more completely understand the interrelationships between these factors. Additionally, to effectively manage elasmobranch stocks, it is essential to determine size and age at maturity, fecundity and reproductive cycles for all common and/or exploited elasmobranch species. Studies must include discerning the habitat requirements for both mating and pupping grounds as well as focusing on social dynamics and mating behavior.

Acknowledgements

Grateful appreciation is extended to R. Ring, W. Landrum and P. and J. Taylor of the National Park Service, to Frank Murru, Ray Davis and the veterinarians and research staff from SeaWorld Adventure Parks, Orlando, Florida. We thank B. Causey, J. Halas, S. Baumgartner of NOAA's Florida Keys National Marine Sanctuary Program, the Newfound Harbor Marine Institute at Seacamp and the United States Coast Guard. N. Kohler of NOAA/NMFS Narragansett Laboratory provided logistical support and we thank T. Duncan and C. Carrier for many hours of shark observations and field assistance. We thank Nick Caloyianis for his efforts to popularize our research, making his photographs available to us for study and for his kindness in allowing us to use Figure 15j. We are very grateful to F. Sato whose photograph appears in Figure 12d. Figure 4 is reproduced with permission of the Journal of Experimental Biology and the Company of Biologists. Figure 8 is reproduced with permission from Kluwer Academic Publishers. This project was supported in part by funds to HLP from NOAA's Highly Migratory Species Management Division and to JCC from the Hewlett-Mellon Faculty Development funds and also from the A. Merton Chickering Endowed Professorship of Albion College. We thank Reviewer 1, S. Michael, F. Almeida, R. Merrick and A. Stoner who all reviewed the manuscript and made many improvements, and T. Tricas for convening the Nelson symposium and shepherding our efforts into final form.

References cited

Agassiz, L. 1871. On the method of copulation among selachians. Proc. Boston Soc. Nat. Hist. 14: 339–341.

Bigelow, H.B. & W.C. Schroeder. 1953. Fishes of the Gulf of Maine. U.S. Fish Wildl. Serv. Fish. Bull. 53: 351–357.

Bolau, H. 1881. Über die Paarung and Fortpflanzung der Scyllium-Arten. Z. wiss. Zool. 35: 321–325.

Bres, M. 1993. The behaviour of sharks. Rev. Fish Biol. Fisheries. 3: 133–159.

Brockmann, F.W. 1975. An observation of mating behavior of the southern stingray, *Dasyatis americana*. Copeia 1975: 784–785.

Callard. I.P., L. Klosterman & G.V. Callard. 1988. Reproductive physiology. pp. 277-317. *In:* T. J. Shuttleworth (ed.) Physiology of Elasmobranch Fishes, Springer Verlag, New York.

Compagno, L.J.V. 1984. Sharks of the world. FAO species catalogue, Vol. 4, Pt.1, FAO Fish. Syn.125: 1–655.

Carrier, J.C. 1985. Nurse sharks of Big Pine Key: comparative success of three types of external tags. J. Fl. Acad. Sci. 48: 146–154.

Carrier, J.C. & H.L. Pratt, Jr. 1998. Habitat management and closure of a nurse shark breeding and nursery grounds. Fish. Res. 39: 209–213.

Carrier, J.C., H.L. Pratt, Jr. & L.K. Martin. 1994. Group reproductive behavior in free-living nurse sharks, *Ginglymostoma cirratum*. Copeia 1994: 646–656.

Casey, J.G. & N.E. Kohler. 1992. Tagging studies on the shortfin mako shark (*Isurus oxyrinchus*) in the Western North Atlantic. Aust. J. Mar. Freshwater Res. 43: 45–60.

Castro, J.I. 1983. The sharks of North American waters, Texas A & M University Press, College Station. 180 pp.

Castro, J.I. 1993. The biology of the finetooth shark, *Carcharhinus isodon*. Env. Biol. Fish. 36: 219–232.

Castro, J.I. 1996. Biology of the blacktip shark, *Carcharhinus limbatus*, off the southeastern United States. Bull. Mar. Sci. 59: 508–522.

Castro J.I., P.M. Bubucis & N.A. Overstrom 1988. The reproductive biology of the chain dogfish, *Scyliorhinus retifer*. Copeia 1988: 740–46.

Clark, E. 1963. The maintenance of sharks in captivity, with a report on their instrumental conditioning. pp. 115-49. *In:* P. Gilbert (ed.) Sharks and Survival, D. C. Heath and Co., Boston.

Clark, E. & K. von Schmidt. 1965. Sharks of the central Gulf Coast of Florida. Bull. Mar. Sci. 15: 13–83.

Clark, E. 1975. The strangest sea. Nat. Geo. Mag. 148: 338–343.

Dempster, R.P. & E.S. Herald. 1961. Notes on the hornshark, *Heterodontus francisci*, with observations on mating activities. Occ. Papers Calif. Acad. Sci. 33: 1–7.

Demski, L.S. 1990a. Neuroendocrine mechanism controlling the sexual development and behavior of sharks and rays. J. Aquariculture and Aquatic Sci. 5: 53–67.

Demski, L.S. 1990b. Elasmobranch reproductive biology: implications for captive breeding. J. Aquariculture and Aquatic Sci. 5: 84–95.

Dobson, S. & J.M. Dodd. 1977. The role of temperature and photo period in the response of the testis of the dogfish, *Scyliorhinus canicula*, L. to partial hypophysectomy (ventral lobectomy). Gen. Comp. Endocrinol. 32: 114–115.

Dodd, J.M. 1983. Reproduction in cartilaginous fishes (Chondrichthyes). pp. 31–95. *In:* W.S. Hoar, D.J. Randall & E.M. Donaldson (ed.) Fish Physiology, Vol. IXA, Academic Press, New York.

Dral, A.J. 1980. Reproduction en aquarium du requin de fond tropical, *Chiloscyllium griseum* Müll. et Henle (Orectolobidés). Rev. Fr. Aquariol. 7: 99–104.

Dugger, A.J. 1987. Mating stingrays. Sea Frontiers 33: 352–354.

Eberhard, W.G. 1985. Sexual selection and animal genitalia. Harvard University Press, Cambridge. 244 pp.

Ebert, D.A. 1996. Biology of the sevengill shark, *Notorynchus cepedianus* (Peron, 1807), in the temperate coastal waters of Southern Africa. S. Afr. J. Mar. Sci. 17: 93–103.

Economakis, A.E. & P.S. Lobel. 1998. Aggregation behavior of the grey reef shark, *Carcharhinus amblyrhynchos*, at Johnston Atoll, central Pacific Ocean. Env. Biol. Fish. 51: 129–139.

Feder, H.M., C.H. Turner & C. Limbaugh. 1974. Observations on fishes associated with kelp beds in Southern California. Calif. Fish. Bull. 160: 1–144.

Francis, M.P. 1996. Observations on a pregnant white shark with a review of reproductive biology. pp. 157–172. *In:* A.P Klimley & D.G. Ainley (ed.) Great White Sharks: the Biology of *Carcharodon carcharias*, Academic Press, San Diego.

Fowler, H.W. 1906a. Some cold-blooded vertebrates of the Florida Keys. Proc. Acad. Sci. Phila. 58 : 77–113.

Fowler, H.W. 1906b. The fishes of New Jersey. Ann. Rep. N. J. State Mus. 1905, Pt. 2: 35–477.

Gilbert, P.W. 1981. Patterns of shark reproduction. Oceanus 24: 30–39.

Gilbert. P.W. & G.W. Heath. 1972. The clasper-siphon sac mechanism in *Squalus acanthias* and *Mustelus canis*. Comp. Biochem. Physiol. 42A: 97–119.

Gilmore, R.G. 1993. Reproductive biology of lamnoid sharks. Env. Biol. Fish. 38: 95–114.

Gilmore, R.G., J. W. Dodrill & P. A. Linley. 1983. Reproduction and embryonic development of the sand tiger shark, *Odontaspis taurus* (Rafinesque). U.S. Fish. Bull. 81: 201–225.

Gordon, I. 1993. Pre-copulatory behavior of captive sand tiger sharks, *Carcharias taurus*. Env. Biol. Fish. 38: 159–164.

Gudger, E.W. 1912. Summary of work done on the fishes of Tortugas. Carnegie Inst. Washington Yearbook 11: 148–150.

Hamilton, W.D. 1971. Geometry for the-selfish herd. J. Theoretical Biol. 31: 295–311.

Hardy, A. 1959. The open sea: its natural history. Houghton Mifflin Co., Boston. 322 pp.

Harvey-Clark, C.J., W.T. Strobo, E. Helle & M. Mattson. 1999. Putative mating behavior in basking sharks off the Nova Scotia Coast. Copeia 1999: 780–782.

Hoenig, J.M. & S.H. Gruber. 1990. Life history patterns in the elasmobranchs: implications for fisheries management. pp. 1–16. *In:* H.L. Pratt, Jr., S.H. Gruber & T. Taniuchi (ed.) Elasmobranchs as Living Resources: Advances in Biology, Ecology, Systematics and Status of the Fisheries, NOAA Tech. Rep. NMFS 90.

Johnson, R.H. & D.R. Nelson. 1973. Agonistic display in the gray reef shark, *Carcharhinus menisorrah*, and its relationship to attacks on man. Copeia 1973: 76–84.

Johnson, R.H. & D.R. Nelson. 1978. Copulation and possible olfaction-mediated pair formation in two species of carcharhinid sharks. Copeia 1978: 539–542.

Jones, S.N. & R.C. Jones. 1982. The structure of the male genital system of the Port Jackson shark, *Heterodontus portusjacksoni* with particular reference to the genital ducts. Aust. J. Zool. 30: 523–541.

Kajiura, S.M. & T.C. Tricas. 1996. Seasonal dynamics of dental sexual dimorphism in the Atlantic Stingray, *Dasyatis sabina*. J. Exp. Bio. 199: 2297–2306.

Klimley, A.P. 1980. Observations of courtship and copulation in the nurse shark, *Ginglymostoma cirratum*. Copeia 1980: 878–882.

Klimley, A.P. 1985. Schooling in the large predator, *Sphyrna lewini*, a species with low risk of predation: a non-egalitarian state. Zeitschrift fur Tierpsychology 70: 297–319.

Klimley, A.P. 1987. The determinants of sexual segregation in the scalloped hammerhead shark, *Sphyrna lewini*. Env. Biol. Fish. 18: 27–40.

Krebs, J.R. & N.B. Davies. 1993. An introduction to behavioural ecology. Blackwell Science Ltd., Cambridge. 420 pp.

Kohler, N.E., J.G. Casey & P.A. Turner. 1998. NMFS Cooperative Shark Tagging Program, 1962-93: an atlas of shark tag and recapture data. Mar. Fish. Rev. 60: 1–87.

LaMarca, M.J. 1964. The functional anatomy of the clasper and clasper gland of the yellow stingray, *Urolophus jamaicensis* (Cuvier). J. Morph. 114: 303–324.

Leigh-Sharpe, W.H. 1920. The comparative morphology of the secondary sexual characters of elasmobranch fishes. Mem. 1, J. Morph. 34: 245–265.

Lent, R. 1999. Final fishery management plan for Atlantic tunas, swordfish and sharks, Vol. 2. U.S. Dep. Commer., National Oceanic and Atmospheric Administration, NMFS, Washington, D.C. 302 pp.

Libby, E.L. & P.W. Gilbert. 1960. Reproduction in the clearnosed skate, *Raja eglanteria*. Anat. Rec. 138:365.

Luer, C.A. & P.W. Gilbert. 1985. Mating behavior, egg deposition, incubation period and hatching in the clearnose skate, *Raja eglanteria*. Env. Biol. Fish. 13: 161–171.

Maruska, K.P., E.G. Cowie & T.C. Tricas. 1996. Periodic gonadal activity and protracted mating in elasmobranch fishes. J. Exp. Zool. 276: 219–232.

Matthews, L.H. 1950. Reproduction in the basking shark, *Cetorhinus maximus* (Gunner). Phil. Trans. Royal. Soc. Lond., Ser. B 234: 247–316.

McCourt, R.M. & A.N. Kerstitch. 1980. Mating behavior and sexual dimorphism in dentition in the stingray, *Urolophus concentricus* from the Gulf of California. Copeia 1980: 900–901.

McKibben, J.R. & D.R. Nelson. 1986. Patterns of movements and grouping of grey reef sharks, *Carcharhinus amblyrhynchos* at Enewetak, Marshal Islands. Bull. Mar. Sci. 38: 89–110.

Metten, H. 1941. Studies on the reproduction of the dogfish. Phil. Trans. Royal Soc. Lond., Ser. B 230: 217–238.

Michael, S. 1993. Reef sharks and rays of the world. Sea Challengers, Monterey. 107 pp.

Myrberg Jr., A.A., & S.H. Gruber. 1974. The behavior of the bonnethead shark, *Sphyrna tiburo*. Copeia 1974: 358–374.

Nordell, S.E. 1994. Observations of the mating behavior and dentition of the round stingray, *Urolophus halleri*. Env. Biol. Fish. 39: 219–229.

Northcutt, R.G. 1977. Elasmobranch central nervous system organization and its possible evolutionary significance. Amer. Zool. 17: 411–429.

Northcutt, R.G. 1978. Brain organization in the cartilaginous fishes. pp. 117–194. *In:* E.S. Hodgson & R.F. Mathewson (ed.)

Sensory Biology of Sharks, Skates and Rays, Office of Naval Research, Arlington.

Olsen, A.M. 1954. The biology, migration, and growth rate of the school shark, *Galeorhinus australis* (Macleay) (Carcharhinidae), in south eastern Australian waters. Aust. J. Mar. Freshwater Res. 5: 353–410.

Pratt, H.L., Jr. 1979. Reproduction in the blue shark, *Prionace glauca*. U.S. Fish. Bull. 77: 445–470.

Pratt, H.L., Jr. 1993. The storage of spermatozoa in the oviducal glands of Western North Atlantic sharks. Env. Biol. Fish. 38: 139–149.

Pratt, H.L., Jr. & J.C. Carrier. 1995. Wild mating of the nurse sharks. Nat. Geo. Mag. 187: 44–53.

Pratt, H.L., Jr. & J.G. Casey 1990. Shark reproductive strategies as a limiting factor in directed fisheries, with a review of Holden's method of estimating growth parameters. pp. 97–109. *In:* H.L. Pratt, Jr., S.H. Gruber & T. Taniuchi (ed.) Elasmobranchs as Living Resources: Advances in Biology, Ecology, Systematics and Status of the Fisheries, NOAA Tech. Rep. NMFS 90.

Pratt, H.L, Jr., J.G. Casey & R.E. Conklin. 1982. Observations on large white sharks, *Carcharodon carcharias*, off Long Island, New York. U.S. Fish. Bull. 80: 153–156.

Pratt, H.L., Jr. & S. Tanaka. 1994. Sperm storage in male elasmobranchs: a description and survey. J. Morph. 219: 297–304.

Price, K.S. 1967. Copulatory behavior of the clearnose skate, *Raja eglanteria*, in lower Chesapeake Bay. Copeia 1967: 854–855.

Rasmussen, L.E.L. & S.H. Gruber. 1990. Serum levels of circulating steroid hormones in free- ranging carcharhinoid sharks. pp. 143–155. *In:* H.L. Pratt, Jr., S.H. Gruber & T. Taniuchi (ed.) Elasmobranchs as Living Resources: Advances in Biology, Ecology, Systematics and Status of the Fisheries, NOAA Tech. Rep. NMFS 90.

Rasmussen, L.E.L., D.L. Hess & S.H. Gruber. 1992. Serum steroid hormones during reproduction in elasmobranchs. pp. 19–42. *In:* W.C. Hamlett (ed.) Reproductive Biology of South American Vertebrates, Springer-Verlang, New York.

Reed, J.K. & R.G. Gilmore. 1981. Inshore occurrence and nuptial behavior of the roughtail stingray, *Dasyatis centroura* (Dasyatidae), on the continental shelf, east central Florida. Northeast Gulf Sci. 5: 59–62.

Rouse, N. 1985. Nurse shark's mating ballet. Sea Frontiers 31: 54–57.

Schensky, F. 1914. Tier-und Pflanzleben der Nordsee. Werner Klinkhardt, Leipzig (cited by Matthews 1950).

Springer, S. 1960. Natural history of the sandbar shark, Eulamia milberti. U.S. Fish. Bull. 61: 1–38.

Springer, S. 1967. Social organization of shark populations. pp. 149–174. *In:* P. W. Gilbert, R. F. Mathewson & D. P. Rall (ed.) Sharks, Skates and Rays, John Hopkins Press, Baltimore.

Stevens, J.D. 1974. The occurrence and significance of tooth cuts on the blue shark, *Prionace glauca* (L.) from British waters. J. Mar. Biol. Assoc. U.K. 54: 373–378.

Sumpter, J.P. & J.M. Dodd. 1979. The annual reproductive cycle of the female lesser spotted dogfish, *Scyliorhinus canicula* L., and its endocrine control. J. Fish Biol. 15: 687–695.

Tricas, T.C. 1980. Courtship and mating-related behaviors in myliobatid rays. Copeia 1980: 553–556.

Tricas T.C. & E.M. LeFeuvre. 1985. Mating in the reef white-tip shark, *Triaenodon obesus*. Mar. Biol. 84: 233–237.

Tricas, T.C., S.W. Michael & J.A. Sisneros. 1995. Electrosensory optimization to conspecific phasic signals for mating. Neuroscience Letters 202: 129–132.

Uchida, S., M. Toda & Y. Kamei. 1990. Reproduction of elasmobranchs in captivity. pp. 211-237. *In:* H.L. Pratt, Jr., S.H. Gruber & T. Taniuchi (ed.) Elasmobranchs as Living Resources: Advances in Biology, Ecology, Systematics and Status of the Fisheries, NOAA Tech. Rep. NMFS 90.

West, J.G. & S. Carter. 1990. Observations on the development and growth of the epaulette shark, *Hemiscyllium ocellatum* (Bonnaterre) in captivity. J. Aquaricult. Aquat. Sci. 5: 111–117.

Wourms, J.P. 1977. Reproduction and development in chondrichthyan fishes. Amer. Zool. 17: 379–410.

Wourms, J.P. 1981. Viviparity: the maternal-fetal relationship in fishes. Amer. Zool. 21: 473–515.

Wright, D.E. & L.S. Demski. 1993. Gonadotropin-releasing hormone (GnRH) pathways and reproductive control in elasmobranchs. Env. Biol. Fish. 38: 209–218.

Yano, K., F. Sato & T. Takahashi. 1999. Observation of the mating behavior of the manta ray, *Manta birostris*, at the Ogasawara Islands, Japan. Ichthyological Res. 46: 289–296.

Appendix. A glossary of elasmobranch reproductive behaviors.

The following is a list of species-typical patterns of reproductive behavior noted by researchers. Following Myrberg & Gruber (1974) we have tried to include records of postures and patterns of movement that were sufficiently stereotyped in form and orientation to allow observers to consistently note their occurrence.

Behaviors of females

Accept – accepting females, permit the approach and precopulatory behaviors of sexually active males. They may engage in 'parallel swimming', and exhibit 'submissive postures', body 'arching', 'cupping' and 'flaring' pelvic fins and will usually permit copulation'(Klimley 1980, Gordon 1993, Carrier et al. 1994) See 'back arching' and 'pectoral fin undulations'.

Arch – female sharks that 'avoid' males may first attempt to escape a 'pectoral grasp' by 'arching' away from the male often twisting the body and cloaca away from the male and, at times, out of the water (Carrier et al. 1994). Accepting females may also arch toward the male (Carrier et al. 1994).

Avoid – females retreating from sexually active pursuing males may 'avoid' an approach or copulation. If engaged, the 'avoiding' female will struggle to escape (Carrier et al. 1994).

Back arching and pectoral fin undulations – female skate precopulatory behavior to attract the attention of males (Luer & Gilbert 1985).

Cup – a female folds her pelvic fin margins to form a cup shape (Gordon 1993).

Flare – a female curves and spreads her pelvic fins exposing the cloaca (Gordon 1993).

Refuge – a female retreating behavior to limit male reproductive access (this study).

Lay on back – the female is motionless and rigid with back to the substrate and pectoral fins outstretched ('lying on back', Klimley 1980).

Shield – the female swims very close to the substrate 'shielding' to prevent males from approaching her cloaca (Gordon 1993).

Submissive behavior – the female swims slowly with the head lowered about 15 degrees exposing the pelvic region just before copulation (Gordon 1993).

Behaviors of males

Bite – see 'courtship bite', similar to the 'grasp' (Carrier et al. 1994).

Block – a single male positions his body in front of the heads of the mating pair of sharks and thus presumably aids the 'pectoral grasp' and copulation by keeping the pair from moving forward (Carrier et al. 1994).

Bob and sway – male rays swim in a sinuous path and 'bob' vertically or 'sway' horizontally in pursuit of a female (Tricas 1980).

Carry – after a successful 'pectoral grasp' in shallow water (< 0.5 m) the male will attempt to carry the 'accepting' female to water deep enough to permit copulation (this study).

Clasper flexion – the erection of claspers. It may be the movement of individual claspers dorsally before a possible insertion attempt by a male bat ray (Tricas 1980) or backward and forward by a 'patrolling' *Carcharias taurus* (Gordon 1993). Myrberg & Gruber (1974) noted that *Sphyrna tiburo* could pivot its claspers separately, alternately, and sometimes rapidly while swimming alone. They also noted 'clasper flexion with thrust', a side roll with clasper extension and body arching in *S. tiburo*. See 'splaying'.

Courtship bite – a non-feeding bite usually made by a male shark on a female's body to facilitate mating, often leaving mating scars (Stevens 1974, Pratt 1979). The bite may act as a releaser for female mating behaviors and may stimulate ovulation and other physiological changes (Maruska et al. 1996). Biting in some species may also be an expression of reproductive hierarchy (Gordon 1993).

Crossing claspers – a male overlapping both claspers until they point toward opposite side (Dral 1980, Gordon 1993).

Follow - often a 'precoupling' or precopulatory behavior (Carrier et al. 1994) in which the male closely follows the female, usually within one body length changing direction frequently as needed. It may be brief or prolonged (Carrier et al. 1994). 'Following' is similar to the 'close follow' of Johnson & Nelson (1978), 'parallel swimming' (Klimley 1980), 'chasing' (Yano et al. 1999). Following is not always of a reproductive nature. Myrberg & Gruber (1974) observed that the largest bonnetheads, *S. tiburo*, frequently follow large members of the opposite sex within 1 m without reproductive intent. Sharks of the same sex will also closely follow each during the mating season (this study).

Gouge – male spotted eagle rays, *Aetobatus narinari*, dive upon and 'gouge' the female's back with their lower toothplate and bite the caudal margins of the female's pectoral fins (Tricas 1980), called 'nibbling' by Uchida et al. (1990).

Grasp – a bite used for holding rather than feeding or defense. It may be directed to the female pectoral fin or body (Carrier et al. 1994).

Group behavior – multiple males compete to grasp a female pectoral fin (Tricas 1980, Uchida et al. 1990, Carrier et. al. 1994) or possibly cooperate to aid other male's attempts to copulate by 'blocking' the movement of the mating sharks (Carrier et. al. 1994).

Insertion and copulation – if a male is successful in 'positioning and alignment' clasper insertion and copulation will usually follow (Carrier et al. 1994).

Nosing – a response to 'cupping'. A male comes from behind and beneath a female and places his snout just below the cloaca (Gordon 1993). Similar to 'nudging' (Klimley 1980).

Nudge – a male moves the female from the perpendicular to the parallel position by placing his head under and in contact with the female (Klimley 1980).

Pectoral bite – a type of 'courtship bite', the male bites and holds a female's pectoral fin. A common male maneuver to hold and position a female while mating (Springer 1960, Klimley 1980, Carrier et al. 1994). It may lead to the 'grasp' (Carrier et al. 1994) and is similar to pectoral nipping (Yano et al. 1999).

Pivot and roll – an avoidance behavior originated by the female to roll over the back of a male and escape his grasp (Klimley 1980).

Positioning and alignment – a prelude to 'insertion and copulation' (Carrier et al. 1994) in which the male successfully aligns the female's body for copulation. The resultant body positions are similar to 'nudging' (Klimley 1980).

Post-copulation – the male removes the clasper from the cloaca, but may briefly maintain his oral hold of the female pectoral (Yano et al. 1999). Carrier et al. (1994) noted that the pair sometimes remains together quiescently, more typically, one or both rapidly depart in different directions.

Precoupling – may include 'following' behavior (Carrier et al. 1994) and 'parallel swimming' (Klimley 1980)

Lay on back (male) – after the 'grasp', the male rolls over on his back, then both sharks remain motionless, side-by-side on the substrate. At this time in Klimley's observations, the right clasper, the one closest to the female, is inserted into the female's cloaca ('lying on back (male)' Klimley 1980, Uchida et al. 1990).

Separating – the male releases the pectoral fin of the female, setting her free (Yano et al. 1999).

Snapping – a male aggressive behavior. A quick physical bite directed toward other resident species, then retraction from them (Gordon 1993).

Splayed claspers – a male contorts and opens his claspers laterally up to ninety degrees to the body axis (Gordon 1993).

Stalking – an aggressive display. Males circle and closely pass other fish and divers in a captive environment (Gordon 1993).

Tailing – one male follows another male so closely that the lead shark's tail movement is restricted (Gordon 1993).

Tail tuck – a male tucks his ventral post-pelvic surface behind the female's first dorsal fin to guide clasper alignment.

Torso thrust (with clasper flexion) – an exaggerated swimming behavior in scalloped hammerheads, perhaps to communicate with schooling females and possibly to fill the siphon sacs with sea water prior to copulation (Klimley 1985).

Wrap – males of small, limber shark species may wrap their body around the female to facilitate copulation (Bolau 1881, Hardy 1959, Dral 1980, Uchida et al. 1990).

Behaviors of males and females

Copulation – the male typically inserts one clasper into the cloaca of the female and may transfer sperm to her uteri (Carrier et al. 1994, Yano et al. 1999) (Figure 12).

Lay parallel on substrate – sharks lay with bodies abreast, less than two pectoral widths apart, prior to a 'grasp' ('lying parallel on substrate' per Klimley 1980).

Mating event – a 'precopulatory' encounter usually starting with a pectoral 'grasp', which may or may not lead to and include 'copulation' (Carrier et al. 1994).

Mating scars – wounds left by a 'courtship bite'. Lesions may be fresh and open, with or without bleeding, or healed marks typified by disrupted denticle patterns that may persist for some time. Scars are usually present on the pectoral fins and often on the trunk around the pelvic fins of females. Scars may occur anywhere on body but are rare around the head of most species (Stevens 1974, Pratt 1979). Similar to 'scarring' (Gordon 1993). See 'courtship bite'.

Parallel swimming – male and female swim together less than two pectoral fin widths apart (Klimley 1980). Occurs in 'precoupling' (Carrier et al. 1994). See 'following'.

Patrolling – relatively straight line swimming. The most common mode of swimming for *S. tiburo* (Myrberg & Gruber 1974). Patrolling may or may not have reproductive consequences and is included here because it is used by several authors in a reproductive context (Springer 1967, Gordon 1993). Contrast with 'following'.

Stall – an individual shark stops all forward movement and hovers above the bottom. This may only be a behavior of those few sharks like *C. taurus*, whose body densities are very near that of sea water which permits them to hover without sinking (Gordon 1993).

Part 3. New techniques–tagging, telemetry and movements

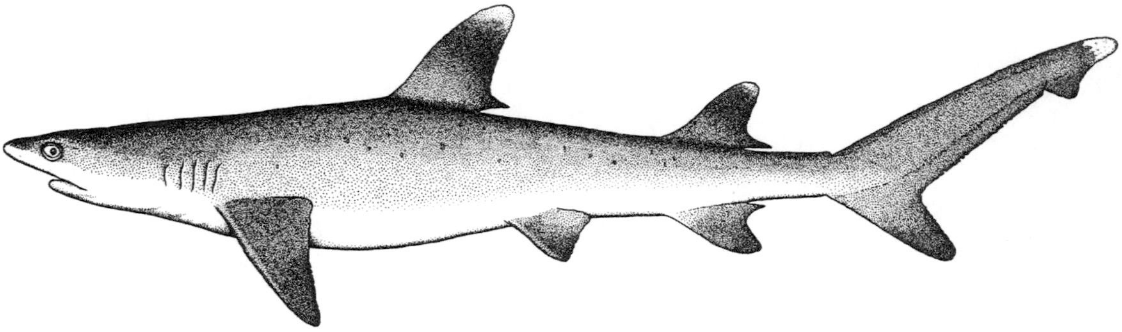

Female of the whitetip reef shark, *Triaenodon obesus*, 66 cm long from Indonesia (PV).

Environmental Biology of Fishes **60:** 191–223, 2001.

Shark tagging: a review of conventional methods and studies

Nancy E. Kohler & Patricia A. Turner
NOAA/NMFS, Northeast Fisheries Science Center, 28 Tarzwell Drive, Narragansett, RI 02882, U.S.A.
(e-mail: nancy.kohler@noaa.gov)

Received 15 July 1999 Accepted 28 April 2000

Key words: elasmobranch, mark, tag, recapture, migration, movement

Synopsis

The tagging of sharks using conventional tags has long been recognized as a valuable means for studying various aspects of their life history, migrations and movements, and population structure. Conventional tags are defined as those that can be identified visually without the use of special detection equipment. Tagging studies specifically targeting sharks began in the late 1920's, and today numerous cooperative shark tagging programs exist worldwide. Cooperative programs depend on the joint participation of scientists and public volunteers to accomplish research objectives. Benefits and problems of these programs are discussed using the tagging methodologies, protocols, and results of the National Marine Fisheries Service Cooperative Shark Tagging Program. An additional 63 shark tagging studies and programs of all types are reviewed. Information useful for behavioral, biological, and fishery management studies can be derived from data resulting from these studies, including species and size composition, sex ratios, spatial and temporal distribution, migrations, movement patterns, rates of travel, delineation of pupping grounds, distribution of maturity intervals, indices of relative abundance, and recognition of individuals. Specific tagging experiments can be designed to provide additional data on age and growth, homing and site fidelity, dispersal rate, residence time, movement rates, tag shedding, and population parameters (e.g. size, mortality, recruitment, exploitation, interaction rates, and stock identity). Sources of bias inherent in tagging and recapture data include mortality, variation in tagging effort and fishing pressure, non-recovery and non-reporting of tags, and tag shedding. Recent advances in tagging methodologies that complement and extend conventional tagging studies will further our knowledge on shark movements and migrations, particularly in the areas of resource utilization and management, space utilization, and population dynamics.

Introduction

External and internal tags have been used for centuries as markers on marine and freshwater fishes for the purposes of identification and information retrieval. From the earliest application on salmonids to the more recent use on sharks, these markers can be divided into five general categories. Overall, physical tags have been used most extensively, followed by dyes and pigments, mutilation, branding, and meristic and morphometric characters including ageing structures (Arnold[1], McFarlane et al. 1990). More than twenty-five general types of physical tags (both external and internal) can be defined, as described by Rounsefell & Everhart (1953), Jakobsson (1970), and McFarlane et al. (1990). These tags are made of various materials, are sometimes designed for specific species or groups of animals, and have been used with varied success. The most common physical tag is external, with some form of anchor that penetrates the skin or other tissues, and includes a component for recognition of individuals (Bergman et al. 1992).

Tagging is a valuable technique for investigating the behavior of sharks (Sciarrotta & Nelson 1977). Oceanic and deep-water sharks are difficult to study, and most shallow water species are either large, active

[1] Arnold, D.E. 1966. Marking fish with dyes and other chemicals. Technical paper No. 10, Bureau of Sport Fisheries and Wildlife, U.S. Dept. of Interior. 44 pp.

predators not easily observed for any length of time, or are species that are rarely seen and in small numbers (McLaughlin & O'Gower 1971). Sharks in general are particularly suited to tagging studies because of their extensive movements, schooling behavior, and relatively large size which allows them to carry most tags successfully (Bonham et al. 1949).

Tagging studies provide valuable information on a wide variety of aspects of elasmobranch biology, including life history parameters, stock status, behavioral and distribution patterns, and migration patterns. The ability to account for the presence of a particular fish or group of fish in time and space by marking provides an important tool to the fishery manager (Rounsefell & Everhart 1953) and may be the most cost-effective, reliable, and direct means to obtain data for studying populations (Everhart & Youngs 1981, Gordon 1990).

The purpose of this paper is to summarize the results of the major shark tagging studies that have used conventional tags. Conventional tags are defined as those that can be identified visually without the use of special detection equipment. We review methods and data protocols used by the National Marine Fisheries Service (NMFS) Cooperative Shark Tagging Program (CSTP) which typify those of many large scale game fish tagging programs around the world. We use results from the CSTP and other tagging studies to review objectives of tagging programs, types of information derived from such programs, and possible problems and biases in these data. In addition, we explore the future use of conventional tags and tagging techniques in combination with new tagging technologies.

Materials and methods

This review contains a comprehensive compilation of 64 past and present worldwide shark tagging studies and programs. Appendix 1 includes the scientific or common name as initially reported in the study, original units for the distance data converted to kilometers and time at liberty data converted to years. Distance traveled is customarily reported as a minimum straight line distance between point of tagging and recapture. An attempt was made to keep study results discrete. It was often difficult, however, to clearly distinguish between studies from the information provided in the reports we reviewed, since multiple authors have reported results from the same studies at different times, or performed different analyses on the same data, and other authors updated results from long term studies. For some multi-year studies, we relied on personal communications from the program managers for the most current results.

Historical review

Development of tagging technology

In a comprehensive review of the historical development of external tags and marks documenting approximately 900 published studies, McFarlane et al. (1990) traced the earliest occurrence of animal marking to between 218 and 201 BC. Much of the initial fish tagging was on salmonids. As early as the 1600s, juvenile *Salmo salar* were marked with colored wool ribbons on their tails to document their return from the sea to natal rivers. In the 1800s, salmon were marked using a variety of methods including fin clipping, fin and jaw wires (ring tags), tail and opercle tags (collar tags and numbered label tags), and dangler-type tags attached through the muscles in front of the dorsal fin (Archer and Atkins tags) (Rounsefell & Everhart 1953, McFarlane et al. 1990). Pelagic herring were first tagged with barbed hooks (Jakobsson 1970) during this same time period. With the introduction of the Petersen disc tag in 1894, the range of studies that could be carried out through tagging was greatly expanded. Demersal species such as plaice were first marked using this extremely popular and successful tag type (Rounsefell & Everhart 1953, Jakobsson 1970).

Until about 1900, the tagging of fish developed very slowly and only about 100 000 fish were tagged by 1910 (Rounsefell & Everhart 1953). Between 1900 and 1929, salmonids and some demersal species such as plaice, cod, haddock, pollock, and halibut were marked with a variety of newly developed tag types. These included Lofting, safety pin, strap, wire loop, Heinicke ring and stud, and bachelor button tags attached through the muscle, fin, jaw, and opercle. By the 1930's, the number of fish tagged increased to over 400 000 as other types of markers, including body cavity, sturgeon, hydrostatic, barbed, and internal anchor, came into use primarily on anadromous and demersal marine species (see Rounsefell & Everhart 1953, Jakobsson 1970, McFarlane et al. 1990).

The first attempt to tag a pelagic species of fish was in 1893 when the Fishery Board of Scotland tagged 600 herring using barbed hooks with a rigid numbered plate on the shaft (Jakobsson 1970). In

1911, Sella tagged *Thunnus thynnus* by using a collar tag consisting of a piece of copper chain fastened around the caudal peduncle. Both methods yielded no returns (Rounsefell & Everhart 1953, Jakobsson 1970). Tuna, herring and mackerel were tagged in subsequent decades using celluloid collar and disc, opercle clip, barb, strap, Petersen disc, plastic strip, body cavity, and hook tags with varied success (Rounsefell & Everhart 1953, Jakobsson 1970, McFarlane et al. 1990).

The internal anchor tag was developed by Rounsefell in 1936 to satisfy the need for an externally visible tag that fish could carry successfully, even as they increased greatly in size, without the high shedding rate of early body cavity tags (Rounsefell & Everhart 1953). By the early 1950's, relatively inexpensive vinyl tubing was readily available in a variety of sizes and colors for use as components of new types of external tags. With the development of the spaghetti loop tag in 1952 by Wilson (Jakobsson 1970) and the dart and streamer tag in 1954 by Mather[2], tagging of pelagic species became widespread. Both tag types were composed of plastic vinyl tubing with a printed number and legend and had little drag in flowing water. Dart tags, developed for larger fish, could be applied with a minimum handling time while the fish remained in the water. The harpoon-like head of the dart tag was initially made of stainless steel and later fitted by Yamashita & Waldron (1958) with a nylon barb. This nylon and stainless steel barb was further modified in size and shape in subsequent years for specific uses on target species and life history intervals like school tuna and *Scomberomorus cavalla* (Mather[2], Bayliff & Holland[3], McFarlane et al. 1990).

The modern anchor tag, first used in the 1960's, is similar to the dart tag except that the barbed head has been replaced by a nylon T-bar. These streamer tags are used widely because of the ease with which large numbers of fish can be tagged with individual serial numbers, their low cost, and their relatively high retention rate (McFarlane et al. 1990).

Since the early 1960's, a variety of subcutaneous tags has been developed and used initially on salmonids. Many of these involve the use of magnetic and electronic detection devices. While they fall outside of our definition of conventional tags, we have included a brief description of their development here. The first of these are small magnetized stainless steel coded wire tags (CWT) that are hypodermically implanted into suitable tissue such as the snout or cheek (Monan 1982). Advantages of CWT include easy identification of large numbers of experimental groups, low tag mortality, and a high tag retention rate. Disadvantages include high initial cost of application and detection equipment, non-visibility of tags, and the necessity of having to kill the fish before removal of the tag to decipher the binary coding (etching) (Raymond 1974, Pepper & O'Connell 1983, Bergman et al. 1992). A more recent coded wire tag (rare earth) allows data to be read with x-ray fluorescent spectroscopy without killing the fish (Monan 1982). Another tag, the passive integrated transponder (PIT) tag, was developed in the early 1980s and has proven very effective. The PIT tag consists of an integrated microchip and antenna encapsulated in ceramic material. The tag is energized by an external detection decoder system to transmit a unique 10-digit alphanumeric identification code (Prentice et al. 1990). Advantages of this tag type include: *in situ* detection of individual fish through soft and hard tissue, glass, plastic, and freshwater and saltwater; no need to anesthetize, handle or restrain the fish during data gathering; and nearly 100% tag retention in the body cavity with no effect on growth, survival, or behavior for salmonids (Prentice et al. 1990). A drawback to all of these tag types is that they must be surgically inserted (which limits the number of fish that can be tagged within a given time) and must be recovered by personnel using the proper equipment to read the tag. A more recent subcutaneous tag that is individually discernable and externally visible without the use of an external device, is the visible implant (VI) tag that is implanted beneath the integument in transparent tissue (Haw et al. 1990, Bergman et al. 1992).

Early elasmobranch tagging

The successful tagging of elasmobranchs has been a relatively recent event. By 1936, 700 skates and rays were tagged and released around the British Isles, and only 1005 had been tagged in all European waters up to 1940 (Olsen 1953). Beginning in the 1940's, however, the tagging of sharks, primarily *Squalus acanthias* and *Galeorhinus galeus*, took place in both the Pacific and Atlantic Oceans.

[2] Mather, F.J., III. 1963. Tags and tagging techniques for large pelagic fishes. pp. 288–293. *In:* International Commission of Northwest Atlantic Fisheries Special Publication 4, North Atlantic Fish Marking Symposium.

[3] Bayliff, W.H. & K.N. Holland. 1986. Materials and methods for tagging tuna and billfishes, recovering the tags and handling the recapture data. FAO Fish. Tech. Pap. 279. 36 pp.

In the Pacific, a *G. galeus* fishery developed during World War II for the vitamin A oils from its liver. Tagging of this species in southeastern Australia commenced in 1942 when 22 fish were internally tagged (Olsen 1953). In 1947, a comprehensive tagging program was undertaken by the Commonwealth Scientific and Industrial Research Organization (CSIRO) (Olsen 1953, 1954). This program, part of the general investigations on the biology of the species, was started in response to concern about declining catches. Up until 1956 when the program ceased, a total of 6502 *G. galeus* were tagged with a combination of internal and external tag types (Olsen 1984). In addition, 587 *Mustelus antarcticus* were also released. Tagging resumed between 1973 and 1976 by the then Fisheries and Wildlife Division of Victoria and again in 1990 by the Fisheries Research and Development Corporation Southern Shark Tagging Project. By 1996, a total of 20 185 sharks of 22 species (Appendix 1) had been tagged in both the Pacific and Indian Oceans (Walker et al.[4]). A male *G. galeus* tagged with an internal plastic tag, was recaptured after 41.8 years from the early CSIRO tagging program off Southern Australia. This is the longest recorded period at large for any species of tagged fish (Coutin 1992).

Due to the demand for vitamin A in liver oil of *S. acanthias*, an intensive fishery for this species developed on the Pacific coast of the United States (US) between 1941 and 1950. The development of this fishery prompted the need for life history information (Holland 1957). A few *S. acanthias* (176) were tagged in 1940 incidental to a study of the otter-trawl fishery, but a sustained tagging program began in 1942. This was a cooperative project involving commercial fishermen, the Washington State Fisheries Department, and the Technological Laboratory of the US Fish and Wildlife Service (USFWS) at Seattle, and was designed to determine the extent and pattern of migration, rate of growth, and fishing mortality. The primary tagging areas were off the coasts of Washington and British Columbia. A total of 9705 *S. acanthias* were tagged and released by August 1946 (Bonham et al. 1949, Holland 1957). *S. acanthias* in the eastern North Pacific were tagged with celluloid Petersen-type disk tags. By the end of 1953, 6.7% of the tagged sharks had been recaptured, with a maximum time at liberty of 10 years and a maximum distance traveled of 8704 km (Appendix 1).

Additional early North American Pacific tagging included the release of 564 *S. acanthias* off British Columbia reported by Foerster[5] and 16 588 tagged *S. acanthias* off Japan between 1929 and 1957 (Taniuchi personal communication). An additional 427 tagged *G. zyopterus* were released in Oregon waters (Westrheim in Herald & Ripley 1951) and another 118 *G. zyopterus* tagged in California waters reported by Herald & Ripley (1951).

In the Atlantic, one of the first shark tagging experiments was on *Somniosus microcephalus*, carried out by the Greenland Fisheries Investigations in 1936, 1939, 1948, and 1949 to determine the growth rate of the species. Hansen (1963) reported a total of 411 sharks tagged using Petersen disc tags with silver and stainless steel wire attachments. Twenty-eight sharks were recaptured up to 1296 km away and after 16 years at liberty. While these returns yielded little or no growth information, they did provide important longevity information for this species.

Most of the initial tagging in the North Atlantic was also on *S. acanthias*. Unlike in the Pacific Ocean, however, *S. acanthias* was not fished commercially on the Atlantic coast, and its large population was considered a hindrance to cod fishing (Templeman 1954). A tagging study was initiated by the government of Newfoundland to determine if the stock was local rather than international, and thus could be locally managed. In 1942, Templeman (1944, 1954) released 279 *S. acanthias* near St. John's Newfoundland that were tagged with Petersen disc tags attached with a nickel wire. This study continued until 1965 with a total of 2855 *S. acanthias* tagged, of which 8.2% were recaptured. The longest time at liberty was 11.2 years with one trans-Atlantic recapture reported (Templeman 1976). Other early North Atlantic *S. acanthias* tagging included 907 sharks tagged in the New England–Gulf of Maine area between 1956 and 1964 and over 20 000 *S. acanthias* tagged in the northeast Atlantic between 1957 and 1963 (Aasen 1960, 1962, Holden 1965, Templeman 1976).

The first tagging of elasmobranchs off South Africa in the Indian Ocean was initiated in 1964 by the

[4] Walker, T.I., L.P. Brown & N.F. Bridge. 1997. Southern shark tagging project. Final report to Fisheries Research and Development Corporation (FRDC Project 93/066), Client report, (November 1997), Marine and Freshwater Resources Institute, Queenscliff, Victoria, Australia. 61 pp.

[5] Foerster, R.E. 1942. Dogfish tagging – preliminary results. Canada Fish. Res. Board. Pacific Coast Sta., Prog. Rep. No. 53: 12–13.

Oceanographic Research Institute (ORI) to help devise protective measures for swimmers against shark attacks in the area (Davies & Joubert 1967). A total of 1001 sharks were initially tagged off Durban with Jumbo Rototags and Petersen disc tags (39% were recaptured in the first year) and later with a modified Jumbo Rototag called the ORI tag (Davies & Joubert 1967). Bass et al. (1973) and Bass (1977) reported the further early tagging of several species of sharks off the Durban coast.

Cooperative tagging programs

Most of the aforementioned tagging studies were initiated and carried out by scientists to answer specific biological questions. All tagging, and in some cases returns, were done by investigators directly involved in the study. Another type of program is a cooperative tagging program, i.e. one that depends on joint participation of scientists and public volunteers to accomplish research objectives (Scott et al.1990). One of the first cooperative marine gamefish tagging programs that utilized assistance from volunteer fishermen was initiated in the US in 1951 by Frank Mather of the Woods Hole Oceanographic Institution (Mather[2]). Since that time, numerous small and large scale programs have developed worldwide and have expanded to include a variety of shark species. Some of these cooperative programs that specifically target sharks include: the ORI Program and the Southern Shark Tagging Project mentioned previously; a tagging study begun by the Ministry of Agriculture, Fisheries and Food in England in conjunction with a local shark angling club in 1970 (Stevens 1990); the Australian cooperative game-fish tagging program begun in 1973 by the Fisheries Research Institute of the New South Wales Department of Agriculture and Fisheries (Pepperell 1990); and the Ireland Central Fisheries Board's cooperative tagging efforts primarily on *Prionace glauca* and *G. galeus*, which began in 1970 (Fitzmaurice 1994, Green personal communication). In the US, these programs include the California Department of Fish and Game Pelagic Shark Tagging Program, which was initiated in 1983 (Ugoretz[6]) the Center for Shark Research at Mote Marine Laboratory in Sarasota, Florida (Hueter personal communication), and the NMFS CSTP described in more detail below.

[6] Ugoretz, J. 1999. Shark tagging news, a newsletter of the California Department of Fish and Game shark tagging program (January 1999). 4 pp.

Numerous other cooperative fish tagging programs have also included sharks as a component of their tagged fish. These include the New Zealand Ministry of Agriculture and Fisheries highly migratory tuna and game-fish efforts (Murray 1990, Hartill & Davies[7]) the American Littoral Society in Highlands, New Jersey, and various US state tagging programs.

NMFS Cooperative Shark Tagging Program

The NMFS CSTP is an extensive Atlantic shark tagging program, which is part of continuing research directed to the study of the biology of Atlantic sharks. The CSTP was started in 1962 by John G. Casey at the Department of Interior's USFWS Sandy Hook Laboratory, Highlands, NJ, in response to several shark attacks along the New Jersey coast. Volunteer participation began with a group of 74 anglers involved in tagging feasibility studies in 1963. The program expanded in subsequent years, coming under the auspices of NMFS in 1970, and currently includes over 6500 volunteers distributed along the Atlantic and Gulf coast of North America, and Europe. An overview of the early history of the CSTP is included in Casey (1985) and Kohler et al. (1998).

Sampling procedures described below for NMFS CSTP are similar to those of other tagging studies and cooperative tagging programs worldwide (Sluczanowski 1988). The two principal tag types used in the CSTP are a fin tag and a dart tag. The fin tags are variously colored, two piece, nylon cattle ear tags (Jumbo Rototags and Rototags) inserted through the first dorsal fin. Rototags are currently in use on neonate and juvenile sharks. Jumbo Rototags were first used by staff biologists on small sharks during the early years of the CSTP when the sharks were taken on board the boat, tagged, measured and released. The fin tags are applied with an applicator through a hole punched in the leading edge of the first dorsal fin by a leather punch. Today these tags are still primarily used by NMFS or other cooperating biologists.

As the program expanded to include thousands of volunteer fishermen, the dart tag was developed to be easily and safely applied to sharks in the water. Originally developed by Frank Mather for tunas (Mather[2]), the dart tags were modified for use on sharks (Casey 1985). A variety of stainless steel dart tags were used,

[7] Hartill, B. & N.M. Davies. 1999. New Zealand billfish and gamefish tagging 1997–98. NIWA Technical Report 57. 39 pp.

beginning in 1965. These included an F1 Streamer tag, NMFS M(s) Stainless tag, a Floy F66 tag, and a NMFS M tag. Recently, a small Hallprint nylon barbed tag is being used on neonates and small sharks. The M tag is the standard dart tag used by volunteer anglers. It has a stainless steel dart head, monofilament line, and a Plexiglas capsule containing a vinyl legend with return instructions printed in English, Spanish, French, Japanese and Norwegian (Kohler et al. 1998). The dart tags are used by biologists and volunteer recreational and commercial fishermen, and are applied using a sharpened stainless steel needle that is mounted in a variable length tagging pole. The head of the dart tag fits into the slotted point of the needle and the entire tag is held in place on the pole by a rubber band near the tag capsule. The dart head is sharpened and curved so that the two rear points will face downward into the muscle when the tag is inserted. Tags are implanted in the dorsal musculature near the base of the first dorsal fin. The dart is inserted at an angle toward the head end of the shark so that the capsule assumes a trailing position on the body. Hallprint tags are also applied with a stainless steel applicator that is affixed in a pole of suitable length. The tip of the applicator is sharpened and is inserted under the first dorsal fin until the nylon barb anchors in the basal cartilage.

Numbered dart tags are sent to volunteer participants on self-addressed return post cards for recording tagging information. The information requested from the volunteers is species, size, and sex of the shark, date, location, and capture gear, as well as the tagger's name and address. A remarks section is included so that the angler can add additional biological or environmental data. Size is estimated or measured and reported in total length, fork length, and/or weight. A card is filled out immediately following a tagging episode. First time taggers are sent instructions on proper catching and handling procedures to minimize injuring the shark before releasing it. The condition of the tagged shark is assessed before and after release and noted on the tag card. The assessment is based on how lively the fish is during the tagging process and after release and whether any injuries are evident. These data are numerically coded by assigning a shark to a general fish condition category: good, fair, or poor, with a specific fish condition category enumerating more details (e.g. bleeding, gut hooked).

There was an initial one-dollar reward sent as an incentive for returning tags, which after a few years was increased to five dollars. Since 1988, a hat with an embroidered logo has been sent as a reward. The tagging program is publicized through various articles in sport fishing magazines and other venues.

Literature such as newsletters, which detail biological information on shark life history parameters, shark identification, results of scientific studies, and management issues, are sent directly to all tagging program participants. In addition, staff frequently travel to fishing centers and to fishermen's forums to further educate constituents on shark conservation and the benefits of tag and release. Most tagging program participants, in turn, are very receptive to scientific information and are gratified to help further knowledge of the various species of sharks that they encounter. Tagging studies have been mostly single release events in which recoveries are made opportunistically by recreational and commercial fishermen. When a tagged shark is re-caught, information similar to that obtained at tagging is requested from the recapturer. Minimum straight-line distance between tagging and recapture (T/R) sites (distance traveled), the number of days at liberty, and the rate of travel are determined for all tags returned. All distance data are converted to kilometers from their original reported units.

Benefits and problems of cooperative programs

The aim of cooperative tagging programs is to gather various types of information on the target species. These data fall into three general categories: (1) life history information, which can be derived from tagging data and to some extent from the results of recaptures; (2) population dynamics information, which comes primarily from the recapture data; and (3) strategies for management, which can be determined from a combination of analysis of both types of data.

The cost/benefit ratio for cooperative tagging programs is extremely low. It would be nearly impossible for an individual or an individual institution or agency to mark and recapture the large quantity of fish over the extensive areas covered by utilizing literally thousands of knowledgeable volunteer recreational and commercial fishermen. Aside from the overhead of the over-seeing institution, this research is accomplished for basically the cost of the tags.

Cooperative shark tagging programs have resulted in large numbers of fish tagged, and time at liberty and distance records for many species of sharks, including *P. glauca*, *Isurus oxyrinchus*, *Carcharhinus plumbeus*, and *Lamna nasus* (Table 1). For example, one cooperative program, the CSTP, was responsible

Table 1. Distance traveled and time at liberty maximums for 101 species of sharks from a review of 64 shark tagging studies using conventional tags (x = data unavailable).

Species	Number of studies	Number tagged	Maximum distance traveled (km)	Maximum time at liberty (yr)
Alopias pelagicus	1	18	x	x
A. superciliosus	3	429	2767	6.5
A. vulpinus	8	150	1556	8.8
Aprionodon isodon	1	127	x	x
Apristurus brunneus	1	23	x	x
Carcharhinus acronotus	2	1331	315	9.2
C. albimarginatus	4	195	< 9	0.7
C. altimus	2	175	3343	11.2
C. amblyrhynchoides	1	122	173	8.8
C. amblyrhynchos	4	202	x	4.2
C. amboinensis	3	177	242	4.7
C. brachyurus	5	6729	1320	3.1
C. brevipinna	5	1806	1665	4.5
C. dussumieri	2	81	4	1.4
C. falciformis	7	1245	1339	7.1
C. fitzroyensis	1	55	150	0.2
C. galapagensis	6	659	2859	5.1
C. isodon	1	108	6	0.8
C. leucas	7	4883	643	7.9
C. limbatus	8	6289	2146	7.3
C. longimanus	4	723	2811	3.3
C. macloti	1	1610	711	10.5
C. maculipinnis	1	183	1383	0.6
C. melanopterus	3	1049	5	3.9
C. obscurus	7	15074	3800	15.8
C. perezi	1	630	48	4.4
C. plumbeus	7	19846	3776	27.8
C. porosus	2	106	> 37	x
C. sealei	1	124	367	x
C. signatus	2	225	2669	13.8
C. sorrah	2	2941	1116	9.9
C. spallanzi	1	1	x	x
C. tilstoni	1	4846	1348	12.9
C. wheeleri	1	33	x	x
Carcharodon carcharias	8	578	1445	2.6
Centrophorus granulosus	1	3	x	x
C. uyato	1	6	x	x
Cetorhinus maximus	3	216	x	x
Chlamydoselachus anguineus	1	1	x	x
Echinorhinus brucus	1	1	x	x
Eusphyra blochii	1	34	21	1
Furgaleus macki	1	79	x	x
Galeocerdo cuvieri	9	8132	6747	10.9
Galeorhinus galeus	10	17574	4940	41.8
G. zyopterus	5	913	2037	2.1
Ginglymostoma cirratum	4	1366	541	7.8
Haploblepharus fuscus	1	164	8	x
Hemipristis elongatus	1	4	x	x
Heterodontus francisci	2	333	19	11.2
H. portusjacksoni	3	550	850	1.8
Hexanchus griseus	1	2	x	x
H. vitulus	1	11	x	x
Isurus oxyrinchus	9	10828	4543	12.8
I. paucus	1	90	3430	5.5
Lamna nasus	7	1074	4260	13.0
Loxodon macrorhynchus	1	2	x	x
Mustelus antarcticus	2	8698	x	x
M. canis	3	425	402	4.0
M. henlei	1	1	x	x
M. lenticulatus	1	2234	1159	5.0
M. nigropunctatus	1	1	x	x
M. norrisi	2	57	x	x
M. palumbes	1	47	210	x
Nasolamia velox	1	1	x	x
Nebrius ferrugineus	2	11	43	x
Negaprion acutidens	2	132	5	3.9
N. brevirostris	3	1724	426	4.1
Notorynchus cepedianus	4	1022	539	1.4
Odontaspis taurus	5	4032	1897	11.0
Poroderma africanum	1	421	1964	x
Prionace glauca	16	115177	7871	10.7
Pristiophorus cirratus	1	376	x	x
P. nudipinnis	1	499	x	x
Pseudocarcharias kamoharai	1	21	x	x
Rhincodon typus	2	264	x	x
Rhizoprionodon acutus	3	1052	1004	1.8
R. lalandii	1	2	x	x
R. longurio	1	73	1111	x
R. porosus	1	14	x	x
R. taylori	1	119	92	0.2
R. terraenovae	3	3803	1037	7.3
Scyliorhinus retifer	1	1	x	x
Somniosus microcephalus	2	433	1296	16.0
Sphyrna lewini	11	3278	1671	9.6
S. media	1	1	x	x
S. mokarran	3	220	1180	4.2

Table 1. (*Continued*).

Table 1. (*Continued*).

Species	Number of studies	Number tagged	Maximum distance traveled (km)	Maximum time at liberty (yr)
S. tiburo	3	3885	343	5.6
S. tudes	1	2	x	x
S. zygaena	6	1427	1122	2.1
Squalus acanthias	14	78386	8704	11.2
S. cubensis	1	9	x	x
S. megalops	2	664	39	0.4
S. mitsukuri	1	10	x	x
Squatina africana	1	8	x	x
S. australis	1	4	x	x
S. californica	4	615	61	6.3
S. dumerili	1	106	x	x
Stegostoma fasciatum	2	3	x	x
Triaenodon obesus	2	237	9	3.4
Triakis megalopterus	2	4550	1066	x
T. semifasciata	3	1004	140	2.1

for over 70% of *P. glauca* tagged worldwide. In this program, 142 868 sharks of more than 52 species were tagged by CSTP participants between 1962 and 1997. Most species (30) have more than 100 sharks tagged (Appendix 1). In addition to species composition, information on migrations, movement patterns and rates of travel has resulted from analysis of CSTP data. Distances traveled for 31 species of sharks in the CSTP ranged from negligible to 6926 km. Seven species traveled distances over 3000 km: *P. glauca*, *Galeocerdo cuvier*, *I. oxyrinchus*, *C. obscurus*, *C. plumbeus*, *I. paucus*, and *C. altimus*. Maximum time at liberty for any shark in the CSTP is 27.8 years. Overall, individuals of 6 species of shark have been at liberty for over 10 years (*C. plumbeus*, *C. obscurus*, *C. signatus*, *I. oxyrinchus*, *C. altimus*, and *G. cuvier*) (Figures 1,2).

There are additional educational and conservation benefits of cooperative tagging programs. Tagging is now a socially acceptable component to recreational fishing, and there are many anglers who solely practice catch and release (van der Elst 1990). The old adage 'the only good shark is a dead shark' is no longer widely accepted, due partly to the education received by shark fishermen through volunteer tagging programs. In the past, shark fishing tournaments landed every shark caught. This resulted in the capture of many small sharks that would not qualify for a prize. Through educational efforts on tag and release and the implementation of minimum sizes, far fewer sharks are landed today. Most tournaments have a tag and/or release prize, and a few closely monitored tournaments have become tag and release only.

Along with the benefits are some significant data quality issues with a large volunteer tagging program. New participants sometimes fail to report all the requested information. Since the shark is usually tagged in the water, size estimates are often exaggerated, and species identification errors, particularly of the genus *Carcharhinus*, are common. Even though volunteers are becoming more skilled observers, errors in species identification and measurement are often not discovered until the shark is recaptured. Much time is spent by tagging program personnel contacting individual taggers and recapturers to fill in missing information and to verify data, especially when the tag or recapture is a record of some sort, is on an uncommon species, or when the shark is caught at an unusual location or time of year. Another problem in many cooperative tagging programs is that as the program expands, the demand for tags often exceeds the supply available. Care must be taken to distribute a limited tag supply over as large a geographic area and season as possible, to avoid unequal distribution of tagging effort. In addition, weather, research cruises, opening or closing of a commercial fishery, commercial value of a species, and life history characteristics all play a role in the tag and recapture rates and distribution patterns (Kohler et al. 1998).

Information derived from tagging and recapture data

Historically, initial objectives of most tagging studies were to study movements and migrations, with more recent techniques developed as tools for studying population dynamics (Rounsefell & Everhart 1953). For instance, in a review of 59 marine tagging studies completed in New Zealand waters, 58% dealt with movements, 22% with age and growth, and 20% with population dynamics (Murray 1990).

Data collected in a tagging program can be used to develop information on behavior, species composition, size composition, sex ratios, spatial and temporal distribution, delineation of nursery and pupping areas, distribution of maturity intervals, indices of relative abundance, and recognition of individuals. External tags can be used to readily identify specific sharks

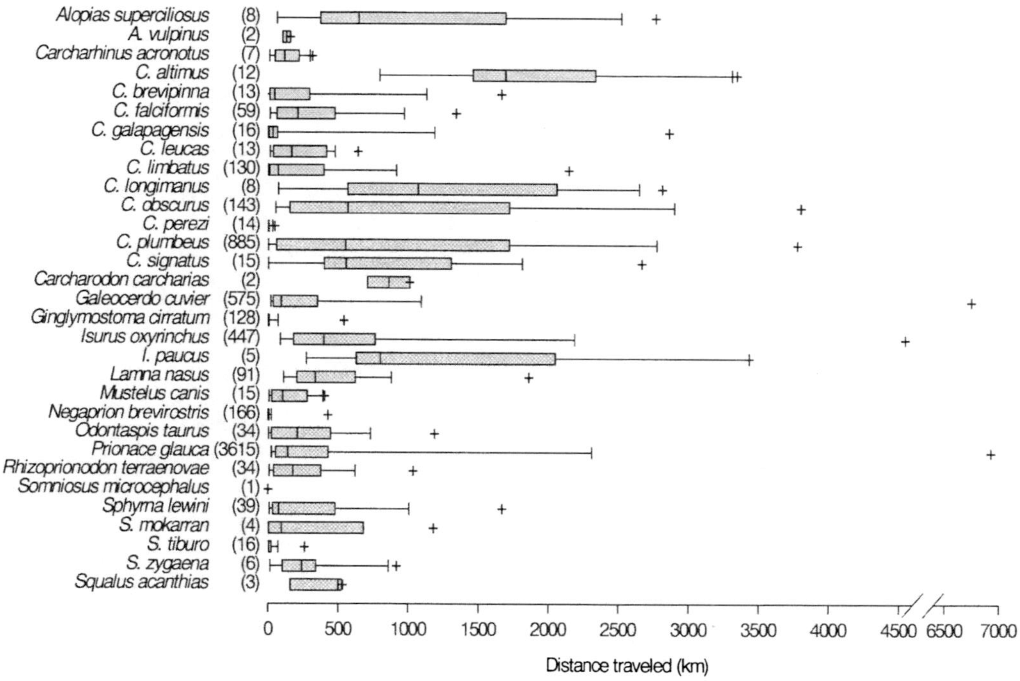

Figure 1. Box and whisker plots of distances traveled for 31 species of sharks from the CSTP, 1962–1997. The box indicates the 25th and 75th percentiles, the line within the box marks the median. Whiskers indicate the 10th and 90th percentiles. Maximum reported value for distance traveled by a species is indicated by +. Sample size in parentheses.

in behavioral studies (Compagno & Fergusson[8]), to determine the stability of a school, and to follow changes in group composition and size over a period of time (Klimley & Nelson 1984). Combinations of tag types, colors, and patterns can be used to distinguish individuals and a particular tagging year or location (Pratt & Carrier 2000). Tagging experiments can also be designed to examine age validation and growth parameters, homing and site fidelity, dispersal rates, residence times, identification and composition of fisheries, fisheries interactions, changes in fishing patterns, gear selectivity and catchability, stock identity, movement rates, population size, abundance, mortality and survival rates, recruitment, interaction rates between areas, catch (harvest) estimates, exploitation rates, tag shedding, survival and growth of hatchery and transplanted fish, and environmental data.

Age and growth studies

A T/R study provides an excellent means for obtaining empirical growth data when the fish are measured at tagging and recapture and the time at liberty is known (Jensen[9], Cailliet 1990). Growth in sharks calculated from tagging experiments is well documented (e.g. Bonham et al. 1949, Holland 1957, Jensen et al. 1961, Hansen 1963, Kato & Carvallo 1967, Tester[10], Clarke 1971, Gubanov 1976, Stevens 1976, 1984, 1990, Bass 1977, Randall 1977, Thorson & Lacy 1982, Pratt & Casey 1983, Pittenger 1984, Tucker 1985). Data from recaptures have been used to determine longevity,

[8] Compagno, L.J.V. & I.K. Fergusson. 1994. Field studies and tagging of great white sharks, *Carcharodon carcharias* (Linnaeus, 1758) off Cape Province, South Africa: a synopsis for 1992–1993. pp. 18–21. *In:* S.L. Fowler & R.C. Earll (ed.) Proceedings of the Second European Shark and Ray Workshop, 15–16 February 1994, Tag and Release Schemes and Shark and Ray Management Plans, Unpublished report.

[9] Jensen, A.C. 1963. Further field experiments with tags for haddock. pp. 194–203. *In:* International Commission of Northwest Atlantic Fisheries Special Publication 4, North Atlantic Fish Marking Symposium.

[10] Tester, A.L. 1969. Cooperative shark research and control program final report 1967–1969. University of Hawaii, Honolulu. 47 pp.

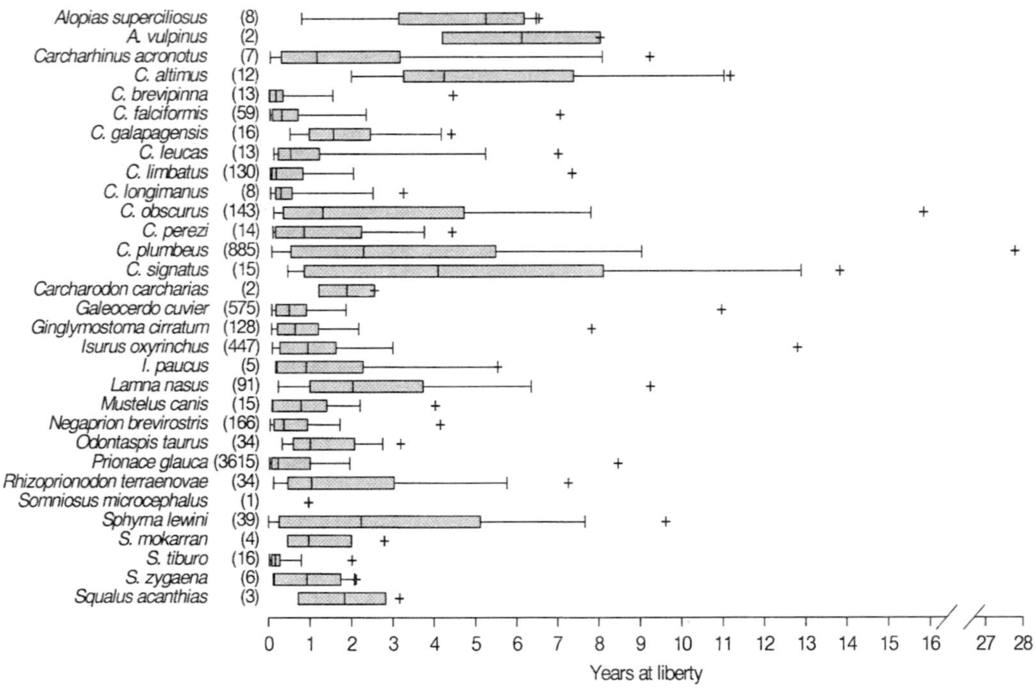

Figure 2. Box and whisker plots of time at liberty for 31 species of sharks from the CSTP, 1962-1997. The box indicates the 25th and 75th percentiles, the line within the box marks the median. Whiskers indicate the 10th and 90th percentiles. Maximum reported value for time at liberty for a species is indicated by +. Sample size in parentheses.

growth rates, and von Bertalanffy growth parameters (Gulland & Holt 1959, Fabens 1965, Francis 1988b,c). In sharks, calculation of growth parameters and/or age verification using T/R data have been accomplished for *C. plumbeus* (Casey & Natanson 1992), *Negaprion brevirostris* (Gruber & Stout 1983), *S. tiburo* (Parsons 1987), *G. cuvier* (Natanson et al. 1999), *Rhizoprionodon terraenovae* (Branstetter 1987), *G. galeus* (Grant et al. 1979, Olsen 1984, Moulton et al. 1992, Francis & Mulligan 1998), *Squatina californica* (Cailliet et al. 1992), *P. glauca* (Skomal 1990), *S. acanthias* (Ketchen 1975, Tucker 1985, McFarlane & Beamish 1987), *Triakis semifasciata* (Smith 1984, Kusher et al. 1992), *M. lenticulatus* (Francis & Francis 1992), *M. antarcticus* (Moulton et al. 1992), *C. tilstoni* and *C. sorrah* (Davenport & Stevens 1988).

Population size

Although marking fish has long been recognized as the most direct means of studying populations for calculation of abundance and size (Ricker 1975, Everhart & Youngs 1981, Jensen 1981), these techniques have infrequently been applied to sharks. Indications of relative population size for juvenile and neonate *Sphyrna lewini* have been inferred from the recapture rate and by the method of Jolly (1965) (Clarke 1971). An estimate of population size from the Jolly-Seber and Peterson methods and an abundance index based on resightings was calculated for *Carcharodon carcharias* (Cliff et al. 1996, Strong et al. 1996). Resighting data were also used to estimate the population size of *S. californica* with the Lincoln–Peterson index, and to calculate size from a simple assumption of proportionality in the numbers of tagged/resighted and untagged/resighted sharks for *Heterodontus portusjacksoni* (McLaughlin & O'Gower 1971, O'Gower & Nash 1978). In addition, population size (Stevens & West[11]) and a framework established for estimating rate of movement (Hilborn 1988, Xiao 1996) were determined for *G. galeus*, and

[11] Stevens, J.D. & G.J. West. 1997. Investigation of school and gummy shark nursery areas in south eastern Australia. Fisheries Research and Development Corporation Project Final Report: Project 93/061. 76 pp.

population density and size have been calculated for *C. melanopterus* (Stevens 1984).

Other life history information

Tag returns from a commercial or recreational fishery in conjunction with tag-recovery and capture–recapture models can be used to determine other life history parameters (Jones 1976, Brownie et al. 1978, Schwarz & Arnason 1990, Pollock et al. 1991, Myers et al. 1997, Brooks et al. 1998, Hoenig et al. 1998a,b). Mortality rates based on recapture data have been calculated for *G. galeus* (Grant et al. 1979, Olsen 1984, Xiao et al. 1999b), *T. semifasciata* (Smith & Abramson 1990, Cailliet 1992), and *C. carcharias* (Cliff et al. 1996). Other fishery parameters derived from shark T/R data include an interaction rate between areas for *M. lenticulatus* (Francis 1988a), an exploitation rate for *M. lenticulatus* (Francis 1989) and *T. semifasciata* (Smith & Abramson 1990, Cailliet 1992), yield per recruit for *G. galeus* (Grant et al. 1979, Olsen 1984) and *T. semifasciata* (Smith & Abramson 1990, Cailliet 1992), gear catchability for *M. antarcticus* (Walker 1992), and stock replacement values for *T. semifasciata* (Smith & Abramson 1990, Cailliet 1992).

Environmental data that are collected ancillary to the tagging of fishes either in a directed study or opportunistically by volunteer taggers can be used for the characterization of pupping areas, definition of essential fish habitat, and habitat preferences of a species or life interval of that species or to determine the causative factors for a particular behavior or movement pattern. Seasonal changes in environmental conditions and fluctuations in temperature, in general, are thought to be chief factors influencing movements of highly migratory species including sharks (e.g. Backus et al. 1956, Strasburg 1958, Springer 1960, Jensen et al. 1961, Kato & Carvallo 1967, Hoey 1983, Talent 1985, Pyle et al. 1996).

There is considerable evidence of seasonal movements by sharks discerned through the use of T/R data. Movement patterns include latitudinal and inshore–offshore migrations in response to increasing water temperatures (Francis 1988a) and have been shown for almost all families of tagged sharks, e.g. heterodontids (McLaughlin & O'Gower 1971), lamnids (Bruce 1992, Casey & Kohler 1992), carcharhinids (Stevens 1976, Tricas 1977, Olsen 1984, Francis 1988a), sphyrnids (Clarke 1971), and squalids (Holland 1957, Jensen et al. 1961, Templeman 1976).

The dispersal rate, residence time, movement patterns, and behavior of tagged fishes can be used to determine the extent of their home range (Harden Jones 1968). Site affinity, homing ability, and/or a home range for sharks have been demonstrated using conventional tagging methods alone or in combination with acoustic tagging techniques for the following species: *N. brevirostris* (Gruber et al. 1988, Morrissey & Gruber 1993), *H. portusjacksoni* (McLaughlin & O'Gower 1971, O'Gower & Nash 1978), *S. californica* (Pittenger 1984), and *C. carcharias* (Strong et al. 1992, Compagno & Fergusson[8]).

Sources of bias in tagging and recapture data

In an ideal world, tagging would be an excellent means to derive the life history and population dynamics information described above. In practice, however, there are a variety of factors that influence the rate of tag returns and limit the validity of the inferences that may be drawn from these data. More than half (55%) of the 191 reported recapture rates for 72 species of sharks from 52 shark tagging studies reviewed in this study reported return rates of less than 5% (Figure 3). Factors that influence this tag return rate include natural mortality of tagged fish (which may differ from year to year and among migration destinations), tagging induced mortality (which includes mortality associated with capture), variation in fishing pressure, emigration out of the principal tagging and fishing area, immigration from non-tagging areas, variation in tagging location and time of release and recovery, changes in the susceptibility to capture over time, experience level of tagging personnel, incorrect species identification, incorrect recording of tag or recapture data, failure to receive recovered tags (including non-recognition, non-recovery, and non-reporting of tags), and tag shedding or loss (Holland 1957, Ricker 1975, Buchanan et al.[12], Grant et al. 1979, Porter 1979, Gulland 1983, Francis 1989, Pepperell 1990, Schwarz & Arnason 1990).

Insuring the reporting of recovered tags in a voluntary program is difficult. The magnitude of this

[12] Buchanan, C.C., F.J. Mather, III & J.M. Mason, Jr. 1977. An overview: cooperative game fish tagging program (Atlantic Ocean). pp. 181–206. *In:* R.A. Barnhart & T.D. Roelofs (ed.) Catch-and-Release Fishing as a Management Tool – A National Sport Fishing Symposium, September 7–8, 1977, Humboldt State University, California.

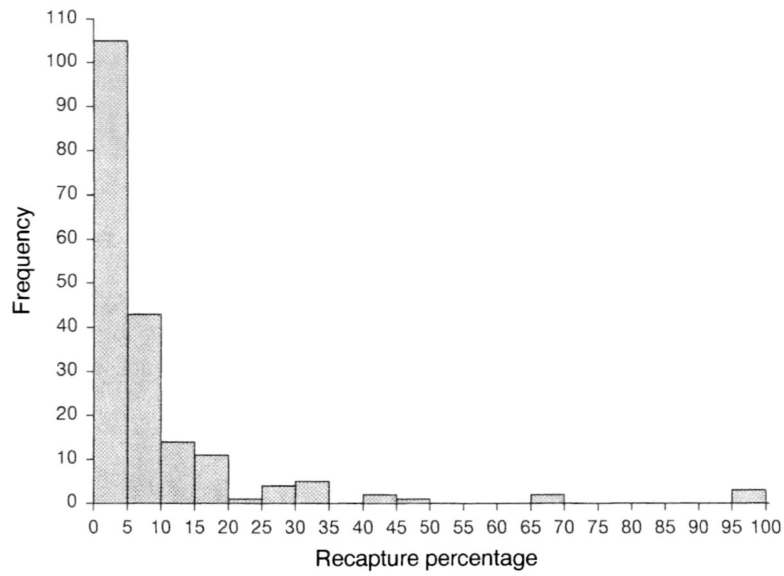

Figure 3. Frequency distribution of 191 reported recapture rates for 72 species of sharks from 52 shark tagging studies using conventional tags.

factor may vary among fisheries and from year to year (Eisner & Ritter 1979). A well-publicized reward and/or lottery system and an easily recognizable tag legend will encourage returns. Stricter management regulations in some fisheries, however, may decrease the likelihood of commercial fishermen sending in information on a recovered tag and cause differences in under-reporting among fisheries (e.g. Harden Jones 1968, Murray 1990). For example, data from the *G. galeus* fishery in Australia showed that only about half of the recaptured sharks have been recorded due to deliberate suppression of information by some fishermen, as well as the fact that some internal tags were accidentally lost during gutting and cleaning operations at sea (Olsen 1984). In contrast, it was found, on average, that only 3% of recreational anglers participating in a cooperative tagging program off Natal, South Africa, did not return tags (van der Elst 1990). Olsen (1953) reported that the return percentage increased three fold after the practice of double tagging was instituted, suggesting that many tags were either shed or missed by fishermen. The color of the tags, however, made no difference in reporting rates (Olsen 1953, Stevens et al. 2000). Various methods have been derived to assess the probability that a tag is returned by fishermen or processors and to calculate the tag reporting rate (Jakobsson 1970, Hilborn 1988, Hoenig et al. 1998a).

When applying conclusions drawn from tagged fish to the overall population, it is often assumed that: tagged fish mix rapidly with the untagged part of a population; tagged fish are as likely to be caught as untagged fish; and the act of catching or tagging has no effect on behavior, growth, migration patterns, or subsequent survival (Gulland 1983, Schwarz & Arnason 1990). The effect of capture and tagging of sharks on growth has been investigated by a number of researchers. Some studies have concluded that the normal growth rate of some shark species may be interrupted by the effects of tagging for various periods of time, depending on injuries received by the fish during tagging operations and the amount of irritation inflicted by various types of tags. This has been shown for *Ginglymostoma cirratum* (Carrier 1985), *S. acanthias* (Bonham et al. 1949, Holland 1957, Ketchen 1975, Templeman 1984), *S. tiburo* (Parsons 1987), *N. brevirostris* (Manire & Gruber 1991), *M. lenticulatus* (Francis & Francis 1992), *C. tilstoni* and *C. sorrah* (Davenport & Stevens 1988). Another *N. brevirostris* tagging study by Gruber & Stout (1983), however, showed that tagging per se did not greatly affect the growth rate when food is not limiting. No effect of tagging on growth was found for *G. cuvier* (Natanson et al. 1999) and *I. oxyrinchus* (Pratt & Casey 1983) and no detrimental effect on body size was found for *Hemiscyllium ocellatum* (Heupel & Bennett 1997).

The tagging process may alter behavior patterns. Underwater observations indicate that adult *G. cirratum* tagged with a spear gun did not leave the tagging area or left only briefly to return the next day (Pratt & Carrier 2001). In another study, approximately 50% of *S. californica* tagged in situ near the base of the tail did not flee after tagging (Pittenger 1984). Some tagging disturbance was noted when *H. portusjacksoni* were banded with caudal peduncle tags, in situ, and seen to move away from their initial site of tagging within their reef environment (McLaughlin & O'Gower 1971). It was determined, however, that no marked change in behavior had occurred (McLaughlin & O'Gower 1970).

In addition to direct observations, volunteer taggers in the CSTP often report that tagged *P. glauca* return immediately to the boat after tagging and/or are subsequently recaught on another line. While this anecdotal information does not address any long term effects, multiple recaptures of individual tagged *G. cuvier* over time (Natanson et al. 1999) suggest that the act of capture and tagging produces a minimum of trauma for some species.

An indirect method of determining initial mortality from the tagging operation is to compare the proportion of sharks of different release conditions that have been recaptured. If there were no differences between groups, this would suggest low tagging mortality (Beverton et al. 1959, Jakobsson 1970, Ricker 1975, Gulland 1983). A quantitative assessment of fish condition is commonly recorded in shark studies (e.g. Bonham et al. 1949, Holland 1957, Clarke 1971, Smith & Abramson 1990, Stevens et al. 2000). Results from some studies show that recovery rates are similar for all sharks released except for those released in what is judged to be the poorest condition category (e.g. Holland 1957, Clarke 1971, Francis 1989). Stevens et al. (2000) reported that recapture rates were not related to condition for *C. sorrah* and *C. macloti*, but there were fewer returns from *C. tilstoni* originally released in fair or poor condition. In addition, it has been suggested that shark length (Francis 1989) and capture gear (Stevens et al. 2000) can affect recapture rates.

The mortality of tagged sharks may vary by gear type. Reported recapture rates of trawl caught and tagged *M. lenticulatus* were less than those of set net caught and tagged *M. lenticulatus*, suggesting that trawl caught fish had significant initial mortality (Francis 1989). For *S. lewini* tagged in the laboratory, there were no deaths directly attributable to tagging, even though these sharks had been handled more and had been held in a small container en route to the laboratory (Clarke 1971). Other evidence that tagging-induced mortality may be low in sharks is derived from sonic tagging experiments. Results from these studies suggest that sharks survive the stress of capture, tagging, and release, and have been tracked successfully up to 90 days (Nelson 1990).

A final influence on the rate of tag returns is tag shedding or loss. Tag loss is possibly the most critical issue that must be considered in the evaluation of results from tagging experiments (McFarlane et al. 1990). In many studies, a tagged animal is assumed to retain its tag permanently, which is clearly not a valid assumption for certain types of tags (Xiao et al. 1999a). Double tagging experiments (e.g. Kato & Carvallo 1967, Francis 1989, Xiao et al. 1999a, Stevens et al. 2000) and aquarium studies (Davies & Joubert 1967) are two ways to estimate tag loss (Harden Jones 1968, Bayliff & Holland[3], McFarlane et al. 1990). The presence or absence of wounds or scars left from shed tags may also give some anecdotal information on tag shedding (Clarke 1971, Randall 1977, Klimley & Nelson 1984, Pittenger 1984).

Tag shedding information in the literature is published in two forms, percentages and rates. Early studies used a tag shed loss percentage to evaluate the effectiveness of various tag types (e.g. Davies & Joubert 1967, Kato & Carvallo 1967). One very recent study has calculated an instantaneous rate of tag shedding for *G. galeus* (range = 0.2829 – 4.5992) (Xiao et al. 1999). Tag shedding rates in sharks, however, must be evaluated by species (Pepperell 1990, Walker et al.[4]) and within species (Holden & Horrod 1979). Tag shedding rates have been found to vary with tag type (e.g. Davies & Joubert 1967, Xiao et al. 1999a), sex (e.g. Xiao et al. 1999a), capture gear (e.g. Hurst et al. 1999), and tagging position (e.g. Davies & Joubert 1967, Kato & Carvallo 1967, Xiao et al. 1999a). For example, spaghetti loop tags placed in front of the first dorsal fin of *M. lenticulatus* were shed more often than those placed between the dorsal fins (18% and 6%, respectively) (Francis 1989). Rototags applied closer to the tip of the first dorsal fin were lost more frequently than those placed towards the base of the leading edge of the fin (Stevens et al. 2000). The shedding rate of dart tags anchored in the basal cartilage of the dorsal fin was about half that of dart tags anchored in the dorsal musculature for *G. galeus* and *M. antarcticus*.

In addition, tag shedding rates for these two species did not vary with length at release or time at liberty (Xiao et al. 1999a).

Tissue irritation was found in some species of sharks marked with dart tags (Davies & Joubert 1967, Manire & Gruber 1991). These studies, however, looked at the external appearance of the wound. When tissue response and healing rates were investigated, it was reported that a spaghetti dart tag inserted into the dorsal musculature of *H. ocellatum* produced only localized tissue disruption and was not detrimental to long term health. In addition, tagging did not appear to predispose the sharks to infection (Heupel & Bennett 1997). A similar result was found using fin tags on juvenile *C. melanopterus*, *C. plumbeus*, and *C. obscurus* (Heupel et al. 1998). The authors caution that the fin tags should be large enough to allow for fin growth over time.

Tag choice

Many of the early shark tagging studies utilized Petersen discs (e.g. Holland 1957, Kato & Carvallo 1967, Bane 1968) or Rototags (e.g. Davies & Joubert 1967, Kato & Carvallo 1967) as their primary tags. Petersen disc tags were found to have a higher return percentage than dangler tags for *S. acanthias* (Holden 1965, Templeman 1984) but have been replaced by Rototags as the principal fin tag used on sharks because of their much lower shedding rate (Kato & Carvallo 1967, Thorson 1971, Holden & Horrod 1979, Walker et al.[4], Xiao et al. 1999a). Other tagging studies on sharks have used internal tags (Olsen 1953) and miscellaneous dangler, collar, loop, and hydrostatic tags and, in some cases, multiple tagging protocols on each fish (Gruber 1982, Gruber & Stout 1983, Pratt & Carrier 2001). These general tag types have been used in a variety of shapes, sizes, materials, and colors and have had varying success rates. With the advent of cooperative shark tagging programs, nylon and stainless steel dart tags were more commonly used (Casey 1985, Pepperell 1990, van der Elst 1990, Hartill & Davies[7], Ugoretz[6]). Overall, of the 54 studies reviewed that included information on tag type, the dart tag was the most widely used, followed by the Rototag (Figure 4).

There are disadvantages to using all of these tag types. External tags generally have a shorter life expectancy, while internal tags have been recovered after many years (Olsen 1953, Grant et al. 1979, Hurst et al. 1999). For recovery purposes, however, an internal tag is not highly visible to fishermen and may be difficult and time consuming to insert (Walker et al.[4]). In general, dart tags have an incidence of higher shedding than fin tags (Davies & Joubert 1967, Kato & Carvallo 1967, Walker et al.[4], Xiao et al. 1999a) but have been retained for longer periods on sharks than other tag types (Carrier 1985). A drawback to fin tags (Rototags, Petersen disc, and strap tags) is their increased susceptibility to fouling in capture gear and vegetation (Olsen 1953, Davies & Joubert 1967, van der Elst 1990) and the restriction of substantial growth in body and fin thickness during long-term experiments (Davies & Joubert 1967, Carrier 1985, McFarlane et al. 1990, Stevens et al. 2000). This can lead to subsequent splitting and deterioration of the fin and is especially true when used on immature sharks (Kato & Carvallo 1967, Carrier 1985) whose cartilaginous dorsal rays are softer (Olsen 1953) and the shark has more dramatic growth over time.

Many sources of bias can be associated with, or mitigated by, the choice of tags. The performance of various tags can be evaluated by comparing the differences in recovery rate and in the rate of decline of reported recoveries among different tags over time. Results from the Australian Cooperative Gamefish Program from 1973 to 1987 showed that higher proportions of sharks tagged with steel anchor tags were likely to be recaptured after long periods compared with various nylon barbed tuna dart tags, which suggests a slower shedding rate for the steel anchor tags. The longest times at liberty were also from fish tagged with this tag type (Pepperell 1990). Results from other studies on pelagic teleosts and some shark species have shown that return rates were similar for the nylon head and stainless steel barb (Mather[2], Lenarz et al. 1973, Baglin et al. 1980) or higher for the stainless steel dart (Beckett[13], Mather et al. 1974, Carrier 1985) depending on the species of fish (McFarlane et al. 1990, Pepperell 1990) and the method of tagging used (Bayliff & Holland[3]). The ORI fin tag was returned after longer times at liberty than the dart tag and the sheep ear tag when different tag types were tested on sharks in the ORI Tagging Program in South Africa (van der Elst 1990). In the NMFS CSTP, sharks have been at liberty for a maximum of 22.6 and 27.8 years for stainless steel dart and Jumbo Rototags, respectively, with similar overall return rates (5.2%, 6%).

[13] Beckett, J.S. 1970. Swordfish, shark and tuna tagging 1961–1969. Fish. Res. Board Can. Tech. Rep. 193: 1–32.

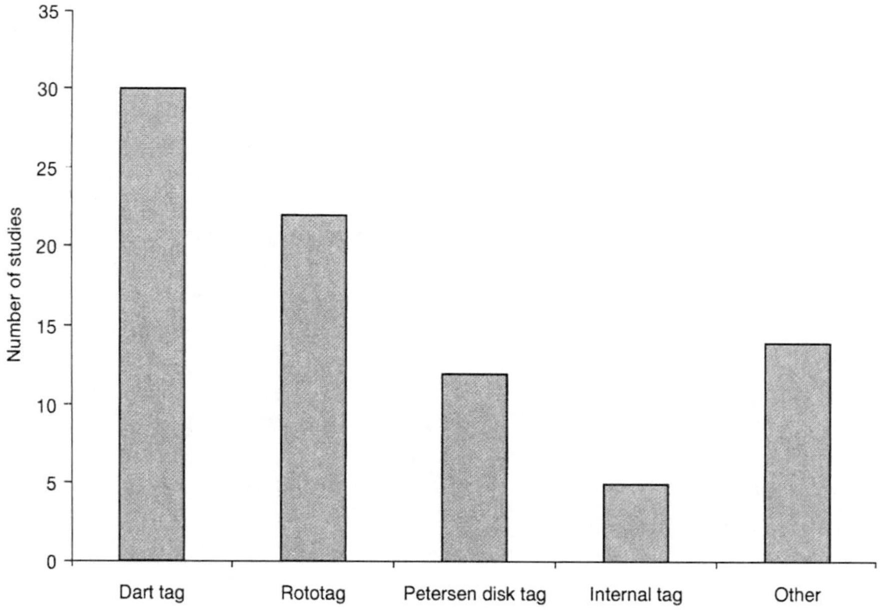

Figure 4. Relative uses of tag types from 54 shark tagging studies using conventional tags. Some studies used more than one tag type. Other tag category includes hydrostatic, freeze brand, loop, dangler, strap, Carlin disc, stainless steel bridle, caudal, and disk.

Choosing a tag type is thus a major factor to consider when a tagging program is planned. No single tag type is appropriate for all shark species or even for all life intervals of a species. No single technique is completely acceptable from a biological or technical standpoint (Prentice et al. 1990). It is, therefore, critical to consider any known life history information about the species to be marked in determining what form of tag to use (McFarlane et al. 1990). Factors determining the selection of an external tag are: the objectives of the study; the effect of the tag on behavior, survival, growth, or other life history characteristics; the stability of the mark; the number and size of the fish to be tagged; the stress of capture, handling, and marking of the fish; the ease of skin penetration or application of the tag; the length of time that the tag should remain on the fish; the availability and skills required by tagging personnel; the cost of conducting the experiment and recovery of tagged fish; the amount of coordination required among agencies, states, or countries; the species of fish the tag is to be used on; and the methods of recovery and reporting of the tags (Rounsefell & Everhart 1953, Arnold[1], Jakobsson 1970, McFarlane et al. 1990). The most important concerns in selecting a tag type for the CSTP (and for most cooperative programs) were that the external tag must be visible, be simple to use with inexpensive equipment, be easily and safely applied by volunteer fishermen, and contain detailed return instructions in several languages (Casey 1985).

Beyond conventional tagging studies: future directions

Historically, data analyses from tagging programs primarily involved descriptions of distributions and movements. Reports on these programs generally relied on descriptive statistics, such as averages (numbers tagged, recapture percentages), and maximums (days free, distances traveled), length frequencies, and distributions by species. Though some current researchers have begun to move beyond conventional data analyses, the time has come to more thoroughly investigate the practical application of T/R data to crucial management issues. Future tagging programs should include studies designed to investigate research questions in the following areas: resource utilization, space utilization, and population dynamics. In addition, the wealth of information from existing programs provides an opportunity for further critical analysis of these research areas.

Resource utilization and management hypotheses

Currently, one of the major challenges to fishery managers is the management and allocation of transboundary or migratory stocks (Hilborn et al. 1990). The migratory (nomadic) nature of many shark species requires international cooperation for management. For example, mating grounds for *P. glauca* occur in the western North Atlantic, whereas pupping grounds exist in the eastern North Atlantic (Casey 1985). Tagging programs can continue to provide data on stock structure, distribution of life history intervals, the exploitation of a resource by multinational fisheries, and direct evidence of fish movements across national and international boundaries. Traditional tagging techniques can be effective tools to test predicted migratory pathways and determine utilization of the resource.

Space utilization hypotheses

Another challenge to fishery managers is developing strategies for the recovery of over fished stocks including establishing minimum sizes, delineating known pupping areas, and determining essential fish habitat. For example, a directed tagging study on juvenile *C. plumbeus* in Delaware Bay has delineated the extent of the pupping area and established baseline neonate life history parameters (Merson 1999). Conventional tag data can serve as a cost-effective research tool (Smith & Abramson 1990) to test the predictions of management models that assume specific movement patterns or space utilization of the study species. Answers to these questions can then provide important information to develop stock maintenance strategies and test the success of these management initiatives.

Population dynamics hypotheses

An important application of conventional tagging methods includes the design of specific studies with experimental components to estimate critical population parameters (e.g. population size, exploitation/recruitment rates), develop fishery models (Xiao et al. 1999a), and to further investigate the possible bias in T/R data (Xiao 1994). Many of the questions and hypotheses on which these experimental studies are based can be answered best by conventional tagging techniques. In addition, quantitative evaluation of the potential bias and errors inherent in tagging data and assuring that the tagged fish are representative of the entire population would vastly improve the analysis of data available from tagging programs worldwide. Although techniques for estimating these parameters from tag data have existed for some time, relatively few programs have focused on experimental studies to investigate the population dynamics of large sharks.

The study of fish migrations includes both the description of movements, and the determination of the causative factors for the movements (Harden Jones 1968). Detailed examination of tag-recapture data can add significantly to both of these kinds of studies. While conventional tagging studies have traditionally described movement patterns, the next step in many programs is to determine what influences these migrations or behaviors through use of a combination of tagging methodologies and state of the art technology. New developments in internal, chemical, electronic, satellite, and archival tags, and genetic and biochemical stock analysis methods have occurred (McFarlane et al. 1990, Eckert & Stewart 2001, West & Stevens 2001). The future of tagging studies lies in the integration of these techniques with traditional methods to help answer some of these questions and further our knowledge on the movements and migrations of these highly migratory sharks.

Acknowledgements

We are indebted to the personnel of the Apex Predators Program, in particular, to L. Natanson who provided vital and timely assistance to B. Wetherbee who provided us with many tagging references and a valuable review of the manuscript, to W. Pratt who made additional comments, to J. Mello who assisted in preliminary preparations and ongoing research, and to R. Briggs for continued support in the NMFS CSTP. Our sincere thanks to the many researchers who contributed unpublished results from their shark tagging programs to render a more comprehensive review of worldwide programs. We apologize if some studies were inadvertently omitted. We are also grateful for critical reviews by T. Tricas, F. Almeida, W. Gabriel, and J. Kocik, and to K. Zewinski for providing valuable technical assistance.

References cited

Aasen, O. 1960. The Norwegian taggings of spiny dogfish (*Squalus acanthias*). Annales Biologiques 17: 85–95.

Aasen, O. 1961. Pigghaundersokelsene. Fisken Havet 1: 1–9.

Aasen, O. 1962. Norwegian dogfish tagging. Annales Biologiques 18: 106–107.

Backus, R.H., S. Springer & E.L. Arnold, Jr. 1956. A contribution to the natural history of the white-tip shark, *Pterolamiops longimanus* (Poey). Deep-Sea Res. 3: 178–188.

Baglin, R.E., Jr., M.I. Farber, W.H. Lenarz & J.M. Mason, Jr. 1980. Shedding rates of plastic and metal dart tags from Atlantic bluefin tuna, *Thunnus thynnus*. U.S. Fish. Bull. 78: 179–185.

Bane, G.W. 1968. The great blue shark. Calif. Currents 1: 3–4.

Bass, A.J. 1977. Long-term recoveries of tagged sharks. Copeia 1977: 574–575.

Bass, A.J., J.D. D'Aubrey & N. Kistnasamy. 1973. Sharks of the east coast of southern Africa. I. The genus *Carcharhinus* (Carcharhinidae). Invest. Rep. Oceanogr. Res. Inst. Durban 33. 168 pp.

Bergman, P.K., F. Haw, H.L. Blankenship & R.M. Buckley. 1992. Perspectives on design, use, and misuse of fish tags. Fisheries 17: 20–25.

Beverton, R.J.H., J.A. Gulland & A.R. Margetts. 1959. Whiting tagging: how the tag return rate is affected by the condition of fish when tagged. J. du Cons. 25: 53–57.

Bonham, K., F.B. Sanford, W. Clegg & G.C. Butcher. 1949. Biological and vitamin A studies of dogfish landed in the State of Washington (*Squalus suckleyi*). Wash. Dept. Fish. Biol. Bull. 49A: 83–114.

Branstetter, S. 1987. Age and growth validation of newborn sharks held in laboratory aquaria, with comments on the life history of the Atlantic sharpnose shark, *Rhizoprionodon terraenovae*. Copeia 1987: 291–300.

Brooks, E.N., K.H. Pollock, J.M. Hoenig & W.S. Hearn. 1998. Estimation of fishing and natural mortality from tagging studies on fisheries with two user groups. Can. J. Fish. Aquat. Sci. 55: 2001–2010.

Brownie, C., D.R. Anderson, K.P. Burnham & D.S. Robson. 1978. Statistical inference from band recovery data – a handbook. U.S. Fish and Wildl. Serv. Resour. Publ. No. 131. 212 pp.

Bruce, B.D. 1992. Preliminary observations on the biology of the white shark, *Carcharodon carcharias*, in South Australian waters. Aust. J. Mar. Freshwater Res. 43: 1–11.

Burnett, C.D., J.S. Beckett, C.A. Dickson, P.C.F. Hurley & T.D. Iles. 1987. A summary of releases and recaptures in the Canadian large pelagic fish tagging program 1961–1986. Can. Data Rep. Fish. Aquat. Sci. 673: 1–99.

Cailliet, G.M. 1990. Elasmobranch age determination and verification: an updated review. pp. 157–165. *In:* H.L. Pratt, Jr., S.H. Gruber & T. Taniuchi (ed.) Elasmobranchs as Living Resources: Advances in the Biology, Ecology, Systematics, and the Status of the Fisheries, NOAA Tech. Rep. NMFS 90.

Cailliet, G.M. 1992. Demography of the central California population of the leopard shark (*Triakis semifasciata*). Aust. J. Mar. Freshwater Res. 43: 183–193.

Cailliet, G.M., H.F. Mollet, G.C. Pittenger, D. Bedford & L.J. Natanson. 1992. Growth and demography of the Pacific angel shark (*Squatina californica*), based upon tag returns off California. Aust. J. Mar. Freshwater Res. 43: 1313–1330.

Carrier, J.C. 1985. Nurse sharks of Big Pine Key: comparative success of three types of external tags. Florida Sci. 48: 146–154.

Casey, J.G. 1985. Transatlantic migrations of the blue shark: a case history of cooperative shark tagging. pp. 253–68. *In:* R.H. Stroud (ed.) World Angling Resources and Challenges: Proceedings of the First World Angling Conference, Cap d'Agde, France, September 12–18, 1984, International Game Fish Association, Ft. Lauderdale.

Casey, J.G. & N.E. Kohler. 1992. Tagging studies on the shortfin mako shark (*Isurus oxyrinchus*) in the Western North Atlantic. Aust. J. Mar. Freshwater Res. 43: 45–60.

Casey, J.G. & L.J. Natanson. 1992. Revised estimates of age and growth of the sandbar shark (*Carcharhinus plumbeus*) from the western North Atlantic. Can. J. Fish. Aquat. Sci. 49: 1474–1477.

Clarke, T.A. 1971. The ecology of the scalloped hammerhead shark, *Sphyrna lewini*, in Hawaii. Pac. Sci. 25: 133–144.

Cliff, G., R.P. van der Elst, A. Govender, T.K. Witthuhn & E.M. Bullen. 1996. First estimates of mortality and population size of white sharks on the South African coast. pp. 393–400. *In:* A.P. Klimley & D.G. Ainley (ed.) Great White Sharks: the Biology of *Carcharodon carcharias*, Academic Press, San Diego.

Coutin, P. 1992. Sharks... and more sharks. Australian Fisheries 51: 41–42.

Davenport, S. & J.D. Stevens. 1988. Age and growth of two commercially important sharks (*Carcharhinus tilstoni* and *C. sorrah*) from Northern Australia. Aust. J. Mar. Freshwater Res. 39: 417–433.

Davies, D.H. & L.S. Joubert. 1967. Tag evaluation and shark tagging in South African waters, 1964–65. pp. 111–140. *In:* P.W. Gilbert, R. F. Mathewson & D.P. Rall (ed.) Sharks, Skates, and Rays, Johns Hopkins Press, Baltimore.

Ebert, D.A. 1996. Biology of the sevengill shark, *Notorynchus cepedianus* (Peron, 1807) in the temperate coastal waters of southern Africa. S. Afr. J. Mar. Sci. 17: 93–103.

Eckert, S.A. & B.S. Stewart. 2001. Telemetry and satellite tracking of whale sharks, *Rhincodon typus*, in the Sea of Cortez, Mexico, and the north Pacific Ocean. Env. Biol. Fish. 60: 299–308 (this volume).

Eisner, R.A. & J.A. Ritter. 1979. The Canadian Atlantic salmon (*Salmo salar*) smolt tag and its recovery system: a review. ICES, C.M. 1979/M: 24. 12 pp.

Everhart, W.H. & W.D. Youngs. 1981. Principles of fishery science. Cornell University Press, Ithaca. 349 pp.

Fabens, A.J. 1965. Properties and fitting of the von Bertalanffy growth curve. Growth 29: 265–289.

Ferreira, C.A. & T.P. Ferreira. 1996. Population dynamics of white sharks in South Africa. pp. 381–391. *In:* A.P. Klimley & D.G. Ainley (ed.) Great White Sharks: the Biology of *Carcharodon carcharias*, Academic Press, San Diego.

Fitzmaurice, P. 1994. Tagging studies of blue sharks and tope off the Irish coast. p. 17. *In:* S.L. Fowler & R.C. Earll (ed.) Proceedings of the Second European Shark and Ray Workshop, 15–16 February 1994, Tag and Release Schemes and Shark and Ray Management Plans.

Francis, M.P. 1988a. Movement patterns of rig (*Mustelus lenticulatus*) tagged in southern New Zealand. N.Z. J. Mar. Freshwater Res. 22: 259–272.

Francis, M.P. 1989. Exploitation rates of rig (*Mustelus lenticulatus*) around the South Island of New Zealand. N.Z. J. Mar. Freshwater Res. 23: 239–245.

Francis, M.P. & R.I.C.C. Francis. 1992. Growth rate estimates for New Zealand rig (*Mustelus lenticulatus*). Aust. J. Mar. Freshwater Res. 43: 1157–1176.

Francis, M.P. & K.P. Mulligan. 1998. Age and growth of New Zealand school shark, *Galeorhinus galeus*. N.Z. J. Mar. Freshwater Res. 32: 427–440.

Francis, R.I.C.C. 1988b. Are growth parameters estimated from tagging and age-length data comparable? Can. J. Fish. Aquatic Sci. 45: 936–942.

Francis, R.I.C.C. 1988c. Maximum likelihood estimation of growth and growth variability from tagging data. N.Z. J. Mar. Freshwater Res. 22: 42–51.

Gordon, W.G. 1990. Fish marking and the Magnuson Act. Amer. Fish. Soc. Symp. 7: 1–4.

Grant, C.J., R.L. Sandland & A.M. Olsen. 1979. Estimation of growth, mortality, and yield per recruit of the Australian school shark, *Galeorhinus australis* (Macleay), from tag recoveries. Aust. J. Mar. Freshwater Res. 30: 625–637.

Gruber, S.H. 1982. Role of the lemon shark, *Negaprion brevirostris* as a predator in the tropical marine environment: a multidisciplinary study. Florida Sci. 45: 46–75.

Gruber, S.H., D.R. Nelson & J.F. Morrissey. 1988. Patterns of activity and space utilization of lemon sharks, *Negaprion brevirostris*, in a shallow Bahamian lagoon. Bull. Mar. Sci. 43: 61–76.

Gruber, S.H. & R.G. Stout. 1983. Biological materials for the study of age and growth in a tropical marine elasmobranch, the lemon shark, *Negaprion brevirostris* (Poey). pp. 193–205. *In:* E.D. Prince & L.M. Pulos (ed.) Proceedings of the International Workshop on Age Determination of Oceanic Pelagic Fishes: Tunas, Billfishes, and Sharks, NOAA Tech. Rep. NMFS 8.

Gubanov, Y.P. 1976. The first catch of a tagged thresher shark (*Alopias vulpinus*). J. Ichthyol. 16: 497–498.

Gulland, J.A. 1983. Fish stock assessment. FAO/Wiley Series on Food and Agriculture, New York. 223 pp.

Gulland, J.A. & S.J. Holt. 1959. Estimation of growth parameters for data at unequal time intervals. J. Con. Explor. Mer 25: 47–49.

Hansen, P.M. 1963. Tagging experiments with the Greenland shark (*Somniosus microcephalus* (Bloch and Schneider)) in subarea 1. pp. 172–175. *In:* International Commission of Northwest Atlantic Fisheries Special Publication 4, North Atlantic Fish Marking Symposium.

Harden Jones, F.R. 1968. Fish migration. Edward Arnold, London. 325 pp.

Haw, F., P.K. Bergman, R.D., Fralick, R.M. Buckley & H.L. Blankenship. 1990. Visible implanted fish tag. Amer. Fish. Soc. Symp. 7: 311–315.

Herald, E.S. & W.E. Ripley. 1951. The relative abundance of sharks and bat stingrays in San Francisco Bay. Calif. Fish Game 37: 315–329.

Heupel, M.R. & M.B. Bennett. 1997. Histology of dart tag insertion sites in the epaulette shark. J. Fish Biol. 50: 1034–1041.

Heupel, M.R., C.A. Simpfendorfer & M.B. Bennett. 1998. Analysis of tissue responses to fin tagging in Australian carcharhinids. J. Fish Biol. 52: 610–620.

Hilborn, R. 1988. Determination of tag return from recaptured fish by sequential examination for tags. Trans. Amer. Fish. Soc. 117: 510–514.

Hilborn, R., C.J. Walters & D.B. Jester, Jr. 1990. Value of fish marking in fisheries management. Amer. Fish. Soc. Symp. 7: 5–8.

Hoenig, J.M., N.J. Barrowman, W.S. Hearn & K.H. Pollock. 1998a. Multiyear tagging studies incorporating fishing effort data. Can. J. Fish. Aquat. Sci. 55: 1466–1476.

Hoenig, J.M., N.J. Barrowman, K.H. Pollock, E.N. Brooks, W.S. Hearn & T. Polachek. 1998b. Models for tagging data that allow for incomplete mixing of newly tagged animals. Can. J. Fish. Aquat. Sci. 55: 1477–1483.

Hoey, J.J. 1983. Analysis of longline fishing effort for apex predators (swordfish, shark, and tuna) in the western North Atlantic and Gulf of Mexico. Ph.D. Thesis, University of Rhode Island, Kingston, 288 pp.

Holden, M.J. 1962. English spurdogfish tagging. Annales Biologiques 18: 107–108.

Holden, M.J. 1965. The stocks of spurdogs (*Squalus acanthias* L.) in British waters, and their migrations. Fish. Invest. Minist. Agric. Fish. Food G.B. Ser. II Salmon Freshwater Fish. 244: 1–20.

Holden, M.J. 1967. Transatlantic movement of a tagged spurdogfish. Nature 214: 1140–1141.

Holden, M.J. & R.G. Horrod. 1979. The migrations of tope, *Galeorhinus galeus* (L), in the eastern North Atlantic as determined by tagging. J. Cons. Int. Explor. Mer 38: 314–317.

Holland, G.A. 1957. Migration and growth of the dogfish shark, *Squalus acanthias* (Linnaeus), of the eastern North Pacific. Wash. Dept. Fish. Fish. Res. Pap. 2: 43–59.

Hurst, R.J., N.W. Bagley, G.E. McGregor & M.P. Francis. 1999. Movements of the New Zealand school shark, *Galeorhinus galeus*, from tag returns. N. Z. J. Mar. Freshwater Res. 33: 29–48.

Jackobsson, J. 1970. On fish tags and tagging. Oceanogr. Mar. Biol. Ann. Rev. 8: 457–499.

Jensen, A.L. 1981. Sample sizes for single mark and single recapture experiments. Trans. Amer. Fish. Soc. 110: 455–458.

Jensen, A.C., R.L. Edwards & G.C. Matthiessen. 1961. The spiny dogfish – a review. Woods Hole Lab Report No. 61-7: 1–42.

Jolly, G.M. 1965. Explicit estimates from capture–recapture data with both death and immigration-stochastic model. Biometrika 52: 225–247.

Jones, R. 1976. The use of marking data in fish population analysis. FAO Fish. Tech. Pap. 153: 1–42.

Kato, S. & A.H. Carvallo. 1967. Shark tagging in the eastern Pacific Ocean, 1962–65. pp. 93–109. *In:* P.W. Gilbert, R.F. Mathewson & D.P. Rall (ed.) Sharks, Skates, and Rays, Johns Hopkins Press, Baltimore.

Ketchen, K.S. 1975. Age and growth of dogfish *Squalus acanthias* in British Columbia waters. J. Fish. Res. Board Can. 32: 43–59.

Ketchen, K.S. 1986. The spiny dogfish (*Squalus acanthias*) in the northeast Pacific and a history of its utilization. Canadian Special Publication of Fisheries and Aquatic Sciences 88: 1–78.

Klimley, P.A. & D.R. Nelson. 1984. Diel movement patterns of the scalloped hammerhead shark (*Sphyrna lewini*) in relation to El Bajo Espiritu Santo: a refuging central-position social system. Behav. Ecol. Sociobiol. 15: 45–54.

Kohler, N.E., J.G. Casey & P.A. Turner. 1998. NMFS cooperative shark tagging program, 1962–1993: an atlas of shark tag and recapture data. Mar. Fish. Rev. 60: 1–87.

Kusher, D.I., S.E. Smith & G.M. Cailliet. 1992. Validated age and growth of the leopard shark, *Triakis semifasciata*, with comments on reproduction. Env. Biol. Fish. 35: 187–203.

Lenarz, W.H., F.J. Mather, III, J.S. Beckett, A.C. Jones & J.M. Mason, Jr. 1973. Estimation of rates of tag shedding by northwest Atlantic bluefin tuna. U.S. Fish. Bull. 71: 1103–1105.

Manire, C.A. & S.H. Gruber. 1991. Effect of M-type dart tags on field growth of juvenile lemon sharks. Trans. Amer. Fish. Soc. 120: 776–780.

Mather, F.J., III, D.C. Tabb, J.M. Mason, Jr. & H.L. Clark. 1974. Results of sailfish tagging in the western North Atlantic Ocean. pp. 194–210. *In:* R.S. Shomura & F. Williams (ed.) Proceedings of the International Billfish Symposium, Kailua-Kona, Hawaii, 9–12 August 1972, Part 2, Review and Contributed Papers, U.S. Dep. Comm. NOAA Tech. Rep. NMFS SSRF-675.

McFarlane, G.A. & R.J. Beamish. 1987. Validation of the dorsal spine method of age determination for spiny dogfish. pp. 287–300. *In:* R.C. Summerfelt & G.E. Hall (ed.) Age and Growth of Fish, Iowa State University Press, Ames.

McFarlane, G.A., R.S. Wydoski & E.D. Prince. 1990. Historical review of the development of external tags and marks. Amer. Fish. Soc. Symp. 7: 9–29.

McLaughlin, R.H. & A.K. O'Gower. 1970. Underwater tagging of the Port-Jackson shark Heterodontus portusjacksoni. Bull. Inst. Oceanogr. (Monaco) 69: 3–10.

McLaughlin, R.H. & A.K. O'Gower. 1971. Life history and underwater studies of a heterodont shark. Ecol. Monogr. 41: 271–289.

Merson, R.R. 1998. Nursery grounds and maturation in the sandbar shark in the Western North Atlantic. Ph.D. Thesis, University of Rhode Island, Kingston, 150 pp.

Monan, G.E. 1982. Advances in tagging and tracking hatchery salmonids: coded wire tags, multiple-coded and miniature radio tags, and the passive integrated transponder tag. pp. 33–37. *In:* C.J. Sindermann (ed.) Proceedings of the Eleventh U.S.–Japan Meeting on Aquaculture, Salmon Enhancement, Tokyo, Japan, October 19–20, 1982, NOAA Tech. Rep. NMFS 27.

Morrissey, J.F. & S.H. Gruber. 1993. Home range of juvenile lemon sharks, *Negaprion brevirostris*. Copeia 1993: 425–434.

Moulton, P.L., T.I. Walker & S. Saddlier. 1992. Age and growth studies of gummy shark, *Mustelus antarcticus* Gunther, and school shark, *Galeorhinus galeus* (Linnaeus), from southern Australian waters. Aust. J. Mar. Freshwater Res. 43: 1241–1267.

Murray, T. 1990. Fish-marking techniques in New Zealand. Amer. Fish. Soc. Symp. 7: 737–745.

Myers, R.A., M.O. Hammill & G.B. Stenson. 1997. Using mark-recapture to estimate the numbers of a migrating stage-structured population. Can. J. Fish. Aquat. Sci. 54: 2097–2104.

Natanson, L.J. & G.M. Cailliet. 1990. Vertebral growth zone deposition in Pacific angel sharks. Copeia 1990: 1133–1145.

Natanson, L.J., J.G. Casey, N.E. Kohler & T. Colket, IV. 1999. Growth of the tiger shark (*Galeocerdo cuvier*) in the western North Atlantic, based on tag returns and length-frequencies; and a note on the effects of tagging. U.S. Fish. Bull. 97: 944–953.

Nelson, D.R. 1990. Telemetry studies in sharks: a review, with applications in resource management. pp. 239–256. *In:* H.L. Pratt, Jr., S.H. Gruber & T. Taniuchi (ed.) Elasmobranchs as Living Resources: Advances in the Biology, Ecology, Systematics, and the Status of the Fisheries, NOAA Tech. Rep. NMFS 90.

O'Gower, A.K. & A.R. Nash. 1978. Dispersion of the Port Jackson shark in Australian waters. pp. 529–544. *In:* E.S. Hodgson & R.F. Mathewson (ed.) Sensory Biology of Sharks, Skates, and Rays, Office of Naval Research, Department of the Navy, Arlington.

Olsen, A.M. 1953. Tagging of school shark, *Galeorhinus australis* (Macleay) (Carcharhanidae) in south-eastern Australian waters. Aust. J. Mar. Freshwater Res. 4: 95–104.

Olsen, A.M. 1954. The biology, migration, and growth rate of the school shark, *Galeorhinus australis* (Macleay) (Carcharhanidae) in south-eastern Australian waters. Aust. J. Mar. Freshwater Res. 5: 353–410.

Olsen, A.M. 1984. Synopsis of biological data on the school shark *Galeorhinus australis* (Macleay 1881). FAO Fisheries Synopsis No. 139, FIR/S139. 42 pp.

Parsons, G.R. 1987. Life history and bioenergetics of the bonnethead shark, *Sphyrna tiburo* (Linnaeus): a comparison of two populations. Ph.D., University of South Florida, Tampa. 51 pp.

Pepper, V.A. & M.F. O'Connell. 1983. Coded wire microtag applications in Newfoundland, the first three years. ICES C.M. 1983/M 15: 1–14.

Pepperell, J.G. 1990. Australian cooperative game-fish tagging program, 1973–1987: status and evaluation of tags. Amer. Fish. Soc. Symp. 7: 765–774.

Pittenger, G.G. 1984. Movements, distribution, feeding and growth of the Pacific angel shark, *Squatina californica*, at Catalina Island, California. M.S. Thesis, California State University, Long Beach. 83 pp.

Pollock, K.H., J.M. Hoenig & C.M. Jones. 1991. Estimation of fishing and natural mortality when a tagging study is combined with a creel survey or port sampling. Amer. Fish. Soc. Symp. 12: 423–434.

Porter, T.R. 1979. A review of factors affecting the rate of return of tagged Atlantic salmon and effects of tagging. ICES C.M. 1979/M 15: 1–17.

Pratt, H.L., Jr. & J.C. Carrier. 2001. A review of elasmobranch reproductive behavior with a case study on the nurse shark, *Ginglymostoma cirratum*. Env. Biol. Fish. 60: 157–188 (this volume).

Pratt, H.L., Jr. & J.G. Casey. 1983. Age and growth of the shortfin mako, *Isurus oxyrinchus*, using four methods. Can. J. Fish. Aquat. Sci. 40: 1944–1957.

Prentice, E.F., T.A. Flagg, C.S. McCutcheon, D.F. Brastow & D.C. Cross. 1990. Equipment, methods, and an automated data-entry station for PIT tagging. Amer. Fish. Soc. Symp. 7: 335–340.

Pyle, P., S.D. Anderson, A.P. Klimley & R.P. Henderson. 1996. Environmental factors affecting the occurrence and behavior of white sharks at the Farallon Islands, California. pp. 281–291. *In:* A.P. Klimley & D.G. Ainley (ed.) Great White Sharks: the Biology of *Carcharodon carcharias*, Academic Press, San Diego.

Randall, J. E. 1977. Contribution to the biology of the whitetip reef shark (*Triaenodon obesus*). Pacific Sci. 31: 143–164.

Raymond, H.L. 1974. Marking fishes and invertebrates. I. State of the art of fish branding. Mar. Fish. Rev. 36: 1–6.

Ricker, W.E. 1975. Computation and interpretation of biological statistics of fish populations. Bull. Fish. Res. Board Can. 191. 382 pp.

Rounsefell, G.A. & W.H. Everhart. 1953. Fishery science – its methods and applications. John Wiley & Sons, New York. 444 pp.

Schwarz, C.J. & A.N. Arnason. 1990. Use of tag-recovery information in migration and movement studies. Amer. Fish. Soc. Symp. 7: 588–603.

Sciarrotta, T.C. & D.R. Nelson. 1977. Diel behavior of the blue shark, *Prionace glauca*, near Santa Catalina Island, California. U.S. Fish Bull. 75: 519–528.

Scott, E.L, E.D. Prince & C.D. Goodyear. 1990. History of the cooperative game fish tagging program in the Atlantic Ocean, Gulf of Mexico, and Caribbean Sea, 1954–1987. Amer. Fish. Soc. Symp. 7: 841–853.

Shafer, T.C. 1970. Migration and distribution of the spiny dogfish (*Squalus acanthias* L.) in the western North Atlantic. M.S. Thesis, University of Rhode Island, Kingston, 45 pp.

Skomal, G.B. 1990. Age and growth of the blue shark, *Prionace glauca*, in the North Atlantic. M.S. Thesis, University of Rhode Island, Kingston, 103 pp.

Sluczanowski, P.W.R. 1988. The development of fish tagging projects. pp. 134–140. *In:* D.A. Hancock (ed.) Australian Society for Fish Biology Tagging Workshop, Sydney, 21–22 July 1988, Bureau of Rural Resources Proceedings No. 5.

Smith, S.E. 1984. Timing of vertebral-band deposition in tetracycline-injected leopard sharks. Trans. Amer. Fish. Soc. 113: 308–313.

Smith, S.E. & N.J. Abramson. 1990. Leopard shark *Triakis semifasciata* distribution, mortality rate, yield, and stock replenishment estimates based on a tagging study in San Francisco Bay. U.S. Fish. Bull. 88: 371–381.

Springer, S. 1960. Natural history of the sandbar shark *Eulamia milberti*. U.S. Fish. Bull. 61: 1–38.

Stevens, J.D. 1976. First results of shark tagging in the north-east Atlantic, 1972- 1975. J. Mar. Biol. Ass. U.K. 56: 929–937.

Stevens, J.D. 1984. Life-history and ecology of sharks at Aldabra atoll, Indian Ocean. Proc. R. Soc. Lond. B222: 79–106.

Stevens, J.D. 1990. Further results from a tagging study of pelagic sharks in the north-east Atlantic. J. Mar. Biol. Ass. U.K. 70: 707–720.

Stevens, J.D., G.J. West & K.J. McLoughlin. 2000. Movements, recapture patterns, and factors affecting the return rate of carcharhinid and other sharks tagged off northern Australia. Mar. Freshwater Res. 51: 127–141.

Strasburg, D.W. 1958. Distribution, abundance, and habits of pelagic sharks in the central Pacific Ocean. 1958. U.S. Fish. Bull. 58: 335–361.

Strong, W.R., Jr., R.C. Murphy, B.D. Bruce & D.R. Nelson. 1992. Movements and associated observations of bait-attracted white sharks, *Carcharodon carcharias*: a preliminary report. Aust. J. Mar. Freshwater Res. 43: 13–20.

Strong, W.R., Jr., B.D. Bruce, D.R. Nelson & R.D. Murphy. 1996. Population dynamics of white sharks in Spencer Gulf, South Australia. pp. 401–414. *In:* A.P. Klimley & D.G. Ainley (ed.) Great White Sharks: the Biology of *Carcharodon carcharias*, Academic Press, San Diego.

Talent, L.G. 1985. The occurrence, seasonal distribution, and reproductive condition of elasmobranch fishes in Elkhorn Slough, California. Calif. Fish Game 71: 210–219.

Templeman, W. 1944. The life-history of the spiny dogfish (*Squalus acanthias*) and the vitamin A values of dogfish liver oil. Newfoundl. Dept. Nat. Res. Bull. (Fish.) 15: 1–102.

Templeman, W. 1954. Migrations of spiny dogfish tagged in Newfoundland waters. J. Fish. Res. Board. Can. 11: 351–354.

Templeman, W. 1976. Transatlantic migrations of spiny dogfish (*Squalus acanthias*). J. Fish. Res. Board Can. 33: 2605–2609.

Templeman, W. 1984. Migrations of spiny dogfish, *Squalus acanthias*, and recapture success from tagging in the Newfoundland area, 1963–65. J. Northw. Atl. Fish. Sci. 5: 47–53.

Thorson, T.B. 1971. Movement of bull sharks, *Carcharhinus leucas*, between Caribbean Sea and Lake Nicaragua demonstrated by tagging. Copeia 1971: 336–338.

Thorson, T.B. & E.J. Lacy, Jr. 1982. Age, growth rate and longevity of *Carcharhinus leucas* estimated from tagging and vertebral rings. Copeia 1982: 110–116.

Tricas, T.C. 1977. Food habits, movements, and seasonal abundance of the blue shark, *Prionace glauca* (Carcharhinidae), in southern California waters. M.S. Thesis, California State University, Long Beach. 79 pp.

Tucker, R. 1985. Age validation studies on the spines of the spurdog (*Squalus acanthias*) using tetracycline. J. Mar. Biol. Assoc., U.K. 65: 641–651.

van der Elst, R.P. 1990. Marine fish tagging in South Africa. Amer. Fish. Soc. Symp. 7: 854–862.

Walker, T.I. 1992. Fishery simulation model for sharks applied to the gummy shark, *Mustelus antarcticus* Gunther, from southern Australian waters. Aust. J. Mar. Freshwater Res. 43: 195–212.

West, G.J. & J.D. Stevens. 2001. Archival tagging of school sharks, *Galeorhinus galeus*, in Australia: initial results. Env. Biol. Fish. 60: 283–298 (this volume).

Williams, H. & A.H. Schaap. 1992. Preliminary results of a study into the incidental mortality of sharks in gill-nets in two Tasmanian shark nursery areas. Aust. J. Mar. Freshwater Res. 43: 237–250.

Xiao, Y. 1994. Growth models with corrections for the retardative effects of tagging. Can. J. Fish. Aquat. Sci. 51: 263–267.

Xiao, Y. 1996. A framework for evaluating experimental designs for estimating rates of fish movement from tag recoveries. Can. J. Fish. Aquat. Sci. 53: 1272–1280.

Xiao, Y., L.P. Brown, T.I. Walker & A.E. Punt. 1999a. Estimation of instantaneous rates of tag shedding for school shark, *Galeorhinus galeus*, and gummy shark, *Mustelus antarcticus*, by conditional likelihood. U.S. Fish. Bull. 97: 170–184.

Xiao, Y., J.D. Stevens & G.J. West. 1999b. Estimation of fishing and natural mortalities from tag experiments with exact or grouped times at liberty. Can. J. Fish. Aquat. Sci. 56: 868–874.

Yamashita, D.T. & K.D. Waldron. 1958. An all-plastic dart-type fish tag. Calif. Fish Game 44: 311–317.

Appendix 1. A summary of 64 past and present worldwide shark tagging studies and programs using conventional tags. For studies not including scientific names, the reported common names were used.

Source	Species	Tagged number	Recaptured number	%	Max. speed (km day^{-1})	Max. distance (km)	Max. time (yr)	Number of species tag	recap.	Location	Tag type
Aasen in Templeman (1976), Aasen (1961)	*Squalus acanthias*	8122		10.8				1	1	Norway, Shetland Islands	yellow polyethylene film with stainless steel bridle
Bane (1968), Bane (unpublished) in Tricas (1977)	*Prionace glauca*	250	3	1.2		2561	2.1	1	1	California	Petersen disc tag, roto snap tag, stainless steel cattle tag
Bass (1977)	Summary information									South Africa – Durban	
	Carcharhinus galapagensis						5.1				
	Carcharhinus leucas						7.7				
Bass et al. (1973)	Summary information									South Africa – Durban	
	Carcharhinus obscurus	2174	98	4.5							
	Carcharhinus brevipinna	960	31	3.2							
Beverton et al. (1959)	*Squalus acanthias*	75	2	2.7			0.6	1	1	Irish Sea	yellow plastic flag on braided nylon loop
Burnett et al. (1987)	Summary information	2514	25	1.0		2228	9.5	> 1.3	5	Atlantic ocean	dart dag
	Prionace glauca	2003	17	0.8		1072					
	Isurus oxyrinchus	110	5	4.5		1028					
	Carcharhinus obscurus	80	1	1.3		6					
	Carcharhinus falciformis	74	0	0.0							
	Carcharhinus longimanus	73	0	0.0							
	Sphyrna lewini	60	0	0.0							
	Carcharhinus plumbeus	34	1	2.9		2228					
	Carcharhinidae	18	0	0.0							
	Galeocerdo cuvieri	18	1	5.6							
	Sphyrnidae	16	0	0.0							
	Lamnidae	11	0	0.0							
	Lamna nasus	8	0	0.0							
	Carcharhinus signatus	5	0	0.0							
	Cetorhinus maximus	2	0	0.0							
	Alopias superciliosus	1	0	0.0							
	Rhizoprionodon terraenovae	1	0	0.0							
Carrier (1985)	*Ginglymostoma cirratum*	70	14	20.0			2.2	1	1	Big Pine Key, Florida	stainless steel barb, plastic barb, Carlin disc

Appendix 1. (*continued*).

Source	Species	Tagged number	Recaptured number	%	Max. speed (km day^{-1})	Max. distance (km)	Max. time (yr)	Number of species tag	recap.	Location	Tag type
Caunter in Stevens (1976)	*Prionace glauca*	200+	1					1	1	Great Britain	
Clarke (1971)	*Sphyrna lewini*	410	76	18.5			0.3	1	1	Hawaii	numbered plastic dart tag
Davies & Joubert (1967)	Summary information	1062	385	36.3	59.1	1383	0.6	16	5	South Africa – Durban, Port Elizabeth	Petersen disc, WHOI nylon barb, WHOI stainless dart, Rototag, Jumbo Rototag, ORI tag
	Carcharhinus obscurus	727	322	44.3	59.1		0.6				
	Carcharhinus maculipinnis	183	47	25.7		1383	0.6				
	Rhizoprionodon acutus	83	6	7.2		14	0.3				
	Carcharhinus sorrah	22	0	0.0							
	Mustelus canis	13	9	69.2		< 2	0.3				
	Carcharhinus albimarginatus	11	0	0.0							
	Squalus cubensis	9	0	0.0							
	Carcharhinus limbatus	3	0	0.0							
	Carcharhinus tjutjot	2	0	0.0							
	Galeocerdo cuvieri	2	0	0.0							
	Loxodon macrorhynchus	2	0	0.0							
	Carcharhinus galapagensis	1	0	0.0							
	Carcharhinus leucas	1	1	100.0			1 day				
	Carcharhinus spallanzi	1	0	0.0							
	Mustelus nigropunctatus	1	0	0.0							
	Sphyrna lewini	1	0	0.0							
Ebert (1996)	*Notorynchus cepedianus*	614	26	4.2	3.5	539	1.4	1	1	Southern Africa	rototag (ORI tag)
Ferreira & Ferreira (1996)	*Carcharodon carcharias*	147						1		South Africa, Dyer Island	plaque attached to ORI spaghetti dart tag
Foerster[1]	dogfish (grayfish)	564	34	6.0				1	1	British Columbia	colored celluloid disk
Francis (1988, 1989)	*Mustelus lenticulatus*	2234	382	17.1	21.0	1159	> 5	1	1	Southern New Zealand – South Island & SW North Island	Floy FT4, 12 cm tube length (yellow plastic)

Govender et al.[2]	Summary information	29 564	1451	4.9	1964	> 36	> 33	South Africa	steel head dart tag, round cattle ear tag
	dusky shark	5065	422	8.3	1835				
	blacktail	4976	148	3.0	788				
	spotted gulley shark	4548	299	6.6	1066				
	copper/bronze shark	3958	137	3.5	1320				
	smooth blackspot houndshark	2158	80	3.7	1407				
	spotted ragged-tooth	1637	79	4.8	1416				
	smooth hammerhead	1216	31	2.5	1122				
	cow and frill sharks	784	29	3.7	497				
	milk shark	692	31	4.5	1004				
	soupfin shark	459	17	3.7	453				
	striped/pyjama catshark	421	20	4.8	1964				
	blacktip shark	402	21	5.2	380				
	hammerhead sharks	384	4	1.0	233				
	sandbar shark	295	6	2.0	347				
	broadnose sevengill shark	284	8	2.8	539				
	scalloped/bronze hammerhead	278	7	2.5	254				
	bluntnose/spiny dogfish	262	4	1.5	87				
	longnose/spinner shark	222	11	5.0	164				
	shorttail catshark	215	24	11.2	1010				
	tiger shark	168	8	4.8	145				
	great white shark	165	7	4.2	1445				
	brown shyshark	164	4	2.4	8				
	Zambezi shark	158	11	7.0	125				
	blackspot shark	124	8	6.5	367				
	flapnose shark	85	5	5.9	54				
	sliteye shark	70	4	5.7	575				
	banded catshark	53	10	18.9	63				
	hardnose/smooth houndshark	53	1	1.9	228				
	whitespotted/smooth houndshark	47	4	8.5	210				
	spotted/spiny dogfish	43	1	2.3	47				
	catsharks	29	2	6.9	0				
	shortfin/mako shark	28	1	3.6	6				
	brown catshark	23	1	4.3	0				
	blue shark	19	0	0.0	0				
	Java shark	15	2	13.3	61				
	blackspot catshark	14	2	14.3	38				

[1]Foerster, R.E. 1942. Dogfish Tagging – preliminary results. Canada Fish. Res. Bd. Pacific Coast Sta., Prog. Rep. No. 53: 12–13.
[2]Govender, A., E. Bullen & R. van der Elst. 1995. Tagging news No. 11 Newsletter from the Oceanographic Research Institute, Durban, South Africa.

Appendix 1. (*continued*).

Source	Species	Tagged number	Recaptured number	%	Max. speed (km day^{-1})	Max. distance (km)	Max. time (yr)	Number of species tag	recap.	Location	Tag type
	Galapagos shark	14	0	0.0		0					
	thintail/thresher shark	13	0	0.0		0					
	grey reef shark	12	1	8.3		0					
	requiem sharks	11	1	9.1		568					
Green (personal communication)	Summary information	17 768	763	4.3		7871	15.2	4	3	Ireland	Jumbo Rototag, dart tags, Petersen disc
	blue shark	14 990	511	3.4		7871	3.9				
	tope	2722	246	9.0		4047	15.2				
	porbeagle shark	55	6	10.9		4260	10.8				
	thresher shark	1	0	0.0							
Gruber & Stout (1983)	*Negaprion brevirostris*	~1500	70+	5.0				1	1	Florida Keys, Bahamas	freeze band, mini-rototag, plastic dart tag, internal tag
Gubanov (1976)	Summary information	225	2	0.9				12		Sumatra	hydrostatic tag (polyethylene cylindrical ampule)
	Alopias vulpinus					1556	2.0				
Hansen (1963)	*Somniosus microcephalus*	411	28	6.8		1296	16.0	1	1	Greenland	Petersen disc
Hartill & Davies[3]	Summary information	11 986	285	2.4	42.4		11.9	10	5	New Zealand waters	stainless steel dart tag
	mako	9143	224	2.4	42.4		6.5				
	blue shark	2411	30	1.2	27.0		1.7				
	hammerhead	165	0	0.0							
	school shark	95	25	26.3	8.0		11.9				
	bronze whaler	75	3	4.0	14.1		3.1				
	thresher	55	0	0.0							
	sevengill	24	3	12.5	9.8		1.9				
	porbeagle	16	0	0.0							
	carpet	1	0	0.0							
	white tip	1	0	0.0							
Herald & Ripley (1951)	*Galeorhinus zyopterus*	118	4	3.4	44.0	2037	2.1	1	1	California waters	
Holden (1962, 1965, 1967)	*Squalus acanthias*	11 996	1044	8.7			3.8	1	1	England, Scotland, Norway, Faroe Islands	Petersen disc, internal and external polythene flag tag

Holden & Horrod (1979)	*Galeorhinus galeus*	491	32	6.5		2526	10.8	1	1	England	Petersen disc, Rototag
Holland (1957)	*Squalus acanthias*	9705	655	6.7		8704	10.0	1	1	Washington	celluloid Petersen tag
Hueter (personal communication)	Summary information	8028	350	4.4	13.9	519	5.6	16	11	Gulf of Mexico	Hallprint PDB plastic-tipped dart tag
	Sphyrna tiburo	2966	96	3.2	3.2	343	5.6				
	Carcharhinus limbatus	2375	184	7.7	13.9	519	2.9				
	Rhizoprionodon terraenovae	1263	17	1.3	3.2	102	4.0				
	Carcharhinus acronotus	805	22	2.7	0.9	111	4.0				
	Carcharhinus brevipinna	150	7	4.7	1.3	370	0.8				
	Carcharhinus leucas	131	14	10.7	1.5	194	0.5				
	Carcharhinus isodon	108	2	1.9	< 0.1	6	0.8				
	Mustelus norrisi	42	0	0.0							
	Carcharhinus plumbeus	41	1	2.4	0.1	7	0.2				
	Sphyrna mokarran	39	2	5.1	0.4	213	2.0				
	Negaprion brevirostris	34	3	8.8	0.1	4	0.4				
	Sphyrna lewini	28	0	0.0							
	Ginglymostoma cirratum	25	0	0.0							
	Galeocerdo cuvier	11	2	18.2	1.4	482	1.0				
	Mustelus canis	8	0	0.0							
	Carcharhinus falciformis	2	0	0.0							
Hurst et al. (1999)	*Galeorhinus galeus*	3950	207	5.2	22.8	4940	9.6	1	1	New Zealand	Floy dart tag, loop tag, plastic Nesbit internal tag
Jensen in Templeman (1976)	*Squalus acanthias*	907		3.0				1	1		
Kato & Carvallo (1967)	Summary information	860	83	9.7		1111	> 0.6	> 14	> 8	Mexican coast, Revillagigedo Islands, offshore Southern California to Peru	Petersen disc, Jumbo Rototag, strap (cattle size ear tag), Floy dart tag
	Carcharhinus galapagensis	215	29	13.5		> 59					
	Carcharhinus limbatus	184	19	10.3		> 37					
	Carcharhinus albimarginatus	139	19	13.7		< 9					
	Carcharhinus falciformis	123	5	4.1		176					
	Carcharhinus porosus	77	6	7.8		> 37					
	Rhizoprionodon longurio	73	2	2.7		1111					

[3]Hartill, B. & N.M. Davies. 1999. New Zealand billfish and gamefish tagging 1997–98. NIWA Technical Report 57. 39 pp.

Appendix 1. (continued).

Source	Species	Tagged number	Recaptured number	%	Max. speed (km day^{-1})	Max. distance (km)	Max. time (yr)	Number of species tag	recap.	Location	Tag type
	Sphyrna lewini	25	1	4.0		> 37					
	Prionace glauca	5	0	0.0							
	Carcharhinus altimus	4	0	0.0							
	Sphyrna zygaena	4	0	0.0							
	Carcharhinus longimanus	3	0	0.0							
	Alopias vulpinus	2	0	0.0							
	Mustelus spp.	2	0	0.0							
	Carcharhinus velox	1	0	0.0							
	Ginglymostoma cirratum	1	1	100.0		> 37					
	other	2	1	50.0							
Ketchen (1986)	*Squalus acanthias*	24 079	1150	4.8		7890		1	1	Straits of Georgia, Canada, Puget Sound	
Klimley & Nelson (1984)	*Sphyrna lewini*	100						1	1	Gulf of California	color coded plastic streamer (dart tip)
Kohler & Turner (this study)	Summary information	142 868	7352	5.1	82.4	6926	27.8	> 52	> 31	Atlantic Ocean, Mediterranean Sea, Gulf of Mexico	stainless steel dart tag, Jumbo Rototag, Rototag
	Prionace glauca	82 080	4001	4.9	82.4	6926	8.5				
	Carcharhinus plumbeus	19 344	987	5.1	21.6	3776	27.8				
	Galeocerdo cuvier	7161	620	8.7	61.4	6747	10.9				
	Carcharhinus obscurus	6707	155	2.3	41.3	3800	15.8				
	Isurus oxyrinchus	4419	483	10.9	66.1	4543	12.8				
	Carcharhinus limbatus	3239	142	4.4	30.4	2146	7.3				
	Rhizoprionodon terraenovae	2539	39	1.5	10.6	1037	7.3				
	Sphyrna lewini	2240	45	2.0	11.1	1671	9.6				
	Negaprion brevirostris	1690	171	10.1	7.4	426	4.1				
	Ginglymostoma cirratum	1270	134	10.6	13.6	541	7.8				
	Carcharhinid sharks	1182	101	8.5							
	Carcharhinus falciformis	974	61	6.3	59.7	1339	7.1				
	Lamna nasus	942	96	10.2	40.7	1861	9.2				
	Hammerhead sharks	867	20	2.3							
	Sphyrna tiburo	849	19	2.2	5.8	261	2.0				
	Misc. sharks	784	83	10.6							
	Odontaspis taurus	684	39	5.7	5.4	1187	3.2				

Carcharhinus perezi	630	16	2.5	1.5	48	4.4
Carcharhinus leucas	621	15	2.4	20.1	643	7.0
Carcharhinus longimanus	597	8	1.3	32.4	2811	3.3
Carcharhinus acronotus	526	9	1.7	1.5	315	9.2
Carcharhinus brevipinna	494	14	2.8	6.1	1665	4.5
Mustelus canis	404	15	3.7	9.7	402	4.0
Carcharhinus galapagensis	403	16	4.0	1.8	2859	4.4
Alopias superciliosus	346	8	2.3	17.3	2767	6.5
Squalus acanthias	237	5	2.1	0.8	532	3.2
Carcharhinus signatus	220	16	7.3	11.2	2669	13.8
Sphyrna zygaena	185	6	3.2	4.8	919	2.1
Carcharhinus altimus	171	12	7.0	4.6	3343	11.2
Cetorhinus maximus	157	0	0.0			
thresher sharks	144	0	0.0			
Sphyrna mokarran	133	4	3.0	1.2	1180	2.8
Aprionodon isodon	127	0	0.0			
Squatina dumerili	106	0	0.0			
Isurus paucus	90	5	5.6	9.6	3430	5.5
Alopias vulpinus	72	3	4.2	0.1	159	8.0
mackerel sharks	42	1	2.4			
Carcharodon carcharias	38	2	5.3	1.6	1011	2.5
Carcharhinus porosus	29	0	0.0			
Somniosus microcephalus	22	1	4.5		0	1.0
Pseudocarcharias kamoharai	21	0	0.0			
Galeorhinus galeus	16	0	0.0			
Mustelus norrisi	15	0	0.0			
Rhizoprionodon porosus	14	0	0.0			
Hexanchus vitulus	11	0	0.0			
Triakis semifasciata	5	0	0.0			
Centrophorus granulosus	3	0	0.0			
Hexanchus spp.	3	0	0.0			
Rhincodon typus	3	0	0.0			
Hexanchus griseus	2	0	0.0			
Rhizoprionodon lalandii	2	0	0.0			
Sphyrna tudes	2	0	0.0			
Chlamydoselachus anguineus	1	0	0.0			
Echinorhinus brucus	1	0	0.0			

Appendix 1. (continued).

Source	Species	Tagged number	Recaptured number	%	Max. speed (km day^{-1})	Max. distance (km)	Max. time (yr)	Number of species tag	recap.	Location	Tag type
	Mustelus henlei	1	0	0.0							
	Scyliorhinus retifer	1	0	0.0							
	Sphyrna media	1	0	0.0							
	Squatina californica	1	0	0.0							
Levine (personal communication)	*Rhincodon typus*	261						1		Indian Ocean, Caribbean Sea	custom dart tag applied *in situ*
Matsunaga (personal communication)	Summary information	1670	14	0.8	35.2	1550	0.4	> 7	1	North Pacific	Floy stainless dart
	blue shark	1394	14	1.0	35.2	1550	0.4				
	bigeye thresher	82	0	0.0							
	silky shark	69	0	0.0							
	oceanic whitetip	50	0	0.0							
	porbeagle	24	0	0.0							
	pelagic thresher	18	0	0.0							
	mako shark	14	0	0.0							
	other	19	0	0.0							
McLaughlin & O'Gower (1971)	*Heterodontus portusjacksoni*	295	4	1.4	1.8	400	0.6	1	1	Bondi, Sydney, Australia	caudal peduncle collar tag applied *in situ*
Natanson & Cailliet (1990)	*Squatina californica*	105	6	5.7				1	1		Jumbo Rototag
Nelson (personal communication), Nelson (unpublished) in Tricas (1977), Sciarrotta & Nelson (1977)	*Prionace glauca*	16	2	12.5		3589	3.2	1	1		Floy FH-69 stainless steel dart tag
Nelson (personal communication)	*Heterodontus francisci*	21	7	33.3		19	11.2	1	1		
Nelson (personal communication)	Summary information	218	117	53.7				3	3		
	Triakis obesus	113	78	69.0		9	3.4				
	Carcharhinus amblyrhynchos	65	28	43.1			4.2				
	Carcharhinus melanopterus	40	11	27.5			0.3				
O'Gower & Nash (1978)	*Heterodontus portusjacksoni*	230	14	6.1	6.5	850	1.8	1	1	Bondi, Sydney, Australia	caudal peduncle collar tag applied *in situ*
Parsons (1987)	*Sphyrna tiburo*	70	5	7.1				1	1	Tampa Bay, Florida	Dalton Rototag

Pepperell (1990), Henry (personal communication)	Summary information	17 401	351	2.0	24.1	5504	7.3	> 6	> 6	Australia	plastic streamer, nylon barbed dart, T-anchor, steel anchor dart
	Isurus oxyrinchus	3581	69	1.9	13.0	3293	5.0				
	Sphyrna spp.	3542	37	1.0	13.0	556	2.7				
	Carcharhinus spp.	3188	57	1.8	9.3	1796	7.3				
	Carcharhinus brachyurus	2518	50	2.0							
	Prionace glauca	2308	40	1.7	24.1	5504	1.6				
	Galeocerdo cuvier	452	14	3.1							
	Carcharodon carcharias	96	7	7.3							
	Eugomphodus taurus	76	4	5.3							
	Alopias spp.	75	1	1.3							
	Misc. sharks	1565	72	4.6			6.0				
Pittenger (1984)	*Squatina californica*	402	111	27.6		24	3.0	1	1	Catalina Island, California	Floy FH-69 stainless steel dart tag applied *in situ*
Randall (1977)	*Triaenodon obesus*	124	7	5.6	< 0.1	3	2.0	1	1	Johnston Island	monel cattle ear tag, streamer dart tag with bright yellow vinyl
Schwartz (personal communication)	Misc. sharks	203 000	3103	1.5		2126	1.5	28	10	North Carolina	metal strap, Petersen, Hallprint dart tag
Shafer (1970)	*Squalus acanthias*	3583	61	1.7	21.1	1578	1.9	1	1	Rhode Island, North Carolina, Maine	Jumbo Rototag
Smith & Abramson (1990)	*Triakis semifasciata*	948	108	11.4		140				San Francisco Bay, California	plastic Rototag
Stevens (1984)	Summary information	1218	274	22.5		11	3.9	6	5	Aldabra Atoll	Jumbo Rototag
	Carcharhinus melanopterus	1003	233	23.2		5	3.9				
	Negaprion acutidens	131	19	14.5		5	3.9				
	Carcharhinus albimarginatus	44	15	34.1		7	0.7				
	Carcharhinus wheeleri	33	6	18.2							
	Nebrius concolor	5	0	0.0							
	Carcharhinus falciformis	2	1	50.0		11	0.4				
Stevens (1990)	Summary information	2883	102	3.5	7.5	7176	13.0	4	4	North East Atlantic	Jumbo Rototag
	Prionace glauca	2585	51	2.0	7.5	7176	10.7				
	Galeorhinus galeus	271	42	15.5	3.0	2461	6.7				
	Lamna nasus	26	8	30.8		2370	13.0				
	Isurus oxyrinchus	1	1	100.0		390	4.6				

Appendix 1. (continued).

Source	Species	Tagged number	Recaptured number	%	Max. speed (km day^{-1})	Max. distance (km)	Max. time (yr)	Number of species tag	recap.	Location	Tag type
Stevens et al. (2000)	Summary information	10 489	579	5.5	18.6	1348	12.9	23	16	Northern Australia	Jumbo Rototag, Rototag
	Carcharhinus tilstoni	4846	402	8.3	24.7	1348	12.9				
	Carcharhinus sorrah	2919	83	2.8	6.8	1116	9.9				
	Carcharhinus macloti	1610	52	3.2	18.6	711	10.5				
	Rhizoprionodon acutus	277	4	1.4	0.1	45	1.8				
	Carcharhinus amboinensis	131	13	9.9	18.3	242	4.7				
	Carcharhinus amblyrhynchoides	122	13	10.7	8.3	173	8.8				
	Rhizoprionodon taylori	119	1	0.8		92	0.2				
	Sphyrna lewini	93	1	1.1		113	< 0.1				
	Carcharhinus dussumieri	79	1	1.3		4	1.4				
	Carcharhinus brevipinna	59	1	1.7		19	1.7				
	Carcharhinus fitzroyensis	55	1	1.8		150	0.2				
	Galeocerdo cuvier	55	2	3.6	3.3	156	5.7				
	Sphyrna mokarran	48	2	4.2	0.8	385	4.2				
	Eusphyra blochii	34	1	2.9		21	1.0				
	Carcharhinus amblyrhynchos	12	0	0.0							
	Carcharhinus melanopterus	6	0	0.0							
	Nebrius ferrugineus	6	1	16.7		43					
	Carcharhinus limbatus	5	1	20.0							
	Carcharhinus plumbeus	5	0	0.0							
	Hemipristis elongatus	4	0	0.0							
	Stegostoma fasciatum	2	0	0.0							
	Carcharhinus falciformis	1	0	0.0							
	Negaprion acutidens	1	0	0.0							
Strong et al. (1992, 1996), Bruce (1992)	*Carcharodon carcharias*	40	4	10.0		220	0.2	1	1	South Australia	Hallprint stainless steel dart tag
Strong in Nelson (personal communication)	*Heterodontus francisci*	312	46	14.7		4	2.1	1	1		
Sutcliffe[4]	*Galeorhinus galeus*	74	12	16.2		3200	12.0	1	1	Scotland	Dalton Rototag, Floy FT-1 dart tag

Taniuchi (personal communication)	*Squalus acanthias*	16 588	340	2.0	25.9		5.9	1	1	Tail region of Tsushima Warm Current, Japan	caudal tags
Templeman (1976)	*Squalus acanthias*	2855		8.2			11.2	1	1	Newfoundland	Petersen disc, flat dangler tag, hydrostatic tag
Tester[5]	Summary information	279	17	6.1		83	1.6	7	4	Hawaii	metal strap tag (small & large), internal plastic tag
	gray reef	113	11	9.7			1.5				
	sandbar	86	1	1.2			1.6				
	tiger	41	4	9.8		83	0.8				
	galapagensis	26	1	3.8			0.5				
	scalloped hammerhead	11	0	0.0							
	blacktip	1	0	0.0							
	smooth hammerhead	1	0	0.0							
Thorson & Lacy (1982), Thorson (1971)	*Carcharhinus leucas*	3859	623	16.1			7.9	1	1	Nicaragua, Costa Rica	Petersen disc, Jumbo Rototag
Tricas (1977)	Summary information	123	4	3.2	2.8	~160	0.9	2	2	Santa Catalina Island, California	Floy FH-9 stainless steel dart tag with plastic spaghetti streamer, telemetry tag
	Prionace glauca	120	3	2.5	2.8	119	0.9				
	Isurus oxyrinchus	3	1	33.3		~160	3 days				
Ugoretz[6], Ugoretz (personal communication)	Summary information	10 105	172	1.7		6147	6.3	> 9	> 5	California – Monterey Bay to Cabo San Lucas, Baja	Floy FH-69 steel dart tag, yellow and red
	Prionace glauca	6958	49	0.7		6147	3.8				
	Isurus oxyrinchus	2674	108	4.0		3728	2.6				
	Alopias spp.	143	2	1.4		19	0.2				
	Squatina californica	107	6	5.6		61	6.3				
	sevengill	65	6	9.2		352	2.1				
	basking shark	57	0	0.0							
	leopard	51	1	2.0		6	2.1				
	Carcharodon carcharias	16	0	0.0							
	soupfin	5	0	0.0							
	Squalus acanthias	4	0	0.0							
	Mustelus spp.	3	0	0.0							
	other	22	0	0.0							

[4]Sutcliffe, R. 1994. Twenty years of tagging common skate and tope off the west coast of Scotland. pp. 14–16. In: S.L. Fowler & R.C. Earll (ed.) Proceedings of the Second European Shark and Ray Workshop, 15–16 February 1994. Tag and Release Schemes and Shark and Ray Management Plans. Unpublished report.
[5]Tester, A.L. 1969. Cooperative shark research and control program final report 1967–69. University of Hawaii, Honolulu. 47 pp.
[6]Ugoretz, J. 1999. Shark tagging news, a newsletter of the California Department of Fish and Game shark tagging program (January 1999). 4 pp.

Appendix 1. (*continued*).

Source	Species	Tagged number	Recaptured number	%	Max. speed (km day^{-1})	Max. distance (km)	Max. time (yr)	Number of species tag	recap.	Location	Tag type
Vas (personal communication)	Summary information	337	0	0.0				3	0	Southern coast of England	Floy dart tag
	Galeorhinus galeus	298	0	0.0							
	Prionace glauca	36	0	0.0							
	Lamna nasus	3	0	0.0							
Walker et al.[7], Coutin (1992)	Summary information	20 185	2753	13.6				> 21	12	Australia	Rototag, Jumbo Rototag, nylon-headed dart tag, Petersen fin tag, internal Nesbit tag, T-bar tag
	Galeorhinus galeus	9638	1011	10.5			41.8				
	Mustelus antarcticus	8647	1659	19.2							
	Squalus megalops	617	9	1.5							
	Pristiophorus nudipinnis	499	21	4.2							
	Pristiophorus cirratus	376	30	8.0							
	Notorynchus cepedianus	116	8	6.9							
	Furgaleus macki	79	0	0.0							
	Carcharhinus obscurus	75	7	9.3							
	Carcharhinus brachyurus	46	2	4.3							
	Heterodontus portusjacksoni	25	3	12.0							
	Cephaloscyllium sp.	16	1	6.3							
	Squalus mitsukuri	10	1	10.0							
	Sphyrna zygaena	7	0	0.0							
	Squalus acanthias	7	0	0.0							
	Centrophorus uyato	6	0	0.0							
	Alopias vulpinus	5	0	0.0							
	Squalus sp.	4	0	0.0							
	Squatina australis	4	0	0.0							
	Carcharodon carcharias	3	1	33.3							
	Carcharias taurus	2	0	0.0							
	Prionace glauca	2	0	0.0							
	Isurus oxyrinchus	1	0	0.0							
Westrheim in Herald & Ripley (1951)	*Galeorhinus zyopterus*	309				167	0.2		1	Oregon Waters	
G.P. Whitley in Olsen (1953)	*Galeorhinus zyopterus*	22							1	Tasmanian waters	internal tag
Williams & Schaap (1992)	Summary information	302	2	0.7		39	~0.4	4	1	Tasmania	cattle ear tag, T-bar tag with streamer

	Squalus acanthias	185	0	0.0							
	Mustelus antarcticus	51	0	0.0							
	Squalus megalops	47	2	4.3		39	~0.4				
	Galeorhinus galeus	19	0	0.0							
Wintner (personal communication), Cliff et al. (1996)	Summary information	2799	159	5.7	39.4	1897	11.0	> 19	> 11	South Africa, KwaZulu-Natal	steel head dart tag, round cattle ear tag (ORI tag)
	Carcharias taurus	1633	121	7.4	39.4	1897	11.0				
	Carcharhinus obscurus	246	6	2.4	10.6	360	4.8				
	Galeocerdo cuvier	224	9	4.0	1.2	143	2.2				
	Carcharhinus brevipinna	143	2	1.4	9.0	45	5 days				
	Carcharhinus brachyurus	132	2	1.5	53.0	53	1.2				
	Carcharhinus leucas	113	4	3.5	14.0	123	0.6				
	Carcharhinus limbatus	80	3	3.8	0.9	6	2.8				
	Carcharodon carcharias	73	6	8.2	28.7	1409	2.6				
	Carcharhinus plumbeus	41	0	0.0							
	Sphyrna lewini	32	2	6.3	1.7	267	0.4				
	Carcharhinus amboinensis	31	2	6.5	0.3	84	4.4				
	Sphyrna zygaena	14	0	0.0							
	Isurus oxyrinchus	11	2	18.2	1.5	6	2.2				
	Notorynchus cepedianus	8	0	0.0							
	Squatina africana	8	0	0.0							
	Mustelus sp.	4	1	25.0							
	Alopias vulpinus	2	0	0.0							
	Triakis megalopterus	2	0	0.0							
	Carcharhinus albimarginatus	1	0	0.0							
	Stegostoma fasciatum	1	0	0.0							

[7]Walker, T.I., L.P. Brown & N.F. Bridge. 1997. Southern shark tagging project. Final report to Fisheries Research and Development Corporation (FRDC Project 93/066). Client report. (November 1997). Marine and Freshwater Resources Institute, Queenscliff, Victoria, Australia. 61 pp.

A female of the great white shark, *Carcharodon carcharias*, 3 m long (PV).

Environmental Biology of Fishes **60:** 225–250, 2001.

Review of elasmobranch behavioral studies using ultrasonic telemetry with special reference to the lemon shark, *Negaprion brevirostris*, around Bimini Islands, Bahamas

L. Fredrik Sundström[a], Samuel H. Gruber[b], Susi M. Clermont[c], João P.S. Correia[d], Jean R.C. de Marignac[e], John F. Morrissey[f], Courtney R. Lowrance[g], Lori Thomassen[h] & Miguel T. Oliveira[d]
[a]*Department of Zoology, Göteborg University, Animal Ecology, Box 463, SE-405 30 Göteborg, Sweden (e-mail: fredrik.sundstrom@zool.gu.se)*
[b]*Bimini Biological Field Station and University of Miami, Rosenstiel School of Marine and Atmospheric Science, Miami, FL 33149-1098, U.S.A.*
[c]*Zoological Institute, University of Copenhagen, Tagensvej 16, 2200 Copenhagen, Denmark*
[d]*Oceanário de Lisboa, Esplanada D. Carlos I, 1998 Lisboa, Portugal*
[e]*Moss Landing Marine Laboratories, 8272 Moss Landing Road, CA 95039, U.S.A.*
[f]*Department of Biology, 114 Hofstra University, Hempstead, NY 11549, U.S.A.*
[g]*Department of Biological Sciences, Florida Institute of Technology, 150 West University Blvd., Melbourne, FL 32901-6975, U.S.A.*
[h]*SeaWorld of Florida, Orlando, Florida, U.S.A.*

Received 7 February 2000 Accepted 3 August 2000

Key words: movement pattern, swimming speed, transmitter, homing, rate-of-movement

Synopsis

A review of past behavioral ultrasonic telemetry studies of sharks and rays is presented together with previously unpublished material on the behavior of the lemon shark, *Negaprion brevirostris*, around the Bimini Islands, Bahamas. The review, focusing on movement behaviors of 20 shark and three ray species, reveals that elasmobranchs exhibit a variety of temporal and spatial patterns in terms of rates-of-movement and vertical as well as horizontal migrations. The lack of an apparent pattern in a few species is probably attributable to the scarcity of tracking data. Movements are probably governed by several factors, some still not studied, but data show that food, water temperature, bottom type, and magnetic gradient play major roles in a shark's decision of where and when to swim. A few species exhibit differences in behavior between groups of sharks within the same geographical area. This interesting finding warrants further research to evaluate the causes of these apparent differences and whether these groups constitute different subpopulations of the same species. The lack of telemetry data on batoids and some orders of sharks must be addressed before we can gain a more comprehensive understanding of the behavior of elasmobranch fishes. Previously unpublished data from 47 smaller and 38 larger juvenile lemon sharks, collected over the decade 1988–1998, provide new results on movement patterns, habitat selection, activity rhythms, swimming speed, rate-of-movement, and homing behavior. From these results we conclude that the lemon shark is an active predator with a strong, apparently innate homing mechanism. This species shows ontogenetic differences in habitat selection and behavior, as well as differences in movements between groups of individuals within the same area. We suggest three hypotheses for future research on related topics that will help to understand the enigmatic behavior of sharks.

Introduction

Studying sharks and rays in their natural environment is complicated because of their size, free-ranging behavior, and the fact that they live in a relatively inaccessible and concealing environment. Direct observation through various techniques (e.g. Gruber 1982, Carey & Clark 1995, Anderson & Goldman 1996, Gunn et al. 1999) can give important information about elasmobranch behavior in the temporal and spatial realm while tagging (Casey & Kohler 1990) can provide information on a longer time and geographical scale. Still, there is a definite need to reveal the daily activities of animals that range out of sight of an underwater observer, and to obtain information on behavior beyond release and recapture sites. Ultrasonic telemetry appears to be well suited for such studies as animals can be followed from boats or by other arrangements (Voegeli et al. 2001) for up to several kilometers distance. Although, telemetry studies often involve baiting to attract an animal as well as handling during capture and transmitter attachment, subjects apparently return to normal behavior within a few hours (e.g. Sciarotta & Nelson 1977, Holland et al. 1993, Nelson et al. 1997, Lowe et al. 1998) up to a day (e.g. McKibben & Nelson 1986, Gruber et al. 1988, Sundström et al. 1998).

In this paper we review behavioral data from telemetry studies on elasmobranch fishes and summarize a decade of research on the behavior of the smaller and large juvenile lemon shark, *Negaprion brevirostris*, in the subtropical marine environment of the Bimini Islands, Bahamas. The new data on lemon sharks add to and complement previous publications by Gruber et al. (1988), Morrissey & Gruber (1993a,b), and Sundström & Gruber (1998).

Past studies

We present behavioral data from 34 published field studies comprising 20 shark and three ray species (Appendix 1). However, at least 33 shark and seven ray species have been studied using various telemetry techniques (L.F. Sundström unpublished data). Studies included in this review employed equipment from the simplest type of ultrasonic transmitter, the so called 'pinger', that emits pulses of a given ultrasonic frequency and repetition rate, to transmitters with sensors for monitoring swimming-speed, tail-beat frequency, water depth, and water temperature. Most pingers were augmented with a coded output that allowed for recognition of individual animals. Further technical information on transmitters and signal monitoring is given by Nelson[1] and Voegeli et al. (2001).

Telemetering the behavior of elasmobranchs

Behavior of animals is governed by a multitude of internal and external stimuli (e.g. Becker et al. 1992, Alcock 1993, Bradbury & Vehrencamp 1998). The movements of sharks are influenced by factors such as general condition of the individual (Bass & Rascovich 1965), ambient water temperature (Morrissey & Gruber 1993a), water currents (Medved & Marshall 1983), light levels (Nelson et al. 1997), time of year (Sciarotta & Nelson 1977), geographic location (Holland et al. 1999), water mass, bottom type, geomagnetic gradient (Klimley 1993), presence or absence of prey (Sims 1999), and oxygen levels (Parsons & Carlson 1998). This complexity makes it difficult to explain why sharks behave in a certain way under certain conditions. Added to this, telemetry studies are relatively difficult to conduct and so only a few species have been successfully followed for a full diel cycle or longer.

We expect that future studies will reveal even more causal factors as determinants of elasmobranch behavior. So, our present knowledge can be considered a mere beginning. Thus, the statement by Myrberg (1976) more than two decades ago that the behavior of sharks is 'a continuing enigma' holds true even today.

Rate-of-movement and instantaneous swimming speed

Movement rates of sharks are often estimated from the distance traveled by an animal over a certain time period, where the shark's position is obtained from a telemetry device – usually a simple pinger. This resultant speed is referred to as rate-of-movement (ROM), or point-to-point swimming speed. Instantaneous swimming speed requires more complicated and expensive transmitters to be attached externally. For the remainder of the paper we will use the acronym ROM when

---

[1] Nelson, D.R. 1990. Telemetry studies of sharks: a review, with applications in resource management. pp. 239–256. *In*: H.L. Pratt Jr., S.H. Gruber & T. Taniuchi (ed.) Elasmobranchs as Living Resources: Advances in the Biology, Ecology, Systematics, and the Status of the Fisheries, Proc. 2nd US–Japan Workshop, NOAA Tech. Rep. NMFS 90.

referring to the rate-of-movement and swimming speed when referring to the instantaneous velocity.

In fact, ROM is sometimes mistakenly assumed to be a close approximation of swimming speed. However, ROM is the speed over ground while swimming speed is the speed through water. Yet their relation can yield useful information on a shark's behavior (see Discussion). As the time between position fixes approaches zero, the difference between ROM and swimming speed approaches zero. In reality, however, the two will differ when: (1) the shark is not swimming in a perfectly straight line between two fixes, (2) position of the shark cannot be precisely fixed, (3) vertical movements are made, and (4) currents are present or cannot be accounted for.

Values of ROM indicate that activity patterns of sharks vary greatly between species (Table 1), which is also the case for teleost fishes (Helfman 1993). Blue sharks, *Prionace glauca* (Sciarotta & Nelson 1977), grey reef sharks, *Carcharhinus amblyrhynchos* (McKibben & Nelson 1986), and leopard sharks, *Triakis semifasciata* (Ackerman et al. 2000), are reported to be nocturnal. One shortfin mako shark, *Isurus oxyrhinchus*, tracked by Holts & Bedford (1993) appeared to be crepuscular, while a second was nocturnal and the third showed no particular period of increased ROM. These results were from single tracks of 18–25 h duration and may not be representative of the general behavior of this species.

That activity patterns cannot be determined from a single track of only 24 h is indicated by a 48-h track on one tiger shark, *Galeocerdo cuvier* (Tricas et al. 1981). This individual swam at 1.04 m s^{-1} during the first 24-h period and slowed to 0.94 m s^{-1} during the second 24-h period which was almost equal to the difference between overall ROM during the day (1.05 m s^{-1}) and night (0.92 m s^{-1}). Similar average ROMs were found by Holland et al. (1999) for six other tiger sharks tracked in deep, offshore Hawaiian waters where they displayed extended periods of straight-line swimming. ROM for these tiger sharks was not calculated when they were in shallow water because they exhibited a meandering swimming pattern, which was often too fine for the tracking vessel to follow. However, the sharks were not necessarily swimming slower in shallower waters as Tricas et al.'s (1981) results suggested. This indicates a weakness in the ROM measure because, although it may increase at times, this change might be attributable to straighter swimming rather than increased movement rate. Holland et al. (1993) thought this could have been the case for scalloped hammerheads, *Sphyrna lewini*. They noted that the hammerheads showed a slight, although significantly, lower ROM during the day when they were milling around in a core area compared to the night ROM when they moved out, swimming more directly for extended periods.

For a number of species, ROM was found not to change over the diel period. The filter-feeding whale shark, *Rhincodon typus*, swam at an average diel ROM of 0.7 m s^{-1} with no difference between day and night rates (Gunn et al. 1999). Another filter-feeder, the megamouth shark, *Megachasma pelagios*, showed a higher ROM immediately after release but this was probably due to the stress of capture (Nelson et al. 1997). The needle dogfish, *Centrophorus acus*, swam at depths of more than 220 m and was probably little influenced by light conditions. Accordingly, its ROM did not vary with any particular pattern (Yano & Tanaka 1986). Sandbar sharks, *Carcharhinus plumbeus*, showed no diel differences in ROM. Rather, they appeared to use currents for passive drift which caused an apparent increase in ROM compared to swimming against the current or at slack tide (Medved & Marshall 1983). However, all of the above species were tracked for limited periods, using a limited number of individuals, and thus, some of these species may yet be shown to be rhythmic.

Despite its weaknesses, ROM can yield repeatable values for the same species. Several independent studies reported the same ROM for the same species, e.g. scalloped hammerheads (Holland et al. 1993, Lowe et al. 1998), blue sharks (Sciarotta & Nelson 1977, Carey & Scharold 1990), and white sharks, *Carcharodon carcharias* (Carey et al. 1982, Strong et al. 1992, but see Goldman & Anderson 1999). Disparities in ROM within the same species may be explained by size differences between individuals studied.

The only species where both juveniles and adults were extensively tracked is the lemon shark. Gruber et al. (1988) reported crepuscular peaks in the activity of adult sharks, but Morrissey & Gruber (1993a) found that juveniles (<1 m) swam faster at night. This is further discussed under Lemon shark studies below.

Instantaneous swimming speed has been monitored with speed-sensing transmitters or inferred from a tail-beat transmitter (see Lowe & Goldman 2001). Yet, few attempts have been made to monitor swimming speed in any elasmobranch (Appendix 1). During such

Table 1. Rates-of-movement (m s^{-1}) for elasmobranch species monitored in the field using ultrasonic telemetry.

	Time interval (min)	Size (m)	Rate-of-movement					Reference
			Sunrise	Day	Sunset	Night	Overall[1]	
Carcarhodon carcharias	15	3.0–5.0	–	–	–	–	0.89	Strong et al. (1992)
	60?	4.6	–	–	–	–	0.89	Carey et al. (1982)
	1	3.7–4.9	–	–	–	–	0.64	Goldman & Anderson (1999)
Carcharhinus amblyrhynchos	15, 30, 60	1.3–2	–	0.47	–	0.92	0.69	McKibben & Nelson (1986)
Carcharhinus obscurus	15	0.65–0.96	–	–	–	–	0.22	Huish & Benedict (1978)
Carcharhinus plumbeus	15	0.71–0.99	–	–	–	–	0.34[2]	Medved & Marshall (1983)
Centrophorus acus	30, 60[3]	1.2, 1.36	lower	0.25	lower	0.22		Yano & Tanaka (1986)
Cetorhinus maximus	120–240	7.0	–	–	–	–	0.45[4]	Priede (1984)
Galeocerdo cuvier	60	2–4.2					1.07[5]	Holland et al. (1999)
	15	4	–	1.05	–	0.92	1.0	Tricas et al. (1981)
Isurus oxyrinchus	–	1.36, 2.4	–	–	–	–	0.67	Carey et al. (1981)
	15	1.7–1.8	–	–	–	–	0.31[6]	Holts & Bedford (1993)
Megachasma pelagios	15	4.9	0.35	0.32[7]	0.35	0.31	0.32	Nelson et al. (1997)
Myliobatis californica	–	–	–	–	–	–	0.15	Matern et al. (2000)
Negaprion brevirostris	15	1.5–2.6	0.56	0.50[7]	0.70	0.55	0.53	Gruber et al. (1988)
	5	1.5–1.8	–	–	–	–	0.33	This study
Prionace glauca	30	1.8–2.6	–	0.33	–	0.50	0.42	Sciarotta & Nelson (1977)
	–	1.98–3.0	–	–	–	–	0.42[7,8]	Carey & Scharold (1990)
Rhincodon typus	10	5–6	–	0.70	–	0.77	0.74	Gunn et al. (1999)
Sphyrna lewini	15	0.64	–	–	–	–	0.18	Lowe et al. (1998)
	15	0.38–0.43	–	0.17	–	0.19	0.18[9]	Holland et al. (1993)
Triakis semifasciata	–	0.91–1.19	–	0.11	–	0.18	0.14	Ackerman et al. (2000)

[1]When overall ROM was not given it was calculated as the average from the four diel periods.
[2]ROM given both for sharks with ultrasonic transmitters and floats attached.
[3]30 min at daytime, 60 min at nighttime.
[4]Speeds were calculated for one shark over approximately 4 h and four sharks over approximately 2-h intervals from 6:16–18:44 h.
[5]Only from sharks swimming in deep, offshore water.
[6]Based on one of three tracked sharks.
[7]No significant difference between day and night ROM.
[8]The ROM for a shark in currents was 1.02 m s^{-1}.
[9]Speeds different among periods, one shark moving in a straight line swam at 0.34 m s^{-1}.

attempts, typically, only a few individuals were tracked and most tracks were of short duration rarely completing a full diel cycle. Generally, speeds were found to vary greatly over time. For example, swimming speeds for two bonnethead sharks, *Sphyrna tiburo*, tracked in an artificial pool were higher during the night than during the day but peaked immediately after sunset and sunrise (Parsons & Carlson 1998). Swimming speed corresponded well with oxygen levels, which suggested that the sharks compensated for low oxygen concentration by increasing swimming speed. A higher speed during the night was also found for blue sharks

(Sciarotta & Nelson 1977). Swimming speed peaked a few hours after sunset and stayed comparatively high until a few hours before sunrise. The same pattern was seen for Pacific angel sharks, *Squatina californica*, which were inactive for most of the day, spending much of their time lying on the bottom (Standora & Nelson 1977). Although, swimming speeds indicate that several shark species are more active during the night, the paucity of data make any general conclusion premature. Burst of speed events have been reported for a few shark species (Table 2, see also the data on swimming speed and burst speed under the lemon shark section below).

Three studies monitored both ROM and swimming speed and the latter was found to be generally about 50% higher (Sciarotta & Nelson 1977, Gruber et al. 1988), although in one study, scalloped hammerheads swam 2.3 times faster than their ROM (Lowe et al. 1998).

Vertical movement patterns

Sharks are spatially and temporally distributed throughout the water column and species can roughly be grouped into several categories based on their vertical movements. For example, deep-sea sharks swim at depths of several hundred meters and typically follow close to or parallel to bottom contours. Occasionally, they venture into shallower water or swim closer to the surface. Sixgill sharks, *Hexanchus griseus* tracked by Carey & Clark (1995) spent most of their time at depths of 600–1000 m with the deepest recording being 1500 m. Movements were generally parallel to bottom contours and no particular vertical movement pattern was apparent during the day. Yet, depth could vary by about a hundred meters within a 20-min period. Another deep-sea shark, the needle dogfish, was tracked at depths between 220–680 m and normally swam within 10–50 m of the sea bed (Yano & Tanaka 1986).

Preference for staying either close to the surface or the bottom was found for two large, wide-ranging, predatory species. Tiger sharks did not show a diel difference in average swimming depth, but followed the bottom contour as long as the water depth did not exceed 300 m (Holland et al. 1999). Once in open water they ascended to the surface or near the surface. However, Tricas et al. (1981) reported that tiger sharks may make deep excursions, especially at night. Similarly, white sharks seemed to prefer to stay close to either the surface or the bottom unless the water was deeper than 30 m (Strong et al. 1992, Goldman & Anderson 1999). However, Carey et al. (1982) found that white sharks may dive to depths of at least 50 m. The Pacific angel shark also stayed close to the bottom but rarely rose far above the seabed (Standora & Nelson 1977). The angel sharks appeared to seek shallower waters at dusk and return to deeper waters at dawn.

Several species were observed to make repeated dives from the surface down to 100+ meters. Such dives lasted from a few minutes to several hours. Blue sharks in the Atlantic off New England made dives more regularly during the day with greater depths compared to night dives (Carey & Gibson 1987, Carey & Scharold 1990). In contrast, blue sharks in the Pacific off California swam close to the surface for most of the time, and tended to dive more frequently and deeper at night (Sciarotta & Nelson 1977). Similarly, a shortfin mako shark in the Atlantic undertook deep dives to about 500 m, but rarely swam above 100 m (Carey et al. 1981, given in Carey & Scharold 1990). However,

Table 2. Burst speeds or highest reported speeds for various shark species. ROM = rate-of-movement.

	Speed (m s^{-1})	Method	Reference
Carcharhinus amblyrhynchos	1.47	ROM	McKibben & Nelson (1986)
Galeocerdo cuvier	3.05	ROM	Tricas et al. (1981)
Isurus oxyrhynchus	10	not given	Carey & Teal in Bone (1988)
	13.9	speed of boat	see Wood (1982)
Negaprion brevirostris	2.39	rod and reel	Gero (1952)
	8	speed of boat	L.F. Sundström unpublished data
	5.57	speed transmitter	This study
Prionace glauca	2.0	speed transmitter	Carey & Scharold (1990)
	1.4	speed transmitter	Sciarotta & Nelson (1977)
	19.2	–	see Wood (1982)
Squatina californica	>0.75	speed transmitter	Standora & Nelson (1977)

smaller shortfin mako sharks off California, spent most of their time above 20 m and made infrequent dives (Holts & Bedford 1993). Both whale sharks and scalloped hammerheads made frequent dives throughout the diel cycle. Whale sharks spent about half of their time within about 100 m of the surface (Gunn et al. 1999) while the hammerheads were found throughout the entire water column diving as deep as 500 m (Klimley 1993, Klimley et al. 1993).

The most conspicuous regularity in vertical movements found was for a megamouth shark tracked by Nelson et al. (1997). Depth changes occurred as single, smooth events closely associated with the times of sunrise and sunset (Figure 1). Nocturnal hours were spent at shallow depths of 20 m. The megamouth shark descended to 150 m during day following the 0.4-lux isolume. A night shark, *Carcharhinus signatus*, made similar diurnal dives over a 2-day period in the Straits of Florida off Miami (S.H. Gruber unpublished data).

Several functions for vertical migrations have been postulated. Klimley (1993) suggested that both vertical and horizontal movements of scalloped hammerheads were guided by geomagnetic gradients. He further suggested that changes in local magnetic intensity could produce the oscillating diving pattern he observed. The use of magnetic gradients had previously been demonstrated for round stingray, *Urolophus halleri*, and inferred for leopard shark, lemon shark, southern stingray, *Dasyatis americana*, and spotted eagle ray, *Aetobatus narinari* (Kalmijn 1978, 1988). Wiltschko & Wiltschko (1995) believed that as more elasmobranchs are tested, we will come to understand that the use of magnetic cues will be a general feature of chondrichthyan orientation. Carey & Scharold (1990) discussed the influence of the magnetic gradient on the diving behavior of blue sharks, but found no apparent correlation between behavior and magnetic contours. Instead they suggested that blue sharks dive as an energy saving and hunting behavior.

Gunn et al. (1999) proposed that whale sharks dive as part of their foraging behavior. Feeding could also explain the diving pattern of the filter-feeding megamouth, which may be following the vertical migration of plankton. However, Nelson et al. (1997) believed that

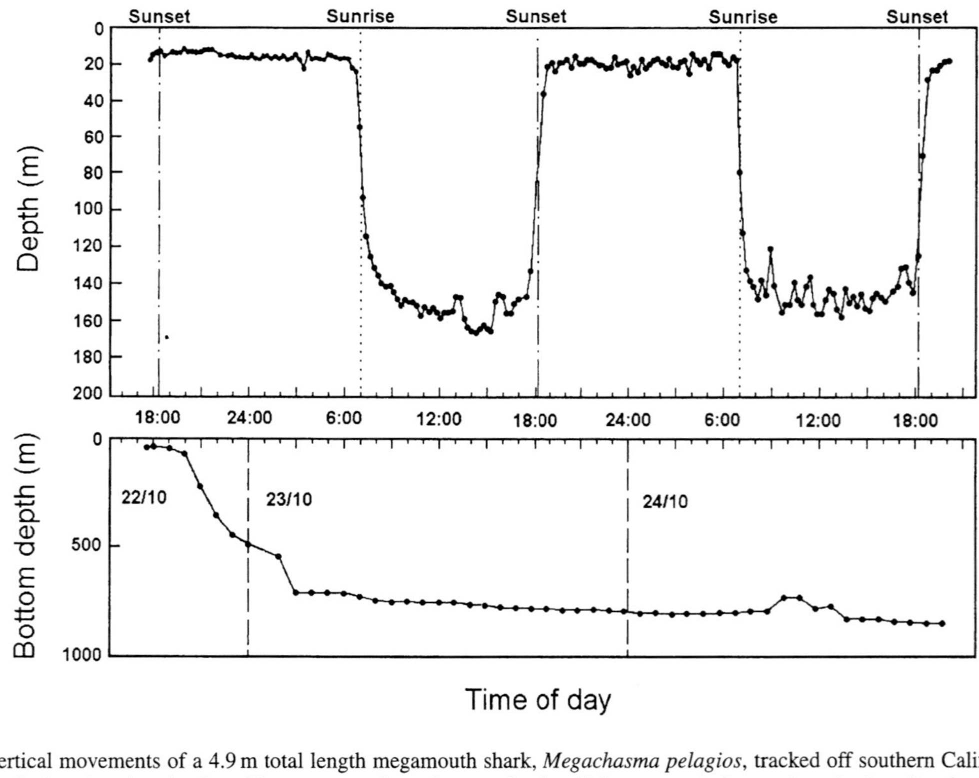

Figure 1. Vertical movements of a 4.9 m total length megamouth shark, *Megachasma pelagios*, tracked off southern California. Upper curve shows shark swimming depth and lower curve shows bottom depth which was much deeper than shark swimming depth except for the start of the track. Diving depth followed the 0.4-lux isolume. From Nelson et al. (1997) with permission from Kluwer Academic Publishers.

vertical movements permitted the megamouth to maintain itself in a constant light regime. A similar explanation was given for vertical movements of the swordfish, *Xiphias gladius* (Carey & Robinson 1981). Temperature, salinity, and oxygen levels can be expected to influence vertical movements mainly by restricting the depth to which a shark will dive. The effect of water temperature on shark diving has been investigated and is discussed below. However, similar studies on salinity and oxygen levels are lacking.

Horizontal movement patterns

A number of different horizontal movement patterns have been found for elasmobranchs. Some species prefered a 'core area' which can be defined as a part of the home range (see 'Activity space' for definition of home range) in which most activities are concentrated (Goodenough et al. 1993). During the day such species were found milling within this area. After sunset sharks typically ventured out on nightly excursions. This pattern was most pronounced in reef whitetip sharks, *Triaenodon obesus*, that spent much of the day resting in caves (Nelson & Johnson[2]). At dusk these sharks ranged far beyond the daytime core area to forage, only to return at dawn. Some reef whitetip sharks used the same cave for several nights in a row or even up to a year. Similarly, several juvenile scalloped hammerhead sharks used the same core area at daytime, followed by nightly excursions at which time larger areas were used (Holland et al. 1993). Klimley & Nelson (1984) found that adult scalloped hammerheads in Baja California did not use a pronounced core area but rather swam back and forth, singly or in groups, along a seamount ridge throughout the day. In the late afternoon or at night they undertook long excursions up to 20 km into the pelagic environment. Juvenile lemon sharks largely swam within a confined area, executing repeated, attenuated loops while remaining in a constant location, but this pattern was not correlated with any particular time of day (Morrissey & Gruber 1993b). From the confined area they made occasional brief excursions called 'jaunts' into little used areas far from their home range. At Johnston Atoll, Central Pacific Ocean, Economakis & Lobel (1998) observed grey reef sharks aggregating during the day but never at night. Although they did not follow sharks during the night when they left the atoll, Economakis & Lobel (1998) still felt that the site of aggregation was a daytime core area.

A related movement pattern was found for blue sharks in the eastern Pacific (Sciarotta & Nelson 1977). Rather than having a daytime core area, these sharks migrated between day and night areas (Figure 2, left). At twilight the blue sharks moved in a rather straight line, from deeper to shallower waters near Catalina Island, then swam along the shoreline all night until just before sunset when they returned to the pelagic zone. Interestingly, this behavior was observed between March and June, but between late June to October the sharks remained offshore (Figure 2, right). Sciarotta & Nelson (1977) suggested that this was a change in response to prey abundance. Some large juvenile lemon sharks also migrate between day and night activity spaces (see lemon shark studies).

[2] Nelson, D.R. & R.H. Johnson. 1980. Behavior of the reef sharks of Rangiroa, French Polynesia. Natl. Geogr. Soc. Res. Rep. 12: 479–499.

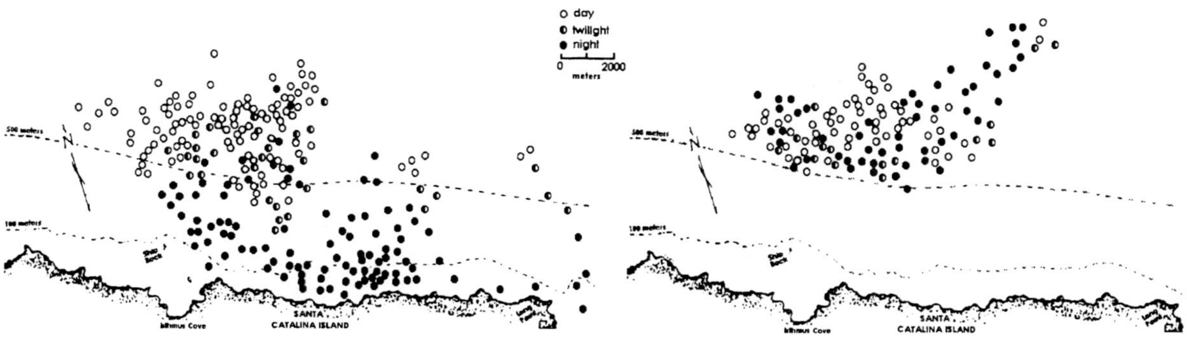

Figure 2. Distribution plots of blue sharks, *Prionace glauca*, at Catalina Island, CA, from late March through early June (left, N = 7) and late June to early October 1972 (right, N = 3). Open circles are day positions and closed circles are night positions and half filled circles indicate twilight. Blue sharks displayed a clear seasonal difference in the distribution of positions, possibly in response to prey behavior. Courtesy of Sciarotta & Nelson (1977) and U.S. Fishery Bulletin.

In the lemon shark section we present data indicating that different groups of sharks within the same lagoon can exhibit different movement patterns. No other sharks are known to partition their activity space in this way. However, two groups of grey reef sharks at Enewetak atoll, Marshall Islands showed different movement patterns depending on where they swam (McKibben & Nelson 1986). Sharks residing in the lagoon typically were site-attached, swimming in areas close to where they were the previous day, while sharks residing on the ocean side were more nomadic, moving up to 2.6 km between day positions (Figure 3).

Three specific movement patterns were described for white sharks around Dangerous Reef, Australia (Strong et al. 1992, 1996). The first, observed at the beginning of each track was called 'downstream circling'. It was characterized by criss-cross swimming, possibly in response to a bait odor corridor created by currents. 'Island patrolling' consisted of white sharks freely ranging around the reef but remaining within 8.4 km of the island. This behavior was also possibly affected by baiting. A third pattern termed 'inter-island cruising', involved more linear paths with occasional excursions directed toward islands along the shark's route. Inter-island cruising involved longer movements away from the tagging area and was probably not affected by baiting.

Goldman & Anderson (1999) and Goldman et al. (1996) reported movement patterns similar to the island patrolling in four white sharks at the Farallon Islands, California. Unlike the sharks around Dangerous Reef, the Farallon white sharks remained near the release site. Such differences could be due to variations in island density as well as prey type and availability. The two larger sharks tracked by Goldman & Anderson (1999) confined their movements to one side of the

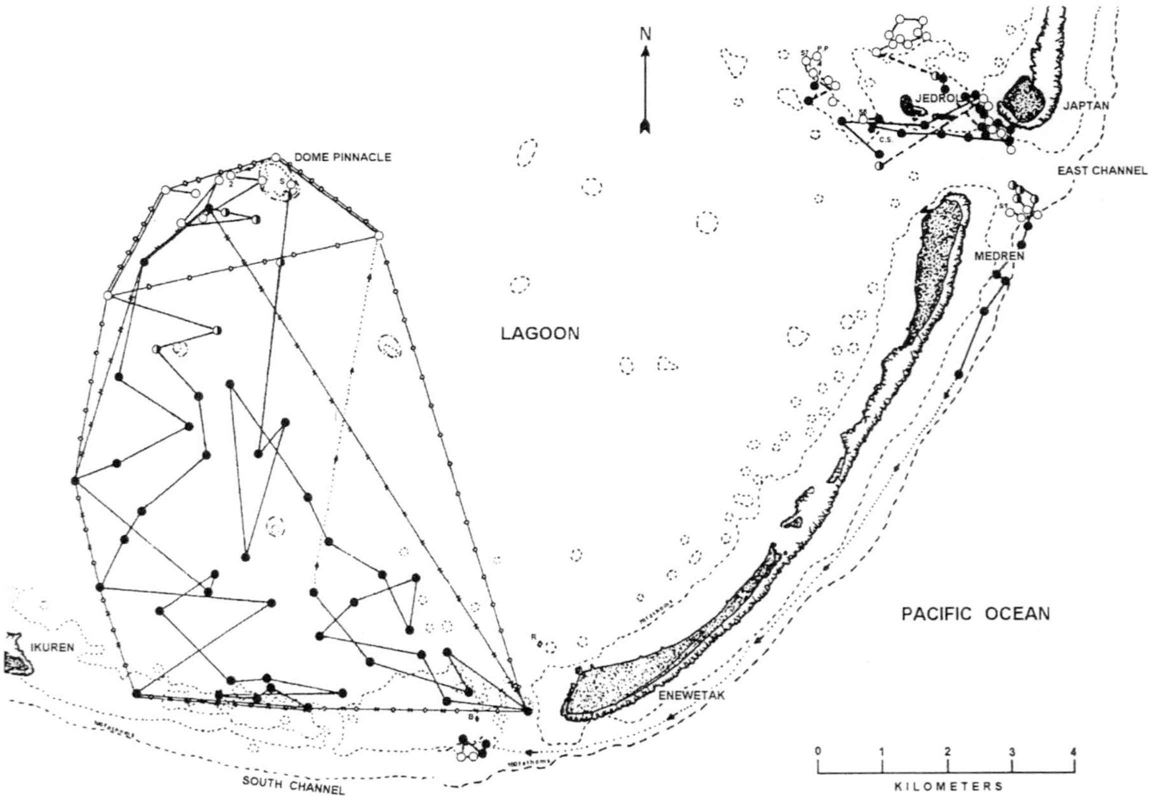

Figure 3. Movement patterns of four gray reef sharks, *Carcharhinus amblyrhynchos*, at Enewetak, Marshall Islands. The encircled track was on a 168 cm total length 'home-ranging' lagoon shark. Along the fringing reef are tracks of three 'nomadic' sharks (139, 152, and 183 cm total length). See the text for explanation of the behavioral differences between the two groups. For the lagoon track activity space is indicated by lines with letters; for daytime areas – D, nighttime areas – N, and overall area – O. It was not meaningful to calculate activity space for the 'nomadic' sharks (modified from McKibben & Nelson 1986 with permission from Bulletin of Marine Science).

islands while the two smaller sharks could be found on all sides. White sharks were also studied by Tricas & McCosker (1984) and Carey et al. (1982) but the movements of these sharks were less complicated. One of the white sharks tracked by Tricas & McCosker (1984) first moved westward, parallel to the north shore of Dangerous Reef, Australia, then once past the island, it moved offshore in a northwesterly direction at which time it undertook a behavior possibly related to island patrolling or inter-island cruising. The white shark studied by Carey et al. (1982) stayed 1.5 days close to a decaying whale carcass before moving inshore to within 11 km of the beach. It then turned southwesterly for the remainder of the 83-h track which was characterized by steady movement down the coast.

Holland et al. (1999) described three general activity patterns for tiger sharks. Those tracked between Oahu and the Penguin Banks, Hawaii, showed extensive offshore movements (when leaving the capture site near Oahu) possibly as a result of capture and handling. When crossing the deep Kaiwi Channel they undertook periods of highly directed swimming and finally made movements toward and subsequent orientation to the Penguin Banks.

Pacific angel sharks provide an example of a species swimming in a confined area with no apparent pattern. These sharks spent most of the day lying on the bottom with only occasional straight-line movements (Standora & Nelson 1977). At night, activity increased as they ranged up to 9 km, still within a relatively confined area. Similarly, two large sixgill sharks stayed in the vicinity of the release site for a few days, venturing off at night between 4.5 and 6 km (Carey & Clark 1995). Whether these two sharks would stay in the same area much beyond the four days they were followed is not certain.

The duration of many tracking studies is too short to adequately characterize movement patterns. This is especially true for pelagic sharks, which may roam over very large areas and need to be tracked for weeks or even longer to reveal any regular movement pattern. Although the blue sharks off California showed a repeated day and night migration (Sciarotta & Nelson 1977) the same species tracked off New England swam with a remarkably constant heading (Carey & Scharold 1990). Three shortfin makos followed between 18–25 h only swam with fairly constant heading throughout the brief track (Holts & Bedford 1993), as did a megamouth shark tracked for 50 h (Nelson et al. 1997). A needle dogfish tracked for 21 h began by circling the capture site for 7 h, then moved in a rather straight line, made a loop and continued straight swimming until the track was ended (Yano & Tanaka 1986).

The importance of long tracks is illustrated by the movement of a basking shark, *Cetorhinus maximus*, tracked by satellite. This individual stayed within the Firth of Clyde in Scotland, for the 17 days while it was tracked (Priede 1984). During the five days directly after tagging this shark moved 55 km south, probably swimming in an almost straight line, then continued straight for another three days before turning north.

Horizontal movements are governed by many factors. For example, the aggregation of sharks in core areas could represent refuging behavior as has been suggested for adult scalloped hammerhead sharks (Klimley & Nelson 1984). On the other hand, movements away from core areas often involve feeding. Whale sharks tracked at Ningaloo Reef, Australia, swam back and forth in a north–south direction following the contours of the reef edge while performing repeated dives probably related to foraging (Gunn et al. 1999). Occasionally, sharks would swim in wide circles when being close to passes in the reef. These events occurred at times when water was flowing out from the reef lagoon, and Gunn et al. (1999) suggested that this water might be transporting prey outside the reef.

Following coastlines or underwater formations is apparently an orientation mechanism. A particular seamount at El Esperito Santo appeared to be an important schooling site for scalloped hammerheads. However, horizontal excursions away from the seamount were highly directional and thought to be a response to changes in water mass and temperature rather than currents or visual cues (Klimley 1993).

Currents such as tidal flow also influence shark movements. Young sandbar sharks usually drifted passively with the current except when feeding (Medved & Marshall 1983). In contrast to those in a current, the behavior of two individuals, apparently feeding in schools of menhaden, *Brevoortia tyrannus*, was characterized by non-directional swimming. Thus, feeding interrupted passive drifting which suggests that drifting may be used as an energy saving, searching behavior. Leopard sharks also used currents to move to and away from muddy littoral zones where extensive food was present (Ackerman et al. 2000).

Response to changing temperature

Several telemetry studies have focused on thermal physiology of sharks (reviewed by Lowe & Goldman 2001), involving telemetered measures of stomach temperature, muscle temperature and ambient water temperature. In addition to physiological data, such studies may yield information on temperature preferanda. Obviously the normal habitat of an animal determines the temperature it swims in. Simply put, deep-sea sharks and polar sharks encounter cold water while tropical shark live in warmer water. A needle dogfish spent most of its time near the bottom in waters 4–13°C (Yano & Tanaka 1986), while whale sharks spent time in waters of 25–28°C (Gunn et al. 1999). These whale sharks did not appear to be orienting to the thermocline but spent most of their time within the mixed layer. Similarly, in the eastern Pacific, shortfin makos spent 90% of their time in the mixed water layer only infrequently plunging below the thermocline (Holts & Bedford 1993).

Several species made vertical migrations through the thermocline and appeared not to be concerned about changes in water temperature. A blue shark preferred temperatures between 14–16°C and made dives from the surface to depths of 250 m experiencing a temperature change of 7–9°C as it passed through the thermocline (Carey & Gibson 1987). None of several blue sharks tracked by Carey & Scharold (1990) ever ventured below 7°C, but they found no evidence that sharks would not tolerate lower temperatures. Instead they believed that depths of dives were adjusted depending on internal temperature and that the return to the warmer surface layer was to regain lost heat.

Another technique to gain information on thermal preference is to plot the position of the shark in conjunction with manually monitored temperatures. This was done by Morrissey & Gruber (1993a) who found that juvenile lemon sharks preferred warmer areas of the nursery in the North Sound of the Bimini Lagoon. Similarly, grey reef sharks moved to and from an aggregation area by following changes in water temperature and light levels (Economakis & Lobel 1998). Klimley et al. (1988) thought that changing temperature and water masses played the most important role in bringing scalloped hammerheads back from nocturnal excursions to the seamount where they remained for large parts of the day. Except for these three studies, temperature was not reported to affect horizontal movement in sharks.

Activity space

The area in which an animal carries out 'normal activities' is called its home range (Goodenough et al. 1993). To be meaningful, a certain minimal amount of tracking time is required to properly estimate the true home range. As for the core area (see 'Horizontal movement patterns' for definition), the use of home range is probably too broad a term for most shark telemetry studies where the studied time period rarely exceed a full diel cycle. Instead we use the term activity space, which denotes the area where the sharks were tracked. In addition, for many sharks, home range will differ in time. For example, the mean daily shift in the center of the activity space for juvenile lemon sharks, in a confined area in the Bimini lagoon, was 327 m day^{-1}, but only 56 m differed between the centers of the initial and final (9–153 days later) trackings (Morrissey & Gruber 1993b). This return to the initial area allowed for a home range to be calculated for these juveniles. However, there was a positive correlation between home range size and body size, and as the lemon sharks in Bimini lagoon grow they begin utilizing larger portions of the lagoon and eventually move into areas around Bimini Islands (L.F. Sundström personal communication). With further growth the large juvenile lemon sharks venture out of the lagoon (Gruber et al. 1988) and are thought to leave the islands at maturity for deeper, offshore waters (Gruber 1982). Based on tag returns, the home range of an individual adult lemon shark must encompass tens of thousands of square km (S.H. Gruber personal communication).

From tag and recapture studies, several other species have been shown to migrate many thousands of kilometers. Tag returns for blue sharks suggest that they use the entire north Atlantic gyre to make ocean crossings (Casey & Kohler 1990). The home range for such sharks could include major areas of the world's oceans. However, during certain periods this same species may use limited areas. Between March and June seven blue sharks tracked by Sciarotta & Nelson (1977) stayed within an area of about 200 km^2 (estimated from Figure 2, left) while during June to October three blue sharks were tracked within an area less than half of that (estimated from Figure 2, right). However, the number of sharks tracked differed, and blue sharks in different oceans may behave differently. Thus such comparisons may be inappropriate.

Differences between day and night activity levels also yield lower daytime activity space than nighttime activity space (e.g. McKibben & Nelson 1986, Holland et al. 1993). Relative to their size, Pacific angel sharks had smaller activity spaces than other species. This may be explained by their sedentary life style as they spend large parts of the day lying still on the bottom (Standora & Nelson 1977).

Contrary to all other shark studies the relationship between size and activity space for white sharks off the Farallon Islands, California, was reversed, that is, larger sharks utilized smaller spaces. Goldman & Anderson (1999) believed this to be a result of the particular hunting behavior at this site. Juvenile white sharks mainly prey on fishes, switching to pinnipeds as they mature (Tricas & McCosker 1984). Young, inexperienced white sharks may need to learn hunting locations and capture methods to feed successfully on elephant seals, *Mirounga angustirostris*, and therefore roamed the Farallon Islands in search of suitable foraging areas. Larger, more experienced white sharks presumably had knowledge about suitable feeding areas and were thus able to limit their search patterns, resulting in a smaller activity space. The tracking data were not, however, lengthy or conducted over complete diel periods so the actual home range was not calculated.

Ray telemetry studies

Very few telemetry studies have been published on batoids. Because of this no general conclusion on their behavior is attempted here. Instead, we summarize the major findings of these studies. Silliman & Gruber (1999) investigated movement patterns and social behavior of a population of spotted eagle rays for 8 months at Bimini Islands, Bahamas. Preliminary results indicated that rays exhibited tidally-mediated diel behavior and tended to refuge during the low-rising phase of the tidal cycle in three specific core areas around the islands. Past mid-rising tide, activity increased as the rays moved on to feeding grounds. On a seasonal basis, rays maintained core areas during the fall and winter but began to disperse in the spring. By late June, most of the approximately 200 individuals had moved offshore but began returning to Bimini by late July. By November 1999, most of the rays had returned. Silliman & Gruber (1999) found that social groups were not segregated by gender or size. They also reported on schooling patterns and produced a catalogue of individual action patterns.

Blaylock (1990) studied the effects of attaching an external transmitter on the behavior of the cownose ray, *Rhinoptera bonasus*. Working in the laboratory, Blaylock (1990) fitted rays with a hybrid transmitter emitting both radio frequency (RF) and ultrasonic signals. The transmitter was tethered to the ray with a harness of braided nylon line inside a nylon tube and passed through its spiracles. Rays less than 2 kg increased their wing beat frequency by 47–93%, yet they could not maintain the same swimming speed as a control group. In contrast, rays weighing 4.2–7.7 kg did not appear to be affected by the transmitter and maintained swimming speed relative to the control group. Blaylock (1990) concluded that if the ratio of transmitter-to-ray mass was < 0.03, the behavior of the ray would be unaffected by towing the package.

Movement rates of California cleftnose eagle rays, *Myliobatis californica*, were not affected by size, tidal stage or time of day, while direction of movement showed a diel pattern (Matern et al. 2000). Between early morning and early afternoon, rays moved toward the warmer, shallower water of the inner Tomales Bay, California. For the remainder of the day they moved back toward the cooler waters of the outer bay. Matern et al. (2000) argued that this movement pattern resulted from the animal's thermoregulatory behavior.

Lemon shark studies

This section presents a series of studies on smaller and large juvenile lemon sharks around the Bimini Islands, undertaken between May 1988 and August 1998. The Biminis form a cluster of subtropical Bahamian islands on the western edge of Great Bahama Bank, approximately 86 km east of Miami, Florida (Figure 4). They are low in elevation, have undeveloped areas covered with mangroves and are arranged in the shape of a triangle that encloses Bimini Lagoon. The lagoon covers an area of about 21 km^2 with a tidal range of approximately 0.75–1.0 m giving an average depth of one meter at mid-tide (Bathurst 1967).

The lemon shark is easily distinguished from most other carcharhinids because of its two large, almost equal sized, dorsal fins (Garrick & Schultz 1963). It used to be common in the western Atlantic, from New Jersey to Brazil but lately their numbers have been depleted, especially around Florida (S.H. Gruber personal communication). The lemon shark is also encountered rarely in the tropical eastern Pacific, from southern Baja California to Ecuador, and sporadically along the West African coast (Compagno 1984). The species is found in shallow waters down to depths of 100 m and is often seen around bays, piers, mangrove islands, seagrass flats, and river mouths.

At birth, lemon sharks measure 60 to 65 cm in total length (TL) and attain a maximum of 340 cm TL.

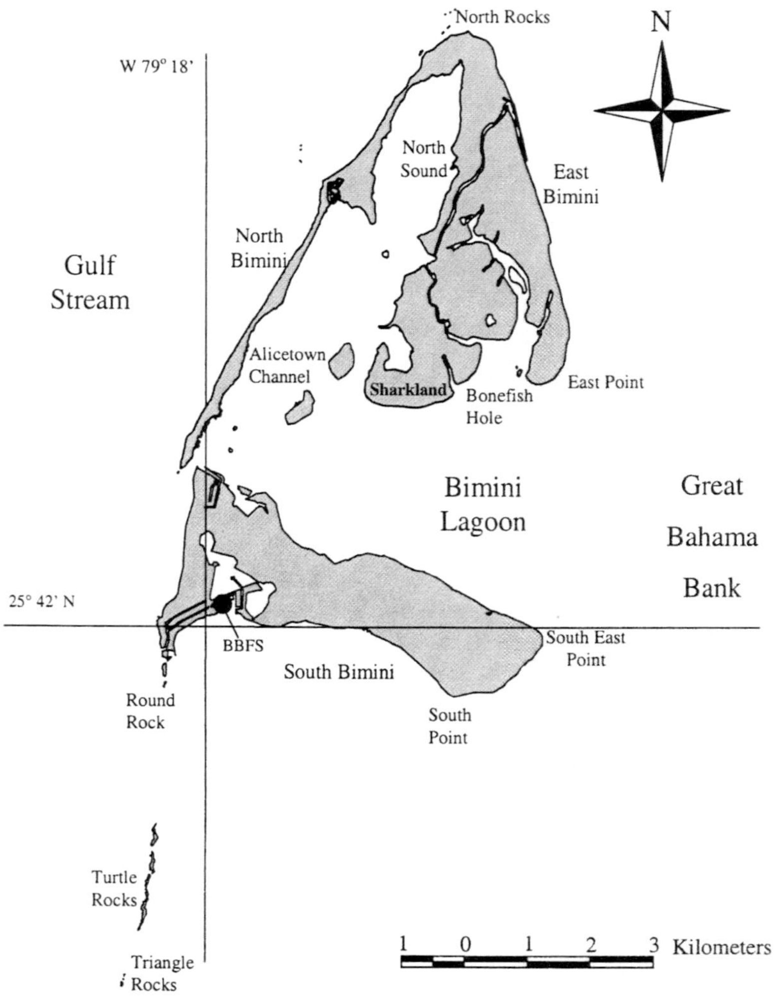

Figure 4. Map of the Bimini Island complex. Bimini is located in the Bahamas, about 86 km east of Miami, Florida and is the primary study area where smaller and large juvenile lemon sharks, *Negaprion brevirostris*, were tracked between February 1989 and August 1998. BBFS shows the location of the Bimini Biological Field Station where scientists were housed.

Maturity of lemon sharks around Bimini is reached at about 235 cm TL equivalent to about 11 to 13 years of age (Brown & Gruber 1988). Lemon sharks may be opportunistic feeders but selectivity is also observed (Cortés & Gruber 1990). The diet of young lemon sharks consists mainly of teleosts and a few percent of crustaceans and mollusks. As they mature the diet changes and becomes even more dominated by teleost and cartilaginous fishes. Gruber (1982) discussed the role of the lemon shark at Bimini and the Florida Keys and established this shark's status as a top predator.

Materials and methods

In the early studies, juvenile lemon sharks (33 females and 24 males) of 60–100 cm TL were captured with 90-kilogram-test, monofilament long lines. Later, sharks were caught in gillnets, set perpendicular to the shoreline in the North Sound, Sharkland, and Bonefish Hole (Figure 4). At times they were pursued by an airboat or outboard skiff and captured with a dipnet. Large juvenile sharks (18 females and 20 males) of 144–198 cm TL, were caught on 1800 m long lines often set at dusk, either on the east side of the Alicetown

Table 3. Lemon sharks, *Negaprion brevirostris*, used in telemetry studies around Bimini Islands, Bahamas, from May 1988 to August 1998. Shark ID: J = smaller juvenile, S = large juvenile, F = female, M = male, and number denotes sequence in female/male capture. TL = total length. Capture location (see Figure 4): ATC = Alice Town Channel, EBL = East Bimini Lagoon, NS = North Sound, SB = South Bimini, SL = Sharkland. Transmitter attachment: I = internal, E = external, St = speed transmitter.

Shark ID	Average TL (cm)	Capture location	Capture date	# fixes	Attachment
JF-1–JF-10	71	NS	16 May 1988–11 Dec 1989	1302	I
JF-11–JF-17	78	SB	16 Oct 1997–12 Feb 1998	459	E
JF-18–JF-21, JF-25–JF-28, JF-30	84	EBL	3 Mar 1998–31 Jul 1998	946	E
JF-22–JF-24, JF-29	81	SB	8 May 1998–25 Jul 1998	447	E
JM-1–JM-8	74	NS	11 Jul 1988–19 Aug 1989	979	I
JM-9–JM-13	74	SB	26 Oct 1997–28 Jan 1998	177	E
JM-14–JM-15, JM-18–JM-21	78	EBL	3 Mar 1998–7 Aug 1998	692	E
JM-16–JM-17	67	SB	13 May 1998–22 Jun 1998	350	E
SF-1–SF-7	168	ATC	3 Nov 1992–25 Jan 1994	1907	E
SF-8–SF-11	160	ATC	16 Jun 1994–3 Aug 1994	2641	I
SF-12–SF-13	163	EBL	29 Aug 1994–22 Jan 1995	1602	I
SF-14–SF-18	164	EBL	16 Dec 1995–23 Jul 1997	3830	I/E (St)
SM-1–SM-8	164	ATC	3 Nov 1992–24 Jan 1994	1309	E
SM-9–SM-12	179	ATC	17 Jun 1994–25 Aug 1994	1558	I
SM-13–SM-15	173	EBL	28 Aug 1994–17 Apr 1995	2830	I
SM-16–SM-20	177	EBL	12 Nov 1995–21 Aug 1997	3055	I/E (St)

Channel or east of the Bimini Lagoon. Detailed data on the sharks used in the tracking studies are given in Table 3.

Between 1988–1991, juvenile sharks were equipped with simple pingers surgically implanted into the celom (see Morrissey & Gruber 1993b). During the years 1997–1998 transmitters were attached to one side of the first dorsal fin by running a monofilament line through the fin and securing on the other side with a neoprene-covered, plastic backing-plate. Transmitters implanted in the shark's celom were left in place while the externally attached packages were secured with a corrosable steel link which would slowly corrode causing the transmitter to dislodge in about one week.

For large juveniles, pingers[3] with a three month battery life were attached to the shark by inserting two steel darts through the skin into the epaxial musculature between the dorsal fins. Pingers operating on 18-month batteries were surgically inserted into the celom as explained above. Speed-sensing transmitters[4] were either secured to the base of the first dorsal fin using a plastic cradle bolted to the fin or fixed to the back between the dorsal fins with four steel darts (see Sundström & Gruber 1998). In this way the expensive transmitter could be recovered by recapturing the shark and unbolting the package. Alternately, the transmitter could be relocated at the sea floor once it was dislodged from the shark.

Tracking

Using shallow-draft skiffs, tracking by a crew of two or three people began immediately after a shark was released and continued throughout day and night as long as the trackers did not loose the signal. Transmitter pulses monitored through stereo-headphones

[3] Pingers (model XTAL 87, Sonotronics, Tucson, AZ, USA). Output frequency = 68.1–79.0 kHz, life span = 3 or 18 months, type of signal = variable pulses between 748 and 1053 ms or pulses combined in different orders as an identity code. Transmitters were 10 × 100 mm plastic cylinders and weighed 40 g in air.

[4] Speed sensing transmitters (model V32S-4H, Vemco Engineers). Output frequency = 69 kHz, life span = 21 days, sensor type = paddle wheel with magnets attached to the blades. A reed switch sensor inside the transmitter detected the magnetic field generated by a passing magnet. Number of pulses per unit time indicated the speed at which the transmitter was traveling (calibration required before first use). Transmitters were 33 × 200 mm and weighed 300 g in air.

were detected using a staff mounted, directional hydrophone[5] connected to a manual, ultrasonic receiver[6]. The signal range of the transmitters was between 200–300 m under normal conditions. A shark's position during the 1988–1989 tracking study was determined by optical triangulation combined with preinstalled mapping landmarks placed around the North Sound (see Morrissey & Gruber 1993a,b). During subsequent studies, hand-held Loran or Global Positioning System (GPS) receivers were used. When possible, position fixes were taken at either 5- or 15-min intervals.

During the 1996–1997 speed trials, the number of signals from a speed-sensing transmitter was counted for 60 s, every fifth minute and signal rate converted to swimming speed using a calibration curve obtained from the individual transmitters. During a burst of speed, the total number of signals was counted and the duration timed with a stopwatch. To be considered as a burst event the duration could not exceed 60 s. Signals from the speed transmitters were consistent so that small changes in signal rate were easily detected. This permitted an accurate evaluation of burst swimming events.

Speed transmitters used during the speed trials differed in their ability to measure low speed due to friction at their axels. The lowest speed measurable by individual transmitters was between $0.36\,\mathrm{m\,s^{-1}}$ and $0.50\,\mathrm{m\,s^{-1}}$. Transmitters had no theoretical upper limit. The subjective accuracy of pulse counting varied between $\pm 0.015\,\mathrm{m\,s^{-1}}$ and $\pm 0.023\,\mathrm{m\,s^{-1}}$, equivalent to a change of one transmitter pulse during any measured minute. To evaluate whether the transmitter's inability to detect speeds below the minimum threshold affected the results, two estimation procedures were done. First, time spent with transmitter at default value was exchanged to speed at $0\,\mathrm{m\,s^{-1}}$ under the assumption that the shark was lying on the bottom. Next, calculations were done by linearly decreasing the speed under the assumption that the shark was swimming below the transmitter's default rate.

Rate-of-movement and swimming speed

ROM was calculated using 5-min intervals between fixes for large juvenile sharks and 15-min intervals for smaller juveniles. The quotient (Q) between median instantaneous swimming speed and the corresponding ROM was calculated for large juvenile sharks. Previous analysis of the swimming-speed data revealed that the large juvenile lemon sharks swam at elevated speeds during the first 24-h period after release (Sundström et al. 1998). Data for this period were therefore eliminated before analysis, which in one case, resulted in only 2 h of data remaining for shark SM-16. This shark was excluded from analysis as were sharks SM-17 and SM-18 because inexperienced trackers stayed too close to the animals causing elevated speeds. Diel differences in swimming speed were evaluated for the periods sunrise (5:00–7:59 h), day (8:00–16:59 h), sunset (17:00–19:59 h), and night (20:00–4:59 h).

Geographic positions

Diel movements of large juvenile sharks caught in the Alicetown channel (Alicetown sharks) were analyzed by comparing the mean of longitudinal (east–west) positions during the same diel periods as for ROM and swimming speed above. Data were separated into individual tracks, ranging from one to 24 h for each shark. Schoener's Ratio (SR), an iterative procedure, which consists of eliminating positions (i.e. increasing the time interval between locations) until they become independent, was calculated for each shark and tested for time-to-independence (see Swihart & Slade 1985). This allowed parametric statistics to be applied.

Differences in the distribution between day and night positions for sharks caught at night, east of the Bimini Lagoon (East Bimini sharks) and those caught at night in the Alicetown channel (Alicetown sharks) were compared by arbitrarily dividing the lagoon into two portions at the western border of Bonefish Hole (Figure 4). The number of day and night fixes was counted for each shark and the proportion of time spent in each of the two parts of the lagoon calculated. Data were arcsine transformed so that parametric statistics could be used.

Homing

The homing ability of 41 juvenile lemon sharks was tested by translocating them in a closed, 100-liter container filled with aerated sea water. Sharks were transported away from their previously determined activity space to 18 randomly chosen release sites 4–15 km offshore from Bimini Islands (Figure 5). The sharks likely never visited these sites before translocation (Morrissey & Gruber 1993b). Immediately after

[5] Sonotronics model DH-2, sensitivity = −84 dBV, beam width = ±6 degrees at half power points.

[6] Sonotronics models USR-D, USR-4D, and USR-5.

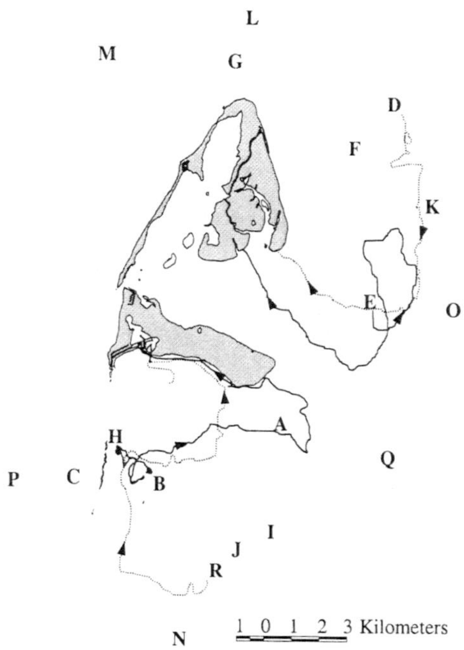

Figure 5. Translocation sites (A–R) around the Bimini Islands, Bahamas and homing paths for four juvenile lemon sharks, *Negaprion brevirostris*, released at sites B, D, E and R.

release, each shark was followed by in-water observers for approximately 30 min, both to confirm that it was swimming properly and to record its initial orientation and behavior. Homing success, i.e. the ability to successfully return to the previously determined activity space, was calculated. Total time required for a shark to reach the shore and it's previous activity space was recorded. Finally, ROM before and after reaching the shore was also calculated.

Results

Rate-of-movements of juvenile sharks

Median ROM based on 2281 positions for sharks JM-1 to JM-8 and JF-1 to JF-10 was $0.15\,m\,s^{-1}$ (range = 0–$1.06\,m\,s^{-1}$). Separation of the diel period into dawn, day, dusk, and night revealed crepuscular peaks in ROM (Kruskal-Wallis test, $p < 0.0001$). To assure that the result was not an artifact of the selected times, the diel period was separated into three additional groupings (four 6-h periods, eight 3-h periods and 24 one-h intervals, Figure 6) and recalculated. Calculations resulted in crepuscular ROM peaks (Kruskal–Wallis tests, $p < 0.001$) for all three groupings.

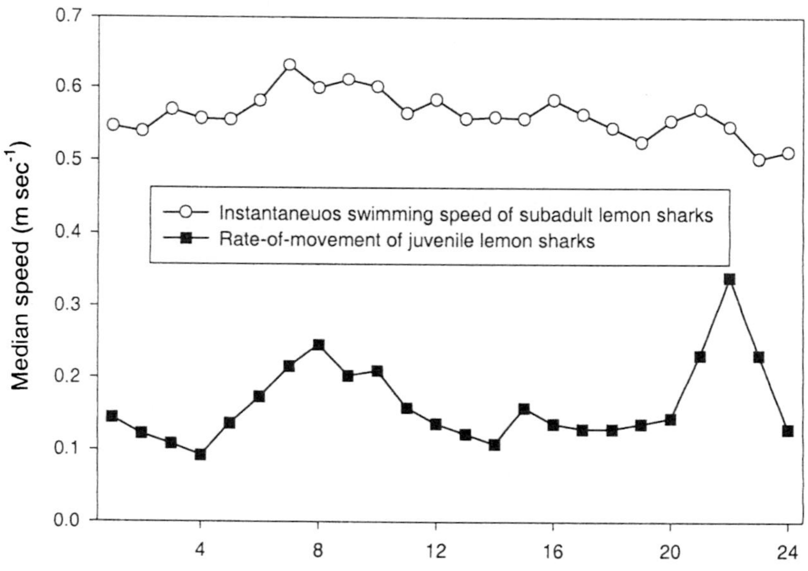

Figure 6. Rate-of-movement of smaller juvenile lemon sharks, *Negaprion brevirostris*, and instantaneous swimming speed measured with a speed-sensing transmitter of large juvenile lemon sharks in relation to time of day. Crepuscular peaks in median rates of movement are highly significant.

Three types of activities were observed in the juveniles. The most common was non-linear movement but two other patterns were observed where the sharks made no forward motion. When sharks were 'parked' the signal direction and amplitude did not change over time suggesting that they were motionless on the bottom. If a tracking boat approached too closely, a parked shark would bolt and dash away. A 'milling' shark was swimming in slow circles within an area of about 25 m in diameter. Milling often occurred away from the mangroves in open areas on the flats, while sharks often parked in the mangroves that fringed the islands around Bimini Lagoon. Parking and milling episodes usually lasted 20–90 min but one shark (J-8) remained parked for several hours. These behaviors caused many periods of 'zero' movement and removing such intervals increased ROM to $0.22\,m\,s^{-1}$ (Student's t-test, $p < 0.0001$).

Both milling and parking were unevenly distributed through the diel period and were significantly less likely to occur during crepuscular periods (ANOVA, $p < 0.01$). At dawn and dusk, sharks exhibited stationary behavior only 20% of the time, compared to both day and night periods when milling and parking comprised about 45% of the shark's activity. Recalculation of ROM after removing all 15-min periods showing 'zero' movements showed that the twilight peaks of ROM was due to faster swimming and not less resting (Kruskal–Wallis test, $p < 0.0001$).

Instantaneous swimming speed

The median swimming speed of $0.55\,m\,s^{-1}$ (time spent with transmitters at default excluded) for seven sharks was not correlated with body length (Spearman $r = 0.57$, $p > 0.05$). Speeds varied greatly within short time periods and the variation was greater within sharks than among. Yet, at times, sharks maintained approximately the same speed for periods exceeding 2 h. Disregarding burst speeds, the highest speed, maintained for one minute or more, ranged from $0.95–1.14\,m\,s^{-1}$.

Of 3671 swimming-speed measurements, 7.1% were below the level detectable by the transmitter. Some default measurements were from sharks lying on the bottom motionless while others were from sharks swimming below the transmitter's critical speed. However, the exact amount of each is unknown. Yet the overall time of default signaling only produced a deviation of $\pm 0.01\,m\,s^{-1}$ from the median value when tested, which is well within the error of overall accuracy of speed measurements. Two sharks (SF-17 and SM-20), never swam slower than the critical limit while the remaining five spent between 6.1–11.5% of their time below the limit.

Except for two sharks (SF-16 and SF-17) differences in swimming speeds were found over the diel cycle (Kruskal–Wallis test, $p < 0.001$). However, the analysis did not reveal any consistent pattern among sharks (Figure 6).

Comparison of ROM with swimming speed in large juvenile sharks

The median ROM, $0.33\,m\,s^{-1}$, of seven large juvenile sharks was based on 3524 position fixes and showed no correlation with body size (Spearman $r = -0.61$, $p > 0.10$). Maximum ROM ranged from $1.44–1.85\,m\,s^{-1}$. Swimming speed and ROM differed for all sharks (Wilcoxon matched pairs, $p < 0.001$, except shark SF-15 where $p < 0.05$). The quotient (Q) between swimming speeds and ROM was 1.67 (SD = 1.2) and the linear correlation between swimming speed and rate-of-movement was positive (Spearman $r = 0.17$, $n = 3524$, $p < 0.0001$).

Burst speeds

A total of 141 burst swimming events was recorded over 469 h of tracking, equivalent to 7.2 burst events per day (weighed average frequency 15.6 events per day). The average burst duration was 7.3 s at a speed of $1.7\,m\,s^{-1}$ (Table 4). Average maximum burst speed was $3.4\,m\,s^{-1}$ and lasted 2 s, while the maximum burst speed ever recorded during a tracking was $5.6\,m\,s^{-1}$ lasting for 2 s. Bursts of speed were negatively correlated with burst duration (Spearman $r = -0.53$, $n = 141$, $p < 0.001$). No clear diel pattern could be seen from the data.

Geographic positions

Using Schoener's ratio, time-to-independence for data of three Alicetown sharks (SF-6, SM-7, and SM-8) was determined to be 3 h. This reduced the number of usable fixes from 1134 to 208, distributed over 34 individual tracks. Plots of the 208 positions showed that these sharks were active on the western side of the lagoon at night and on the eastern side during the day, crossing the lagoon from east to west during dusk and from west to east at dawn (Figure 7). Nocturnal positions were

Table 4. Bursts of speed duration and magnitude (±SD) for seven large juvenile lemon sharks, *Negaprion brevirostris*, fitted with a speed-sensing transmitter.

	Frequency (# events h^{-1})	Mean duration (s)	Mean burst speed ($m\,s^{-1}$)	Max burst speed ($m\,s^{-1}$)
SF-14	1.0 (n = 35)	6.6 ± 5.0	1.24 ± 0.37	2.26
SF-15	1.4 (n = 33)	8.9 ± 6.6	1.22 ± 0.48	3.57
SF-16	0.4 (n = 16)	4.6 ± 3.7	1.83 ± 0.26	2.32
SF-17	0.2 (n = 4)	4.3 ± 2.2	1.65 ± 0.51	2.39
SF-18	0.5 (n = 19)	10.0 ± 14	1.81 ± 0.67	3.81
SM-19	0.7 (n = 26)	10.0 ± 12	1.94 ± 0.97	5.57
SM-20	0.4 (n = 8)	6.9 ± 6.3	2.39 ± 1.06	4.18
All sharks	0.7 ± 0.4	7.3 ± 2.4	1.72 ± 0.41	3.44 ± 1.2

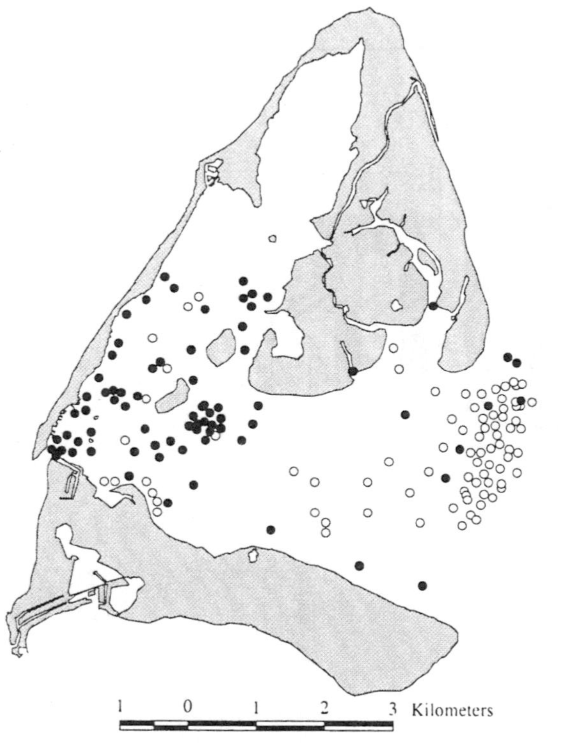

Figure 7. Day (open circles) and night positions (closed circles) of three lemon sharks, *Negaprion brevirostris*, tracked in Bimini Lagoon, Bahamas during the winter of 1994. Note the clear predominance of western positions during night and eastern positions during day (n = 208 independent positions).

significantly westward of diurnal positions for all three sharks (Student's t-test, $p < 0.05$), while sunrise and sunset positions did not differ significantly.

There was a significant difference between the activity spaces of the Alicetown and East Bimini sharks. Based on 1825 position fixes, an apparent westward distribution of night fixes for four Alicetown sharks (SM-10, SM-11, SM-12, and SF-11) and one East Bimini shark (SF-12), was found when compared to three East Bimini sharks (SM-13, SM-14, and SF-13), which did not go as far west as the Alicetown channel but remained near Bonefish Hole and East Point during the night. Night positions were significantly west of day positions in Alicetown sharks and shark SF-12 (Student's t-test, $p < 0.05$), but there was no difference between the day and night positions of the East Bimini sharks (Figure 8).

Homing behavior

Activity spaces of all the juvenile sharks used in the homing study were restricted to areas along the mangroves in depths of less than 1 m and never more than 250 m offshore. Because of their health status, three sharks were eliminated before their first translocation. Ten sharks from South Bimini were released at site B, far more than for any other site. But to make the distribution of the releases at the various sites more equal, data from these sharks were pooled so that only the average data were used when comparing with sharks released at others sites. Therefore, only 22 data points are illustrated in Figure 9, which shows the sharks' orientation patterns in relation to geographic north (N) as well as the direction of their previously determined home range (H). Homing success was 76% (29 sharks). Six sharks resettled in new areas around the islands and contact was lost with the remaining two sharks after they had reached the shore. Only a single shark (JM-20) released in the Gulf Stream, failed to return to the Bimini Islands while 71% (27) returned to the Bimini Islands within less than 12 h.

Immediately after release, 22 of the sharks were observed by divers for 30 min. Sharks typically swam straight to the bottom and moved along the sea floor

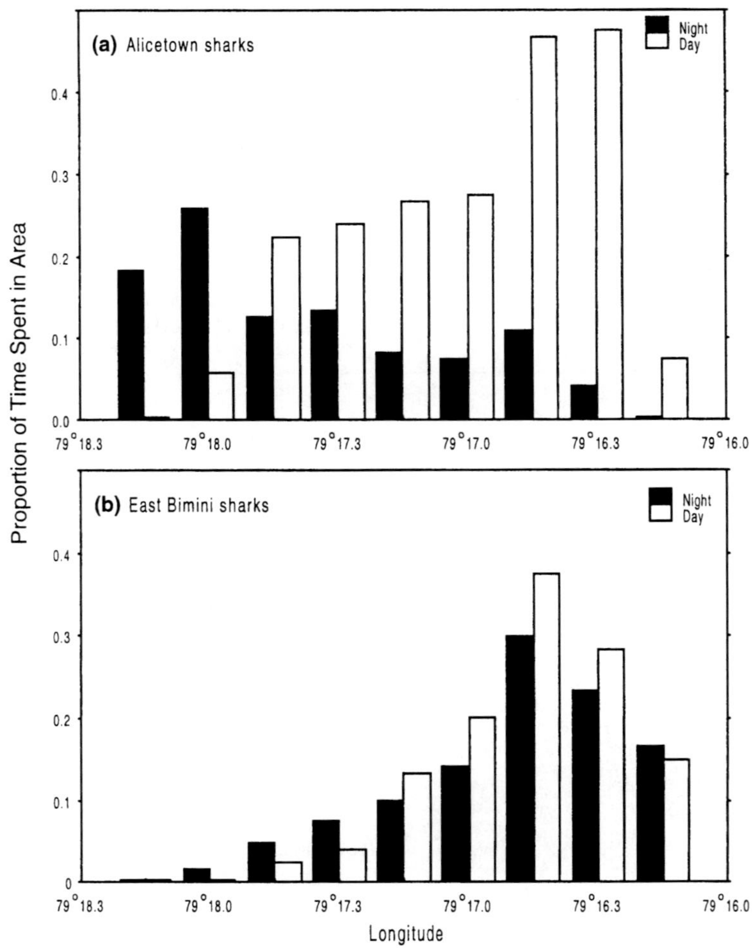

Figure 8. Proportion of time spent by eight large juvenile lemon sharks, *Negaprion brevirostris*, in longitudinal transects of equal area in the Bimini Lagoon, Bahamas. Sharks caught in Alice Town channel (a) moved west at night ($n = 4$, total obs. = 856) while sharks caught in the east of Bimini Lagoon (b) did not exhibit diel movements ($n = 4$, total obs. = 969). The two groups apparently partitioned the lagoon into separate activity spaces by using different movement patterns.

(except for the two sharks released in the Gulf Stream, at sites M and P). The preferred compass bearing at release was toward the east with no direct homeward component (Rayleigh-test, $p < 0.05$, Figure 9a). Subsequent bearings, between release site and the position of the shark 1 h after release (Figure 9b) show a random orientation with respect to both geographic north (N) and the shark's activity space (H). However, the first straight bearing (defined as the first course taken within 30 min that did not deviate more than 5% from a straight line, Figure 9c) showed a significant homeward orientation with an angular deviation of $26.5^\circ \pm 50^\circ$, 95% CI from their activity space. Returning sharks were observed swimming in a zigzag pattern along a particular axis. The duration of each cross over the axis increased with each pass. Sharks maintained this oscillating pattern for two to ten minutes before selecting a heading along a new axis. They typically repeated this 'search' pattern three or four times, before choosing a final bearing and making a concerted move toward shore, with few directional changes.

Most sharks released from sites south of the islands returned to the shore slightly west of the southeasternmost point of South Bimini (Figure 4). In addition to this, the homing paths (Figure 5) demonstrated that sharks were not retracing their outbound path and thus not using route reversal (see Wiltschko & Wiltschko 1995). ROM was significantly higher before reaching the shore (0.37–0.98 m s^{-1}) than after (0.21–0.70 m s^{-1}, Wilcoxon's Matched Pairs test, $p < 0.01$).

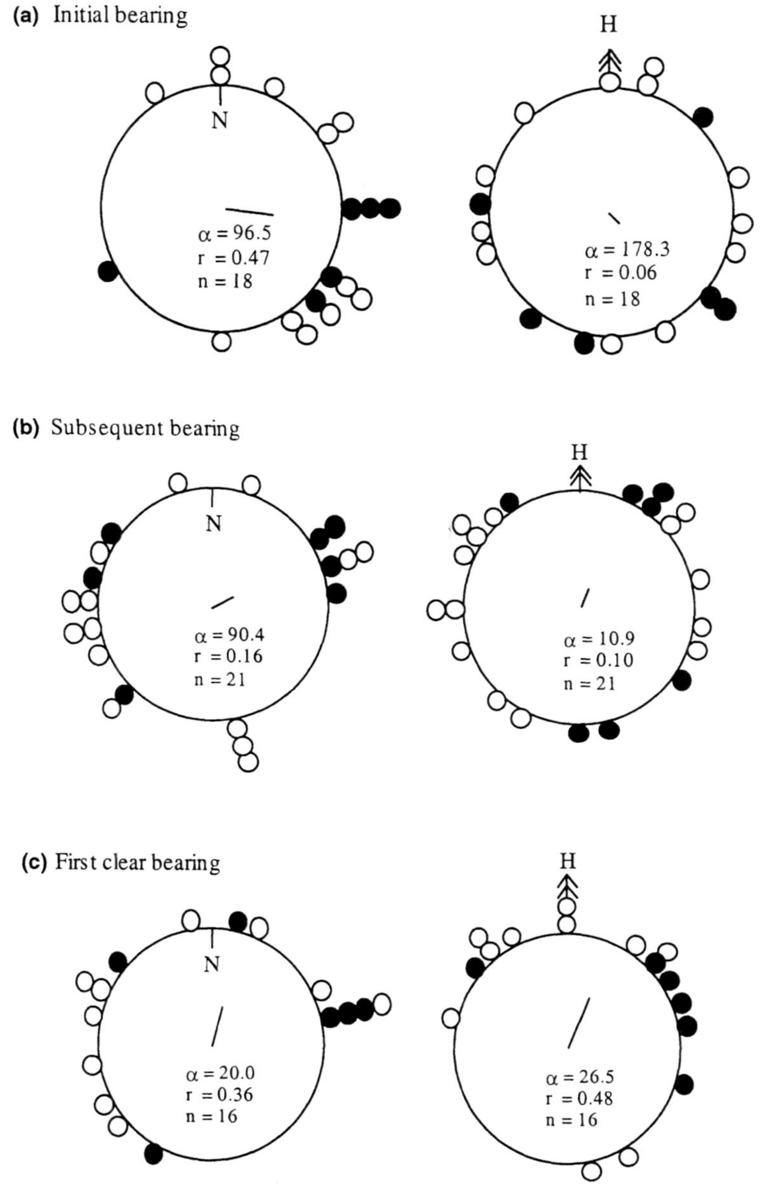

Figure 9. Orientation behavior of juvenile lemon sharks, *Negaprion brevirostris*, in relation to geographic north (N, left diagrams) and activity space (H, right diagrams) with mean vector (central line) after being translocated to unfamiliar locations several kilometers away from their normal activity space around the Bimini Islands, Bahamas. The very first bearing (a) monitored by in-water observation showed a preferred compass direction (PCD) toward east (Rayleigh-test, $p < 0.05$), whereas there was no homeward component. The subsequent orientation within the first hour of release (b) no longer showed a PCD but a very random orientation in relation to both geographic north and home direction. The first straight bearing followed by the sharks (c) showed a significant homeward orientation with an average angular degree of 26° (±50, 95% CI) from the sharks previously determined activity space. Open circles represent sharks caught at East Bimini and closed circles represent sharks caught at South Bimini. See text for more detail on pooled data and definition of first straight bearing.

Discussion

The present study has expanded previous knowledge on the field behavior of lemon sharks (Gruber et al. 1988, Morrissey & Gruber 1993a,b) by providing new information based on detailed studies of diel activity patterns, spatial requirements, swimming behavior, and homing of smaller juvenile as well as large juvenile lemon sharks found in the waters around Bimini Islands.

Movement speed and body size

ROM differed between smaller and large juvenile lemon sharks, although the swimming speed of the large juvenile sharks tracked at Bimini was not directly correlated with body length. Perhaps the size range of the large juveniles was too small and variation in speed too great to produce a statistically significant difference. However, a relation between body size and rate of swimming certainly exists. Laboratory studies on juvenile lemon sharks report average swimming speeds of around 0.2–0.3 m s^{-1} (Nixon & Gruber 1988, Bushnell et al. 1989, Scharold & Gruber 1991, Cortés & Gruber 1994) which is below the 0.5–0.6 m s^{-1} found for the large juvenile lemon sharks in this study. However, preliminary findings by Gruber et al. (1988) showed that a 178 cm and a 210 cm lemon shark swam at 0.74 m s^{-1} and 0.69 m s^{-1}, respectively, and Webb & Keyes (1982) reported that a > 200 cm lemon shark swam at 1.06 m s^{-1} in a large pool. When expressing the swimming speeds in terms of body lengths (bl) per second, all speeds range from 0.2 to 0.5 bl s^{-1}, suggesting that swimming speed for the large juveniles were indeed good estimates of field swimming speed.

In comparison with other species, the lemon shark appears to follow the description of Thomson & Simanek (1977) who placed it in a group together with the blue shark, hammerhead and tiger shark. These animals are characterized by a more heterocercal tail, flattened ventral surface, a relatively less fusiform body and a more blunt head. This grouping also correspond relatively well to actual field data reported for these shark species as well as theoretical swimming speed (Sundström et al. 1998).

Rate-of-movement versus swimming speed

The quotient, Q, between instantaneous swimming speed and ROM can reveal interesting information on a fish's behavior, especially over short periods. A high Q-value ($Q \gg 1$), as found for sharks SF-16, SF-17, SF-18, and SM-19, for an animal in water with insignificant currents, would indicate that the fish is searching around in a small area perhaps looking for hiding prey. However, currents are present in Bimini Lagoon and these may also contribute to a large Q. A migrating shark is more likely to swim in a straight line, and consequently the ratio between swimming speed and ROM would be low in small currents. Such straight swimming was proposed by Sciarotta & Nelson (1977) for blue sharks during crepuscular periods when the sharks migrated between epipelagic diurnal habitats and shallow nocturnal habitats. Some of the lemon sharks also migrated between their day and night habitats, possibly explaining the low Q-values found for SF-14 and SF-15 during crepuscular periods.

Another situation, $Q < 1$, was found for shark SF-15 tracked at sunset. In still waters this result could be explained by an error of the shark's estimated position. A more likely explanation is that SF-15 was swimming, or passively drifting, with currents, possibly as a means of saving energy (Metcalfe et al. 1990). In similar studies dusky sharks, *Carcharhinus obscurus*, were found to swim with currents 61% of the time (Huish & Benedict 1978) and sandbar sharks passively drifted in strong currents (Medved & Marshall 1983).

Evaluating the Q-value is straightforward when a shark is not in currents. However, as explained, the situation is more complicated when currents are involved. It is therefore important to have knowledge about both the magnitude and the direction of a current in which a fish is swimming to adequately describe its behavior. Presently, the tidal phase and current situation in the Bimini Lagoon is under analysis (L. Thomassen personal communication). With a full understanding of the tidal dynamics in the Bimini Lagoon, it will be possible to describe the lemon shark's swimming behavior in greater detail.

Burst swimming speeds

Since territorial behavior has not been observed for lemon sharks, bursts of speed likely indicate either an escape response or an attack on prey. It may be possible to distinguish between successful and unsuccessful feeding or predator avoidance by implanting temperature sensitive transmitters into the stomach (Carey et al. 1981, McCosker 1987, Goldman et al. 1996), however we were not able to make this distinction in the present study. One explanation of the negative correlation between burst speed and burst duration is muscle fatigue. Graham et al. (1990) found the critical swimming speed of a 70 cm lemon shark to be 0.77 m s^{-1}, i.e. it was able to maintain this speed for 30 min. Therefore, it is unlikely that the 150–200 cm sharks used in the present study reached muscle fatigue at speeds below

$1\,m\,s^{-1}$. However, it is difficult to estimate how long a two meter lemon shark could maintain high speeds without reaching fatigue.

The large second dorsal fin and the relatively large caudal fin of lemon sharks indicate that they are strong swimmers capable of short bouts of high acceleration (Webb 1984). Most burst events observed in the present study lasted only a few seconds. Unfortunately, these events were also associated with the highest error. A two-second burst (3 pulse counts) would yield a burst speed of $1.39 \pm 0.44\,m\,s^{-1}$, while a 20-s duration burst (30 pulse counts) would yield a burst speed of $1.39 \pm 0.044\,m\,s^{-1}$. The difference is attributed to an error of counting just one too many or few pulses during a burst event. However, the initial change in signal rate accompanying a burst was easily detected, immediately alerting the tracker. A counting error was more likely to occur at the end of a burst as the signal rate decelerated more slowly.

Diel movement behavior

Gruber et al. (1988) previously reported diel activity for large juvenile lemon sharks but their results were limited to incomplete trackings of a few individuals for a few hours duration only. They suggested that sharks refuge on the east side of the Bimini Lagoon during the day and moved west at night to feed in the Alicetown channel bordering North Bimini.

The repeatable east–west movement pattern found for large juvenile sharks in the present study confirms these early observations. Beginning at sunrise and sunset our sharks regularly traveled a distance of 7–8 km, to and from shallow flats east of Bonefish Hole. Sharks undertook movements toward east Bimini after apparently foraging in the western part of the lagoon, a behavior consistent with the pattern Hamilton & Watts (1970) termed refuging. A similar behavior was found for blue sharks in the Pacific that migrated at twilight, in a rather straight line, to shallower waters near an island, then swam along the coastline all night until just before sunset when they returned to the pelagic zone (Sciarotta & Nelson 1977).

The sharks captured at night east of Bimini Lagoon did not exhibit the same repeatable east–west movement pattern suggesting that there are different groups of lemon sharks in the lagoon and that they might partition the lagoon for more efficient foraging. Such behavioral differences among conspecifics in the same area were reported for grey reef sharks (McKibben & Nelson 1986). One group of grey reefs resided on the inside of a lagoon showing typical home-ranging behavior while another group swam along the outer rim of the lagoon with typically nomadic behavior (Figure 3). There exists no adequate explanation why these sharks differed in behavior.

Activity rhythms of smaller and large juveniles

Activity rhythms of lemon sharks have been studied, with conflicting results. Gruber (1982) first suggested that lemon sharks were crepuscular. Preliminary fieldwork on adult lemon sharks (Gruber et al. 1988) supported this theory but laboratory studies of activity and metabolism showed juveniles to be nocturnal (Nixon & Gruber 1988). Morrissey & Gruber (1993a) also showed that juveniles are more active at night. However, Cortés & Gruber (1990) found no pattern in diel feeding activity. The crepuscular patterns of the large juvenile lemon sharks tracked in this study are supported by nearly 2300 estimates of movement rates measured year round at all times of day. Still, swimming speed measured with speed-sensing transmitters did not reveal any particular diel activity peaks.

The conflicting laboratory results may be explained by constant conditions of light, temperature, salinity, and water flow in the respirometer (Nixon & Gruber 1988) that may have caused the normal, endogenous crepuscular peaks to become out of phase prior to experimentation. Feeding studies (Cortés & Gruber 1990) seemed to indicate that juvenile lemon sharks are equally likely to feed at any time. Perhaps this lack of temporal preference is due to prey availability and not predator activity patterns.

Rogers et al. (1984) suggested that light levels and prey availability could be considered preeminent in regulating the activity patterns of most fishes and that activity peaks of predators should be correlated with periods when environmental and biological factors are optimal for exploitation of a selected prey item. Major (1977) believed that predators have a great advantage during twilight periods when vision is in the mesopic range. Gruber & Cohen (1978) reviewed vision in sharks and reported that the lemon shark's visual system is adapted for seeing both in daylight and in dim light. We suggest that juvenile lemon sharks experience increased predatory success at twilight. Therefore, we predict that, although they will feed opportunistically at all times, young lemon sharks should be most active with respect to feeding during crepuscular or nocturnal periods due to their particular visual adaptations. However, the lemon shark shifts to new prey as it matures and moves offshore (Cortés & Gruber 1990). Maturation is accompanied by a shift in visual pigments' sensitivity to different wavelengths (Cohen 1991) and these factors may explain differences in activity between the smaller juvenile and the large juvenile lemon shark.

Homing behavior of juvenile lemon sharks

Juvenile lemon sharks were clearly able to home after being translocated, and typically returned to the Bimini Islands within less than 12 h. Little time was spent in deepwater sites and the higher ROM before reaching the shore may indicate that the juveniles were under stress and increased their speed to enter shallow water to seek refuge. This may be because numerous large sharks that feed on smaller sharks inhabit the water around the Bimini islands. Juvenile lemon sharks may avoid predation by confining their activities to water depths that are too shallow to be entered by their predators (Morrissey & Gruber 1993a).

Most of the sharks returned precisely to the capture site suggesting high site fidelity. The six sharks that resettled at other sites may have done so to avoid being caught multiple times or may have eventually moved to the capture site after tracks ended. Some of these sharks may also have been caught by us while on long-distance excursions (Morrissey & Gruber 1993b) explaining why they did not return to their capture site.

Most sharks displaced south of South Bimini oriented toward the southernmost point of South Bimini, suggesting that this area played an important role in orientation. However, it is uncertain why the sharks often made landfall at that particular point. Magnetic, auditory, olfactory or other sensory cues including depth gradients could have been involved. However, preliminary homing studies performed at Bimini demonstrated the ability of some juvenile lemon sharks to home with their nostrils blocked and others with their eyes patched (S.H. Gruber personal communication). Some sharks also were displaced into the surf zone, near islands south of Bimini to preclude sound as an initial orienting cue. These preliminary experiments do not rule out the use of these sensory cues in homing but rather demonstrate that the shark does not require them for homeward orientation.

The zigzag swimming pattern observed by divers appears as if the sharks were using a gradient for orientation similar to compass navigation by a marine vessel. In a feasibility study in the Bimini Lagoon, Kalmijn (1988) obtained the first experimental evidence in the field that lemon sharks actually orient to electrical fields during unrestricted swimming. At about the same time, Klimley & Butler (1988) were studying the orientation of scalloped hammerhead sharks in the open ocean. In a series of carefully documented field observations, they showed that at least one hammerhead could maintain a relatively straight course for tens of kilometers while swimming away from and apparently retracing its pathway upon returning to an isolated seamount in the Gulf of California. Klimley & Butler (1988) believed that the hammerheads oriented to the seamount utilizing cues provided by local anomalies in the earth's magnetic field. The zigzag swimming pattern observed for the translocated lemon sharks could also be an indication of the use of the electric or magnetic field for orientation.

Future directions

Over the past decade, we have tracked young-of-the-year to about 9-year old (200 cm TL) lemon sharks at Bimini lagoon and elsewhere. While much of their temporal and spatial activity has been revealed, there is still a great deal to be learned about lemon shark behavior by using both standard tracking methods and newer, more technologically advanced techniques. We present three hypotheses, that can be tested on lemon sharks by using conventional telemetry-tracking or newer techniques, which would further our knowledge on shark behavior in general and lemon shark behavior in particular.

The magnetic orientation hypothesis

At least some shark species use cues from the earth's magnetic field to return to their home range (Kalmijn 1978, 1988, Klimley 1993). Experiments using a multi-position receiver system in combination with a large electrode array could be set up whereby a tagged lemon shark could swim through the array. The experimenter could subtly rotate the earth's magnetic field while the receiver system gives an objective, real-time plot of the shark's second by second path. If the shark's migratory pathway can be manipulated and controlled by the experimenter, this would be strong evidence that lemon sharks attend to magnetic cues.

A second experiment might involve a small electrical field simulator that randomly produces appropriate electrical fields around the shark, continuously varying in time, sign, strength, and direction. The simulator is attached to the shark and the animal is transported to a remote site, released and tracked home. If the simulator 'confuses' the shark, and it looses its way, this would provide evidence that a functional ampullary system is needed for homeward orientation. With proper controls (i.e. sham stimulators) and considering the fact that blocking olfactory, acoustic and visual stimuli do not prevent the sharks from homing, by process of elimination, the electrical sense would perhaps be the most likely candidate as the key sensory system providing homing cues.

If magnetic orientation could be concluded to determine the homing of lemon sharks, another experiment would be to compare the homing behaviors of e.g. lemon sharks born at an island nursery compared to a continental nursery. Possibly insular lemon sharks will have a more precise homing mechanism than continental conspecifics. With essentially infinite nursery habitat, a continental lemon shark might not need as precise a homing mechanism. A series of homing releases in Biscayne Bay, Florida would answer this question. This experiment goes to the question of putative behavioral differences between continental and insular populations of lemon sharks.

The nursery habitat limitation hypothesis

We have established that a high proportion of 150 to 200 cm TL lemon sharks leave the lagoon in the fall and many do not return. Where do these sharks take up residence? Using 'pop up' technology, a radio transmitter can be attached to several lemon sharks in the month of September and the pop up release time set for 45 days. On the day of release, a small aircraft can be used to locate the transmitter. This experiment will address the habitat and space requirements of the species.

The philopatry hypothesis

Previous work has raised the question whether adult female lemon sharks return to their natal ground to mate or give birth. Using implanted archival tags in concert with PSAT technology, transmitters are applied in the spring to as many adult lemon sharks as possible. The following year, returning adults are collected and tags recovered. Since only about 30–50 adults use the lagoon, applying 10 transmitters means that a good proportion of the population will be tagged. The results will also answer the important question of where the adults go for the eight months between June and March when they apparently leave the Bimini area.

Acknowledgements

This study was carried out under Bahamas Research Permit MAF/FIS/17 and received financial support from Earthwatch, National Science Foundation, Bimini Biological Field Station (BBFS), the University of Miami, U.S. Office of Naval Research, and University of Linköping, Sweden. Many people have helped in some material way in the preparation of this paper, and we are sincerely grateful to them all. For their various courtesies and support, special thanks are due to the staff and numerous volunteers from all around the world at the BBFS and the Earthwatch Volunteer Corps for their devotion and continual commitment to this research. Special thanks also go M. Marcotte, K.J. Goldman and C.L. Lowe for assistance with literature, R. Billerbeck for computer assistance, D. Brumbaugh for providing telemetry equipment, and D. Capone for cartography. T.C. Tricas is acknowledged for his encouragement and helpful guidance during the writing process. He and two anonymous reviewers are also thanked for many constructive comments that greatly improved the manuscript. For all the support thanks to our parents, wives, families, and friends; without them we would not have made it! We gratefully acknowledge the following corporations for believing in and sponsoring our shark research: Adelino Pereira Santiago Lda. (Portugal), Davie Marine of Florida (USA), Fundação Berardo (Portugal), Kawasaki (USA), Mariner (Brunswick Corporation - USA), Marpro Lda. (Portugal), Mundo Submerso (Portugal), Pelican Products (USA) and Trimble Navigation (USA). Finally, thanks to C. Higgs and the Bahamas Ministry of Fisheries for cooperation and permission to conduct biological research in their waters and for the enthusiastic support of the overall shark research program.

References cited

Ackerman, J.T., M.C. Kondratieff, S.A. Matern & J.J. Cech. 2000. Tidal influence on spatial dynamics of leopard sharks, *Triakis semifasciata*, in Tomales Bay, California. Env. Biol. Fish. 58: 33–43.

Alcock, J. 1993. Animal behavior. Sinauer Associates, Sunderland. 625 pp.

Anderson, S.D. & K.J. Goldman. 1996. Photographic evidence of white shark movements in California waters. Calif. Fish. Game 82: 182–186.

Bass, G.A. & M. Rascovich. 1965. A device for the sonic tracking of large fishes. Zoologica 50: 75–83.

Bathurst, R.G.C. 1967. Oölitic films on low energy carbonate sand grains, Bimini Lagoon, Bahamas. Mar. Geol. 5: 89–109.

Becker, J.B, S.M. Breedlove & D. Crews. 1992. Behavioral endocrinology. The MIT Press, Cambridge. 574 pp.

Blaylock, R.A. 1990. Effects of external biotelemetry transmitters on behavior of the cownose ray *Rhinoptera bonasus* (Mitchill 1815). J. Exp. Mar. Biol. Ecol. 141: 213–220.

Bone, Q. 1988. Muscles and locomotion. pp. 99–141. *In*: T.J. Shuttleworth (ed.) Physiology of Elasmobranch Fishes, Springer-Verlag, Berlin.

Bradbury, J.W. & S.L. Vehrencamp. 1998. Principles of animal communication. Sinauer Associates, Sunderland. 882 pp.

Brown, C.A. & S.H. Gruber. 1988. Age assessment of the lemon shark, *Negaprion brevirostris*, using tetracycline validated vertebral centra. Copeia 1988: 747–753.

Bushnell, P.G., P.L. Lutz & S.H. Gruber. 1989. The metabolic rate of an active, tropical elasmobranch, the lemon shark (*Negaprion brevirostris*). Exp. Biol. 48: 279–283.

Carey, F.G. & E. Clark. 1995. Depth telemetry from the sixgill shark, *Hexanchus griseus*, at Bermuda. Env. Biol. Fish. 42: 7–14.

Carey, F.G. & Q.H. Gibson. 1987. Blood flow in the muscle of free-swimming fish. Physiol. Zool. 60: 138–148.

Carey, F.G. & B.H. Robinson. 1981. Daily patterns in the activities of swordfish, *Xiphias gladius*, observed by acoustic telemetry. U.S. Fish. Bull. 79: 277–292.

Carey, F.G. & J.V. Scharold. 1990. Movements of blue sharks (*Prionace glauca*) in depth and course. Mar. Biol. 106: 329–342.

Carey, F.G., J.W. Kanwisher, O. Brazier, G. Gabrielson, J.G. Casey & H.L. Pratt, Jr. 1982. Temperature and activities of a white shark, *Carcharodon carcharias*. Copeia 1982: 254–260.

Carey, F.G., J.M. Teal & J.W. Kanwisher. 1981. The visceral temperatures of mackerel sharks (Lamnidae). Physiol. Zool. 54: 334–344.

Casey, J.G. & N.E. Kohler. 1990. Long distance movements of Atlantic sharks from the NMFS cooperative shark tagging program. pp. 87–91. *In*: S.H. Gruber (ed.) Discovering Sharks, Underwater Naturalist, Highlands.

Cohen, J.L. 1991. Adaptations for scotopic vision in the lemon shark. J. Exp. Zool. suppl. 5: 76–84.

Compagno, L.J.V. 1984. FAO species catalogue of the world, vol 4: Sharks of the world. An annotated and illustrated catalogue of shark species known to date, part 2: Carcharhiniformes. FAO Fish. Synop. 125, Vol.4, Pt.2: 251–665.

Cortés, E. & S.H. Gruber. 1990. Diet, feeding habits and estimates of daily ration of young lemon sharks, *Negaprion brevirostris* (Poey). Copeia 1990: 204–218.

Cortés, E. & S.H. Gruber. 1994. Effect of ration size on growth and gross conversion efficiency of young lemon sharks, *Negaprion brevirostris*. J. Fish Biol. 44: 331–341.

Economakis, A.E. & P.S. Lobel. 1998. Aggregation behavior of the grey reef shark, *Carcharhinus amblyrhynchos*, at Johnston Atoll, Central Pacific Ocean. Env. Biol. Fish. 51: 129–139.

Garrick, J.A.F. & L.P. Schultz. 1963. A guide to the kinds of potentially dangerous sharks. pp. 3–60. *In*: P.W. Gilbert (ed.) Sharks and Survival, D. C. Heath, Boston.

Gero, D.R. 1952. The hydrodynamic aspects of fish propulsion. Amer. Mus. Novitates 1601: 1–32.

Goldman, K.J., S.D. Anderson, J.E. McCosker & A.P. Klimley. 1996. Temperature, swimming depth, and movements of a white shark at the South Farallon Islands, California. pp. 111–120. *In*: A.P. Klimley & D.G. Ainley (ed.) Great White Sharks: The Biology of *Carcharodon carcharias*, Academic Press, San Diego.

Goldman, K.J. & S.D. Anderson. 1999. Space utilization and swimming depth of white sharks, *Carcharodon carcharias*, at the South Farallon Islands, central California. Env. Biol. Fish. 56: 351–364.

Goodenough, J., B. McGuire & R.A. Wallace. 1993. Perspectives on animal behavior. John Wiley & Sons, New York. 762 pp.

Graham, J.B., H. Dewar, N.C. Lai, W.R. Lowell & S.M. Arce. 1990. Aspects of shark swimming performance determined using a large water tunnel. J. Exp. Biol. 151: 175–192.

Gruber, S.H. 1982. Role of the lemon shark, *Negaprion brevirostris* (Poey) as a predator in the tropical marine environment: a multidisciplinary study. Florida Sci. 45: 4–75.

Gruber, S.H. & J.S. Cohen. 1978. Vision in sharks. pp. 11–105. *In*: E. Hodgson & R. Mathewson (ed.) Sensory Biology of Sharks, Skates, and Rays, Off. Naval Res., Arlington.

Gruber, S.H., D.R. Nelson & J.F. Morrissey. 1988. Patterns of activity and space utilization of lemon sharks, *Negaprion brevirostris*, in a shallow Bahamian lagoon. Bull. Mar. Sci. 43: 61–76.

Gunn, J.S., J.D. Stevens, T.L.O. Davis & B.M. Norman. 1999. Observations on the short-term movements and behaviour of whale sharks (*Rhincodon typus*) at Ningaloo Reef, Western Australia. Mar. Biol. 135: 553–559.

Hamilton, W. J. & K. E. Watts. 1970. Refuging. Ann. Rev. Ecol. Syst. 1: 263–287.

Helfman, G.S. 1993. Fish behaviour by day, night and twilight. pp. 479–512. *In*: T.J. Pitcher (ed.) Behaviour of Teleost Fishes, Chapman & Hall, London.

Holland, K.N., C.G. Lowe, J.D. Peterson & A. Gill. 1992. Tracking coastal sharks with small boats: hammerhead shark pups as a case study. Aust. J. Mar. Freshwat. Res. 43: 61–66.

Holland, K.N., B.M. Wetherbee, C.G. Lowe & C.G. Meyer. 1999. Movements of tiger sharks (*Galeocerdo cuvier*) in coastal Hawaiian waters. Mar. Biol. 134: 665–673.

Holland, K.N., B.M. Wetherbee, J.D. Peterson & C.G. Lowe. 1993. Movements and distribution of hammerhead shark pups on their natal grounds. Copeia 1993: 495–502.

Holts, D.B. & D.W. Bedford. 1993. Horizontal and vertical movements of the shortfin mako shark, *Isurus oxyrinchus*, in the southern California Bight. Aust. J. Mar. Freshwat. Res. 44: 901–909.

Huish, M.T. & C. Benedict. 1978. Sonic tracking of dusky sharks in the Cape Fear River, North Carolina. J. Elisha Mitchell Scien. Soc. 93: 21–26.

Kalmijn, A.J. 1978. Experimental evidence of geomagnetic orientation in elasmobranch fishes. pp. 347–353. *In*: K. Schmidt–Koenig & W.T. Keeton (ed.) Animal Migration, Navigation, and Homing, Springer-Verlag, Heidelberg.

Kalmijn, A.J. 1988. Detection of weak electric fields. pp. 151–186. *In*: A.J. Kalmijn, J. Atama, R.R. Fay, A.N. Popper & W.N. Tavola (ed.) Sensory Biology of Aquatic Animals, Springer-Verlag, Heidelberg.

Klimley, A.P. 1993. Highly directional swimming by scalloped hammerhead sharks, *Sphyrna lewini*, and subsurface irradiance, temperature, bathymetry, and geomagnetic field. Mar. Biol. 117: 1–22.

Klimley, A.P. & S.B. Butler. 1988. Immigration and emigration of a pelagic fish assemblage to seamounts in the Gulf of California related to water mass movements using satellite imagery. Mar. Ecol. Prog. Ser. 49: 11–20.

Klimley, A.P., S.B. Butler, D.R. Nelson & A.T. Stull. 1988. Diel movement of scalloped hammerhead shark, *Sphyrna lewini* Griffith and Smith, to and from a seamouth in the Gulf of California. J. Fish Biol. 33: 751–761.

Klimley, A.P., I. Cabrera-Mancilla & J.L. Castillo-Geniz. 1993. Horizontal and vertical movements of the scalloped hammerhead shark, *Sphyrna lewini*, in the southern Gulf of California, Mexico. Ciencias Marinas 19: 95–115.

Klimley, A.P. & D.R. Nelson. 1984. Diel movement patterns of the scalloped hammerhead shark (*Sphyrna lewini*) in relation to El Bajo Espiritu Santo: a refuging central-position social system. Behav. Ecol. Sociobiol. 15: 45–54.

Lowe, C.G. & K.J. Goldman. 2001. Thermal and bioenergetics of elasmobranchs: bridging the gap. Env. Biol. Fish. 60: 251–266.

Lowe, C.G., K. N. Holland & T.G. Wolcott. 1998. A new acoustic tail beat transmitter for fishes. Fish. Res. 36: 275–283.

Major, P.F. 1977. Predator-prey interactions in schooling fishes during periods of twilight: a study of the silverside *Pranesus insularum* in Hawaii. U.S. Fish. Bull. 75: 425–426.

Matern, S.A., J.J. Cech & T.E. Hopkins. 2000. Diel movements of bat rays, *Myliobatis californica*, in Tomales Bay, California: evidence for behavioral thermoregulation? Env. Biol. Fish. 58: 171–180.

McCosker, J.E. 1987. The white shark, *Carcharodon carcharias*, has a warm stomach. Copeia 1987: 195–197.

McKibben, J.N. & D.R. Nelson. 1986. Patterns of movement and groupings of grey reef sharks, *Carcharhinus amblyrhynchos*, at Eniwetak, Marshall Islands. Bull. Mar. Sci. 38: 89–110.

Medved, R.J. & J.A. Marshall. 1983. Short-term movements of young sandbar sharks, *Carcharhinus plumbeus*. Bull. Mar. Sci. 33: 87–93.

Metcalfe, J.D., G.P. Arnold & P.W. Webb. 1990. The energetics of migration by selective tidal stream transport: an analysis for plaice tracked in the southern North Sea. J. Mar. Biol. Ass. U.K. 70: 149–162.

Morrissey, J.F. & S.H. Gruber. 1993a. Habitat selection by the juvenile lemon shark, *Negaprion brevirostris*. Env. Biol. Fish. 38: 311–319.

Morrissey, J.F. & S.H. Gruber. 1993b. Home range of juvenile lemon sharks, *Negaprion brevirostris*. Copeia 1993: 425–434.

Myrberg, A.A. Jr. 1976. Behavior of sharks – a continuing enigma. Nav. Res. Rev. 29: 1–11.

Nelson, D.R., J.N. McKibben, W.R. Strong, Jr., C.G. Lowe, J.A. Sisneros, D.M. Schroeder & R.J. Lavenberg. 1997. An acoustic tracking of a megamouth shark, *Megachasma pelagios*: a crepuscular vertical migrator. Env. Biol. Fish. 49: 389–399.

Nixon, A.J. & S.H. Gruber. 1988. Diel metabolic and activity patterns of the lemon shark (*Negaprion brevirostris*). J. Exp. Zool. 248: 1–6.

Parsons, G.R. & J.K. Carlson. 1998. Physiological and behavioral responses to hypoxia in the bonnethead shark, *Sphyrna tiburo*: routine swimming and respiratory regulation. Fish Phys. Biochem. 19: 189–196.

Priede, I.G. 1984. A basking shark (*Cetorhinus maximus*) tracked by satellite together with simultaneous remote sensing. Fish. Res. 2: 201–216.

Rogers, S.C., D.W. Church, A.H. Weatherly & D.G. Pincock. 1984. An automated ultrasonic telemetry system for the assessment of locomotor activity in free-ranging rainbow trout, *Salmo gairdneri* Richardson. J. Fish Biol. 25: 697–710.

Scharold, J. & S.H. Gruber. 1991. Telemetered heart rate as a measure of metabolic rate in the lemon shark, *Negaprion brevirostris*. Copeia 1991: 942–953.

Sciarrotta, T.C. & D.R. Nelson. 1977. Diel behavior of the blue shark, *Prionace glauca*, near Santa Catalina Island, California. U.S. Fish. Bull. 75: 519–528.

Silliman, W.R. & S.H. Gruber. 1999. Behavioral biology of the spotted eagle ray, *Aetobatus narinari* (Euphrasen, 1790), in Bimini, Bahamas; an interim report. Bahamas J. Sci. 7: 13–20.

Sims, D.W. 1999. Threshold foraging behaviour of basking sharks on zooplankton: life on an energetic knife-edge? Proc. R. Soc. Lond. B 266: 1437–1443.

Standora, E.A. & D. R. Nelson. 1977. A telemetric study of the behavior of free-swimming Pacific angel sharks, *Squatina californica*. Bull. South. Cal. Acad. Sci. 76: 193–201.

Strong, W.R., Jr., B.D. Bruce, D.R. Nelson & R.D. Murphy. 1996. Population dynamics of white sharks in Spencer Gulf, South Australia. pp. 401–414. *In*: A.P. Klimley & D.G. Ainley (ed.) Great White Sharks: The Biology of *Carcharodon carcharias*, Academic Press, San Diego.

Strong, W.R., Jr., R.D. Murphy, B.D. Bruce & D.R. Nelson. 1992. Movements and associated observations of bait-attracted white sharks, *Carcharodon carcharias*: a preliminary report. Aust. J. Mar. Freshwat. Res. 43: 13–20.

Sundström, L.F. & S.H. Gruber. 1998. Using speed sensing transmitters to model the bioenergetics of subadult lemon sharks, *Negaprion brevirostris* (Poey), in the field. Hydrobiol. 271/272: 241–247.

Sundström, L.F., J. Sterk & S.H. Gruber. 1998. Effects of a speed-sensing transmitter on the swimming speed of lemon sharks. Bahamas J. Sci. 6: 12–22.

Swihart, R.K. & N.A. Slade. 1985. Testing for independence of observations in animal movements. Ecology 66: 1176–1184.

Thomson, K.S. & D.E. Simanek. 1977. Body form and locomotion in sharks. Amer. Zool. 17: 343–354.

Thorson, T.B. 1971. Movement of bull sharks, *Carcharhinus leucas*, between the Caribbean Sea and Lake Nicaragua demonstrated by tagging. Copeia 1971: 336–339.

Tricas, T.C. & J.E. McCosker. 1984. Predatory behavior of the white shark, *Carcharodon carcharias*, with notes on its biology. Proc. Calif. Acad. Sci. 14: 221–238.

Tricas, T.C., L.R. Taylor & G. Naftel. 1981. Diel behavior of the tiger shark, *Galeocerdo cuvier*, at French Frigate Schoals, Hawaiian Islands. Copeia 1981: 904–908.

Voegeli, F.A., M.J. Smale, D.M. Webber, Y. Andrade & R.K. O'Dor. 2001. Ultrasonic telemetry, tracking and automated monitoring technology for sharks. Env. Biol. Fish. 60: 267–281.

Webb, P.W. 1984. Form and function in fish swimming. Sci. Amer. 241: 72–82.

Webb, P.W. & R.S. Keyes. 1982. Swimming kinematics of sharks. U.S. Fish. Bull. 80: 803–812.

Wiltschko, R. & W. Wiltschko. 1995. Magnetic Orientation in Animals. Springer-Verlag, Heidelberg. 297 pp.

Wood, G.L. 1982. The Guinness book of animal facts and feats. G.L. Woods and Guinness Superlatives LTD, New York. 275 pp.

Yano, K. & S. Tanaka. 1986. A telemetric study on the movements of the deep sea squaloid shark, *Centrophorus acus*. pp. 372–380. *In*: T. Uyeno, R. Arai, T. Taniuchi & K. Matsuura (ed.) Indo-Pacific Fish Biology, The Ichthyological Soc. Japan, Tokyo.

Appendix 1. Published telemetry-tracking studies focusing on elasmobranch behavior. Swimming depth was obtained from transmitters with a pressure sensing transducer, or in the case of Carey & Scharold (1990) using water temperature in conjunction with isotherms, and swimming speed was obtained by speed-sensing transmitters or inferred from a tail-beat transmitter. Behavior: ab = aggregation, d = depth, da = diel activity, h = homing, hr = home range, hs = habitat selection, m = movement, n = navigation, and ss = instantaneous swimming speed.

Elasmobranch species	Common name	Behavior	Reference
Aetobatus narinari	spotted eagleray	m	Silliman & Gruber (1999)
Carcharhinus amblyrhynchos	gray reef shark	ab, da, m	Economakis & Lobel (1998)
			McKibben & Nelson (1986)
			Nelson & Johnson[1]
Carcharhinus leucas	bull shark	m	Thorson (1971)
Carcharhinus limbatus	blacktip shark	da, hr	Nelson & Johnson[1]
Carcharhinus melanopterus	blacktip reef shark	da, hr	Nelson & Johnson[1]
Carcharhinus obscurus	dusky shark	m	Huish & Benedict (1978)
Carcharhinus plumbeus	sandbar shark	m	Medved & Marshall (1983)
Carcharodon carcharias	white shark	d, m	Carey et al. (1982)
			Goldman et al. (1996)
			Goldman & Anderson (1999)
			Strong et al. (1992)
Centrophorus acus	needle dogfish	d, m	Yano & Tanaka (1986)
Galeocerdo cuvier	tiger shark	d, m	Holland et al. (1999)
			Tricas et al. (1981)
Hexanchus griseus	bluntnose sixgill shark	d, m	Carey & Clark (1995)
Isurus oxyrhinchus	shortfin mako shark	d, m	Carey et al. (1981)
			Holts & Bedford (1993)
Megachasma pelagios	megamouth shark	d, m	Nelson et al. (1997)
Myliobatis californica	California cleftnose eagle ray	m	Matern et al. (2000)
Negaprion brevirostris	lemon shark	da, h, hr, hs, m, ss	Gruber et al. (1988)
			Morrissey & Gruber (1993a,b)
			This study
Prionace glauca	blue shark	d, da, m, ss	Carey & Gibson (1987)
			Carey & Scharold (1990)
			Sciarotta & Nelson (1977)
Rhincodon typus	whale shark	m	Gunn et al. (1999)
Sphyrna lewini	scalloped hammerhead	da, h, m, n, ss	Holland et al. (1992, 1993)
			Klimley & Nelson (1984)
			Klimley & Butler (1988)
			Klimley et al. (1988)
			Klimley (1993)
			Klimley et al. (1993)
			Lowe et al. (1998)
Sphyrna tiburo	bonnethead shark	ss	Parsons & Carlson (1998)
Squatina californica	angel shark	d, m, ss	Standora & Nelson (1977)
Triaenodon obesus	white tip reef shark	da, hr	Nelson & Johnson[1]
Triakis semifasciata	leopard shark	m	Ackerman et al. (2000)

[1]Nelson, D.R. & R.H. Johnson. 1980. Behavior of the reef sharks of Rangiroa, French Polynesia. Natl. Geogr. Soc. Res. Rep. 12: 479–499.

Environmental Biology of Fishes **60:** 251–266, 2001.

Thermal and bioenergetics of elasmobranchs: bridging the gap

Christopher G. Lowe[a] & Kenneth J. Goldman[b]
[a]*Department of Biological Sciences, California State University, Long Beach, 1250 Bellflower Blvd., Long Beach, CA 90840, U.S.A. (e-mail: clowe@csulb.edu)*
[b]*The College of William and Mary, School of Marine Science, Virginia Institute of Marine Science, P.O. Box 1346, Gloucester Point, VI 23062, U.S.A.*

Received 26 January 1999 Accepted 17 April 2000

Key words: energetics, thermoregulation, ultrasonic telemetry, transmitters

Synopsis

Physiological telemetry is a powerful tool in studying the thermal biology and energetics of elasmobranchs in the laboratory and field. Controlled laboratory studies have increased our understanding of the physiology and behavior of many elasmobranchs, but have focused primarily on small, slow moving species. Extrapolating results from these laboratory studies to free-swimming animals in the field or to other unstudied species may be problematic, due to laboratory constraints or species specific differences. Some elasmobranchs are too large or logistically difficult to maintain in captivity, making them extremely difficult to study in the laboratory, and thus can only be studied in the field. Physiological telemetry offers a 'bridge' between the laboratory and the field providing an opportunity to elucidate similarities and differences. Previous studies have coupled a variety of sensors with ultrasonic transmitters to relay information on epaxial muscle and stomach temperatures of free-swimming lamnid sharks. Even though these studies indicate lamnids exhibit elevated body temperatures, the degree to which these sharks may control body temperature is still not fully understood. Telemetry of heart rate, swimming speed, muscle contraction rate, and tail beat frequency has been used to estimate energy consumption of free-swimming elasmobranchs with varying success. Based on recent advances in technology, several hypotheses regarding thermoregulation, cardiac output, and obligate ram ventilation are discussed. Although many telemetry studies have been restricted by logistical difficulties in conducting long-term tracks, recent developments such as acoustic modems, underwater listening stations and satellite telemetry may significantly increase the amount and types of physiological data that can be collected. These improvements in technology and captive animal husbandry techniques will help to 'bridge the gap' between the laboratory and the field.

Introduction

Due to logistical difficulties in studying marine fishes in their natural environment, much of what is known about the physiological ecology of many elasmobranchs has come from laboratory studies (e.g. Bushnell et al. 1989, Scharold et al. 1989, Lowe 1998). Although controlled laboratory studies have shown how some elasmobranchs respond to changes in their environment, there can be problems in extrapolating results from laboratory studies to free-swimming animals in the field or to other unstudied species (Lowe et al. 1998). For example, early estimates of energy consumption of elasmobranchs were restricted to laboratory studies of small, less active species such as dogfish (e.g. *Squalus acanthias, Scyliorhinus stellaris, Scyliorhinus canicula*) (Winberg 1956, Pritchard et al. 1958, Brett & Blackburn 1978). While these studies provided the first estimates of metabolism for elasmobranchs, they were all restricted to the laboratory and most likely do not reflect actual metabolic rates of free ranging individuals in the field. In some cases, metabolic data from these less active species have been used to predict metabolic rates of more active species such as shortfin mako, *Isurus oxyrinchus*, and sandbar sharks, *Carcharhinus plumbeus* (Stillwell & Kohler 1982, Medved et al. 1988). However, these inter-specific extrapolations may greatly misrepresent the metabolic expenditure of these more active species. Thus, the need for comparable field studies is essential.

Conversely, because of the large size and high mobility of many elasmobranch fishes, controlled laboratory studies are extremely difficult, and thus these animals can only be studied in the field. For example, lamnid sharks (family Lamnidae) are extremely difficult to maintain in captivity, and therefore, determining the thermoregulatory capabilities of these sharks whose body temperatures are warmer than that of ambient water temperature has only been feasible from studies conducted in the field (Carey & Teal 1969, Carey et al. 1985, Goldman 1997). Studying animals in the field poses an equally difficult challenge in examining the physiological ecology of these species. So, how do we 'bridge the gap' between the laboratory and the field, or study the physiological ecology of elasmobranchs not amenable to laboratory study?

The development of ultrasonic telemetry in the 1960's greatly enhanced our ability to quantify the physiological and behavioral responses of elasmobranchs to changes in their natural environments (Nelson 1974, 1978, 1990). Originally used to remotely track the movements of fishes in the wild, researchers began to expand the use of ultrasonic telemetry by incorporating sensors to monitor physiological parameters (Carey & Lawson 1973, Kanwisher et al. 1974). This spawned the development of physiological telemetry, a technique used to 'telemeter' information on the physiological state of a free-ranging organism (Lucas et al. 1993). Since then, a variety of sensors have been used on elasmobranchs ranging from stingrays to white sharks to telemeter information concerning swimming speed (Gruber et al. 1988, Carey & Scharold 1990, Sundström & Gruber 1998, Parsons & Carlson 1998), tail beat frequency (Lowe et al. 1998), heart rate (Scharold et al. 1989, Scharold & Gruber 1991), muscle, cranial and stomach temperature (Carey 1982, Carey et al. 1984, Tricas & McCosker 1984, McCosker 1987, Carey & Scharold 1990, Goldman 1997), and neural activity (Kotchabhakdi et al. 1973, Tricas unpublished data).

Physiological telemetry continues to evolve, as recent developments in sensor technology and miniaturization of electronics increase the diversity and applications of this technology. However, this technique may not always prove the most cost-effective method and is still restrictive in terms of sample size and small craft logistics. The primary goal of this paper is to review the use of physiological telemetry to study the thermal biology and bioenergetics of elasmobranch fishes. In addition, we will examine the costs and benefits of telemetry techniques for studying the physiology, behavior, and ecology of elasmobranchs, and posit several hypotheses to inspire further research in this field.

Thermal telemetry

Elevated body temperature, which can be found in certain teleosts of the family Scombridae (tribe Thunnini) and elasmobranchs of the order Lamniformes, is a highly evolved physiological specialization directly tied to their energetics and ecology (Block & Finnerty 1994, Naylor et al. 1997, Goldman 1997). Heat derived from oxidative metabolism in the internalized red swimming muscle provides the primary source of heat production in these fishes (Hochachka et al. 1978, Carey et al. 1985). Using vascular counter-current heat exchangers called the retia mirabilia, heat is retained, producing body temperatures that are significantly warmer than ambient water (Eschricht & Müller 1835a,b, Burne 1923, Carey & Teal 1969, Carey et al. 1981, Carey et al. 1985, Fudge & Stevens 1996). Retia in sharks can be located in the cranium near the eyes (orbital retia), in locomotor musculature (lateral cutaneous retia), and viscera (suprahepatic and kidney retia). Body shape and the large size of some lamnid species further minimizes conductive heat loss to the surrounding seawater (Neill et al. 1976, Graham 1983, Block & Finnerty 1994).

The first measurements of elevated body temperature in elasmobranchs were taken using thermometers or thermistors (temperature-sensing devices) inserted into the epaxial muscle immediately after being caught. Muscle temperatures ranged from 0–14°C higher than ambient water temperatures, depending on species (Carey & Teal 1969, Carey et al. 1971, Smith & Rhodes 1983, Block & Carey 1985, Carey et al. 1985) (Table 1). Although these data may indicate elevated body temperatures, they may not accurately reflect those of free-swimming animals, because probe measurements were made on stressed, moribund and/or dead sharks. In response to the need to measure body temperatures of

Table 1. Probed and telemetered body temperatures of warm-bodied elasmobranch fishes.

Method Species Body location	Body temperature (°C)	Water temperature (°C)	n	Reference
Probed				
Lamna nasus				
deep muscle	17.2 (13.5–20.2)	8.0 (6.0–11.0)	3	Carey & Teal (1969)
"	19.3	11.5	7	Carey et al. (1971)
"	21.7 (20.8–22.6)	13.5 (12.7–14.0)	2	Carey et al. (1981)
"	20.1 (16.0–24.2)	12.9 (10.2–15.5)	2	Carey et al. (1985)
"	21.1 (20.3–21.8)	15.0 (14.4–15.5)	6	Block & Carey (1985)
shallow muscle	13.3 (12.3–14.2)	15.0 (14.4–15.5)	6	Block & Carey (1985)
stomach	18.3 (18.2–18.3)	13.5 (12.7–14.0)	2	Carey et al. (1981)
Lamna ditropis				
deep muscle	18.0 (16.0–19.0)	8.0	3	Smith & Rhodes (1983)
"	24.1 (18.9–27.5)	13.6 (11.5–15.4)	29	Anderson & Goldman unpublished
shallow muscle	19.6 (17.7–21.0)	13.6 (11.5–15.4)	5	Anderson & Goldman unpublished
Isurus oxyrinchus				
deep muscle	25.1 (19.4–28.6)	21.1 (15.8–27.4)	12	Carey & Teal (1969)
"	24.9	20.4	26	Carey et al. (1971)
"	25.2 (22.0–28.5)	18.9 (11.0–21.9)	5	Carey et al. (1981)
"	24.2 (2.5)[1]	20.2 (17.1–23.3)	38	Carey et al. (1985)
stomach	21.6 (19.0–26.8)	18.9 (11.0–21.9)	5	Carey et al. (1981)
"	22.9 (2.2)[1]	20.2 (17.1–23.3)	10	Carey et al. (1985)
"	24.6 (2.8)[1]	20.2 (17.1–23.3)	11	Carey et al. (1985)
Isurus paucus				
deep muscle	22.7	23.1	1	Carey et al. (1985)
Alopias superciliosus				
deep muscle	21.6 (20.3–23.0)	18.6 (16.0–21.2)	2	Carey etal. (1971)
Telemetered				
Isurus oxyrinchus				
stomach	22.1 (19.1–24.0)[2]	16.0 (9.2–17.2)	1	Carey et al. (1981)
"	25.2 (22.5–27.7)[2]	19.0 (13.5–23.9)	1	Carey et al. (1981)
Carcharodon carcharias				
muscle	21.0 (0.9)	15.0 (4.8–18.9)	1	Carey et al. (1982)
"	24.7 (0.4)	21.1 (20.9–21.5)	1	Tricas & McCosker (1984)
stomach	24.3 (0.9)	18.1	1	McCosker (1987)
"	26.6 (0.5)	13.3 (12.3–13.9)	1	Goldman et al. (1996)
"	26.2 (0.2)	13.7 (12.9–15.3)	1	Goldman (1997)
"	26.3 (0.3)	14.0 (12.9–15.3)	1	Goldman (1997)
"	26.6 (1.07)	17.4 (17.1–17.9)	1	Goldman & Gordon unpublished

[1]Mean and standard deviation. All other probed temperatures are mean and range.
[2]Median and range. All other telemetered temperatures are mean and standard deviation. Water temperatures are mean and range. n = number of animals studied.

unstressed, free-swimming lamniform sharks, swimming muscle temperature and ingestible stomach temperature transmitters were developed (Figure 1a, b). Both were initially designed to answer similar questions about tuna and lamnid endothermy, yet each has specific benefits and costs.

Muscle temperature telemetry

Telemetering fish muscle temperature involves placing a rigid thermistor deep into the internalized epaxial red muscle of the animal, which then measures changes in muscle temperature by changing the transmitter

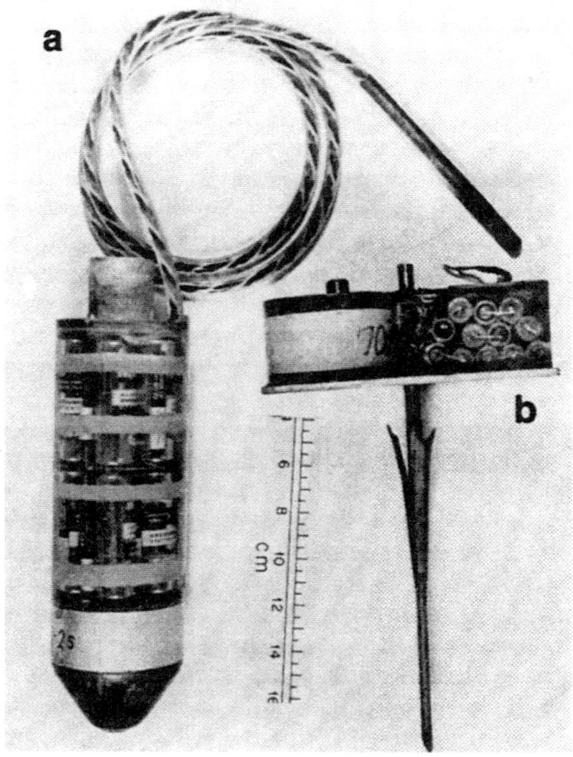

Figure 1. Temperature sensitive ultrasonic transmitters designed for bluefin tuna. a – Stomach temperature ultrasonic transmitter designed to be ingested with bait. b – Muscle temperature transmitter designed to be harpooned into the dorsal musculature where the harpoon tip functions as the thermistor (from Carey & Lawson 1973).

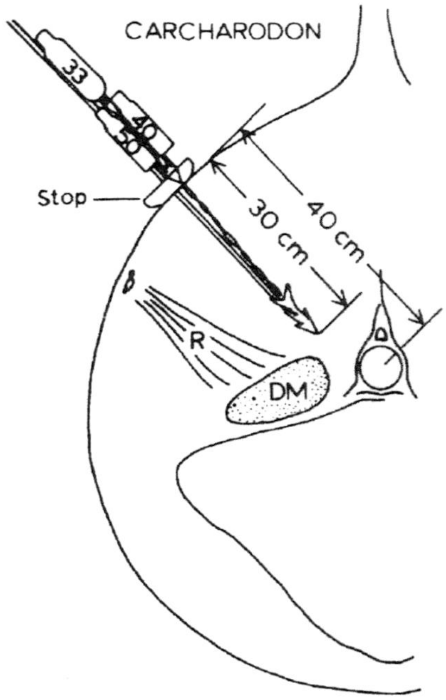

Figure 2. Ultrasonic transmitter package designed to simultaneously monitor epaxial muscle and ambient water temperatures, and depth. Muscle temperature was detected by a thermistor placed at the tip of the dart head and was transmitted at a frequency of 50 kHz. Water temperature was transmitted at a frequency of 40 kHz and swimming depth at 33 kHz (from Carey et al. 1982).

pulse rate (Figures 1b, 2) (Carey & Lawson 1973, Carey & Robison 1981). Carey et al. (1982) designed a multi-transmitter package consisting of an epaxial muscle thermistor, ambient water thermistor, and depth-sensing transmitters that could be harpooned into the dorsal musculature of a shark (Figure 2). Each transmitter produced a different frequency so data could be accumulated simultaneously, thus allowing for direct water and body temperature comparisons as the shark swam at different depths. Initially, these multi-transmitter packages were quite large. Subsequent designs consolidated all of the sensors into one transmitter by multiplexing sensor information at the same frequency, reducing the size of the transmitter package (Carey & Scharold 1990).

Because the transmitters are usually harpooned into the shark, drawbacks of this technique include difficulty in positioning the thermistor into the internalized red muscle, accuracy, and reproducibility. In addition, the connector wires for the muscle thermistors are relatively short (~ 30 cm), therefore limiting the penetration of the thermistor into the epaxial muscle (Carey et al. 1981, Tricas & McCosker 1984). For example, in large white sharks, the thermistor may not extend into the red muscle, thereby underestimating body core temperatures. This method is also intrusive and may irritate muscle activity during swimming, possibly altering the animal's behavior.

Although muscle temperature may underestimate the core body temperatures of fishes, valuable information regarding the thermoregulatory capabilities of sharks has been determined using this technique. Carey (1973) found that the muscle temperature of a dusky shark, *Carcharhinus obscurus*, followed changes in ambient water temperature. This pattern was also observed for muscle and cranial temperatures of blue sharks, *Prionace glauca* (Carey & Scharold 1990) (Table 2). Because both of these sharks lack internalized red swimming muscle and corresponding retia, these field data support the ectothermic nature of these sharks.

Muscle temperature telemetry studies of lamnid sharks have also been conducted. For example, epaxial muscle temperatures have been telemetered from two white sharks, *Carcharodon carcharias*. Carey et al. (1982) found that a large white shark tracked in the northwest Atlantic exhibited a 3–5°C elevation in muscle temperature over ambient temperature (Table 1). This shark showed a preference for the thermocline, and because it moved from warm to cold water, Carey et al. (1982) were able to estimate its rate of metabolism from the rate of change in its muscle temperature. They estimated a metabolic rate of 20 kJ kg^{-1} d^{-1} and concluded that although this metabolic rate may seem low, when scaled properly it would be three times higher than that estimated for a one ton spiny dogfish at 20°C (Table 3). This method of determining metabolism is only accurate for warm-bodied fishes (Carey et al. 1982). Tricas & McCosker (1984) tracked a white shark off South Australia that also exhibited a 3–5°C elevation in muscle temperature over ambient temperature, however, water temperature in this study was an average of 4°C warmer (Table 1). No thermocline was observed and water temperature did not vary over the time of the track, hence water temperature preferences and metabolic rate could not be examined.

Stomach temperature telemetry

In order to obtain more accurate measurements of body core temperature, the aforementioned ingestible stomach temperature transmitters were used (Figure 1a). In these transmitters the thermistor is located on the surface of the transmitter package and measures the temperature of the surrounding environment, which in this case is the stomach. Thus, transmitters can be fed to

Table 2. Elasmobranch studies using physiological sensors in conjunction with ultrasonic telemetry.

Topic	Sensor type	Parameter	Species	Study type	Reference
Energy consumption	Swimming speed	Activity	*Prionace glauca*	Field	Carey & Scharold (1990)
			Negaprion brevirostris	Lab/field	Gruber (1984), Sundström & Gruber (1998)
			Sphyrna tiburo	Lab	Parsons & Carlson (1998)
	Tail beat	Tail beat frequency	*Sphyrna lewini*	Lab/field	Lowe et al. (1998), Lowe (1998)
			P. glauca	Field	Carey & Scharold (1990)
	Electrocardiogram	Heart rate	*N. brevirostris*	Lab	Scharold & Gruber (1991)
			Triakis semifasciata	Lab	Scharold et al. (1989)
Thermal biology	Thermistor	Epaxial muscle temperature	*P. glauca*	Field	Carey & Scharold (1990)
			Carcharodon carcharias	Field	Carey et al. (1982), Tricas & McCosker (1984)
			Carcharhinus obscurus	Field	Carey (1973)
		Cranial temperature	*P. glauca*	Field	Carey & Scharold (1990)
		Stomach temperature	*C. carcharias*	Field	McCosker (1987), Goldman et al. (1996), Goldman (1997)
			I. oxyrinchus	Field	Carey et al. (1981)

Table 3. Routine metabolic rates estimated from elasmobranchs using physiological telemetry methods.

Species	Body weight (kg)	Water temperature (°C)	Routine metabolic rate (kJ kg^{-1} d^{-1})	Telemetry type	Reference
Triakis semifasciata	2.2–5.8	16	43	Heart rate EKG	Scharold et al. (1989)
Negaprion brevirostris	1.1–1.6	25	78	Heart rate EKG	Scharold & Gruber (1991)
Negaprion brevirostris	20–34	22	35	Swimming speed	Sundström & Gruber (1998)
Sphyrna tiburo	4.2–5.5	30–33.5	37.5	Swimming speed	Parsons & Carlson (1998)
Sphyrna lewini	0.6–0.8	26	96	Tail beat frequency	Lowe (1998), Lowe et al. (1998)
Carcharodon carcharias	943	15	20	Muscle temperature	Carey et al. (1982)

sharks in a piece of bait during either induced or natural feeding events (Goldman 1997). Because the stomach is centrally located within the body lying ventral to the subcutaneous lateral retia and dorsal to the suprahepatic rete, it is well insulated from outside temperature changes (Goldman et al. 1996, Goldman 1997). These features indicate that stomach temperature may be a better indicator of body core temperature than epaxial muscle temperature.

Although elasmobranchs can regurgitate indigestible items, treble hooks can be attached to the transmitter so that the transmitter becomes attached to the stomach wall, thereby increasing the retention time (McCosker 1987). Because the use of treble hooks is slightly intrusive, another option to increase retention time in the stomach is to use transmitters that are larger in diameter than that of the stomach's pyloric sphincter. This method has produced retention times of at least three weeks (Goldman 1997).

One of the primary drawbacks of this technique is that it does not allow for simultaneous measurements of surrounding ambient water temperature. As a result, water temperatures must be estimated or measured via other methods during the track. In addition, short-term reductions in stomach temperature through the influx of ingested seawater may temporarily lower the average body core temperature of the shark (McCosker 1987, Goldman 1997).

Telemetered stomach temperatures from shortfin mako sharks showed a 6–8°C elevation over water temperature, while those from white sharks were 7.4–14.1°C higher (Carey et al. 1981, McCosker 1987, Goldman 1997) (Figure 3). Data to date suggest that both species may maintain constant elevated body core temperatures independent of ambient water temperature (Goldman et al. 1996).

One of the primary advantages of using ingestible transmitters is that there is little or no trauma associated with tagging (Goldman 1997, Goldman & Anderson 1999). Another advantage of stomach temperature telemetry is the short-term reduction in stomach temperature observed when these sharks feed (McCosker 1987, Goldman 1997). Although stomach temperature reduction appears to last on the order of 10 to 15 minutes, the overall body core temperatures can be up to 1.7°C lower on feeding days than on non-feeding days (Goldman 1997). This correlation holds potential for determining feeding periodicity in warm-bodied sharks. Additionally, stomach temperatures were higher and showed more consistent (and repeatable) mean values than muscle temperatures in white sharks, providing more evidence that stomach temperature telemetry provides a more accurate measure of body core temperature (Table 1).

Summary of thermal biology

It has been 30 years since Carey and colleagues began the work, which demonstrated that lamnid sharks are

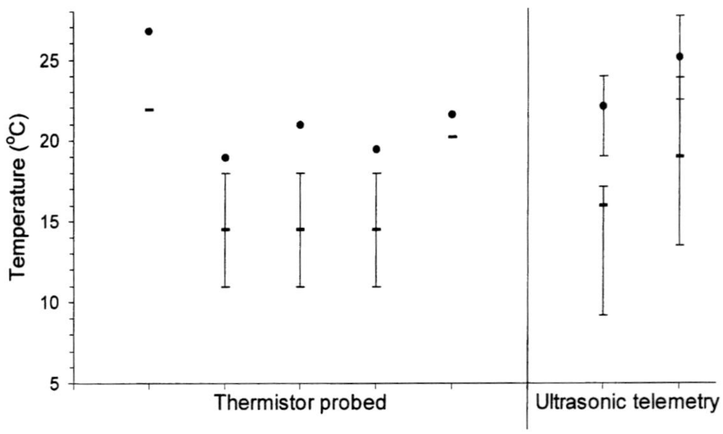

Figure 3. Probed and telemetered stomach temperatures from shortfin mako sharks, *Isurus oxyrinchus*. Closed circles represent stomach temperatures. Dashes represent mean water temperatures. Error bars indicate maximum and minimum values. Although water and stomach temperatures appear to have overlapped on the second telemetered shark, stomach temperature was always at least 2°C above ambient water temperature (from Carey et al. 1981).

endothermic fishes. Anatomical studies have shown that lamnids possess vascular shunts around the suprahepatic rete, which may alter the rete's efficiency (Carey et al. 1981). They possess a mechanism for transporting warm blood from the muscle to the brain and supplementing the orbital retia (Wolf et al. 1988). Additionally, they possess hematological and cardiac characteristics similar to those of other endotherms and homeotherms (Emery 1985,1986, Lai et al. 1989). This, along with results from telemetry studies, suggests they may be able to physiologically thermoregulate, a process presumably controlled by the central nervous system (Brill et al. 1994).

In addition to the lamnids, alopiid sharks and other elasmobranch species have been shown to possess retia. For example, recent dissections have shown the presence of heat exchanging retia in three species of myliobatiod rays (Alexander 1995,1996, and personal communication). However, no reliable temperature measurements have been published, so their thermoregulatory ability is still unknown.

Recent shipboard and laboratory studies of oxygen consumption of small shortfin mako sharks hold promise for a future 'bridge' between controlled laboratory studies of thermoregulation and field investigations (Graham et al. 1990, Lai et al. 1997, Bernal personal communication). Until that time, ultrasonic telemetry of free-swimming sharks provides the only mechanism for studying their thermal physiology.

Heart rate telemetry

In some fishes, heart rate can be a good indicator of activity, reproductive state, stress, feeding, and metabolic rate (Priede & Young 1977, Jones & Randall 1978, Armstrong 1986). As a fish becomes more active it must increase its cardiac output, the product of heart rate and heart stroke volume, in order to meet the increased metabolic demand of the activity (Farrell & Jones 1992). As long as the pulse of oxygen carried by the flow of blood remains constant, the rate of oxygen consumption (metabolic rate) can be measured by determining the product of cardiac output and the difference in arteriovenus oxygen concentration (Webber et al. 1998).

Heart rates of fishes can be easily monitored by the use of electrocardiogram (EKG) electrodes, which are surgically implanted into the fish's pericardium to detect electric impulses generated by the heart chamber contractions. Measuring stroke volume, however, requires the use of more complex techniques and invasive surgery to attach electromagnetic or Doppler flow probes into or around the ventral aorta. Due to these difficulties in measuring stroke volume on free-swimming fishes, most studies have used heart rate as an indicator of metabolism or other behaviors, such as spawning and feeding.

To monitor heart rates of free-swimming fishes, a variety of heart-rate telemetry systems have evolved and have been successfully tested on teleost fishes in the laboratory and field (Wardle & Kanwisher 1974, Armstrong et al. 1989, Lucas et al. 1991, Lucas et al. 1993). EKG electrodes attached to microamplifiers increase the electric impulses produced during the R wave portion (ventricular systole) of the heart beat cycle and trigger the transmitter to produce an output pulse (Armstrong et al. 1989, Lucas et al. 1991). These heart rate sensors have been integrated with a wide variety of carrier signal transmitters including FM radio, FM acoustic, ultrasonic pulse, electromagnetic transmitters, and ultrasonic transponders, further increasing the versatility of this technique (Armstrong et al. 1989).

Unfortunately, one of the main drawbacks of this technique is that it requires extensive handling of the animal for proper surgical placement of electrodes and is more intrusive compared to other telemetry methods. For many species, this may pose serious challenges and limitations to the application. In addition, the success of this technique is entirely dependent on prior laboratory calibration of the relationship between heart rate, metabolism, and other behaviors.

The first applications of heart rate telemetry for estimating metabolic rates of free swimming elasmobranchs were tested on leopard, *Triakis semifasciata*, and lemon sharks, *Negaprion brevirostris* (Scharold et al. 1989, Scharold & Gruber 1991) (Table 2). In these studies, sharks instrumented with EKG ultrasonic transmitters were observed in respirometers to determine the relationship between their heart rates and $\dot{V}O_2$. Scharold et al. (1989) exercised instrumented leopard sharks in a swim tunnel respirometer over a range of aerobic swimming speeds, and found that although heart rate increased at a significant linear rate with increases in swimming speed, heart rate varied considerably with increases in $\dot{V}O_2$ ($r^2 = 0.38$) (Figure 4). As a result, heart rate was found to account for only 32% of the rise in $\dot{V}O_2$. Similar results were obtained from juvenile lemon sharks observed swimming voluntarily in an annular

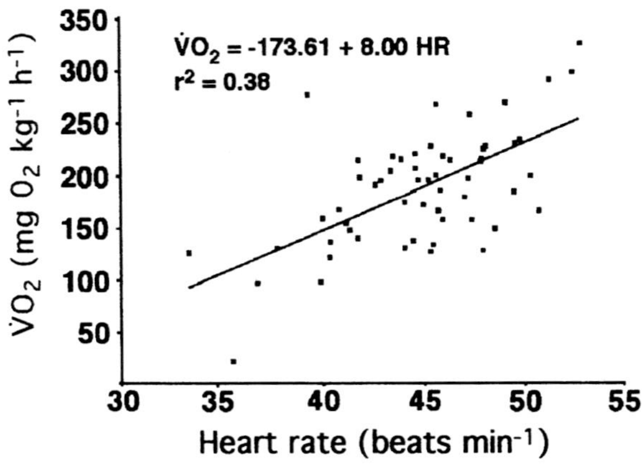

Figure 4. The relationship between heart rate and oxygen consumption rate for leopard sharks instrumented with ultrasonic heart rate transmitters (from Scharold et al. 1989).

respirometer (Scharold & Gruber 1991) (Table 2). Heart rates of lemon sharks increased at a significant linear rate with increases in swimming speed, but also varied considerably with increases in $\dot{V}O_2$. Heart rate only accounted for an 18% rise in $\dot{V}O_2$ for the lemon shark. The mean ($\pm$ SE) resting heart rate and $\dot{V}O_2$ measured for leopard sharks was 37 ± 2 beats min^{-1} and 105 ± 36 mg O_2 kg^{-1} h^{-1}, with a maximum sustained swimming rate of 47 ± 0.9 beats min^{-1} and 229 ± 13 mg O_2 kg^{-1} h^{-1}. Assuming these sharks have a routine (average) swimming speed of 0.4 body lengths per second (BL s^{-1}), based on the relationship between swimming speed and $\dot{V}O_2$ derived from this experiment, they would have a routine metabolic rate of 43 kJ kg^{-1} d^{-1} (Table 3). This is a lower metabolic rate than that measured for the lemon shark, which had a routine heart rate of 56 ± 2 beats min^{-1} and $\dot{V}O_2$ of 240 ± 13 mg O_2 kg^{-1} h^{-1} (78 kJ kg^{-1} d^{-1}). Although it is likely that differences in experimental methodology (swim tunnel vs. annular respirometer), water temperature and shark size account for much of the difference in metabolic rate, leopards sharks still have higher heart rates than lemon sharks.

In both studies it was concluded that heart rate contributes a relatively small percentage to changes in $\dot{V}O_2$, and that cardiac output is likely facilitated by increases in stroke volume and/or arteriovenus oxygen differences with increased activity. As a result, heart rate was considered to be an inadequate indicator of metabolic rate for these two species, and no field experiments were conducted. These findings are supported by those of Lai et al. (1989) who measured changes in heart rate and stroke volume of resting and swimming leopard sharks in a swim tunnel and found that these sharks modulated their stroke volume more than their heart rates. Based on similarities in heart structure, it has been suggested that other ectothermic elasmobranch species may also exhibit this cardiac response (Emery et al. 1985, Farrell 1991, Toda & Gattusa 1996), but this may not be true for endothermic elasmobranchs. Recent studies of shortfin mako shark cardiac physiology indicate that they have hearts more like birds and mammals and may modulate heart rate more than stroke volume (Lai et al. 1997), in which case heart rate alone may provide an adequate field indicator of metabolic rate.

Even though heart rate telemetry may not provide an adequate predictor of the metabolic rates of leopard and lemon sharks in the field, these studies have provided valuable information on the cardiac physiology of elasmobranchs in the laboratory (Scharold et al. 1989, Scharold & Gruber 1991). Another benefit of these studies is that they offered the first quantification of the energetic cost to a shark carrying an externally mounted transmitter (Scharold et al. 1989, Scharold & Gruber 1991). By measuring the $\dot{V}O_2$ of lemon sharks with and without EKG transmitters, Scharold & Gruber (1991) noted that sharks with transmitters had metabolic rates 7% higher than sharks without transmitters. This additional cost of 5.5 kJ kg^{-1} d^{-1} was attributed to drag incurred by the externally mounted transmitter package. Few other studies have attempted to measure the cost of carrying transmitters and most have asserted these costs were negligible. Wardle & Kanwisher (1974) surgically implanted EKG transmitters in the stomach of cod specifically to reduce drag incurred by an externally mounted package. Although 'transmitter effects' pose a valid concern for any telemetry study, they are especially relevant in studies of energetics. Therefore, the energetic costs of carrying packages should be quantified whenever possible.

Nonetheless, heart rate telemetry may also be useful in quantifying behaviors other than those related to activity. For example, heart rate telemetry could be used to determine when elasmobranchs feed, because many fishes experience increased heart rate during and after feeding events (Armstrong 1986). Although monitoring heart rate of elasmobranchs may be useful in determining how elasmobranchs respond to environmental and behavioral stressors, until stroke volume can be monitored remotely, the utility of this technique for quantifying energetic demands of free swimming fishes is limited (Webber et al. 1998).

Telemetry of swimming activity

Almost all studies on metabolic costs of elasmobranchs have come from laboratory respirometer experiments, where the animal is allowed to swim voluntarily or is forced to swim over a range of speeds (Graham et al. 1990, Scharold & Gruber 1991, Lowe 1998). Using this technique, the metabolic rates of these elasmobranchs can be measured indirectly via oxygen consumption and correlated with their activity rate. In some studies, researchers have attempted to estimate energy consumption costs of elasmobranchs in the field by quantifying their swimming speeds, tail beat frequencies, or activity rates in respirometers (e.g. Bushnell et al. 1989, Parsons 1990, Lowe 1998).

Swimming speed telemetry

A number of studies have used speed-sensing transmitters to measure swimming speeds and energy consumption of elasmobranchs in the field (Gruber et al. 1988, Carey & Scharold 1990, Parson & Carlson 1998, Sundström & Gruber 1998). These transmitters incorporate a paddlewheel or turbine, which spins at different rates depending on the velocity of water passing by it. The pulse rate of the transmitter varies depending on the revolution rate of the sensor. By calibrating the transmitter pulse rate over several known water velocities these transmitters can be used to determine the swimming speeds of free-swimming animals. To ensure that the speed sensor is directly in the flow of water passing over the shark, transmitters are externally mounted on or near the dorsal fin (Parson & Carlson 1998, Sundström & Gruber 1998).

A variety of speed-sensing transmitters are commercially available and have been shown to be very useful in quantifying swimming of active fishes in the field, yet there are a number of limitations to these type of transmitter. One limitation is that the transmitters require linear calibration of the sensor with water velocity, and may not always be equally sensitive over a full range of swimming speeds (Sundström & Gruber 1998). In addition, speed sensors may not always be accurate in indicating when a shark is actively swimming. For example, if a shark were gliding through the water, the speed sensor would still indicate that the shark is moving forward even though it is not actually swimming. Another possible drawback is that speed-sensing transmitters also tend to be quite large compared to other transmitters and might incur more drag, thus affecting the swimming speeds of sharks in the field. This restricts the use of these types of transmitters to larger sharks.

In the early 1980's Gruber and colleagues systematically quantified all of the components of the energy budget of the lemon shark in the laboratory, albeit focusing on smaller individuals (Gruber 1984). As part of this effort, Bushnell et al. (1989) determined the relationship between swimming speed and oxygen consumption rates of juvenile lemon sharks in an annular respirometer. Gruber et al. (1988) attached speed sensing (paddlewheel style) ultrasonic transmitters to two mature lemon sharks in Bimini Lagoon to make direct measurements of swimming speeds (Table 2). These sharks (178 and 210 cm TL) exhibited mean routine swimming speeds of 0.33 BL s^{-1}. Later, Sundström & Gruber (1998) tracked three immature lemon sharks (154–188 cm TL) in Bimini Lagoon using similar speed-sensing transmitters and observed mean swimming speeds of 0.37 BL s^{-1}. Using data from Bushnell et al. (1989) they were able to estimate the metabolic rates of the tracked lemon sharks based on the measured swimming speeds (Table 3). From these data they were able to construct the first field-derived energy budget for an elasmobranch, finally 'bridging of the gap' between laboratory- based data and field measurements of free-swimming sharks in their natural environment.

Although these studies have developed the most detailed description of a shark's energy budget to date, these data are only from larger sharks tracked in the field. Smaller sharks can be studied in respirometers, but cannot be tracked using the speed-sensing transmitters because of the large size of the transmitter package, while large sharks can be tracked in the field but not studied in respirometers due to restrictions on respirometer size (Graham et al. 1990). Extrapolating metabolic rate data from juvenile sharks to adults is problematic, requiring mass specific corrections to account for the differences in size between those sharks studied in the laboratory and those tracked in the field.

Parsons & Carlson (1998) used speed sensing (propeller style) ultrasonic transmitters to quantify the swimming speeds of bonnethead sharks, *Sphyrna tiburo*, under normoxic and hypoxic conditions in an artificial lagoon. Oxygen consumption of sharks at different swimming speeds and under different oxygen concentrations was compared in an annular respirometer. Bonnethead sharks swam significantly faster and increased their mouth gape in hypoxic conditions than in normoxic conditions. As a result, sharks experienced higher $\dot{V}O_2$ in hypoxic conditions, which the researchers attributed to the increased swimming speeds (Table 2). Based on laboratory measurements these free-swimming sharks would have metabolic rates of 37.5 kJ kg^{-1} d^{-1} under normoxic conditions at their routine swimming speeds (Table 3). Parsons & Carlson (1998) concluded that the speed-sensing transmitters provided a more accurate measure of activity and also noted that differences in swimming speeds between sharks in situ and those measured in the respirometer may have been attributed to the added stress of handling and confinement.

Because of their negative buoyancy and need for continuous forward movement for adequate ventilation, Weihs (1973) suggested that pelagic sharks and tunas may conserve energy by exhibiting a 'fly and glide' vertical movement pattern. To test this hypothesis, Carey & Scharold (1990) conducted open ocean tracks of blue sharks using a combination of speed, depth, and tail beat sensing transmitters (Table 2). Although only two sharks were tracked with these transmitters, neither shark was observed to glide during their descent. In addition, it was concluded that the angles of decent and ascent were not suitable for achieving this type of energy saving. Because of the lack of calibrating laboratory physiology data, no estimates of energy expenditure were made for these sharks (Carey & Scharold 1990).

The use of speed-sensing transmitters has undoubtedly increased the accuracy of measuring swimming speeds of elasmobranchs in the field (Parsons & Carlson 1998, Sundström & Gruber 1998). Previous methods for quantifying activity, such as measuring speed over ground, have been shown to routinely underestimate the swimming speeds of elasmobranchs because it does not take into account fine scale movements (Gruber et al. 1988, Carey & Scharold 1990, Lowe et al. 1998). However, much like heart rate, using swimming speed to quantify metabolism of free-swimming elasmobranchs in the field requires laboratory calibration, thereby limiting the use of this technique.

Tail beat telemetry

Tail beat frequency (TBF) has also been used as a correlate of energy consumption. Because sharks use their tails as their main propulsion sources, TBF may provide an easily monitored indicator of swimming for fishes in the field. Laboratory studies have shown that most fishes increase their TBF in proportion to increases in swimming speed, while some species modulate their tail beat amplitude or propulsive wavelength in addition to TBF to increase their forward thrust (e.g. Bainbridge 1958, Hunter & Zweifel 1971, Dewar & Graham 1994, Lowe 1996). Tail beat frequency can therefore provide a reliable indicator of activity and exertion, although detailed laboratory calibrations are required to determine these relationships and the energy expenditures of fishes in the field (e.g. Stasko & Horrall 1976, Briggs & Post 1997, Lowe et al. 1998).

A variety of tail beat sensors have been developed to measure TBF in fishes. The most common method uses electromyogram (EMG) electrodes placed in the epaxial swimming muscle to monitor rhythmic body flexing. Because of difficulties in shielding, EMG electrodes have typically been used with radio transmitters in freshwater (Young et al. 1972, Ross et al. 1981, Weatherley et al. 1996, Briggs & Post 1997). However, Dewar et al. (1999) developed an EMG ultrasonic transmitter for use on marine fishes that amplifies and encodes output information via frequency modulation. Some researchers have used continuous-wave ultrasonic telemetry to indicate TBF, where the rhythmic movement of the fish creates a corresponding Doppler effect (Stasko & Horrall 1976). Another type of tail beat sensor developed uses a simple magnetized pivoting vane, which passes over a reed switch on the fish's caudal peduncle with every lateral sweep of the tail (Figure 5) (Lowe et al. 1998).

Some of the drawbacks of EMG transmitters are that they require precise surgical placement of the electrodes, which like muscle thermistors may affect swimming behavior. In addition, transmitter circuits must be properly tuned to adequately respond to muscle action potential (Lowe et al. 1998, Dewar et al. 1999). Although continuous-wave transmitters do not require attachment of intrusive EMG electrodes and can reduce external drag through surgical implantation in the body cavity, their continuous output severely limits battery longevity. Externally mounted pivoting vane transmitters are less invasive than EMG transmitters, but they still incur some drag that may affect swimming behavior and energetics (Lowe et al. 1998). All of these transmitter types require varying amounts of handling and surgical attachment, which may affect the fish's behavior (Lowe et al. 1998).

The first study to use ultrasonic tail beat telemetry to quantify energy expenditure of an elasmobranch

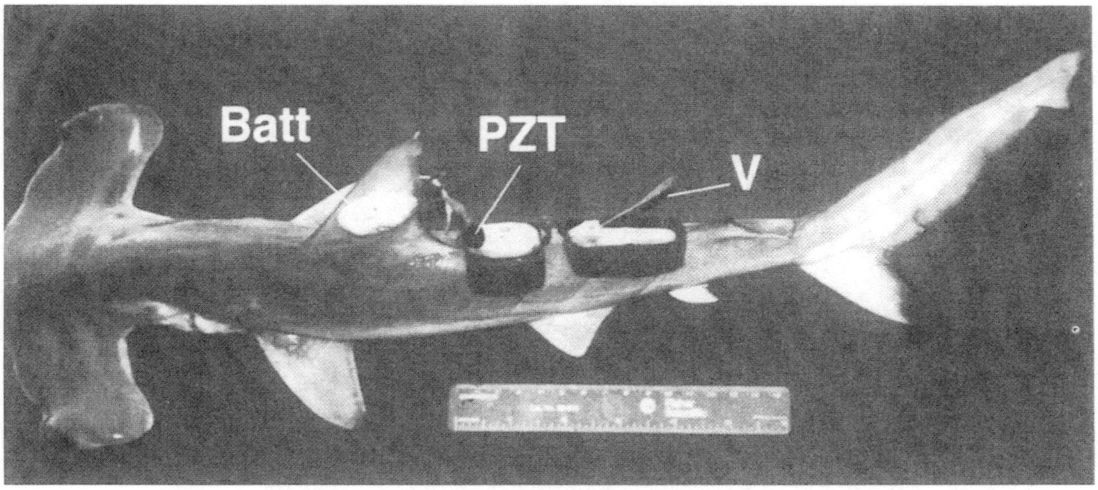

Figure 5. Photograph of the ultrasonic tail beat transmitter attached to the caudal peduncle of a 63 cm TL scalloped hammerhead shark. The posterior unit is the magnetized pivoting vane. Battery pack (Batt) is sandwiched to the dorsal fin and the middle component houses the transmitter circuitry. PZT = transducing element and V = pivoting vane.

was conducted by Lowe (1998). In this study, juvenile scalloped hammerhead sharks, *Sphyrna lewini*, were exercised in a respirometer and the relationship between TBF, swimming speed, and $\dot{V}O_2$ was determined over a range of aerobic swimming speeds (Lowe 1996, Lowe et al. 1998, Lowe 1998) (Table 2). A minimally intrusive pivoting-vane ultrasonic transmitter was used to relay TBF of free-swimming sharks in the field. Sharks were exercised in the respirometer with and without transmitters to account for the energetic costs to those sharks carrying transmitters (Lowe 1998). Although laboratory and field tests indicated that the transmitters accurately relayed TBF of free-swimming sharks, sharks carrying the transmitter packages swam 8–26% slower at a given TBF depending on the velocity (Lowe 1998).

Six hammerhead shark pups instrumented with tail beat transmitters were tracked for periods up to 50 h continuously. Tail beat frequency was recorded from every successive tail beat and averaged over 15 min. periods over the course of the tracks. These data and previously determined laboratory relationships were used to determine swimming speeds and $\dot{V}O_2$ over the course of the track (Figure 6). Tracks indicated that these sharks have higher metabolic requirements than those estimated for other species of tropical elasmobranchs and also swim relatively faster than other species studied (Table 3). Because of the direct coupling of laboratory and field experiments on the same size sharks, this study likely represents the most accurate estimates of field based energy consumption for an elasmobranch to date (Lowe 1998).

Successful use of tail beat telemetry is largely dependent on laboratory-based calibrations, thus restricting its use to smaller elasmobranchs. However, if tail beat sensors are used in conjunction with speed sensors on larger sharks that are too large for laboratory study, they can then be used to more accurately estimate activity and swimming speeds for pelagic sharks. In addition, although the transmitters used in the Lowe (1998) study did affect the sharks' swimming and slowed sharks down at certain velocities, further streamlining of these transmitter packages is needed to reduce drag and transmitter effects on the sharks kinematics.

Future of physiological telemetry of elasmobranchs

Several things become apparent as one examines the history of physiological ecology of elasmobranch fishes. Not only has relatively little work has been done in this area, but physiological research on elasmobranch fishes lags behind that of teleosts. As a result, we pose several hypotheses that can be directly tested using telemetry techniques and how these studies can be used to further advance our understanding of elasmobranch fishes.

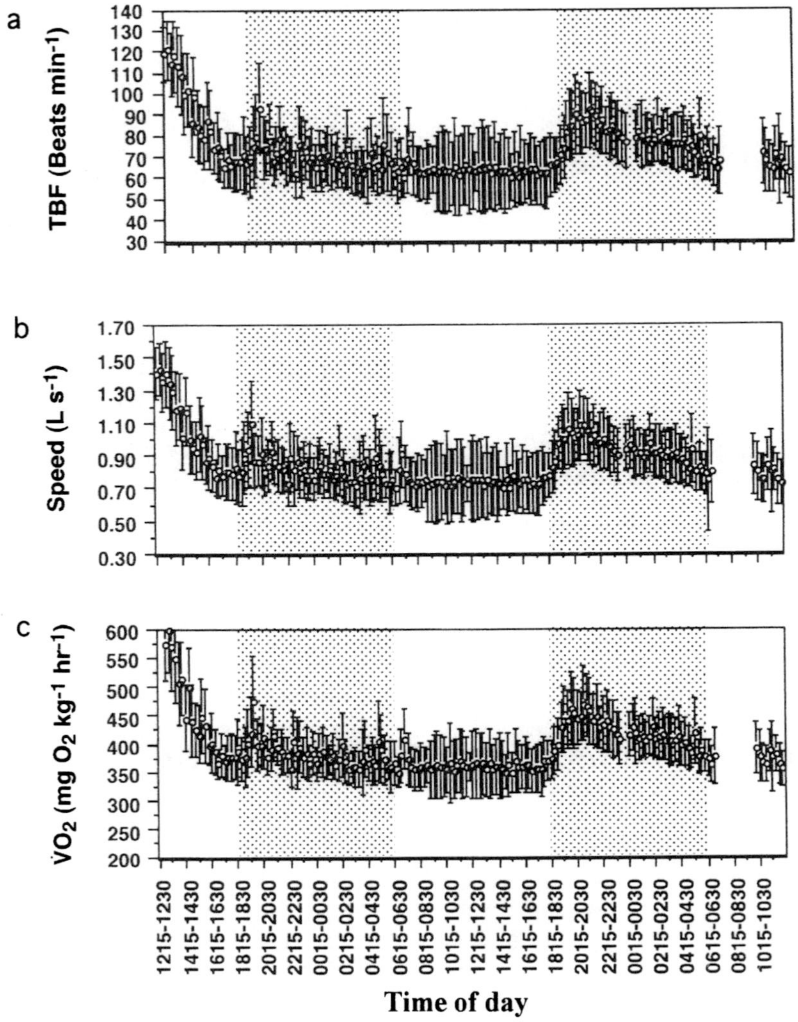

Figure 6. a – The average tail beat frequency (TBF) of a hammerhead shark tracked using the ultrasonic tail beat transmitter over a 48 h period. Circles represent average TBF and bars are standard deviations in 15 min blocks. b – The average instantaneous relative swimming speed estimated using TBF data collected over the 48 h. Circles represent average speed and bars are standard deviations in 15 min blocks. c – The average oxygen consumption rate ($\dot{V}O_2$) estimated using TBF data collected over the 48 h track. Circles represent corrected $\dot{V}O_2$ and bars are standard deviations in averaged in 15 min blocks (from Lowe et al. 1998).

The homeothermy hypothesis

Both temperature probing and field telemetry studies have demonstrated that lamnid sharks exhibit elevated body temperatures (endothermy) (Carey & Teal 1969, Carey et al. 1971, McCosker 1987, Goldman 1996, Goldman 1997). However, limited field data and studies of cardiac and vascular anatomy also indicate that mako and white sharks may be able to maintain an elevated and uniform body temperature that is regulated independent of changes in ambient temperature (Carey et al. 1981, Emery 1985, Goldman et al. 1996, Goldman 1997). If this were true, then these sharks would not only be considered 'endotherms', but may be better defined as 'homeotherms' (Hickman et al. 1984).

One way to test this homeothermy hypothesis would be to use a combination of techniques and transmitters. White sharks could be fed a multiplexing stomach temperature and depth-sensing transponding transmitter during a natural feeding event. Another water temperature-sensing transponding transmitter, operating at a different frequency, could be harpooned into the sharks back while it is feeding. These multiple transmitters would allow for simultaneous measurements of core body temperature in relation to changes in ambient water temperature and depth. By using transponding transmitters, sensor data could be stored on an onboard data logger at programmed intervals. Studies have found that some white sharks show seasonal site fidelity and forage in the same feeding locations (Klimley & Anderson 1996, Goldman & Anderson 1999). Ultrasonic listening stations that periodically emit pulses could be deployed on the ocean floor in areas where sharks are known to patrol. A number of sharks could be instrumented at one time and when the transponding transmitters on these passing sharks detect a listening station pulse, the transmitters would download stored data via an acoustic modem. The use of transponding transmitters would allow for longer monitoring periods while increasing the chances of observing thermoregulation as the sharks moved through different temperatures of water. Additionally, this method could help determine the presence or absence of circadian temperature cycles. Because these cycles occur in other homeotherms such as humans (Vander et al. 1990), this method would provide additional support for homeothermy in these species (Goldman 1997).

The cardiac output hypothesis

Cardiac morphology and physiology studies of some ectothermic sharks indicate that they modulate stroke volume more than heart rate to increase cardiac output. However, most of these studies have focused on non-obligate ram ventilating (swimming) species (Lai et al. 1989, Scharold et al. 1989, Scharold & Gruber 1990). Similar studies of endothermic sharks indicate that they modulate heart rate more than stroke volume, but all endothermic elasmobranch species are obligate swimmers (Lai et al. 1997). So how does an obligate swimming ectothermic shark regulate cardiac output?

To test this hypothesis under natural conditions, heart rate and stroke volume of obligate and non-obligate swimming sharks could be monitored and compared using an EKG sensor and Doppler flow probe integrated in a frequency modulated ultrasonic transmitter. After calibration and surgical attachment, sharks could be placed in a large tank for observation. Activity would then be correlated with heart rate and stroke volume. This would provide a more realistic view of cardiac responses of elasmobranchs than previous studies, because sharks would not be confined in a swim tunnel or small annular respirometer. Further calibration of the cardiac output transmitters in a respirometer would allow for metabolic rate estimates from free-swimming animals in the field.

The obligate ram ventilator hypothesis

Many species of sharks are thought to be obligate swimmers and must continuously swim in order to respire (Lowe 1998). However, because of limited in situ observations or monitoring, it is difficult to determine if some species are truly continuously active. For example, Priede (1984) hypothesized that basking sharks, *Cetorhinus maximus*, in the North Atlantic settle to the sea floor, shed their gill rakers, and enter a torpor-like state to save energy in winter months when plankton abundance is low. Other species such as the reef white tip, *Triaenodon obesus*, can be observed actively swimming or resting on the bottom (Nelson & Johnson 1980). However, it may be difficult to determine whether sharks are actually swimming in a small area or resting on the bottom just using conventional telemetry.

To test this hypothesis, sharks could be fitted with a tail beat-sensing transmitter (Lowe et al. 1998). This transmitter design incorporates a 'resting' pulse rate change to indicate when the shark is no longer swimming. When the shark is actively swimming, the transmitter pulses with every lateral sweep of the shark's tail, but when the sharks stops swimming the transmitter defaults to a fixed pulse rate ($\sim$5 sec pulse interval). To reduce the need to manually track these sharks, the tail beat sensor could be integrated into a transponding transmitter for longer-term data storage. Data derived using this technique could be used to estimate energy costs of both obligate and non-obligate swimming species.

Conclusions

Information on physiological ecology of elasmobranchs has increased significantly in the last 20 years, due to the development and use of telemetry techniques. Although each of the bioenergetics studies have provided valuable information regarding the metabolism of elasmobranchs, it is still very difficult to compare differences in metabolic rates among species due to differences in experimental technique, water temperature, and sizes of animals used (Table 3). Nonetheless, each form of telemetry has provided a tool for assessing physiological correlates of free-swimming fishes. Perhaps some of these methods will be applied to study energetics of other species, particularly on elasmobranchs where direct comparisons can be made between the laboratory and field.

As useful as some of these methods may be, there are still many limitations and restrictions. For example, the high cost of telemetry equipment and transmitters can have a negative impact on funding. As the complexity of multiplexed (multiple sensor) transmitters increases, so does the expense. Alternatively, 'homemade' transmitters can be customized for a specific study species and can be much more cost effective (Lowe et al. 1998, Parsons & Carlson 1998).

Another limitation of telemetry is acquisition of long-term data from open ocean species. Tracking elasmobranchs in offshore waters is difficult and usually requires a large tracking vessel. High vessel costs make it cost prohibitive to follow an individual for more than 3–5 days continuously. However, small vessel tracking has been successfully used for some oceanic species and may offer a cost-effective alternative to using large research vessels (Holland et al. 1985,1992,1999).

Small sample sizes are somewhat inherent to this work due to difficulties in locating and/or tagging target species and tracking more than one animal at a time. Nonetheless, more effort has been made in developing new technologies such as acoustic modems, underwater listening stations and satellite telemetry, which can be used to either accompany or eliminate the need for manual tracking in many situations (Voegeli et al. 2000, Priede 1984, Lutcavage et al. 1999). The continuing development of these technologies will make obtaining larger sample sizes and longer-term data sets easier. These improvements in technology and captive animal husbandry techniques will help to 'bridge the gap' between the laboratory and the field.

Acknowledgements

We thank S. Gruber and T. Tricas for inviting us to participate in this meaningful symposium volume. We dedicate this paper in memory of Don Nelson, who was an innovative and dedicated researcher, patient mentor, and a good friend. We also thank G. Goodman-Lowe and C.C. Goldman for their valuable comments and review of this manuscript. This is VIMS contribution no. 2315.

References cited

Alexander, R.L. 1995. Evidence of a counter-current heat exchanger in the ray *Mobula tarapacana* (Chondrichthyes: Elasmobranchii: Batiodea: Myliobatiformes). J. Zool. Lond. 237: 377–384.

Alexander, R.L. 1996. Evidence of brain-warming in the mobulid rays, *Mobula tarapacana* and *Manta birostris* (Chondrichthyes: Elasmobranchii: Batiodea: Myliobatiformes). J. Linn. Soc. 188: 151–164.

Armstrong, J.D. 1986. Heart rate as an indicator of activity, metabolic rate, food intake, and digestion in pike, *Esox lucius*. J. Fish Biol. 29: 207–221.

Armstrong, J.D., M.C. Lucas, I.G. Priede & L. DeVera. 1989. An acoustic telemetry system for monitoring the heart rate of pike, *Esox lucius* L., and other fishes in their natural environment. J. Exp. Biol. 143: 549–552.

Bainbridge, R. 1958. The speed of swimming of fish as related to size and to the frequency and amplitude of the tail beat. J. Exp. Biol. 35: 1183–1226.

Block, B.A. & F.G. Carey. 1985. Warm brain and eye temperatures in sharks. J. Comp. Physiol. B. 156: 229–236.

Block, B.A. & J.R. Finnerty. 1994. Endothermy in fishes: a phylogenetic analysis of constraints, predispositions, and selection pressures. Env. Biol. Fish. 40: 283–302.

Brett, J.R. & J.M. Blackburn. 1978. Metabolic rate and energy expenditure of the spiny dogfish, *Squalus acanthias*. J. Fish. Res. Board Can. 35: 816–821.

Briggs, C.T. & J.R. Post. 1997. In situ activity metabolism of rainbow trout (*Oncorhynchus mykiss*): estimates obtained from telemetry of axial muscle electromyograms. Can. J. Fish. Aquat. Sci. 54: 859–866.

Brill, R.W., H. Dewar & J.B. Graham. 1994. Basic concepts relevant to heat transfer in fishes, and their use in measuring the physiological thermoregulatory abilities of tunas. Env. Biol. Fish. 40: 109–124.

Burne, R.H. 1923. Some peculiarities of the blood vascular system of the porbeagle shark, *Lamna cornubica*. Phil. Trans. Roy. Soc. London 212: 209–257.

Bushnell, P.G., P.L. Lutz & S.H. Gruber. 1989. The metabolic rate of an active, tropical elasmobranch, the lemon shark (*Negaprion brevirostris*). Exp. Biol. 48: 279–283.

Carey, F.G. & K.D. Lawson. 1973. Temperature regulation in free-swimming bluefin tuna. Comp. Biochem. Physiol. 44A: 375–392.

Carey, F.G. 1982. A brain heater in the swordfish. Science 216: 1327–1329.

Carey, F.G., J.G. Casey, H.L. Pratt, D. Urquhart & J.E. McCosker. 1985. Temperature, heat production, and heat exchange in lamnid sharks. Mem. So. Calif. Acad. Sci. 9: 92–108.

Carey, F.G., J.W. Kanwisher, O. Brazier, G. Gabrielson, J.G. Casey & H.L. Pratt, Jr. 1982. Temperature and activities of a white shark, *Carcharodon carcharias*. Copeia 1982: 254–260.

Carey, F.G., J.W. Kanwisher & E.D. Stevens. 1984. Bluefin tuna warm their viscera during digestion. J. Exp. Biol. 109: 1–20.

Carey, F.G. 1973. Fishes with warm bodies. Sci. Amer. 228: 36–44.

Carey, F.G. & B.H. Robison. 1981. Daily patterns in the activities of swordfish, *Xiphias gladius*, observed by acoustic telemetry. U.S. Fish. Bull. 79: 277–292.

Carey, F.G. & J. Scharold. 1990. Movements of blue sharks (*Prionace glauca*) in depth and course. Mar. Biol. 106: 329–342.

Carey, F.G. & J.M. Teal. 1969. Mako and porbeagle: warm-bodied sharks. Comp. Biochem. Physiol. 28: 199–204.

Carey, F.G., J.M. Teal, J.W. Kanwisher, K.D. Lawson & J.S. Beckett. 1971. Warm-bodied fish. Amer. Zool. 11: 137–145.

Carey, F.G., J.M. Teal & J.W. Kanwisher. 1981. The visceral temperatures of mackerel sharks (Lamnidae). Physiol. Zool. 54: 334–344.

Dewar, H. & J.B. Graham. 1994. Studies of tropical tuna swimming performance in a large water tunnel. III. Kinematics. J. Exp. Biol. 192: 45–59.

Emery, S.H. 1985. Hematology and cardiac morphology in the great white shark, *Carcharodon carcharias*. Mem. So. Calif. Acad. Sci. 9: 73–80.

Emery, S.H. 1986. Hematological comparisons of endothermic vs. ectothermic elasmobranch fishes. Copeia 1986: 700–705.

Emery, S.H., C. Mangano & V. Randazzo. 1985. Ventricle morphology in pelagic elasmobranch fishes. Comp. Biochem. Physiol. 8: 635–643.

Eschricht, D.F. & J. Müller. 1835a. Über die arteriösen and venösen Wundernetze an der Leber und einen merkwürdigen Bau dieses Organes beim Thunfische, *Thynnus vulgaris*. Physikal. Abhandl. d. K. Wissensch. Berlin. 32 pp. (after Fudge & Stevens 1996).

Eschricht, D.F. & J. Müller. 1835b. Nachtrag zu der Abhandlung der Herren Eschricht und Müller über die Wundernetze an der Leber des Thunfisches. Über die Wundernetze am Darmkanal des *Squalus vulpes* L., *Alopecias vulpes* Nob. Physikal. Abhandl. d. K. Wissensch. Berlin. pp. 325–328 (after Fudge & Stevens 1996).

Farrell, A.P. 1991. From hagfish to tuna – a perspective on cardiac function. Physiol. Zool. 64: 1137–1164.

Farrell, A.P. & D.R. Jones. 1992. The heart. pp. 1–88. *In:* W.S. Hoar, D.J. Randall & A.P. Farrell (ed.) Fish Physiology, Vol. 12A, Academic Press, San Diego.

Fudge, D.S. & E.D. Stevens. 1996. The visceral *retia mirabilia* of tuna and sharks: an annotated translation and discussion of the Eschricht & Müller 1835 paper and related papers. Guelph Ichthyol. Rev. 4: 1–92.

Goldman, K.J. 1997. Regulation of body temperature in the white shark, *Carcharodon carcharias*. J. Comp. Physiol. B 167: 423–429.

Goldman, K.J. & S.D. Anderson. 1999. Space utilization and swimming depth of white sharks, *Carcharodon carcharias*, at the South Farallon Islands, central California. Env. Biol. Fish. 56: 353–366.

Goldman, K.J., S.D. Anderson, J.E. McCosker & A.P. Klimley. 1996. Temperature, swimming depth, and movements of a white shark at the South Farallon Islands, California. pp. 111–120. *In:* A.P. Klimley & D.G. Ainley (ed.) Great White Sharks: Ecology and Behavior, Academic Press, San Diego.

Graham, J.B. 1983. Heat transfer. pp. 248–279. *In:* P.W. Webb & D. Weihs (ed.) Fish Biomechanics, Praeger Publishing, New York.

Graham, J.B., H. Dewar, N.C. Lai, W.R. Lowell & S.M. Arce. 1990. Aspects of shark swimming performance determined using a large water tunnel. J. Exp. Biol. 151: 175–192.

Gruber, S.H. 1984. Bioenergetics of captive and free-ranging lemon sharks. AAZPA Ann. Conf. Proc.: 340–373.

Gruber, S.H., D.R. Nelson & J.F. Morrissey. 1988. Patterns of activity and space utilization of lemon sharks, *Negaprion brevirostris*, in a shallow Bahamian lagoon. Bull. Mar. Sci. 43: 61–76.

Hickman, C.P., L.S. Roberts & F.M. Hickman. 1984. Integrated principles of zoology. Times Mirror/Mosbey College Pub., St. Louis. 1065 pp.

Hochachka, P.W., W.C. Hulbert & M. Guppy. 1978. The tuna power plant and furnace. pp. 153–181. *In:* G.D. Sharp & A.E. Dizon (ed.) The Physiological Ecology of Tunas, Academic Press, New York.

Holland, K.N., R.W. Brill, R.K.C. Chang & R. Yost. 1985. A small vessel technique for tracking pelagic fish. Mar. Fish. Rev. 47: 26–32.

Holland, K.N., C.G. Lowe, J.D. Peterson & A. Gill. 1992. Tracking coastal sharks with small boats: hammerhead shark pups as a case study. Aust. J. Mar. Freshwater Res. 43: 61–66.

Holland, K.N., B.M. Wetherbee, C.G. Lowe & C. Meyer. 1999. Movements of tiger sharks in coastal Hawaiian waters. Mar. Biol. 134: 665–673.

Hunter, J.R. & J.R. Zweifel. 1971. Swimming speed, tail beat frequency, tail beat amplitude, and size in jack mackerel, *Trachurus symmetricus*, and other fishes. U.S. Fish. Bull. 69: 253–256.

Jones, D.R. & D.J. Randall. 1978. The respiratory and circulatory systems during exercise. pp. 425–501. *In:* W.S. Hoar & D.J. Randall (ed.) Fish Physiology, Vol. 7, Academic Press, New York.

Kanwisher, J.W., K. Lawson & G. Sundnes. 1974. Acoustic telemetry from fish. U.S. Fish. Bull. 72: 251–255.

Klimley, A.P. & S.D. Anderson. 1996. Residency patterns of white sharks at the South Farallon Islands, California. pp. 365–374. *In:* A.P. Klimley & D.G. Ainley (ed.) Great White Sharks: Ecology and Behavior, Academic Press, San Diego.

Kotchabhakdi, N., J. Kanwisher & C.L. Prosser. 1973. Acoustic telemetry of electrical activity from the brain of free-swimming dogfish. Biol. Bull. 145: 443–444.

Lai, N.C., J.B. Graham, W.R. Lowell & R. Shabetai. 1989. Elevated pericardial pressure and cardiac output in the leopard shark *Triakis semifasciata* during exercise: the role of the pericardioperitoneal canal. J. Exp. Biol. 147: 263–277.

Lai, N.C., K.E. Korsmeyer, S. Katz, D.B. Holts, L.M. Laughlin & J.B. Graham. 1997. Hemodynamics and blood properties of the shortfin mako shark (*Isurus oxyrinchus*). Copeia 1997: 424–428.

Lowe, C.G. 1996. Kinematics and critical swimming speed of juvenile scalloped hammerhead sharks. J. Exp. Biol. 199: 2605–2610.

Lowe, C.G., K.N. Holland & T.G. Wolcott. 1998. A new acoustic tailbeat transmitters for fishes. Fish. Res. 36: 275–283.

Lowe, C.G. 1998. Swimming efficiency and bioenergetics of juvenile scalloped hammerhead sharks in Kaneohe Bay, Hawaii. Ph.D. Dissertation, University of Hawaii, Honolulu. 130 pp.

Lucas, M.C., J.D. Armstrong, I.G. Priede, A.N.Z. Gindy & L. De Vera. 1991. Direct movements of metabolism, activity and feeding behaviour of pike, *Esox lucius* L., in the wild, by the use of heart rate telemetry. J. Fish Biol. 39: 325–345.

Lucas, M.C., A.D.F. Johnstone & I.G. Priede. 1993. Use of physiological telemetry as a method of estimating metabolism of fish in the natural environment. Trans. Amer. Fish. Soc. 122: 822–833.

Lutcavage, M.E., R.W. Brill, G.S. Skomal, B.C. Chase & P.W. Howey. 1999. Results of pop-up satellite tagging of spawning size class fish in the Gulf of Maine: do north Atlantic bluefin tuna spawn in the mid-Atlantic. Can. J. Fish. Aquat. Sci. 56: 173–177.

McCosker, J.E. 1987. The white shark, *Carcharodon carcharias*, has a warm stomach. Copeia 1987: 195–197.

Medved, R.J., C.E. Stillwell & J.G. Casey. 1988. The rate of food consumption of young sandbar sharks (*Carcharhinus plumbeus*) in Chincoteague Bay, Virginia. Copeia 1988: 956–963.

Naylor, G.J.P., A.P. Martin, E.G. Mattison & W.M. Brown. 1997. Interrelationships of lamniform sharks: testing phylogenetic hypotheses with sequence data. pp. 199–217. *In:* T. Kocher & C.A. Stephien (ed.) Molecular Systematics of Fishes, Academic Press, San Diego.

Neill, W.H., R.K. Chang & A.E. Dizon. 1976. Magnitude and ecological implications of thermal inertia in skipjack tuna, *Katsuwonnus pelamis* (Linnaeus). Env. Biol. Fish. 1: 61–80.

Nelson, D.R. 1974. Ultrasonic telemetry of shark behavior. Nav. Res. Rev. 27: 1–21.

Nelson, D.R. 1978. Telemetering techniques for the study of free-ranging sharks. pp. 419–482. *In:* E. Hodgson & R. Mathewson (ed.) Sensory Biology of Sharks, Skates, and Rays, Off. Naval Res., Arlington.

Nelson, D.R. 1990. Telemetry studies of sharks: a review, with applications in resource management. pp. 239–256. *In:* H.L. Pratt, S.H. Gruber & T. Taniuchi (ed.) Elasmobranchs as Living Resources: Advances in the Biology, Ecology, Systematics, and Status of the Fisheries, NOAA Tech. Rep. NMFS 90.

Nelson, D.R. & R.H. Johnson. 1980. Behavior of the reef sharks of Rangiroa, French Polynesia. Natl. Geogr. Soc. Res. Rep. 12: 479–499.

Parsons, G.R. 1990. Metabolism and swimming efficiency of the bonnethead shark *Sphyrna tiburo*. Mar. Biol. 104: 363–367.

Parsons, G.R. & J.K. Carlson. 1998. Physiological and behavioral responses to hypoxia in the bonnethead shark, *Sphyrna tiburo*: routine swimming and respiratory regulation. Fish Physiol. Biochem. 19: 189–196.

Priede, I.G. 1984. A basking shark (*Cetorhinus maximus*) tracked by satellite together with simultaneous remote sensing. Fish. Res. 2: 201–216.

Priede, I.G. & A.H. Young. 1977. Ultrasonic telemetry of cardiac rhythms of wild brown trout (*Salmo trutta,* L.) J. Fish Biol. 10: 299–318.

Pritchard, A.W., E. Florey & A.W. Martin. 1958. Relationship between metabolic rate and body size in an elasmobranch (*Squalus suckleyi*) and in a teleost (*Ophiodon elongatus*). J. Mar. Res. 17: 403–411.

Weatherley, A.H., P.A. Kaselo, M.D. Gare, J.M. Gunn & B. Lipicnik. 1996. Field activity of lake trout during the reproductive period monitored by electrogram telemetry. J. Fish Biol. 48: 647–685.

Ross, L.G., W. Watts & A.H. Young. 1981. An ultrasonic biotelemetry system for the continuous monitoring of tail-beat rate from free-swimming fish. J. Fish Biol. 18: 479–490.

Scharold, J. & S.H. Gruber. 1991. Telemetered heart rate as a measure of metabolic rate in the lemon shark, *Negaprion brevirostris*. Copeia 1991: 942–953.

Scharold, J., N.C. Lai, W.R. Lowell & J.B. Graham. 1989. Metabolic rate, heart rate, and tailbeat frequency during sustained swimming in the leopard shark *Triakis semifasciata*. Exp. Biol. 48: 223–230.

Smith, R.L. & D. Rhodes. 1983. Body temperature of the salmon shark, *Lamna ditropis*. J. Mar. Biol. Ass. U.K. 63: 243–244.

Stasko, A.B. & R.M. Horrall. 1976. Method of counting tailbeats of free-swimming fish by ultrasonic telemetry techniques. J. Fish. Res. Board Can. 33: 2596–2598.

Stillwell, C.E. & N.E. Kohler. 1982. Food, feeding habits, and estimates of daily ration of the shortfin mako (*Isurus oxyrinchus*) in the Northwest Atlantic. Can. J. Fish. Aquat. Sci. 39: 407–414.

Sundström, L.F. & S.H. Gruber. 1998. Using speed-sensing transmitters to construct a bioenergetics model for subadult lemon sharks, *Negaprion brevirostris* (Poey), in the field. Hydrobiologia 371/372: 241–247.

Toda, B. & A. Gattuso. 1996. Heart ventricle pumps in teleosts and elasmobranchs: a morphodynamic approach. J. Exp. Zool. 275: 162–171

Tricas, T.C. & J.E. McCosker. 1984. Predatory behavior of the white shark (*Carcharodon carcharias*), with notes on its biology. Proc. Calif. Acad. Sci. 43: 221–238.

Vander, A.J., J.H. Sherman & D.S. Luciano. 1990. Human physiology: the mechanisms of body function. McGraw-Hill Publishing, New York. 145 pp.

Voegeli, F., M.J. Smale, D.M. Webber, Y. Andrade & R. O'Dor. 2000. Ultrasonic telemetry, tracking and automated monitoring technology for sharks. Env. Biol. Fish. (in press).

Wardle, C.S. & J.W. Kanwisher. 1974. The significance of heart rate in free swimming cod, *Gadus morhua*: some observations with ultra-sonic tags. Mar. Behav. Physiol. 2: 311–324.

Webber, D.M., R.G. Boutilier & S.R. Kerr. 1998. Cardiac output as a predictor of metabolic rate in cod *Gadus morhua*. J. Exp. Biol. 201: 2779–2789.

Weihs, D. 1973. Mechanically efficient swimming techniques for fish with negative buoyancy. J. Mar. Res. 31: 194–209.

Winberg, G.G. 1956. Rate of metabolism and food requirements of fishes. Fish. Res. Board Can. Transl. Ser. No. 194, 1960: 1–253.

Wolf, N.G., P.R. Swift & F.G. Carey. 1988. Swimming muscle helps warm the brain of lamnid sharks. J. Comp. Physiol. B 157: 709–715.

Young, A.H., P. Tyler, F.G.T. Holliday & A. MacFarlane. 1972. A small sonic tag for measurement of locomotor activity in fish. J. Fish Biol. 4: 57–65

Environmental Biology of Fishes **60:** 267–281, 2001.

Ultrasonic telemetry, tracking and automated monitoring technology for sharks

Frederick A. Voegeli[a], Malcolm J. Smale[b], Dale M. Webber[c], Yanko Andrade[c] & Ronald K. O'Dor[c]
[a]*VEMCO Limited, 100 Osprey Drive, Shad Bay, Nova Scotia B3H 4J1, Canada (e-mail: fred@vemco.com)*
[b]*Port Elizabeth Museum, Port Elizabeth, South Africa*
[c]*Dalhousie University, Halifax, Nova Scotia B3H 4J1, Canada (e-mail: odor@is.dal.ca)*

Received 19 January 1999 Accepted 21 February 2000

Key words: acoustic, positioning, remote, pinger, transponder, sensors, code, archival

Synopsis

Sharks were among the first marine animals to carry telemetry systems because of their size and the need to understand their interactions with humans. Modern telemetry systems can gather many kinds of data (limited only by imagination, funding and sensor types), indicating which animals are near telemetry receivers and what they are doing. Receivers now range from simple autonomous detector units for deployment in mid-water in large-scale grids, to sophisticated automated benthic recorders, to triangulating radio-linked buoy systems (RAP), to ship-borne transponders. In addition, archival tags can now gather and store data even while the shark is away, to be downloaded later. Older types had to be recovered, but pop up tags release from sharks automatically, surface and transfer data to satellites, while CHAT tags download whenever queried by a nearby transponding acoustic receiver. Sophisticated animal-borne tags dramatically increase the information gathered about sharks and their environment. The examples provided show the parallel progression of shark biology and acoustic biotelemetry illustrating that telemetry systems are tools for gathering data, which can often be honed to facilitate biological experiments. Future visions include sensors that directly measure shark swimming power and cardiac output, compressing the data so that it can be delivered to RAP systems tracking multiple animals with meter resolution in near real time. CHAT tags as small as 22 mm diameter should be able to return similar data from trips of hundreds of kilometers. Continued communication between biologists and engineers is essential to develop these technologies.

Introduction

Tagging versus telemetry

People have been 'tagging' fish accidentally for centuries; handmade fishhooks would be recognizable to their makers and, no doubt, even stone-age fishers found hooks set years earlier in re-caught fish. Scientific marking of fishes and sharks has been an area of considerable research. The techniques here usually rely on numbered tags that reveal information on movement patterns of individual fish, population information and augment growth studies in the field (Stevens 1976, 1990, Casey 1985, Casey & Kohler 1992, Kohler & Turner 2001). Modern tag and release programs approach telemetry with PIT (passive integrated transponder) tags, so small that they can be implanted in the tissues of even juvenile fish. When activated by powerful radio waves they respond passively with radio signals that can be detected and decoded from up to a meter away as fish pass weirs or are processed in factories after capture. We do not propose to explore this history, but do think it necessary to define some terms.

From its roots, telemetry means 'measuring from a distance', so it is an active process requiring energy output to send a signal over a distance. Electromagnetic energy, particularly at radio frequencies, is rapidly absorbed as it passes through even a few centimeters of seawater (Pincock & Voegeli 1992) essentially eliminating radio waves for telemetry applications in non-surfacing marine elasmobranchs, except when downloading logged data from released tags ('popups', see below). Thus, acoustic energy is the best available technique for marine fishes. The ultrasonic ranges 30–100 kHz are above most animal auditory ranges but transmitted with low energy loss through seawater. The earliest ultrasonic tracking systems for fish appeared in the literature in the mid-fifties (Trefethen et al. 1957) and research for the first decade or so focused on monitoring fish movements on a small scale. Early references to shark biotelemetry are documented in the classic review by Stasko & Pincock (1977), and Stasko[1] starts with Baldwin & Ingle's (1964) olfactory studies on nurse sharks and Bass & Rascovich's (1965) tracking of hammerhead and sandbar sharks. Early elasmobranch tracking studies were summarized by Nelson (1978), as well as some of the pioneering work in the early development of multi-parameter tags that he and his students built themselves. More recent studies are covered by Nelson.[2]

Active tagging basics

Simple pingers broadcast the presence of the transmitter but leave it to the listener to somehow estimate a position. Individual tags may be identified (ID) by unique frequencies, the interval between pings or, in more recent tags, by complex, unique codes based on ping intervals. Typical pingers transmit a fixed rate pulsed tone at one 'ping' per second with a tone duration of about 10 ms because a 1% duty cycle saves battery power. Intervals under 0.5 s cause echo problems and those over 1.5 s do not provide sufficiently rapid audible feedback to take bearings with a directional hydrophone.

Once ID is established data about the animal's physiology or environment can also be encoded. Internal and external temperatures of sharks were used as controls for warm-blooded tuna in the earliest temperature telemetry work (Carey et al. 1971, Carey & Lawson 1973). Complex multi-channel telemetry systems also had their origins in shark studies[3] (Nelson 1978). Scariotta & Nelson (1977) used this approach to monitor swimming speed and turning frequency in near- and offshore as locations illustrated in Figure 1 to demonstrate nocturnal feeding. Such data can now be gathered by automated listening or tracking stations but initially required tremendous effort and risk using manual tracking.

Both ID and data coding can be tricky using acoustic systems. High power requirements and limited bandwidth rule out such radio approaches as continuous amplitude modulation. In the field, there is a trade-off between simple encoding schemes that deliver some information under the worst signal to noise conditions and complex ones that deliver more information under optimal conditions, with limited ranges. Frequency modulation is also not feasible if frequency is being used to identify individuals. Thus, most acoustically telemetered field data are 'interval' encoded, i.e. the interval between pings is proportional to the current datum. Data can also be encoded in the duration of the ping, but power constraints usually mean that 'off-time' is a better way to carry signal than 'on-time', which is minimized for clear signaling. Also, the highly reverberant nature of the underwater channel can extend the apparent length of a 10 ms pulse to 15–20 ms at the receiver. Most standard, commercially available acoustic receiving equipment is designed to decode data in this 'off-interval' format (Voegeli et al. 1983). Maximum update rate (i.e. minimum interval between pulses) determines the amount of data that can be transmitted. In most environments this rate is limited by echoes that arrive later than the most direct signal and are virtually indistinguishable from it. Receivers are usually set to ignore later pulses in a 'blanking interval', typically 200–300 ms. Ironically, more powerful tags echo from further away and, therefore need longer blanking intervals and have lower data rates.

Within these limits, a sensor in the same electronics package as a pinger can modulate the ping interval to

[1] Stasko, A.B. 1975. Underwater biotelemetry, an annotated bibliography. Fish. Mar. Serv. Res. Dev. Tech. Rep. 534. 31 pp.

[2] Nelson, D.R. 1990. Telemetry studies of sharks: a review with applications in resource management. pp. 239–256. *In:* H.L. Pratt, Jr., S.H. Gruber & T. Taniuchi (ed.) Elasmobranchs as Living Resources: Advances in the Biology, Ecology, Systematics, and the Status of the Fisheries, NOAA Technical Report NMFS 90.

[3] Standora, E.A., Jr., T.C. Sciarrotta, D.W. Ferrel, H.C. Carter & D.R. Nelson. 1972. Development of a multichannel, ultrasonic telemetry system for the study of shark behavior at sea. Cal. State Univ., Long Beach Foundation Tech. Rept. 5. 69 pp.

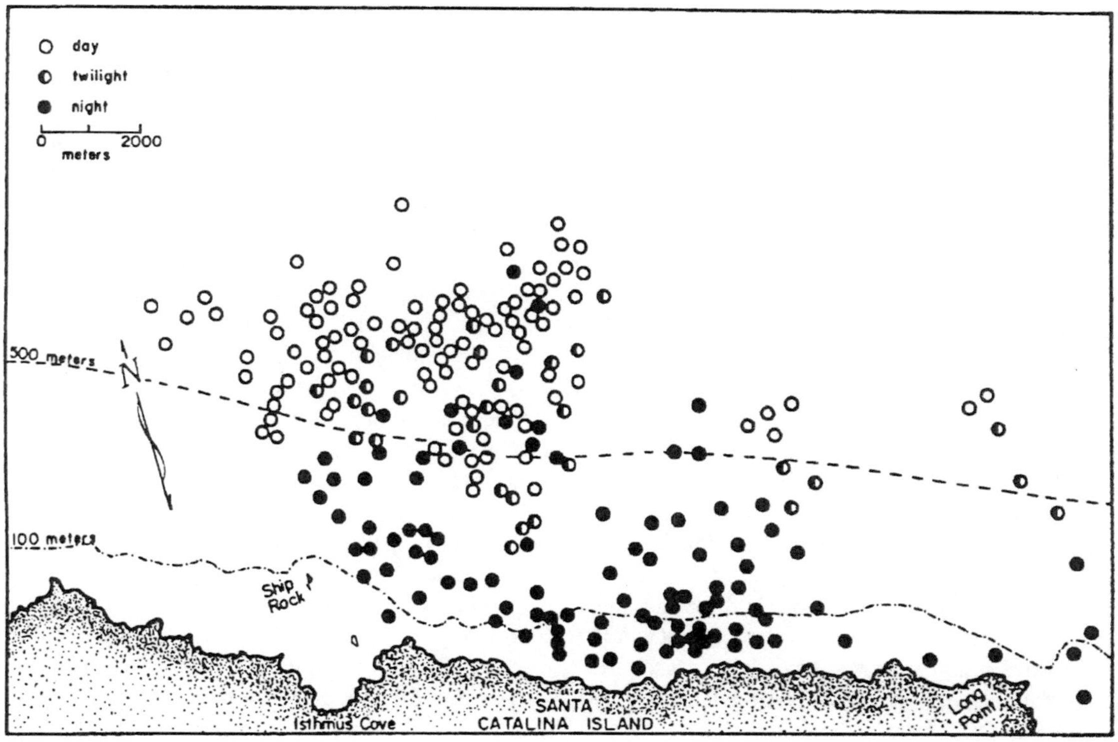

Figure 1. An early record of blue shark, *Prionace glauca*, positions shows clear diel behavior. Multiplexed data indicated active feeding in shallow water at night (Scariotta & Nelson 1977).

be proportional to the sensor value. The relationship between sensor value and ping interval is usually designed to be a linear function, easily described by a slope and an intercept. In this way a decoder can be made to measure the time between received pings and to decode a sensor value in engineering units. Thus, depth can be displayed in meters or fathoms and temperature in °C or °F, etc. In 1975, the first commercial data telemetry decoder for fish-tracking tags was introduced by Communications Associates of Long Island, New York. This CI-40 was capable of decoding both single channel data telemetry tags and dual channel tags. The first dual channel tags combined temperature and depth sensors in one package. Four pings defined a fixed synchronization interval, followed by intervals representing temperature and depth. Repeated continuously this provided a data update of both sensors in approximately 3.5 s. Alternative code schemes have been used by researchers in home-built tags. For example, Carey & Scharold (1990) alternated strings of multiple pings representing temperature with strings representing depth in their study of blue sharks. New technologies are now available (e.g. RCODE, see below) which allow multiple data telemetry transmitters to occupy a single transmission frequency.

Sensors

Researchers have monitored a wide range of biological functions based on several physical sensors, including depth, temperature, swimming speed, mortality, electromyogram, heart rate, tail-beat frequency and differential pressure (Nelson 1978, Wolcott 1995). Many of the more frequently used are available commercially in expendable versions with single or multiple sensors for gathering data from manually tracked sharks. Depth and temperature are the most common variables. Depth sensing transmitters have been commercially available for 20 years and used in numerous studies (Scariotta & Nelson 1977, Nelson et al. 1997). Sizes range down to 12 mm diameter cylinders, but the most common 16 mm diameter size has room for a variety of battery types, silver oxide or lithium, to give a number of life and power output options. Temperature sensors can be used externally or internally. In some waters with distinct thermoclines an external temperature sensor can also give a rough indication of swimming depth. Temperature sensing transmitters applied internally may monitor core body temperature, as in the study of thermoregulation in lamnid sharks (Carey et al. 1982, 1985) or be used as an indication of feeding when the prey is at a different temperature than the predator. For example, warm seals eating cool sharks or cool sharks eating warm seals (Goldman et al. 1996). A more detailed discussion of sensor types is available in this volume (Lowe & Goldman 2000).

Data loggers

Historically, active tags were actively tracked providing continuous access to transmitted data and an approximate position within the limits of tag range. Recent advances in electronic miniaturization and power requirements, particularly for microprocessors and memory circuits, have dramatically increased the options for active tagging and recovering data on complex functions like swimming power: (1) Complex data are averaged or integrated before storage or transmission to save memory, power and bandwidth. (2) Traditional archival tags store data for recovery on recapture. (3) Popup tags release themselves on a signal, at a predetermined time or when they are full, to be recovered or even to automatically transmit their data to satellites. (4) Transponder tags send out stored data in response to a signal (e.g. CHAT tags, see below).

Sensors are read by a microprocessor and the readings stored in semiconductor memory. These can include temperature, depth, light level, swimming speed and compass heading. With flash RAM memory archival tags can be used to gather multiple years of data. Simple archival tags must be recovered in order to read the stored data, often by connecting to a serial port on a personal computer, or sometimes through an optical or electromagnetic interface. They are quite economical, but the need to recapture the shark to read the stored data means that only a low percentage of the loggers are recovered. This must be factored into the cost per data point calculation in a trade-off against the costs of more complex tag communication systems.

The newer techniques offer many advantages over traditional tracking approaches including much longer data sets, but they also create their own unique problems. When one boat tracks one shark transmitting on one frequency the data are expensive, but the context is known. It is automatically clear that a particular shark with a sensor monitoring a specific variable is near the boat. To provide equivalent context, archival tags must not only record the variable, but also identify themselves and record the time and location at which the variable was recorded. Standard archival tags have identifiers on the outside and popups that transmit to satellites start with a unique identifying code. Time can be easily recorded from internal computer clocks, but location is tricky.

Navigation sensors are required if we are to know where an archivally tagged shark has been. The success of geo-location with light sensors on mammals has created a demand for sensors that will work for other animals. The light level sensor has been used for geo-location in much the same way as a navigator would use a sextant. The time of sunrise and sunset is used to estimate local noon, which can be used to compute longitude by comparison to the internal clock. The day length can be used to estimate latitude. Initially, the use of a light sensor meant that the tag had to be attached externally, but the CSIRO fisheries laboratory in Australia developed an implantable archival tag, which could use light levels for geo-location. The light sensor was extended on a thin wire trailing it through a hole in the fish's skin. This same trailing light sensor also contained a temperature sensor so that internal and external temperatures could be recorded. This tag was built commercially by Zelcon and used initially for southern bluefin tuna and recently on school sharks (West & Stevens 2001). Accuracy is on the order of 40 km, but some recovered tags have shown significant migrations even though the tuna was recaptured only a few kilometers from the original tagging site. These techniques can be applied to sharks that swim at depths with adequate light levels and migrate over large enough distances. This requires compensation of light level with depth because vertical movements of sharks cause variations in ambient light level, which may confuse the measurement of sunrise or sunset time. Other sensors, which might aid in the prediction of where sharks have been include: compass, acceleration, magnetic field strength and advanced swimming speed techniques. With these sensors, a form of dead reckoning could be achieved. Temperature and depth data can also be used for positioning when compared to oceanographic data.

Remote release data loggers (popups)

An alternative to recapture of the shark is to provide a method of releasing the data logger by remote command. One of the first release mechanisms used to retrieve a tag from a shark was an electronic timer-triggered release mechanism using the rapid electrolytic corrosion of a metal loop when a battery was applied between the loop and another electrode (Nelson & McKibben 1981). This technique was used to retrieve an ultrasonic depth transmitter and activate a radio beacon to aid in location of the tag. The timer approach required that the researcher be in the area at exactly the right time to retrieve the device. Another more flexible approach is to allow the researcher to release the device at the press of a button on a remote control device. Such a device for release of archival tags was designed for use on sea mammals but could be adapted for use on large sharks. The device uses syntactic foam to make the logger buoyant. A miniature ultrasonic receiver is placed with the logger on a common mounting plate. On receipt of the correct ultrasonic command, the receiver applies a potential between two electrodes. One of these electrodes, a stainless steel pin, is dissolved by rapid galvanic action, which releases a latch to allow the package to float to the surface. The ultrasonic receiver also activates a pinger to aid in the recovery of the device (Hammill et al. 1999).

Current work with devices released from animals uses Argos satellite beacons to signal the release location. This technique can even include telemetry of statistics of archived data to the satellite from a miniature data logger[4].

Automated monitors

The newest technology available for remote monitoring of free-swimming sharks in their natural environment combines acoustic telemetry with a range of automated recorders. In the past, continuous tracking of individuals has been expensive, time consuming and has gathered data from a small number of sharks. The attachment and later retrieval of data storage devices can provide data on where the shark has been, but recapture rates require that many such devices be deployed in the hope of recovering only a few percent. The combination of sensors, data storage and acoustic download of the stored data, without recapture of the shark is now possible. Both mobile tracking receivers and fixed bottom mounted sites have been developed to collect the data from the tagged sharks. Simple fixed sites gather IDs of passing sharks, but more sophisticated ones collect a full data download. Data transfer is also possible in a fine-scale positioning system for studying such questions as mating or predator–prey interaction.

Automated pinger detection and identification

Automated pinger identification sites contain an ultrasonic receiver with a microprocessor and memory. The microprocessor compares received pinger periods with a list of values of actual pingers released. Automated monitors are available in fully submersible packages that can be placed on the sea bottom, on reefs and on fish aggregation devices (FAD). The first commercially available monitor was produced by Ultrasonic Telemetry Systems of Placentia, California. It used two alternating ping frequencies of 38.4 and 40 kHz (see McKibben et al. 1985). One of the important features of an automated monitor is its ability to detect the correct ID of multiple pingers in the same location. The VEMCO VR20 monitor (Figure 2a) developed in 1987 contains a sophisticated algorithm for identifying as many as eight transmitters on the same frequency at the same time. The receiver can also be programmed to search ten or more different frequencies for passing pingers, allowing detection of as many as 80 pingers in the same place and time.

In practice, however, the number of frequencies scanned is reduced to about five. This keeps the time taken to scan all frequencies down to about 3 min or less and allows at least 2 kHz frequency spacing. It is best to place automated monitors in sites with a low probability of multiple transmitters being present on each frequency. Even though it is possible to detect as many as eight pingers on one frequency, there is an increased chance that a pinger that is not there may be reported because there are so many signals in the water.

Simple pingers still have their place in studies that combine automated detection with manual tracking; the continuous 'pings' are easy to follow with directional hydrophones and diver held receivers. However,

---

[4] Hill, R. 1997. Data archival and satellite linked tags from Wildlife Computers. p. 34. *In:* G.W. Boehlert (ed.) Application of Acoustic and Archival Tags to Estuarine, Nearshore, and Offshore Habitat Utilization and Movements by Salmonids, NOAA Tech. Mem. NMFS-SWFSC 236.

a b

Figure 2. a – compares the size of a 34 mm diameter CHAT (Communicating Histogram Acoustic Transponder) tag that stores two types of data to be released on an acoustic query with a bottom-resident VR20T that records CHAT and other tags on multiple frequencies. b – shows the small, economical single frequency VR1 on the same scale, which records RCODE tags on a single frequency as well as the presence of CHAT tags (without data download).

they can be improved on in terms of battery life and number of unique signals to identify sharks by coding the signal instead of a simple fixed-rate ping.

Aural coded ID pingers

Coded ID pingers were developed to increase the number of unique individuals that could be identified. Early attempts at coding such as the Sonotronics aural code pingers used a long string of pulses with missing pulses to create a repeating sequence of pings that can be easily identified by ear without an electronic decoder. The user simply counted the number of pings between breaks and wrote down the three-digit number obtained. This scheme worked until two or more sharks showed up at the same place, with an overlapping and constant series of pings. This overlap in time and space is called a 'code collision'. For continuously repeating codes there is a 100% chance of collision, if two transmitters are present at the same location. For automated detection equipment, a more sophisticated scheme of infrequent code transmissions was developed (Voegeli et al. 1997).

Infrequent codes

Infrequent codes consist of a short burst of pings followed by a long period of silence. The silence provides time for other transmissions to be heard. The ratio of code transmission time to silence can be chosen to suit a particular study. This scheme allows many unique ID codes to exist on the same frequency and allows multiple transmitters to be detected at the same site. A big advantage of infrequent codes is that the transmitter battery life is extended to approximately five times that of a fixed-rate pinger. It is desirable to have the repeat interval between code transmissions different for each tag so that two tags that collide on one occasion will separate on the next. For large numbers of tags in the water this can force some of the repeat intervals to be many minutes which could allow a shark to swim through the detection zone without sending a transmission. To correct this problem a variable silent time is used.

Random repeat infrequent codes (RCODE)

RCODE or random repeat code, uses a pseudo-random number generator to vary the silent period between code transmissions. If two transmissions collide on one occasion, then the next time they transmit they will be apart and easily detected. The big advantage of this scheme over placing fixed-rate pingers on different frequencies is that simple and low cost single-channel detectors and automated monitors can be built. An example is the VR1 receiver (Figure 2b) that can run for six months on a single 'C' cell lithium battery. RCODE transmitters can have either 256 or 4096 different ID codes depending on the level of security required. RCODE transmitters can also carry single sensor data. Single channel data telemetry can be substituted for 14 of the possible ID codes. This allows a mixture of 14 depth/ID tags plus 256 ID tags on the same frequency.

Fixed hydrophone monitors

The use of automated fully submerged monitors is becoming common because they are on the job day and night and in all weather conditions. New RCODE pingers provide up to 4096 different ID codes for individual animals. Low cost automated monitors such as the VR1 have been developed so that large networks of monitors can be deployed for up to six months before battery change. The earlier work with simple pinger identification receivers has shown that some species exhibit a homing behavior and repeatedly visit some sites (Klimley et al. 1988, 1998). This repeated homing behavior can be used to collect data from a device on the shark at each return to a fixed site. Clearly a more sophisticated tag was required along with a more complex communication protocol.

Two-way communication

It is important to note that all transmitters mentioned to this point communicate only in one direction, data and/or ID from the transmitter to the receiver. The VR1 or VR20 automated monitors simply listen for signals from passing sharks. These signals are being transmitted from the sharks constantly, needed or not. The next step in the evolution of fish tracking tags was development of a two-way communicating tag. The simple form of a two-way tag is called a transponder, for each received signal it replies with another. The advantages of transponders are that they do not waste power on transmissions that are not requested, and an accurate distance can be obtained by measurement of time between command and reply (Bagley & Priede 1997). When used for data transmission they have the advantage of verification and if necessary re-transmission of the data.

Remote data telemetry from data loggers (CHAT tags)

The one thing that was obvious from the outset of CHAT tag development was that transmission of megabytes of raw data was out of the question. This would create a communication bottleneck because of the bandwidth of the acoustic channel. It was essential to process as much data on the tag as possible, thereby reducing the amount of stored data. The first data reduction algorithm was a simple set of histograms with minimum, maximum and average values for depth and temperature sensors. Hence the name 'Communicating histogram acoustic transponder' (CHAT). The CHAT tag incorporates sensors, a micro-controller, memory and two-way acoustic communication elements. The first models built in 1997 contained only two sensors, temperature and depth (Figure 2a). In 1998 a swim speed sensor using a paddle wheel was added. The three-sensor paddle wheel version requires external attachment but the simple temperature and depth version can be internally implanted.

The two-way communication link provides a method for verifying the error checking codes and setting pointers within each CHAT tag to keep track of data blocks which have not yet been downloaded. The two-way communication also allows the CHAT tag to be programmed remotely so that its data sampling interval and averaging interval can be changed without recapture. Table 1 serves as an example of some of the tradeoffs in archival tags. Short data-averaging intervals will fill memory more quickly, so it is important to balance averaging interval with expected frequency of downloads when sharks return, to maintain a complete record. It also may not be appropriate to gather large amounts of fine-scale data if the shark does not spend sufficient time at the home site to download the data.

Table 1. Time to fill memory and download data for two channel CHAT tags.

Average interval	Sample interval	Time to fill memory	Download time, 1 week of data
1 week	1 min	39 years*	31 s
1 day	1 min	5.6 years*	3.6 min
12 h	1 min	2.8 years*	7.3 min
6 h	1 min	1.4 years	15 min
2 h	1 min	170 days	44 min
1 h	1 min	85 days	1.5 h
30 min	10 s	42 days	2.9 h
15 min	10 s	21 days	5.8 h
10 min	10 s	14 days	8.7 h
5 min	10 s	7.1 days	17 h

*Battery life of 2 years would expire before memory was filled for these cases.

Another important feature of the two-way communication is the ability to handle multiple CHAT tags in the same area. The communication protocol developed makes sure that each CHAT tag will take its turn to transmit without interrupting an exchange already in progress.

The mobile VR28T tracking system was the first receiver system to communicate with CHAT tags. It has the major advantage of allowing the user to track the shark and stay with it until all the data have been downloaded. It can be used on any vessel that has a sheltered steering station to mount a laptop or larger personal computer. The VR28 receiver system uses four directional hydrophones aimed in forward, aft, port and starboard directions. The hydrophones are each connected to an independent receiver module with a common controller connected by serial port to the user's personal computer (PC), which serves as the interface and data storage facility. The PC screen (Figure 3) provides a graphic signal strength indicator and shows an arrow to guide a helmsman to the target shark. A numeric display shows the distance to the CHAT tag calculated with a high degree of accuracy (~1 m) from speed of sound out and back. This receiver provides a low stress

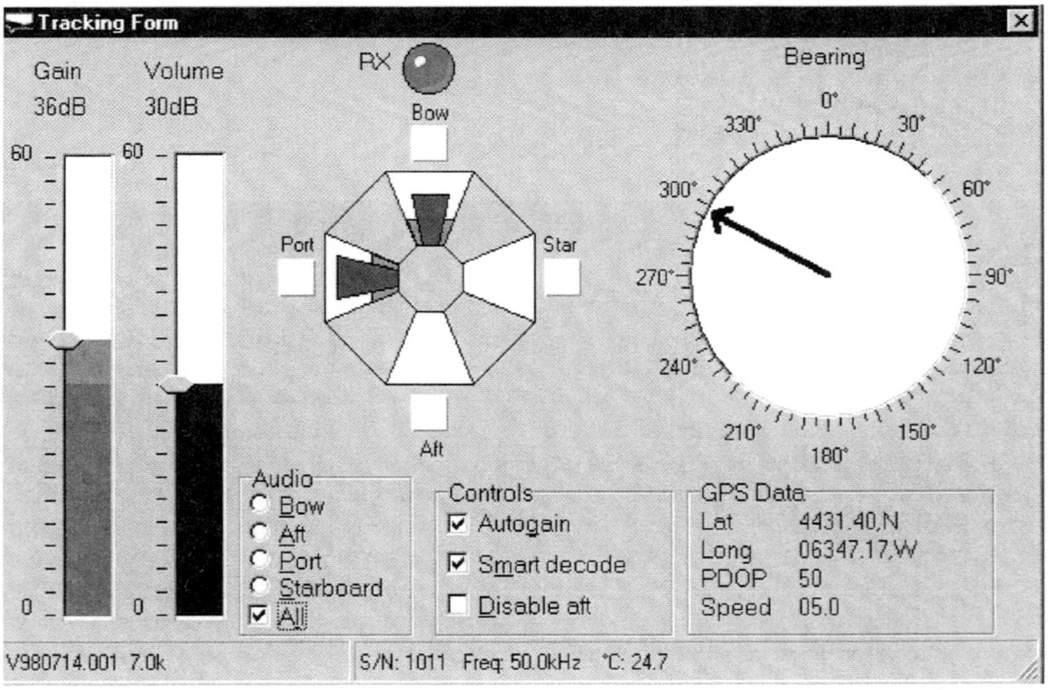

Figure 3. The helmsman's guide display on this four hydrophone VR28 system helps keep the vessel close to the shark for CHAT download.

tracking system, which allows a boat or ship to follow all but the fastest sharks for hours to days, giving ample time to download archived data. For even faster sharks, a high resolution tracking system (VH170) which uses a 17 hydrophone array has also been built that can download data from CHAT tags. Preliminary tests of this receiver on the new 2 494 ton research ship 'Shoyo Maru' moving at 5 kt gave a communication range of 2 km, giving a coverage of 12 km^2.

VR20T automatic download stations provide fixed sites for CHAT tag data (Figure 2a). Given that an individual shark may not stay within the reception zone of a fixed site for more than a few tens of minutes the data sampling and averaging intervals should be set to keep data to a minimum. However, to help increase the chances of a successful download of all the data there is a protocol, which downloads the data piecemeal at a number of sites. For example, if a CHAT tag leaves a fixed site after downloading only 10% of the recorded data, it will start the download process at the next untransmitted block when it encounters another fixed or mobile site. Data reduction programs are available to sort the multiple pieces of data to each tag serial number. At the time of writing of this document tests done on VR20T stations with CHAT tags on noisy inshore locations in Nova Scotia gave a communication distance of up to 1 km or coverage of 3 km^2. To expand the detection network of CHAT tags a small, low-cost VR1 receiver (Figure 2b) now detects the time and date of passage and serial numbers. These units can supplement or predict optimum placement of the full download sites.

Radio acoustic positioning (RAP) buoys

This positioning system is used to obtain detailed information on the movements of pinger-equipped animals, using an approach pioneered by McKibben & Nelson (1981). Other efforts used fixed hydrophone arrays connected by wire to a shore-based receiver system (Voegeli 1988, Wartzok et al. 1992). The fine-scale movements can be correlated with the habitat or movements of other tagged animals of the same species or their predators or prey.

The use of a buoy-based positioning system allows a study to be set up in remote locations where wired hydrophones are not practical (O'Dor et al. 1989). They are used in situations where the monitored sharks are likely to stay within or repeatedly visit an area of a few square kilometers. This is usually an area of feeding or breeding activity that can be many kilometers from the base station site. One application is the study of animal movements in response to human activity, such as trawl-capture efforts (Engas et al. 1996, 1997, Løkkeborg & Fernö 1999), and bait odor plumes (Skajaa et al. 1998). Other possibilities for shark studies include predator/prey interactions. Klimley et al. (2000) have studied sharks as predators using a VRAP system (Figure 4) set up at a sea-mammal breeding colony haul out site at Ano Nuevo Island in California. Figure 5 shows tracks for a stingray and a species of squid it preys upon at the breeding grounds studied by the authors in South Africa.

A VRAP buoy system uses three or more buoys positioned with GPS (preferably differential GPS) containing radio-controlled ultrasonic receivers (Figure 4). The receivers on each buoy select the same transmitter and detect the arrival times of each ping received. The pings from all three buoys are then sent by radio signal to a base station where a PC (DOS or Windows-based) calculates the position of the pinger with respect to the three buoys. The PC is used to show the three buoy positions and the pinger positions on a screen with a different color for each pinger track (Figure 6). The VRAP buoy system performs a self-calibration cycle at user-selected intervals. Calibration is done by turning on each buoy's 'local pinger', measuring the time to adjacent bouys and calculating distances between buoy pairs. The accuracy of pinger positions is best within the buoy triangle and can be to within 1 m. As a pinger is moved outside of the triangle the accuracy degrades to several meters, but a Windows-based plotting-accuracy program is available to show the expected accuracy for a given buoy spacing geometry and uncertainty in signal arrival time at each buoy. It should be noted that a signal must be received by three buoys in order to determine position, but data from any sensor can be recorded as long as the transmitter is received by one buoy. Commercially available software only listens to three buoys at a time, but it is possible to cover larger areas with more buoys by manually switching between buoy triangles. It is also important to know the depth of the sharks in order to obtain an accurate horizontal plane position. If depth telemetry tags are used, the data from each tag can be used in the position equation for a three-buoy system.

With a four-buoy system it is possible to calculate the shark's depth and horizontal plane position. Such a system has been designed and tested by lowering two of four buoy hydrophones on extension cables (Sauer et al. 1997, O'Dor et al. 1998). It was found that the

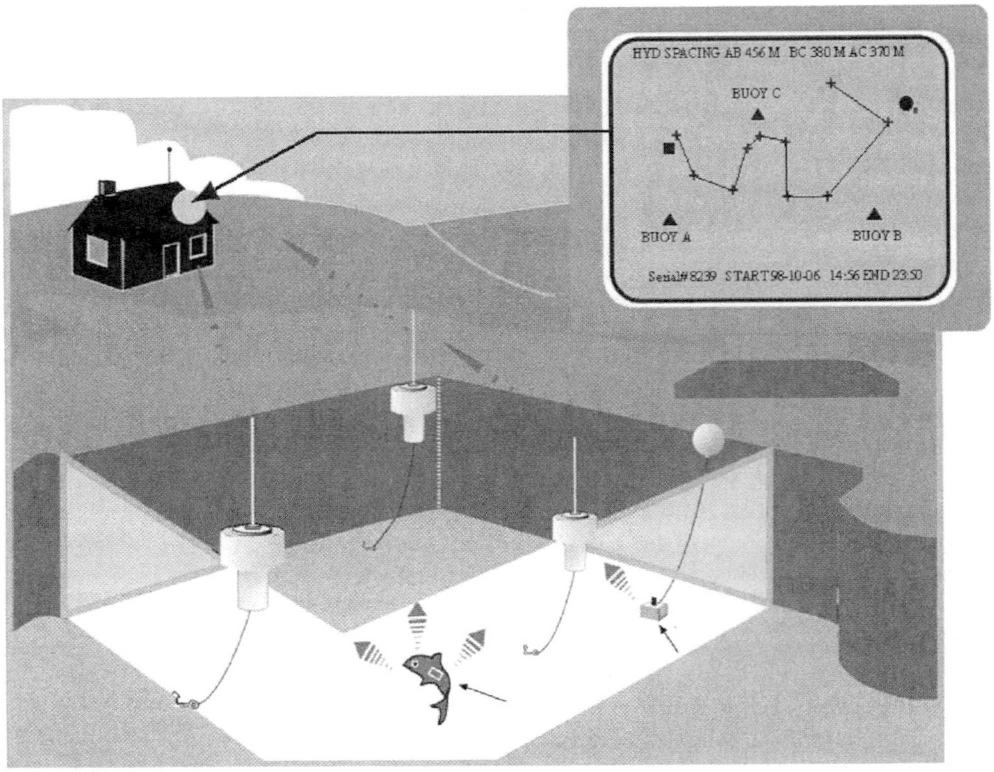

Figure 4. This radio-acoustic positioning system provides continuous triangulated positions and data using three radio-linked, self-positioning buoys and a GPS located reference pinger providing a maximum acoustic range of 1 km and a maximum radio range of 6 km.

depth calculation was sufficiently accurate only if the horizontal to vertical hydrophone spacing was approximately equal. This placed severe limitations on such systems in shallow coastal waters, indicating that such an arrangement is only a last resort for animals too small to carry depth transmitters. However, Figure 5 shows that post-analysis can extract useful information from animals as small as 40 g squid, as well as divers and large rays.

The VRAP buoy system as it now exists can select any ultrasonic tag between 25 and 80 kHz in two ranges 25–50 kHz and 50–80 kHz. The appropriate hydrophone must be selected to match the frequency range. The lower frequency range is best for large sharks because larger transmitters that transmit further can be used. The higher frequency range works for smaller sharks and juveniles but can also be used with larger sharks if detection range is not a concern. Buoy spacing is up to 500 m on each side of an equilateral triangle for the high frequency range, and can approach 1000 m on each side for the low frequency range.

The number of animals that can be tracked at the same time with conventional pingers or telemetry tags is approximately ten. This limitation is one of signal bandwidth. A new version is currently under development which will use the RCODE tags to allow multiple tags on each frequency. With the current single channel heterodyne ultrasonic receivers the buoys must all listen to one frequency at a time. This limits the position update rate to approximately one position every 10 s from each animal. The reason that ten or more seconds are used is so that multiple pings can be received and analyzed before each position is calculated. This means that with ten animals it will take over a minute to gather positions for all of them. Because tracks are viewed in near real time, the user can view individuals selectively. For this there is a 'fast track' algorithm, which allows the user to stay tuned to a particular animal for a longer period of time and obtain a position for every transmitted ping. This can provide position updates about once per second.

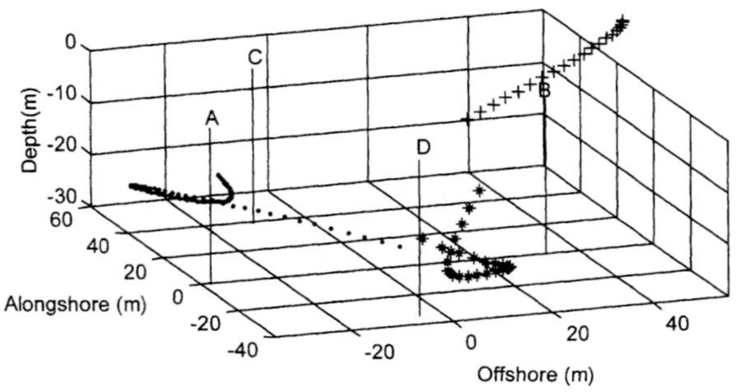

Figure 5. A 1-h '4-D' RAP track of 'predator–prey' interaction. As MJS dives off a boat near the centre (0, 0) of the array of buoys (A, B, C, D) carrying an activated 50 kHz dart tag (*), he scares away a squid, *Loligo vulgaris reynaudii*, swimming near the surface, that is already carrying a 63.7 kHz tag (+). He searches for a squid egg bed near the centre of the array on the spawning grounds in South Africa (Sauer et al. 1997). There he darts a stingray *Dasyatis brevicaudata* (•) hiding among the eggs to prey on mating squid (Sauer & Smale 1991). The ray moves rapidly outside the array, rests near C and then disappears along shore. Although no ray–squid interactions are recorded, the diver–squid, diver–ray interactions demonstrate the capacity of the system. The spacing of symbols on the track at 1 min intervals adds the 4th dimension, time, illustrating velocity. Details of the array and time-spline post-analysis techniques used to produce the figure were described by O'Dor et al. (1998).

Future directions

Future sensors

Speed costs sharks energy, so monitoring it gives an indication both of long-term energy budgets and short-term investments in activity. The most common way to measure it is a paddle wheel, modified propeller or turbine (Voegeli & Pincock 1981). A rotating shaft coupled to an electronic sensor magnetically, allows measurement of speeds down to about $0.2\,\mathrm{ms}^{-1}$, but slower swimming speeds are more difficult to measure (Lowe 1996). These work well on large sharks, but can jam and increase drag dramatically in small ones. Differential pressure sensors used to measure very small variations in pressure without interference from ambient static pressure due to depth are available in 16 mm cylindrical tags. These were originally developed to monitor cephalopod jet pressure (Webber & O'Dor 1986, O'Dor et al. 1993, 1994), but now have many other applications. They have been used to form an electronic speed sensor with no moving parts using the pitot principle, able to resolve speed to approximately the same $0.2\,\mathrm{ms}^{-1}$ lower limit (Voegeli 1990), as shown in Figure 7. Models with increased gain are currently being tested and will take this even lower.

An alternate approach is electromyogram tags for tail-beat frequency, but these are problematic on any size of shark because of the need to place electrodes so as to detect very low potentials. The electrodes tend to move, foul and exhibit varying electrical impedance over time. For those who wish to persevere, such telemetry can yield useful measurements, but there are questions about the value of tail-beat frequency alone as a speed indicator, as beat amplitude is an independent variable. Similar electrodes can measure heart rate to estimate energy consumption directly, but the same problems arise with low potentials and stroke volume has been shown to vary with activity (Scharold et al. 1989, Scharold & Gruber 1991).

Field studies may only augment and validate laboratory studies of small or juvenile sharks but may be indispensable for studies of specimens that are too large or delicate for laboratory studies. Current and future work in speed sensing focuses on direct measurement of overall tail-beat dynamics and the correlation of these measurements with swimming speed (Webber et al. 1999).

Newer techniques for measuring swimming power and cardiac output in teleosts (Webber et al. 1999) are of equal interest for sharks. A key type still under development is the cardiac output sensor using doppler blood flow measurement. Laboratory instruments with electronics packaged in a cabinet the size of a shoe box measure the velocity of blood cells in a vessel acoustically without penetration. An ongoing collaboration between the authors is aimed at producing a cardiac output tag that will fit on an adult cod. This tag will allow a better estimation of the metabolic cost of normal activity by accounting for both heart rate and stroke volume (Webber et al. 1998). These tags will be more informative and more stable than current heart rate tags because they require no direct muscle penetration or invasion of blood vessels. Ultimately, it should be possible to correlate swimming power, measured directly from differential pressure and incorporating both tail-beat frequency and amplitude, with cardiac output for sharks moving known distances and speeds in VRAP systems.

Future improvements in CHAT tag technology

Planned improvements include: (1) making the tag smaller – the next size planned is a 22 mm diameter

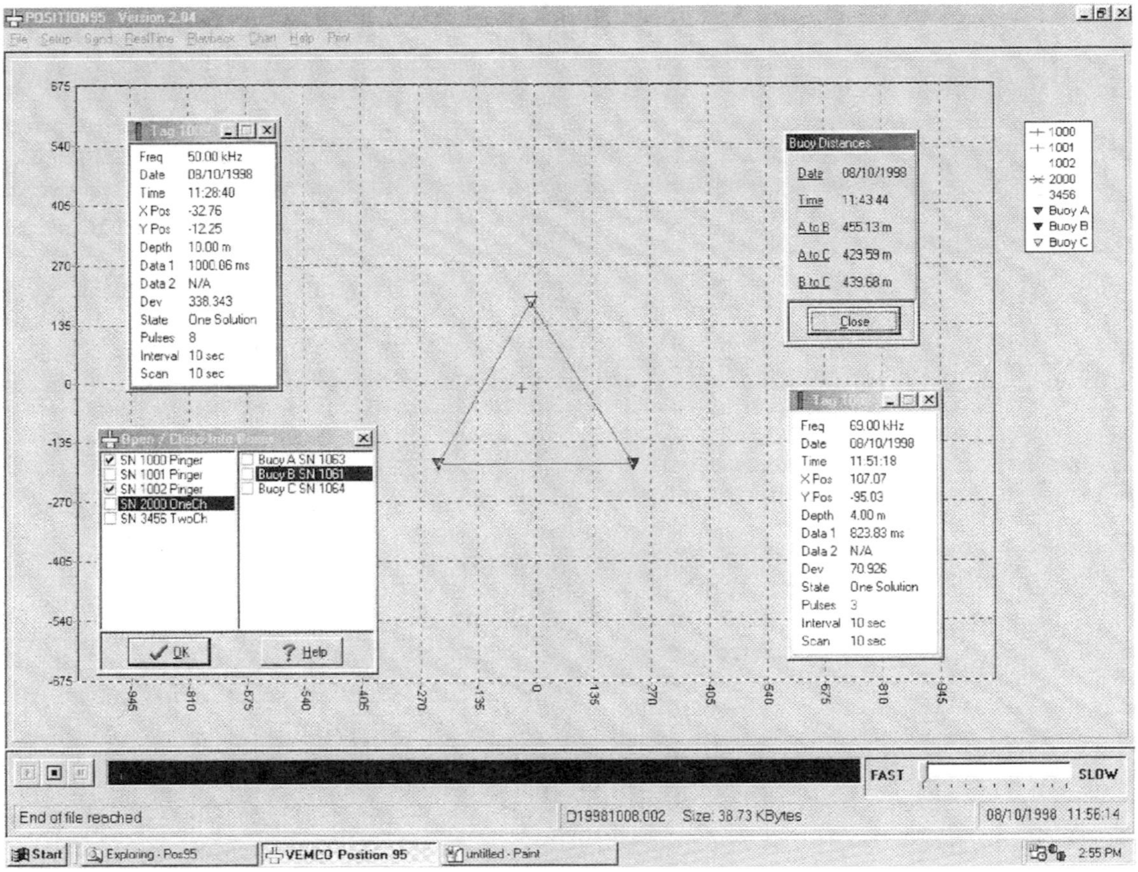

Figure 6. The Windows 95 display for VRAP system shows details about buoy positions and two transmitters currently being positioned while the tracking display remains on screen.

version, (2) adding more sensors with a focus on navigation and geo-location, and (3) increasing data transmission rate. This is not a simple matter because to increase data rate usually means decreasing communication distance which, in turn, results in less time when the shark is within range. Digital signal processing techniques are possible that improve the transmission from CHAT tag to receiver site. It is not as necessary to have a higher data rate in the other direction because there are fewer signals sent.

Future improvements to VRAP buoy technology

The VRAP buoys have battery endurance between charge cycles of about one week. One way to ensure that an array of three buoys can be kept functioning continuously is to have a fourth buoy, which can be charged and then replace one of the operational buoys in rotation. This buoy can also serve as a spare in case one breaks a mooring and goes adrift for a while. There is also a battery charge shuttle device, which consists of two batteries that are charged ashore and then taken out to each buoy to dump the charge through a regulator circuit. Solar panels are under study to provide or at least supplement the battery power.

Versions are under development with multi-channel ultrasonic receivers for applications where it is important to know exactly where two or more animals are at the same instant in time. The penalty for this advance will be that more power will be consumed from the battery so it may not be an advantage for everyone.

Reference of buoy and shark positions to real world co-ordinates is another area under development. In areas where differential GPS is available it may be possible to place GPS receivers on two buoys. The

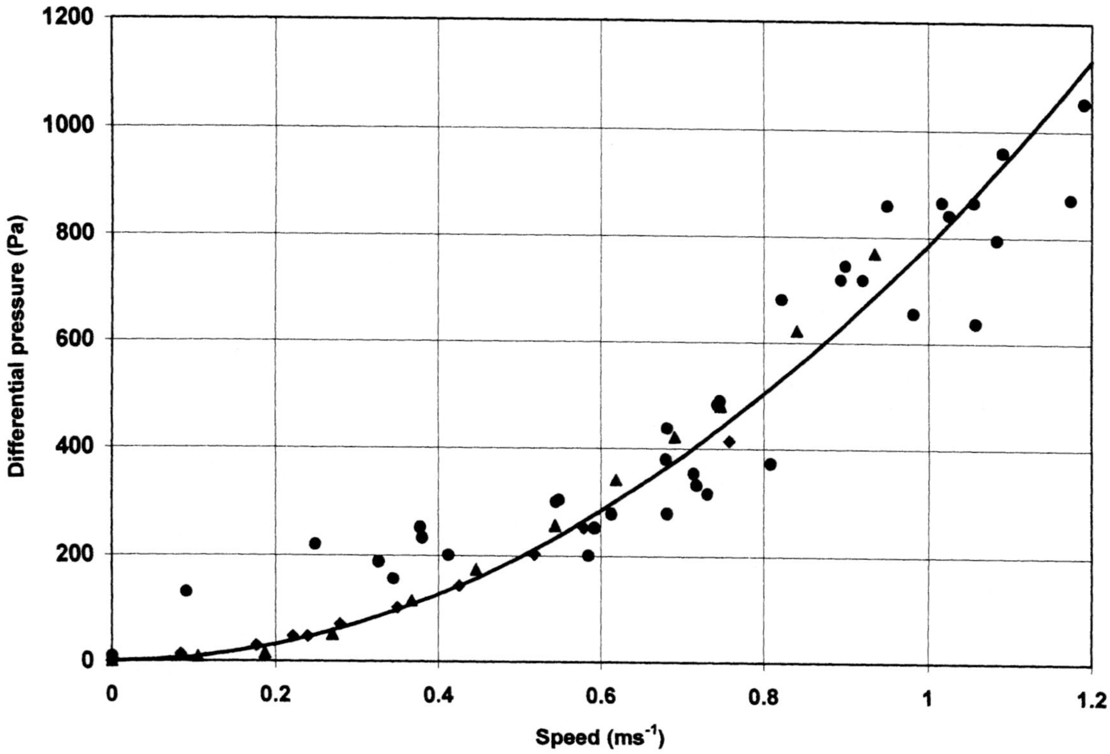

Figure 7. Comparison of speeds in free-swimming, 1.3 m gully shark, *Triakis megalopterus*, to flume calibrations for pitot-type velocimeters with no moving parts. Solid line is a regression of pressure on speed^2 for two static sensors (triangle & diamond) in a flume, oriented with positive port upstream and negative port downstream, to maximize pressure differential at speed. The shark speeds (circles) are estimated from video recordings in the Port Elizabeth Aquarium, S. Africa, and errors are significant at high speeds. Tags respond to lateral as well as forward movements giving elevated pressures at low speeds on undulating sharks (Webber 1999).

use of fixed bottom-mounted references is the alternative, but they must be 'surveyed in' when they are deployed (see Figure 4). When an accurate ground reference mechanism has been established it is possible to superimpose buoy and shark positions on a bit map showing bathymetry and any other graphical information such as bottom type, vegetation, etc. In addition to their normal role as a fine-scale positioning array, individual buoys can be placed to listen to and report the ID of passing RCODE and CHAT tags. This real-time reporting can be used to alert a nearby surface tracking boat that a CHAT tag is in the area.

As the use of acoustic telemetry increases, with both transmitters and receivers proliferating, consideration must be given to the possibility that studies will begin to overlap, particularly for wide-ranging species. On the negative side one scientist's signal becomes another's noise, but on the positive side this could lead to increased data returns through co-operation. Will this constrain a field famous for its individualistic 'telemetry cowboys' (and girls)?

Acknowledgements

Various aspects of the work reported here have been supported by Natural Sciences and Engineering Research Council (Canada) and Foundation for Research Development (South Africa).

References cited

Bagley, P.M. & I.G. Priede. 1997. An autonomous free-fall acoustic tracking system for investigation of fish behaviour at abyssal depths. Aquat. Living Res. 10: 67–74.

Baldwin, H.A. & G.F. Ingle. 1964. A remote control technique for the study of olfaction in sharks. National Biomedical and Scientific Instrumentation Symposium 2: 217–228.

Bass, G.A. & M. Rascovich. 1965. A device for the sonic tracking of large fishes. Zoologica 50: 75–82.
Carey, F.G. & K.D. Lawson. 1973. Temperature regulation in a free-swimming bluefin tuna. Comp. Biochem Physiol. 44A: 375–392.
Carey, F.G. & J.V. Scharold. 1990. Movement of blue sharks (*Prionace glauca*) in depth and course. Mar. Biol. 106: 329–342.
Carey, F.G., J.M. Teal, J.W. Kanwisher, K.D. Lawson & J.S. Beckett. 1971. Warm bodied fish. Amer. Zool. 11: 137–145.
Carey, F.G., J.W. Kaniwisher, O. Brazier, G. Garielson, J.G. Casey & H.L. Pratt Jr. 1982. Temperature and activities of a white shark, *Carcharodon carcharias*. Copeia 1982: 254–260.
Carey, F.G., J.G. Casey, H.L. Pratt, D. Urquhart & J.E. McCosker. 1985. Temperature, heat production and heat exchange in Lamnid sharks. pp. 92–108. *In:* G. Sibley, J.A. Seigel & C.C. Swift (ed.) Biology of the White Shark, Mem. S. Cal. Acad. Sci. 8.
Casey, J.G. 1985. Trans-Atlantic migrations of the blue shark: a case history of cooperative shark tagging. pp. 253–268. *In:* R.H. Stroud (ed.) World Angling Resources and Challenges, International Game Fish Association, Fort Lauderdale.
Casey, J.G. & N.E. Kohler. 1992. Tagging studies on the shortfin mako shark (*Isurus oxyrinchus*) in the Western North Atlantic. Aust. J. Mar. Freshwater Res. 43: 45–60.
Engas, A., S. Løkkeborg & J.T. Ovredal. 1996. Natural and fishing gear induced behaviour of tagged fish, studied by means of a stationary positioning system. pp. 203–211. *In:* E. Baras & J.C. Philippart (ed.) Underwater Biotelemetry, University of Liege, Liege.
Engas, A., E.K. Haugland & J.T. Ovredal. 1997. Reactions of cod (*Gadhus morhua* L.) in the pre-vessel zone to an approaching trawler under different light conditions. Hydrobiologia 371/372: 109–206.
Goldman, K.J., S.D. Anderson, J.E. McCosker & A.P. Klimley. 1996. Temperature, swimming depth and movements of a white shark at the South Farallon Islands, California. pp. 111–120. *In:* A.P. Klimley & D.G. Ainley (ed.) Great White Sharks: The Biology of *Carcharodon carcharias*, Academic Press, San Diego.
Hammill, M.O., V. Lesage, G. Lobb, P. Carter & F.A. Voegeli. 1999. A remote release mechanism to recover time-depth recorders from marine mammals. Mar. Mam. Sci. 15: 584–588.
Klimley, A.P., S.B. Butler, D.R. Nelson & A.T. Stull. 1988. Diel movements of scalloped hammerhead sharks (*Sphyrna lewini* Griffith and Smith) to and from a seamount in the Gulf of California. J. Fish Biol. 33: 751–761.
Klimley, A.P., F.A. Voegeli, S.C. Beavers & B.J. LeBoeuf. 1998. Automated listening stations for tagged marine fishes. Mar. Tech. Soc. J. 32: 94–101.
Klimley, A.P., B.J. LeBoeuf, K.M. Cantara, J.E. Richert, S.F. Davis & S. Van Sommeran. 2000. Radio-acoustic positioning: a tool for studying site-specific behavior of large marine animals. Mar. Biol. (in press).
Kohler, N.E. & P.A. Turner. 2001. Shark tagging: a review of conventional methods and studies. Env. Biol. Fish. 60: 191–223 (this volume).
Løkkeborg, S. & A. Fernö. 1999. Diel activity pattern and food search behaviour in cod, *Gadus morhua*. Env. Biol. Fish. 54: 345–353.
Lowe, C.G. 1996. Kinematics and critical swimming speed of juvenile scalloped hammerhead sharks. J. Exp. Biol. 199: 2605–2610.
Lowe, C.G. & K.J. Goldman. 2000. Physiological telemetry of elasmobranchs: taking the laboratory into the field. Env. Biol. Fish. (in press).
McKibben, J.N. & D.R. Nelson. 1981. A portable real-time, x–y plotting system for ultrasonic tracking of fish. pp. 105–115. *In:* F.M. Long (ed.) Proceedings of the 3rd International Conference on Wildlife Biotelemetry, University of Wyoming, Laramie.
McKibben, J.N., D.R. Nelson & S.H. Gruber. 1985. A data logging monitor for aquatic biotelemetry. pp 90–96. *In:* F.M. Long (ed.) Proceedings of the 5th International Conference on Wildlife Biotelemetry, University of Wyoming, Laramie.
Nelson, D.R. 1978. Telemetering techniques for the study of free-ranging sharks. pp. 419–482. *In:* E.S. Hodgson & R.F. Mathewson (ed.) Sensory Biology of Sharks, Skates and Rays, Office of Naval Research, Department of the Navy, Arlington.
Nelson, D.R. & J. McKibben. 1981. Timed release, recoverable, ultrasonic/radio transmitters for tracking pelagic sharks. pp. 90–104. *In:* F.M. Long (ed.) Proceedings of the 5th International Conference on Wildlife Biotelemetry, University of Wyoming, Laramie.
Nelson, D.R., J.N. McKibben, W.R. Strong, Jr., C.G. Lowe, J.A. Sisneros, D.M. Schroeder & R.J. Lavenberg. 1997. An acoustic tracking of a megamouth shark, *Megachasma pelagios*: a crespuscular vertical migrator. Env. Biol. Fish. 49: 389–399.
O'Dor, R.K., D.M. Webber & F.A. Voegeli. 1989. A multiple buoy acoustic/radio telemetry system for automated positioning and telemetry of physical and physiological data. pp. 444–452. *In:* A. Uchiyama & C.J. Amlaner (ed.) Proceedings of the 10th International Symposium on Biotelemetry, Waseda Univ. Press, Tokyo.
O'Dor, R.K., J. Forsythe, D.M. Webber, J. Wells & M.J. Wells. 1993. The cost of living for *Nautilus* in the wild. Nature 363: 626–628.
O'Dor, R.K., J.A. Hoar, D.M. Webber, F.G. Carey, S. Tanaka, H. Martins & F.M. Porteiro. 1994. Squid (*Loligo forbesi*) performance and metabolic rates in nature. Mar. Fresh. Behav. Physiol. 25: 163–177.
O'Dor, R.K., Y. Andrade, D.M. Webber, W.H.H. Sauer, M.J. Roberts, M.J. Smale & F.A. Voegeli. 1998. Applications and performance of radio-acoustic positioning and telemetry (RAPT) systems. Hydrobiologia 371/372: 1–8.
Pincock, D.G. & F.A. Voegeli. 1992. Quick course in underwater telemetry systems. VEMCO Limited, Shad Bay. 31 pp.
Sauer, W.H.H. & M.J. Smale. 1991. Predation patterns on the inshore spawning grounds of the squid *Loligo vulgaris reynaudii* (Cephalopoda: Loliginidae) off the south-eastern Cape, South Africa. S. Afr. J. Mar. Sci. 11: 513–523.
Sauer, W.H.H., M.J. Roberts, M.R. Lipinski, M.J. Smale, R.T. Hanlon, D.M. Webber & R.K. O'Dor. 1997. Choreography of the squid's 'nuptial dance'. Biol. Bull. 192: 203–207.
Scariotta, T.C. & D.R. Nelson. 1977. Diel behaviour of the blue shark, *Prionace glauca*, near Santa Catalina Island. Calif. Fish. Bull. 75: 519–528.

Scharold, J., N. Chin Lat, W.R. Lowell & J.B. Graham. 1989. Metabolic rate, heart rate, and tail beat frequency during sustained swimming in the leopard shark *Thiakis semifasciata*. J. Exp. Biol. 48: 223–230.

Scharold, J. & S.H. Gruber. 1991. Telemetered heart rate as a measure of metabolic rate in the lemon shark, *Negaprion brevirostris*. Copeia 1991: 942–953.

Skajaa, K., A. Ferno, S. Lokkeborg & E.K. Haugland. 1998. Basic movement pattern and chemo-oriented search towards baited pots in edible crab (*Cancer pagarus* L.) Hydrobiologia 371/372: 143–153.

Stasko, A.B. & D.G. Pincock. 1977. Review of underwater biotelemetry, with emphasis on ultrasonic techniques. J. Fish. Res. Board Can. 34: 1261–1285.

Stevens, J.D. 1976. First results of shark tagging in the north-east Atlantic, 1972–1975. J. Mar. Biol. Ass. UK 56: 929–937.

Stevens, J.D. 1990. Further results from a tagging study of pelagic sharks in the North-East Atlantic. J. Mar. Biol. Ass. UK 70: 707–720.

Trefethen, P.S., J.W. Dudley & M.R. Smith. 1957. Ultrasonic tracer follows tagged fish. Electronics 30: 156–160.

Voegeli, F.A. 1988. Ultrasonic tracking, position monitoring and data telemetry systems. pp. 279–284. *In:* A. Uchiyama & C.J. Amlaner (ed.) Proceedings of the 10th International Symposium on Biotelemetry, Waseda Univ. Press, Tokyo.

Voegeli, F.A. 1990. Ultrasonic telemetry of swimming speed data. pp. 279–284. *In:* C.J. Amlaner (ed.) Proceedings of the 11th International Symposium on Biotelemetry, Yokohama, Japan.

Voegeli, F.A & D.G. Pincock. 1981. Determination of fish swimming speed by ultrasonic telemetry. Biotelemetry and Patient Monitoring 7: 215–220.

Voegeli, F.A., G. Hoar & D.W. Church. 1983. A new generation of ultrasonic receivers for biotelemetry. pp. 63–75. *In:* D.G. Pincock (ed.) Proceedings of the 4th International Conference on Wildlife Telemetry, Applied Microelectronics Inst., Halifax.

Voegeli, F.A, G.L. Lacroix & J.M. Anderson. 1997. Development of miniature pingers for tracking Atlantic salmon smolts at sea. Hydrobiology 371/372: 35–46.

Wartzok, D., S. Sayegh, H. Stone, J. Barchak & W. Barnes. 1992. Acoustic tracking system for monitoring under-ice movements of polar seals. J. Acoust. Soc. Am. 92: 682–687.

Webber, D.M. 1999. Respiratory and cardiovascular correlates of activity metabolism in the Atlantic cod *Gadus morhua*. Ph.D. thesis, Cambridge University, Cambridge. 280 pp.

Webber, D.M. & R.K. O'Dor. 1986. Monitoring the metabolic rate and activity of free swimming squid with telemetered jet pressure. J. Exp. Biol. 126: 205–224.

Webber, D.M., R.G. Boutilier & S.K. Kerr. 1998. Cardiac output as a predictor of metabolism in cod (*Gadus morhua*). J. Exp. Biol. 201: 2779–2789.

Webber, D.M., F.A.Voegeli & M.J. Smale. 2000. The development of differential pressure techniques to measure swimming speed in fish. *In:* J. Eiler (ed.) Proceedings of the 15th International Symposium on Biotelemetry (in press).

West, G.J. & J.D. Stevens. 2001. Archival tagging of school shark, *Galeorhinus galeus*, in Australia: initial results. Env. Biol. Fish. 60: 283–298 (this volume).

Wolcott, T.G. 1995. New options in physiological and behavioural ecology through multichannel telemetry. J. Exp. Mar. Biol. Ecol. 193: 239–256.

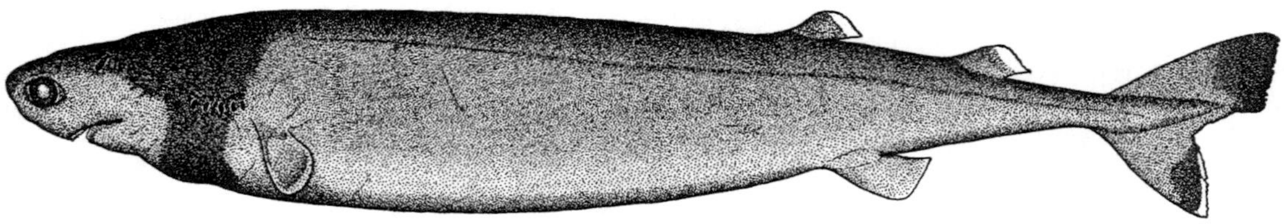

Cookie-cutter shark, *Isistius brasiliensis*, 42 cm long (PV).

Environmental Biology of Fishes **60:** 283–298, 2001.

Archival tagging of school shark, *Galeorhinus galeus*, in Australia: initial results

Grant J. West & John D. Stevens
CSIRO Marine Research, GPO Box 1538, Hobart, Tasmania, Australia 7001
(e-mail: grant.west@marine.csiro.au)

Received 11 January 1999 Accepted 2 November 1999

Key words: data storage tags, swimming depth, movement, diel behaviour, geo-location

Synopsis

Archival tags were used to study the movement and depth behaviour of school sharks, *Galeorhinus galeus*, in southern Australia. Thirty fish were tagged in late 1997, and to date there have been nine recaptures (30% recapture rate). Periods at liberty varied from 8 days to 18 months. The sharks spent about 80% of their time on the continental shelf, and appeared to swim close to the bottom during the day. At night they often ascended for periods of up to several hours, except at times around the full moon. When in deep water, the sharks typically descended at dawn to depths of up to 600 m, before ascending at dusk. It was not possible to use the light data from the tags to estimate position when the sharks were in deep water, because they were often at depths beyond the sensitivity of the tag. In shallower water, longitude was estimated from the light data but latitude was estimated from the maximum daily depth, assuming the fish were on the bottom. The timing of the dives in deepwater appeared sufficiently regular to offer the prospect of using it to estimate longitude. We propose future research using archival tags on this species should address questions about female reproductive migrations, pelagic behaviour and vertical movements.

Introduction

The school shark, *Galeorhinus galeus*, has a widespread but disjunct distribution in temperate waters of the major oceans where it is mainly coastal on continental shelves and upper slopes down to 550 m depth (Compagno 1984, Last & Stevens 1994). However, in the North East Atlantic it is found offshore to the Azores and in Australasia it is recorded across the Tasman Sea between Australia and New Zealand (Holden & Horrod 1979, Hurst et al. 1999).

In southern Australia, school sharks have been fished commercially since the 1920s using longlines and gillnets set on the bottom. The fishery is considered to be over-exploited and current management plans are aimed at rebuilding the stock (Punt & Walker 1998). Conventional tagging of school sharks carried out periodically since the 1940s has shown extensive movements across southern Australia (Olsen 1954, Walker[1]). More recently, tagging in both New Zealand and Australia has demonstrated trans-Tasman Sea migrations (Walker et al.,[2] Hurst et al. 1999). However, our considerable knowledge of school shark migratory patterns could be enhanced if we had more detailed information on depth behaviour, and continuous data on movements over the total period of liberty of a tagged fish.

---

[1] Walker, T.I. 1989. Methods of tagging adopted in the southern shark fishery. pp. 105–108. *In:* Hancock D.A (ed.), Bureau of Rural Resources, Proceedings No. 5, Australian Society for Fish Biology Tagging Workshop (Sydney, 21–22 July 1988), Australian Government Publishing Service, Canberra.

[2] Walker, T.I., L.P. Brown & N.F. Bridge. 1997. Southern shark tagging project. Final report to Fisheries Research and Development Corporation, Marine and Freshwater Resources Institute Queenscliff, Victoria. 61 pp.

Past studies

The development of miniature time-depth recorders since the 1980s led to fascinating insights into the diving behaviour of air-breathing marine animals (Wilson et al. 1995). These types of tags have also been successfully used in the study of fish movements. Metcalfe & Arnold (1997) estimated movements for a demersal fish by combining depth information from the tag with a very detailed knowledge of the tidal cycles of the North Sea. However, this approach is not possible for most studies of fish movement. In a workshop on new technologies that might be used to investigate tuna movements, Hunter et al. (1986) supported the development of an 'archival' tag that could store information on swimming depth, water temperature and light level. The light data could be used to estimate the time of sunrise and sunset, from which longitude could be estimated, and latitude could be determined by depth-specific seawater temperatures (Smith & Goodman[3]). Wilson et al. (1992) reported the successful development of a 'global location sensor' that used light data to estimate both latitude and longitude to an accuracy of about 150 km. DeLong et al. (1992) studied the movements of northern elephant seals, *Mirounga angustirostris*, using a 'geographic location time-depth recorder' (GLTDR) that measured light when the animal was near the surface. The accuracy of their position estimates was considered to be about $\pm 1°$ (Hill 1994). This estimate was supported by the work of Stewart & DeLong (1995), who attached satellite-linked radio transmitters as well as the 'GLTDRs' to northern elephant seals.

Gunn et al. (1994) carried out an experiment in which caged southern bluefin tuna, *Thunnus macoyii*, tagged with archival tags were towed a distance of 450 km to a farm site. They found that the position estimates were accurate to within 1°. By 1994, one company had developed a tag that estimated the time of dawn and dusk on-board, and at depths below the surface. The tag had the ability to calculate light-extinction coefficients on a daily basis, providing that the fish dived during the day, or used appropriate default values. The light-extinction coefficient was used to allow light measured at depth to be corrected to surface intensities. Block et al. (1998) tagged 170 Atlantic bluefin tuna, *Thunnus thynnus thynnus*, ranging in size from 76–234 kg with internal archival tags, two of which have been recovered. Their tests with tags on moorings at 50 m depth indicated that while longitude could be estimated to within 1°, latitude estimates were less accurate. Their experiment was carried out near the equinox, a period when latitude estimates are known to be very poor (Hill 1994). Welch & Eveson (1999) tested a variety of archival tags using both reference light levels and rate of change in light levels to identify dawn and dusk, and a number of techniques to smooth the light signal. They found that the algorithm used to determine dawn and dusk had more influence on the results than the method used to smooth the data. In practice, however, the differences were not large. The average position error was about 140 km (standard deviations of 0.9° of longitude and 1.2° of latitude). Klimley et al.[4] reviewed the state of archival tag development and suggested three general research needs: improved estimates of latitude, additional means of data retrieval other than tag recovery, and improved methods of fish handling and tag attachment.

We used archival tags to learn more about the depth distribution and offshore movements of school sharks, and the reproductive migrations of females. In this paper, we present the initial results from these archival tags, concentrating on the depth distribution of the sharks. An analysis of geo-position based on light and bathymetric data is included for one shark, together with an appraisal of the methods used.

Methods

Prior to application of archival tags on school sharks, we carried out an experiment with 'dummy' tags to ensure that we achieved satisfactory recapture rates. We used three types of tags, two of them external, and one internal. The external tags were cast from epoxy resin, one a rectangular block (52 mm long × 24 mm wide × 12 mm thick), the other a streamlined shape (27 mm maximum diameter and 116 mm long) with fins at the rear. Nylon pins with an expanded arrow head were screwed into the tags, and the tags were pinned through a pre-punched hole in the lower anterior region of the first dorsal fin and secured by the female

[3] Smith, P. & D. Goodman. 1986. Determining fish movements from an 'archival' tag: precision of geographical positions made from a time series of swimming temperature and depth. NOAA Technical Memorandum, NMFS-SWFC, La Jolla 60. 14 pp.

[4] Klimley, A.P, E.D. Prince, R.W. Brill & K. Holland. 1994. Archival tags 1994: present and future. NOAA Technical Memorandum NMFS-SEFSC-357. 31 pp.

half of a cattle ear tag (Daltons Supplies Australia Pty Ltd). The internal tag was simply the rectangular block with a 3 mm diameter plastic-coated copper wire (to simulate the light sensor) projecting 175 mm from the tag. The internal tags were inserted into the coelomic cavity through a 3 cm longitudinal incision made just in front of, and above, the pelvic fin. The incision was closed with sutures leaving the wire protruding through the body wall, alerting fishers to the presence of the tag. The tags were deployed during 1995 as part of an extensive conventional tagging project (Walker et al.[2]).

The archival tags were manufactured by Lotek Marine Technologies Inc. The tags had a depth sensor rated to 1000 m (accuracy 0.5%), a temperature sensor (range 2–30°C, accuracy 0.2°C) and a light sensor with a peak spectral sensitivity at 750 nm (unfiltered in our study). The data from these sensors, together with a time stamp from an internal clock, were stored in a 1 Mb non-volatile RAM chip. We used a 4-min sampling interval for the light and depth sensors, and 2 h for the temperature sensor, which gave a data logging duration of 1.9 y. The circuitry was housed in an air-filled transparent polycarbonate casing (64 mm long × 23 mm diameter, designed for external application) and pressure-rated to 1000 m depth. We chose the same attachment method used in the dummy-tag experiment. The tag housing was fixed to a baseplate using titanium wire. The baseplate incorporated a nylon pin (Figure 1) that was inserted through a pre-punched hole in the first dorsal fin and secured in the same way as the external dummy tags. The tags carried a return address label, and notice of a A$200 reward. The program was publicised extensively, particularly among those involved with the shark fishery.

The software used to calculate position was developed by CSIRO for use with tags applied to southern bluefin tuna (Gunn et al. 1994). It used a reference light value to identify the time of sunrise and sunset. We carried out a number of experiments with tags on moorings to derive a reference light value at specific depths at twilight. With the tags used originally on southern bluefin tuna, light decreased exponentially with depth, and the correction used was:

$$\text{Adjusted light} = \text{rawlight}\,(e^{b\ \text{depth}}),$$

where b is the slope of the linear regression of ln light vs. depth. The light signal in the Lotek tags was, however, logarithmically amplified and this resulted in an approximately linear decrease in light with depth, at least for the first several hundred meters for casts carried out during the day. The correction applied was:

$$\text{Adjusted light} = \text{raw light} + [(\text{reference depth} - \text{actual depth}) \times \text{attenuation slope}].$$

Instead of adjusting light to surface levels, we allowed for a reference depth that would typically be the most common depth occupied by the shark. At greater depths, or for casts carried out at twilight, the response was non-linear and tended towards an exponential relationship at very low light levels.

Latitude estimates based on day-length are generally much less accurate than those for longitude, especially around the time of the equinoxes (Hill 1994). Along the southern coast of Australia, ocean depth increases with latitude. If we could establish that the shark was on the bottom, this offered us the prospect of using the maximum depth from the tag to provide an independent estimate of latitude each day. The charted depth information that was available in digital form was the location of the 50, 100 and 200 m isobaths. For each longitude, the latitude was estimated as the

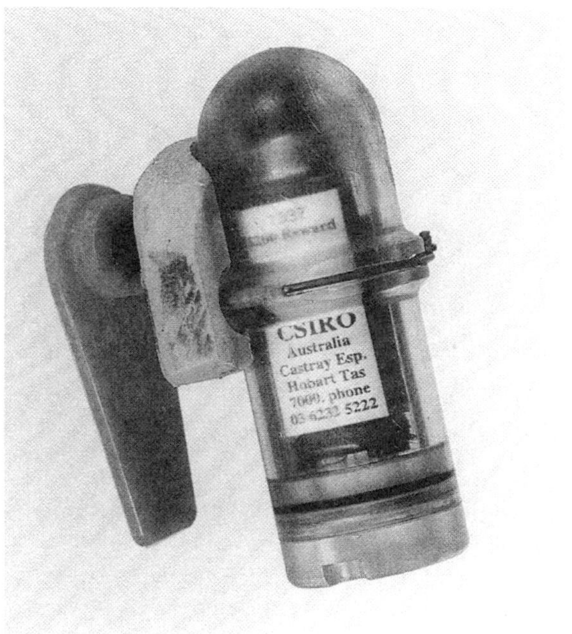

Figure 1. Attachment method used for Lotek archival tags. The archival tag was wired to a polycarbonate block that carried a barbed nylon pin. The pin passed through the shark's first dorsal fin and was secured by a cattle ear-tag.

point midway between the contours above and below the daily maximum depth recorded by the tag.

Sharks were captured during commercial fishing trips using longlines and gillnets set on the bottom. Only specimens that were in a lively condition were tagged. Sharks were measured (total length, TL), sexed, tagged and released.

Results

Dummy tagging

A total of 243 school sharks were tagged in the dummy tagging experiment (Table 1). By mid-1997, the overall recapture rate was 16%, well above the 10% we had set as a minimum criterion to proceed with the use of real archival tags. The recapture rate for the internal tags was higher (21.7%) than for the external streamlined and block tags (16.3% and 11.4%, respectively). Although the results favoured internal tags, at the time of purchase there were none of these available that we considered suitable for our application. By August 1999, the overall recapture rate had risen to 19.3%, with 27.5% of internal tags being returned. However, differences in return rates between the two tag types were not significant at the 5% level ($\chi^2 = 4.35$, 2 df, $p = 0.114$).

Archival tag releases

While it had been intended to concentrate the archival tagging on pregnant sharks, the realities of fishing dictated otherwise. Catches of school shark were small, and very few were pregnant. In all, 23 females (125–165 cm TL) and 7 males (131–141 cm TL) were tagged. The first 15 releases, caught in gillnets, were made in shallow water (40–100 m) in the Great Australian Bight (GAB) (Figure 2). The fifteen longline releases were made at the edge of the shelf, in water 200–350 m deep, off Beachport, South Australia (SA).

In the 18 months following the releases, there have been nine recaptures (30% recapture rate). The release and recapture locations are shown in Figure 2. Six of them were from longline releases and three from gillnet releases. Two fish were at liberty for only 8 and 23 days, another three for 4 months, two for 10 months and two for 17 months. Table 2 summarises release and recapture details for the archival-tagged sharks.

Table 1. Results of the dummy-archival tagging experiment on school shark, *Galeorhinus galeus*, up to August 1999.

Tag type	Number released	Number recaptured	Percent recaptured
Internal	69	19	27.5
External block	88	13	14.8
External streamlined	86	15	17.4
Total	243	47	19.3

Depth behaviour

The depth data from the recaptured sharks showed a general pattern of diel vertical migrations, together with periods spent on the continental shelf, and more occasional periods in deep-water. The daily range in depths for the seven recaptured sharks at liberty for more than one month are shown in Figure 3. The three fish that were released in shallow water in the GAB stayed at depths of < 100 m for the first three to five months, before two of them moved into deeper water. Shark 1396 spent 10.7% of its time (8.4 months) off the continental shelf (> 200 m), while shark 1253 was at depths > 200 m for 8.5% of its 10.7 months at liberty. Shark 1253 was unusual in that it spent about 4 months (February to June 1998) in very shallow (< 50 m) water. There was no obvious pattern in the timing of the excursions off the edge of the shelf. For the sharks tagged at the edge of the continental shelf off Beachport, the two shorter-term recaptures (1391 and 1394) spent only a small amount of time (6.0% and 2.3%, respectively) at depths > 200 m. The two sharks at liberty for more than a year, sharks 1346 and 1259, spent the most time at depths > 200 m (17.9% and 20.7%, respectively), but the timing of these events were different. Shark 1259 did show some indication of an annual pattern, in that it came onto the shelf in November 1997 and 1998.

The general pattern of depth behaviour is illustrated by shark 1394 which was at liberty for 4.4 months (Figure 4). After being tagged, it dived to 335 m before commencing an ascent about 15 min after sunset. It stayed off the shelf for the first six days after release (4–9 November), descending during the day to depths of 185–580 m. At night, it was in depths of 31–66 m. The excursions from deep water towards the surface each night were not always regular, as shown by the depth pattern for 7–9 November (Figure 4 inset). Seven days after release (10 November) it appeared to move onto the shelf where it remained for the next four months at depths from 75–145 m. This shark showed a regular pattern of nightly ascents up to 75 m in extent, from November through January, except during the full

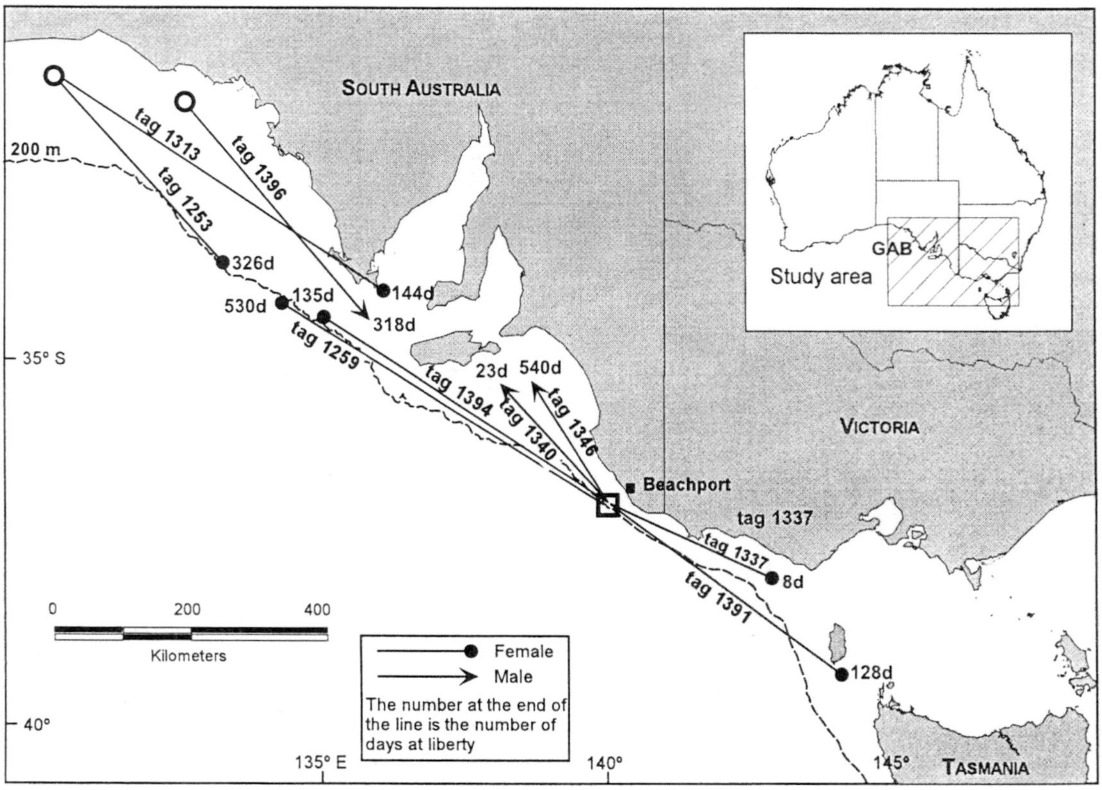

Figure 2. Release and recapture positions for school sharks, *Galeorhinus galeus*, tagged with archival tags. ○ – Gillnet release positions for sharks tagged in October 1997 on the continental shelf. □ – Longline release position for sharks tagged on 4 November 1997 at the shelf edge. Inset shows general study area.

Table 2. Summary of data for recaptures from archival-tagged school shark, *Galeorhinus galeus*.

Tag no.	Release locality	Release date	Recapture date	No. months at liberty	No. months of data	Distance from release (km)[a]	Direction from release (E or W)	Sex	TL (cm) at release
1337	Beachport	4 Nov 97	12 Nov 97	0.3	0.3	284	E	Female	153.5
1340	Beachport	4 Nov 97	27 Nov 97	0.8	0.8	246	W	Male	131.0
1313	GAB	18 Oct 97	11 Mar 98	4.7	4.7	648	E	Female	146.0
1391	Beachport	4 Nov 97	12 Mar 98	4.2	4.2	449	E	Female	128.6
1394	Beachport	4 Nov 97	19 Mar 98	4.4	4.4	534	W	Female	150.7
1396	GAB	25 Oct 97	8 Sep 98	10.4	8.4	438	E	Male	138.3
1253	GAB	18 Oct 97	10 Sep 98	10.7	10.7	420	E	Female	136.1
1259	Beachport	4 Nov 97	18 Apr 99	17.4	14.3	321	W	Female	128.3
1346	Beachport	4 Nov 97	28 Apr 99	17.7	13.0	218	W	Male	130.5

[a]Straight-line distance between release and recapture positions.

moon. We interpret this type of shallow-water depth pattern as the fish swimming close to the bottom during the day, and rising off the bottom at night. In late January through late February the ascents decreased in frequency and duration, and after this virtually ceased.

The timing of the ascents was not always at night. On the continental shelf several sharks were at times more shallow during the day than at night (Figure 5a and b), although the amplitude of the depth change was much less than in deep water. In deep water, the usual pattern

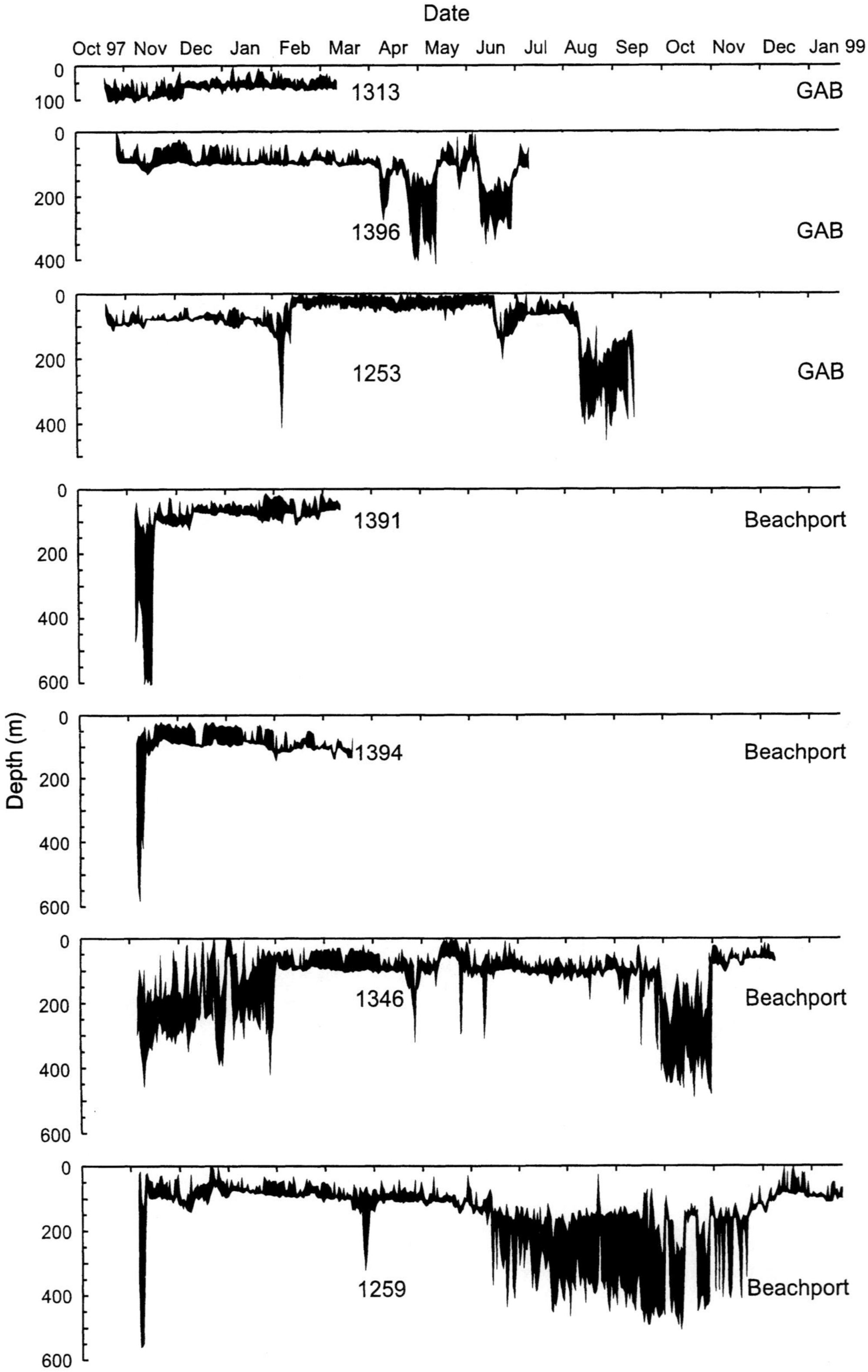

Figure 3. Daily depth-range for seven school sharks at liberty for more than one month released in the GAB (Great Australian Bight) and at Beachport, South Australia.

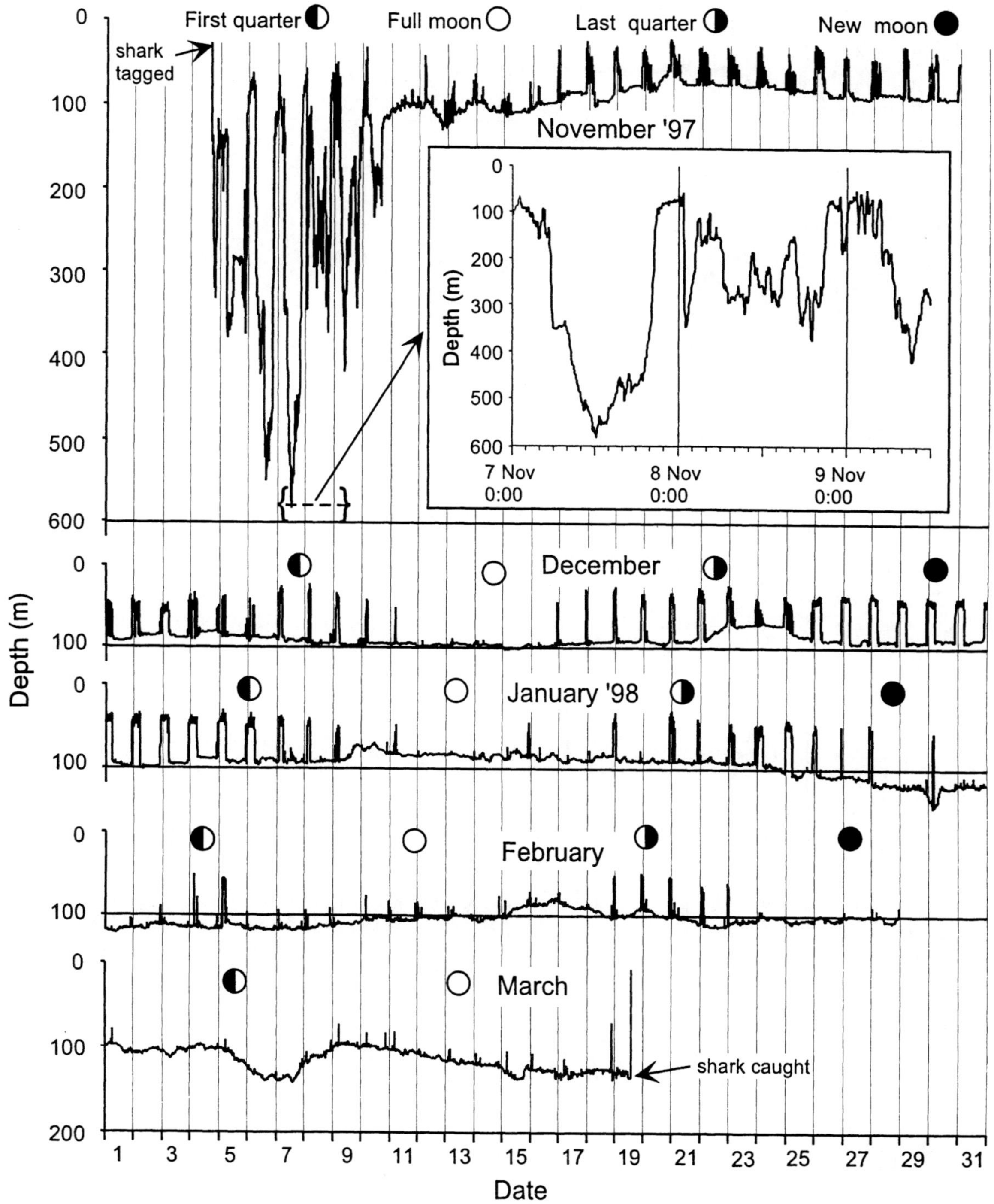

Figure 4. Depth behaviour of shark 1394. Vertical lines indicate midnight, SA daylight saving time. Inset shows expanded depth profile for 7–9 November 1997.

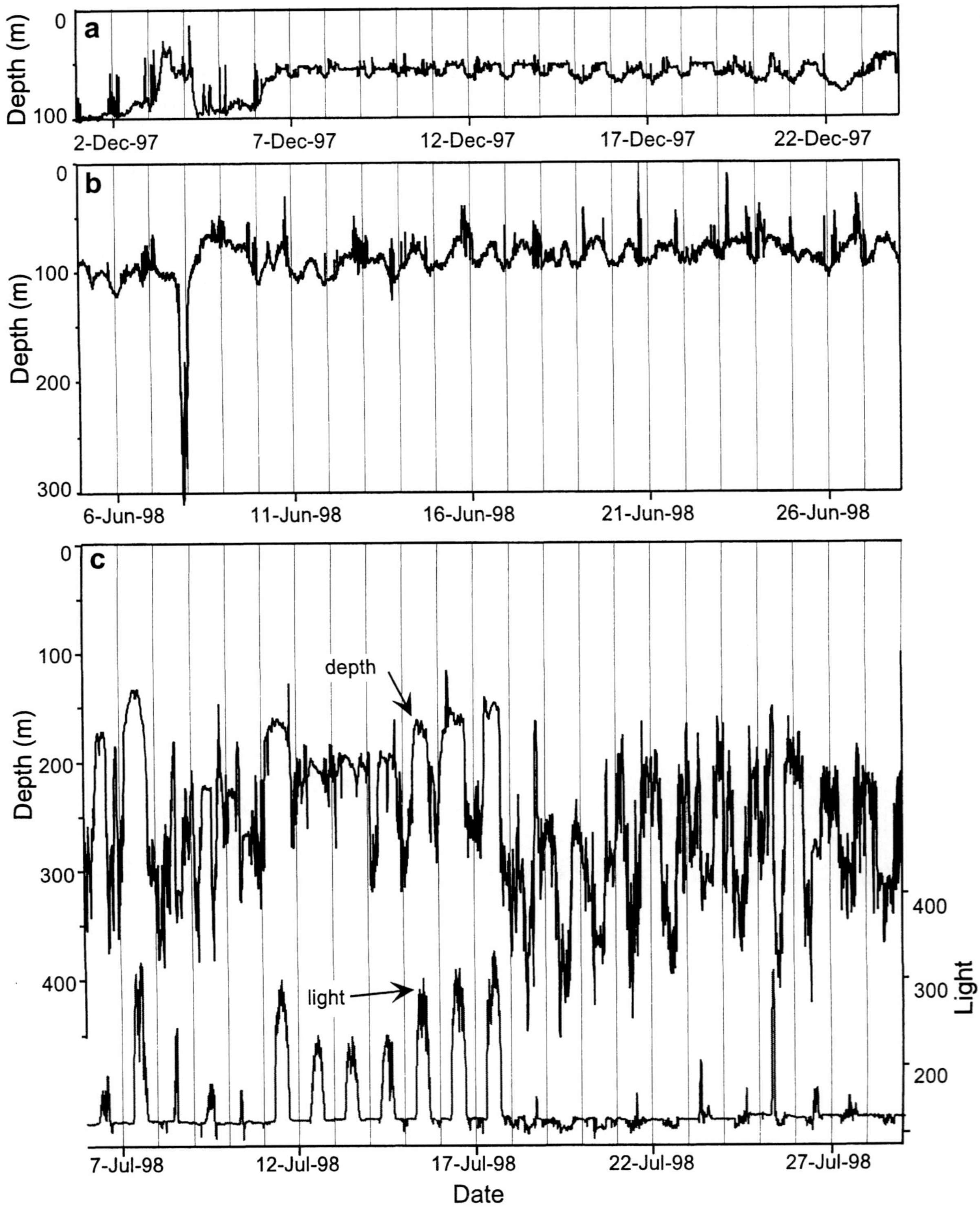

Figure 5. Selected periods from archival tag depth profiles for three school sharks: a – shark 1313, December 1997; b – shark 1346, June 1998; c – shark 1259, July 1998, upper plot is depth profile, lower plot is light. Vertical lines are midnight, SA standard time.

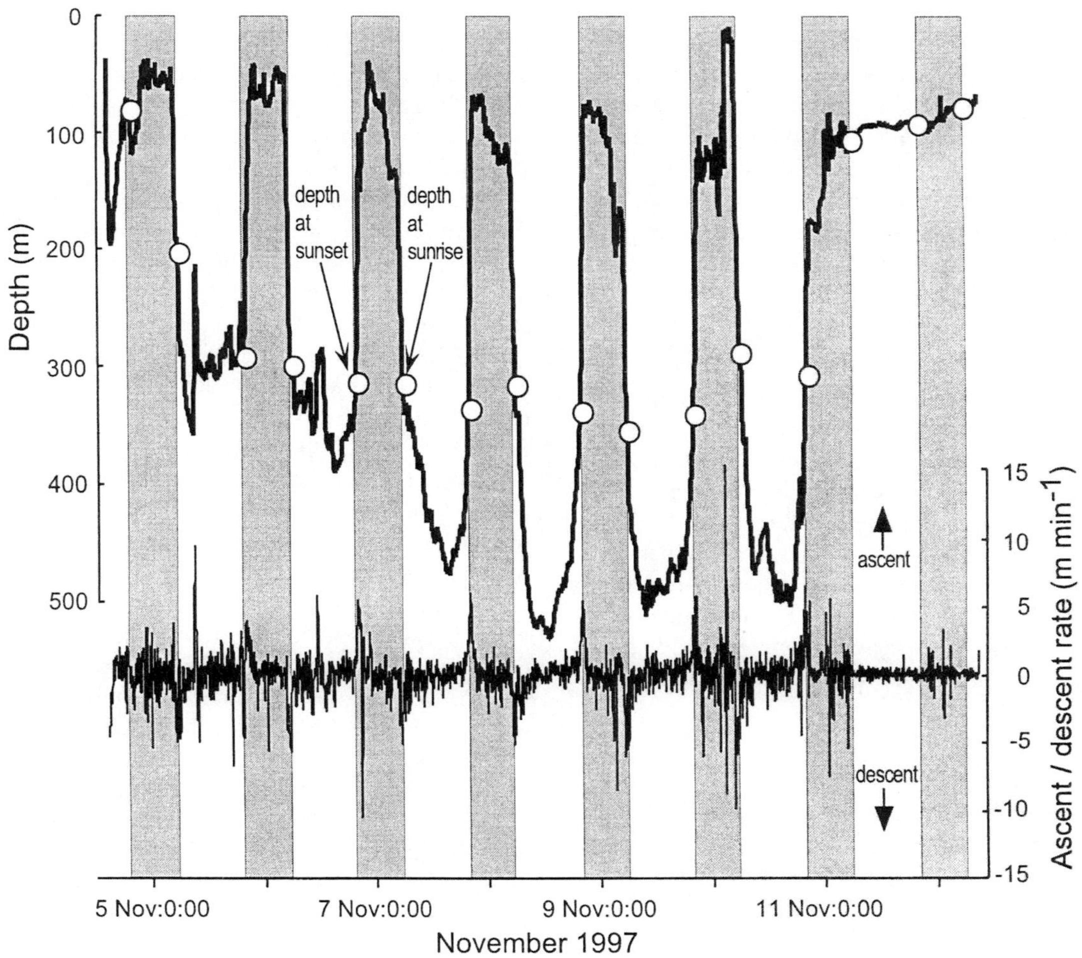

Figure 6. Depth behaviour of shark 1337. Shaded bars represent night (sunset to sunrise, SA daylight saving time). Upper plot is depth profile, lower plot is vertical movement rate.

was for the sharks to ascend during the night. However, shark 1259 spent periods off the shelf during which it ascended during the day (6–17 July), before reverting to the usual pattern (Figure 5c).

We attempted to estimate the timing (relative to sunrise and sunset) of the vertical movements in deep water, but because of the depths the sharks reached during the day we could not use light to estimate this. However, one fortuitous recapture allowed us to estimate daily positions based on rate of travel (i.e. independent of the light data), from which we could calculate time of sunrise and sunset. Shark 1337 was recaptured 284 km from where it was tagged, an average daily distance of 37 km during its 7.7 days at liberty. This is the fastest movement ever recorded for a tagged school shark, being more than twice the maximum rate observed from conventional tagging (L. Brown personal communication). We assumed that this fish moved at a constant rate and estimated its position each day, and the corresponding time of sunrise and sunset (Figure 6). Except for the first night after tagging, this fish was at similar depths at dawn and dusk while it was in deep water (sunrise 286–298 m, sunset 290–305 m). By sunset, the fish had already started to move up into the water column, having covered 22–47% (average 34%) of the ascent (the difference between the daytime maximum depth and the following night's minimum depth). By sunrise, the fish had reached 51–73% (average 60%) of its descent. Most of the tagged sharks showed different or erratic diving behaviour in the first day or so after tagging.

Vertical movement rates

Maximum vertical ascent (21.6 m min^{-1}) and descent (44.7 m min^{-1}) rates were shown by school sharks in deep water. These vertical movements are illustrated by shark 1337 when it was in deep water (Figure 6, lower plot) where it showed regular periods of increased swimming speed around sunrise and sunset that peaked at about 6 m min^{-1}. On other occasions, mainly at night, there were higher movement rates of up to 15 m min^{-1}.

Temperature data

The ambient temperature data from the sharks released off Beachport showed that the thermocline in this area was at about 300 m in November 1979, a depth the sharks regularly exceeded on their daily descents. When the sharks were off the continental shelf, minimum and maximum temperatures varied from 8.1–10.8°C and 13.6–16.3°C, respectively. When the sharks were over the shelf, minimum temperatures varied from 11.3–14.3°C, and maximum temperatures varied from 17.0–20.1°C. The greatest temperature change experienced by the sharks was 5.4°C in 2 h, when diving in deep water.

Geo-location and light data

We used the light data from one recaptured shark that had remained on the continental shelf to estimate position. In depths up to about 200 m there was a clear response from the tag light sensor to ambient light (e.g. Figure 5c, 7–17 July). However, after times varying from 7–55 days, a voltage leak developed in the light circuit of the tags that resulted in the light signal failing to return to zero at night. Our geo-location method used a relatively low light level to identify the time of dawn and dusk, and this was frequently obliterated by the spurious signal. This prevented us from estimating day-length (and hence latitude) for most of the data, but still allowed us to estimate longitude. Shark 1313 had been at liberty for over four months, and the voltage leak had been less severe than in other tags. The shark had not been deeper than 110 m (Figure 3), and we used the median depth of 64 m as the reference depth. The linear correction of light caused considerable problems with our software. When the shark was deeper than 64 m, the correction increased the light level, but it did this to the night-time light levels as well. The program searched for light data between a range of values, in order to estimate the time at which the reference light level occurred, and these values were often lower than the artificially raised night-time levels. The voltage leak that occurred later in this deployment only added to the problem. We reverted to using an exponential correction which, because it was multiplicative rather than additive, did not suffer from this problem. The differences in longitude between consecutive days were calculated, and any that exceeded two degrees were used to reject the outlier. Latitude was estimated for each longitude using the maximum depth for the corresponding day. The results indicated that the shark had moved steadily east, reaching longitude 140° E by January, before turning west. It had spent most of the time in longitudes 135°–137° (Figure 7).

We attempted to estimate longitude for one shark using the time of its daily dives. Other investigators, faced with similar problems in deep-diving pelagic fish have used dive behaviour as a substitute for light data, on the assumption that the fish was responding to changes in light intensity.[5] Shark 1337 (Figure 6) remained too deep during most days for the light sensor to respond, until the day before it was caught when it came onto the shelf. We chose 250 m as a convenient reference depth. The times at which these depths occurred were estimated by linear interpolation between the depths immediately above and below 250 m. These estimates of the time of dawn and dusk were used for geo-location. The longitude estimates varied from 138.8° to 144.5° E (Table 3), compared with the known release and recapture longitudes of 140.0° and 143.0° E, respectively. On the last day at liberty, when the shark was at about 100 m, longitude was also estimated using the light data from the archival tag.

Discussion

The most striking feature of the behaviour of the school sharks we tagged was their extensive vertical movements, which occurred almost every day when they were in deep water. Frequent vertical excursions through the water column appear to be a common feature of large pelagic sharks and teleosts that have been tracked using acoustic telemetry (Carey & Robinson 1981, Carey et al. 1982, Holland

[5] Boggs, C., M. Musyl, C. Curran & R. Brill. 1999. Geolocation by crepuscular diving behavior and lunar influences on depth distribution in a bigeye tuna carrying an archival tag. Proceeedings of the 50th Annual Tuna Conference, Lake Arrowhead, California, May 24–27, 1999, Inter American Tropical Tuna Commission, 8604 La Jolla Shores Drive, La Jolla, CA 92073-1508.

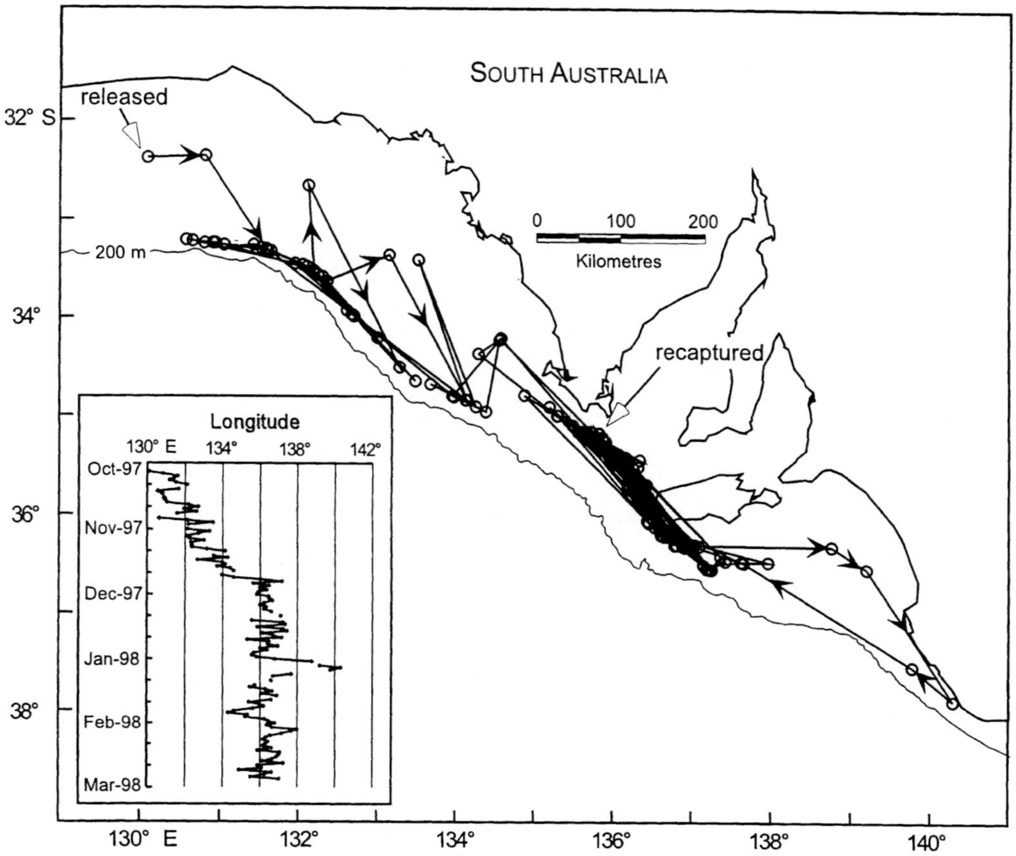

Figure 7. Position estimates for shark 1313, released 18 October 1997, recaptured 11 March 1998. Longitude estimated from archival tag light data, latitude estimated from depth data. Arrows indicate direction of travel. Inset shows the temporal pattern of longitudinal movement.

Table 3. Longitude estimates for shark 1337 based on the timing of its daily dives, and on archival tag light data. Released at 140.0° E on 4 November 1997, recaptured at 143.0° E on 12 November 1997.

Date	Max. depth (m)	Time at 250 m am	Time at 250 m pm	Longitude estimated from time at 250 m		Longitude estimated from light data	
				Dawn-dusk	Dusk-dawn	Dawn-dusk	Dusk-dawn
5 Nov 97	359	6:15	19:29	140.1	144.5	—	—
6 Nov 97	391	5:40	20:11	139.2	140.1	—	—
7 Nov 97	478	5:33	20:15	139.7	139.2	—	—
8 Nov 97	532	5:37	20:17	138.8	140.1	—	—
9 Nov 97	513	5:27	20:26	139.0	137.9	—	—
10 Nov 97	505	5:36	20:09	140.0	—		
11 Nov 97	152	—	—	—	—	141.2	142.5

et al. 1990a, Nelson et al. 1997, Pepperell & Davis 1999, Gunn et al. 2000). These vertical movements have been related to feeding (Carey & Scharold 1990, Nelson 1990) and to thermoregulation in tuna (Hunter et al. 1986, Holland et al. 1992, Holland & Sibert 1994).

The daily vertical excursions made by the tagged school sharks varied from 70 to 550 m. At times the fish were as deep as 600 m, having passed through a thermocline at approximately 300 m. This type of depth pattern, in which animals descend beyond the photic zone, is consistent with a predator that uses olfactory as well as visual cues in search of prey (Carey & Scharold 1990). Information from fishers operating from Beachport during our tagging suggest that squid, *Nototodarus gouldi*, are a common component of the diet of school shark at this time of year. *Nototodarus gouldi* apparently feeds mainly at night, probably moving up from the bottom into the water column (O'Sullivan & Cullen 1983, Nowara & Walker 1998). Certainly, some species of squid undergo similar daily vertical movements to those we observed in school sharks. Nakamura (1993) found that *Ommastrephes bartramii* was near the surface at night, but descended to depths of over 700 m during the day. Tricas (1979) found that pelagic cephalopods were a major prey item of blue sharks, *Prionace glauca*, in California, and suggested they were captured at night when the cephalopods migrate from deeper water to near the surface. Dietary studies in conjunction with archival or sonic tagging might help explain the diving behaviour of the school sharks. Our initial analysis of the diving behaviour revealed instances during which the diel pattern of vertical movement was interrupted by sudden vertical excursions, suggesting predatory behaviour.

Vertical movements provide the opportunity for behavioural thermoregulation. Carey & Scharold (1990) found that blue sharks stayed significantly warmer than the water for several hours at depth, and re-warmed quickly when at the surface. However, school sharks have a more prolonged descent and in general do not come as close to the surface as blue sharks when they ascend at night. In deep water, the school sharks regularly descended through the thermocline, and experienced water temperatures as low as 8.1 °C. One shark spent about 30% of the time in deep water at temperatures of 8–9°C in the first two weeks after release off Beachport. At times, they experienced temperature changes of as much as 5.4°C in 2 h when diving in deep water. These are relatively modest temperature gradients compared with those reported in other studies. Carey & Scharold (1990) reported that one blue shark passed through a 19°C change in water temperature in six minutes, and Carey & Robinson (1981) found that vertical movements in the swordfish, *Xiphius gladius*, were associated with temperature changes of as much as 19°C in 2 h. The use of internal tags in future experiments may allow us an insight into any thermoregulatory behaviour of school sharks.

The maximum rates of ascent and descent we observed were 22 and 45 m min^{-1} respectively, but these were based on a 4-min sampling interval which might not capture the dynamics of short-term dive behaviour (Wilson et al. 1995). The maximum descent rate is similar to the 50 m min^{-1} observed by Carey & Scharold (1990) for the blue shark. The school sharks commenced their diel ascents slowly at first, with vertical movement rates peaking around sunset. Descent rates peaked at sunrise (Figure 6). The diel pattern of diving in the school sharks contrasts with that observed in the pelagic blue and shortfin mako, *Isurus oxyrinchus*. These sharks made dives from the surface to depths of up to 600 m every few hours throughout the entire day, and just as frequent but lower amplitude dives at night (Carey & Scharold 1990). Klimley (1993) also observed frequent vertical oscillations in the scalloped hammerhead shark, *Sphyrna lewini*, both day and night. Swordfish and bigeye tuna, *Thunnus obesus*, have the most similar diel pattern of diving behaviour to the school shark. Swordfish swim deeper during the day (at depths up to 600 m) than at night (Carey & Robinson 1981), although unlike the school shark they often ascend to the surface during the day. In Hawaiin waters, bigeye tuna typically descend to around 250 m during the day, although they too make frequent ascents into warmer water. At night they ascend to near the surface (Holland & Sibert 1994). Carey & Robinson (1981) thought that in some areas the depths may be limited by water with low oxygen concentrations, while in well-oxygenated water the swordfish appeared to follow an isolume. Holland & Sibert (1994) demonstrated large and rapid changes in thermoregulation in the bigeye tuna associated with their diving behaviour.

When tagged school sharks were released, their diving pattern was sometimes quite atypical for periods of several hours to several days. They often came near the surface during daylight, something that was never subsequently observed. Sciarrotta & Nelson (1997) noted 'abnormal' plunge behaviour in about half of the blue sharks they tagged, with the sharks diving to about 95 m within 30 min of being darted with acoustic transmitters. This initial effect disappeared within 90 min of being tagged. Similarly, Carey & Scharold (1990) observed that tagged blue sharks, when first released, commonly went below the thermocline for many minutes to several hours. Their diving pattern was commonly less well developed, and their course heading different for the first half-day or so. Presumably, this different behaviour is related to the stress of capture and tagging. Holts & Bedford (1993) noted that the trauma of capture and release probably lasted only 30–90 min in shortfin mako sharks that they tracked. 'Atypical' behaviour lasting several hours after tagging has been noted in tracked marlin (Holland et al. 1990b, Block et al. 1992).

One question that archival tagging cannot currently answer is whether sharks are in mid-water, or near the bottom but moving up the continental slope, when they ascend at night. It is possible that intensive depth profiling in the release area might yield temperature information that could help to answer this question. Alternatively, acoustic tags could be used to track their movements, although this would need to take into account possible short-term changes in behaviour resulting from the effect of capture and tagging.

On average the sharks spent 86% of their time in depths < 200 m. However, the 79–82% of time at these depths spent by the two fish at liberty for more than a year is probably a better estimate. There was no apparent pattern to the timing of the off-shelf movements.

When the fish were in shallow water, one of the most consistent patterns of vertical behaviour was the nightly ascents, often to near the surface. There were striking differences between fish (and between months) in the frequency and duration of these ascents, but there was a strong lunar influence. At times of full moon, these ascents were depressed. Fishers report better catches with their bottom-set gear at times around the full moon. We have yet to estimate the average amount of time the fish spend off the bottom, when they are less vulnerable to the fishing gear. When the fish were in deep water, there was no obvious lunar effect on their vertical behaviour.

Our estimation of position from light data is still at an early stage of development, but it reflects the current situation that the estimation of longitude appears relatively robust. In deep water, where light data are insufficient, our preliminary results from estimating the longitude of sharks from the temporal pattern in their diving behaviour were encouraging. It should be possible to evaluate this method further by comparing dive-based longitude estimates from other recaptures with light-based estimates from the times immediately before and after the deep-diving periods.

However, estimation of latitude is difficult and usually requires ancillary information. Currently, latitude estimates for southern bluefin tuna from archival-tag data are based on sea-surface temperatures (J. Gunn personal communication) because of the uncertainty surrounding light-based latitude estimates. Three factors in our study made latitude estimation even more difficult: the particular light reference-level we used, problems with the baseline light signal in the tags, and the problem of correcting light for depth attenuation. Our reference level for indicating the time of dawn and dusk relied upon the stable behaviour of the light sensor. The voltage leak in the high impedance light circuit, caused apparently by minute amounts of moisture entering the tag, rendered our method unworkable. Other reference measures such as a percentage of noon light, or the time at which light changes most rapidly (Welch & Eveson 1999), may not have been as sensitive to this problem. Wilson et al. (1992), who used a similar reference light level to ours, showed that the best estimate was obtained when the illuminance was changing most rapidly with time. At the surface, this time is close to the start of civil twilight at dawn, but we found that it is delayed at depth. We established our reference level by deploying a tag on a mooring for a month, to derive an average light level for a particular depth for 1-min intervals throughout twilight. It was necessary to calibrate the tags recovered from sharks against the one used to obtain the reference level. The simplest situation for geo-location is to use the depth at which the reference level has been derived as the 'reference' depth for the correction of light. However, since the light-sensors' attenuation response changes throughout twilight, the correction should be the one appropriate to that zenith angle. Faced with these uncertainties we chose to use a bathymetric method to estimate latitude. These problems do not apply to the estimation of longitude. It requires no specification of zenith angle, it can use an arbitrary reference light level, and there is far more freedom in choosing an appropriate light-attenuation correction. Our local geographic situation was fortuitous in allowing us to use depth to estimate latitude. It was local geographic conditions of tidal amplitude that allowed Metcalfe & Arnold (1997) to geo-locate in the absence of light data. We intend to improve our latitude estimates by obtaining higher resolution bathymetric data, but the method will not work on the eastern coast of Australia where the isobaths run north–south.

Klimley et al.[4] pointed out that day-length and temperature are relatively poor indicators of latitude, and described work in progress on magnetic-field sensors that offered greater accuracy in latitude estimation. However, these have not yet appeared in archival tags, necessitating the use of ancillary data such as bathymetry (this study) or tidal amplitude (Metcalfe & Arnold 1997).

Future direction of archival tag research

The female reproductive migration hypothesis
One of the questions we had hoped to address with the archival tagging of school sharks was the movement patterns of pregnant females. It is not known whether school sharks give birth across their southern Australian range, or whether parturition is restricted to known shallow-water nursery areas in the south-east (Victoria and Tasmania). This is a contentious issue with considerable management implications. The current hypothesis suggests that females spend most of their gestation in SA waters, before migrating to the south-east shortly before parturition. We had hoped to tag pregnant females in the GAB. Subsequent recaptures might indicate an easterly reproductive migration at the hypothesised time. This question will be more easily addressed with the advent of 'pop up' archival tag technology (Klimley et al.[4]). Pregnant sharks could be tagged in the GAB, and the tags timed to release during the peak parturition period when the females would be expected to be in the shallow-water nursery areas in Victoria and Tasmania.

The pelagic vs. slope vertical migration hypothesis
Currently, there is some uncertainty over the deep-water diel behaviour we have documented for school sharks. We suggest that the sharks are making vertical excursions off the bottom, into the water column, and towards the surface at night. However, it is possible that they are remaining close to the bottom and following the continental slope contour up into shallower water at night, returning back down the slope in the day. The extra distances required for the sharks to follow the slope are feasible in regions towards the east of the study area where the shelf is narrow and the depth drops off very quickly. Tracking sharks with acoustic telemetry could resolve this uncertainty. Of particular interest is the swimming depth and behaviour of school sharks when they are offshore in oceanic waters of the Tasman Sea. It is uncertain whether the current archival tags can answer this question. If the sharks stay deep during the day while in the Tasman Sea, there may be insufficient light to allow geo-location and thus prove that they are in oceanic water (unless there is a fortuitous offshore recapture). With 'pop up' archival tags it might be possible to obtain this depth information by translocating a captured school shark to offshore waters. However, there is the problem that the shark might not behave 'normally' under these conditions.

The vertical feeding movement hypothesis
Our results have shown a relatively regular deep-water diel pattern in the vertical movements of school sharks, with ascents into the water column at night. We suggest this is most likely related to feeding, with the sharks following vertically-migrating prey such as squid. Squid are known to be an important item in the diet of school sharks from southern Australia (Walker et al.[6]). Sharks in deep water descended at dawn (reaching depths of up to 600 m during the day) and ascended at dusk. This pattern was most evident in sharks tagged at the shelf edge near Beachport, SA. Examination of school shark stomachs in this area and time period, in conjunction with mid-water sampling of vertically-migrating prey species, could test this hypothesis. Acoustic telemetry could be used to confirm that the school sharks were showing the same diel pattern observed with archival tagging. Internal archival tags, with sensors to measure body and ambient temperature, would clarify any thermoregulatory behaviour.

Acknowledgements

We thank Terry Walker, Lauren Brown and Natalie Bridge from MAFRI, Victoria, and others who were involved in the southern shark tagging project for help with the dummy tagging. Ian Helmond designed the dummy archival tags (constructed by the CSIRO workshop staff) and the moorings we used to test the archival tags. Danny McLaughlan expertly constructed the moorings. John Gunn, Andrew Betlehem, Simon Wotherspoon, Matt Sherlock, Jason Hartog and Ann Cowling generously assisted us in our efforts at understanding the complexities of archival tag technology and software development. Trevor Storeton-West and Geoff Arnold from CEFAS, Lowestoft (where the

[6] Walker, T.I., P.L. Moulton, N.G. Dow & S.R. Saddler. 1989. Southern shark assessment project. Final FIRTA Report, Part B, Internal report no. 175b, Marine Science Laboratories, Queenscliff, Victoria.

Lotek archival tags were designed) gave us invaluable assistance with our technical queries, as did Keith Stoodley from Lotek Marine Technologies Inc. We thank Rob Wilson and Brian Haines, shark fishermen, for allowing us to tag sharks on their boats. Tim Tricas, and two anonymous referees, improved the manuscript through their constructive comments. The project was made possible by funding from FRDC grant 96/128, and support from the fishing industry and AFMA.

References cited

Block, B.A., H. Dewar, T. Williams, E.D. Prince, C. Farwell & D. Fudge. 1998. Archival tagging of Atlantic bluefin tuna (*Thunnus thynnus thynnus*). Marine Technology Society Journal 32: 37–46.

Block, B.A., D.T. Booth & F.G. Carey. 1992. Depth and temperature of the blue marlin, *Makaira nigricans*, observed by acoustic telemetry. Mar. Biol. 114: 175–183.

Carey, F.G. & B.H. Robinson. 1981. Daily patterns in the activities of swordfish, *Xiphius gladius*, observed by acoustic telemetry. U.S. Fish. Bull. 79: 277–292.

Carey, F.G., J.W. Kanwisher, O.B. Brazier, G. Gabrielson, J.G. Casey & H.L. Pratt, Jr. 1982. Temperature and activities of a white shark, *Carcharodon carcharias*. Copeia 1982: 254–260.

Carey, F.G. & J.V. Scharold. 1990. Movements of blue sharks (*Prionace glauca*) in depth and course. Mar. Biol. 106: 329–342.

Compagno, L.J.V. 1984. FAO species catalogue. Vol. 4, Sharks of the world. Annotated and illustrated catalogue of the shark species known to date. FAO Fisheries Synopsis, Part 2 – Carcharhiniformes: i–x, 251–655.

DeLong, R.L., B.S Stewart & R.D Hill. 1992. Documenting migrations of northern elephant seals using day-length. Marine Mammal Science 8: 155–159.

Gunn, J.S., T. Polacheck, T.L.O. Davis, M. Sherlock & A. Betlehem. 1994. The development and use of archival tags for studying the migration, behaviour and physiology of southern bluefin tuna, with an assessment of the potential for transfer of the technology to groundfish research. ICES C.M. Mini Symposium on Fish Migration 2.1. 22 pp.

Gunn, J.S., J.D. Stevens, T.L.O. Davis & B. Norman. 2000. Observations on the short-term movements and behaviour of whale sharks (*Rhincodon typus*) at Ningaloo Reef, Western Australia. Mar. Biol. (in press).

Hill, R.D. 1994. Theory of geolocation by light levels. pp. 227–236. *In*: B.J. Boeuf & R.M. Laws (ed.) Elephant Seals, Population Ecology, Behaviour and Physiology, Uni. Calif. Press, Berkeley.

Holden, M.J. & R.G. Horrod. 1979. The migrations of tope, *Galeorhinus galeus* (L), in the eastern North Atlantic as determined by tagging. J. Cons. int. Explor. Mer 38: 314–317.

Holland, K.N., R.W. Brill & R.K.C. Chang. 1990a. Horizontal and vertical movements of yellowfin and bigeye tuna associated with fish aggregating devices. U.S. Fish. Bull. 88: 493–507.

Holland, K.N., R.W. Brill & R.K.C. Chang. 1990b. Horizontal and vertical movements of Pacific blue marlin captured and released using sportfishing gear. U.S. Fish. Bull. 88: 397–402.

Holland, K.N., R.W. Brill, R.K.C. Chang, J.R. Sibert & D.A. Fournier. 1992. Physiological and behavioural thermoregulation in bigeye tuna (*Thunnus obesus*). Nature 358: 410–412.

Holland, K.N. & J.R. Sibert. 1994. Physiological thermoregulation in bigeye tuna, *Thunnus obesus*. Env. Biol. Fish. 40: 319–327.

Holts, D.B. & D.W. Bedford. 1993. Horizontal and vertical movements of the shortfin mako shark, *Isurus oxyrinchus*, in the southern California Bight. Aust. J. Mar. Freshw. Res. 44: 901–909.

Hunter, J.R., A.W. Argue, W.H. Bayliff, A.E. Dizon, A. Fonteneau, D. Goodman & G.R. Seckel. 1986. The dynamics of tuna movements: an evaluation of past and future research. FAO Fish. Tech. Pap. 277. 78 pp.

Hurst, R.J., N.W. Bagley, G.A. McGregor & M.P Francis. 1999. Movements of the New Zealand school shark, *Galeorhinus galeus*, from tag returns. N.Z J. Mar. Freshw. Res. 33: 29–48.

Klimley, A.P. 1993. Highly directional swimming by scalloped hammerhead sharks, *Sphyrna lewini*, and subsurface irradiance, temperature, bathymetry, and geomagnetic field. Mar. Biol. 117: 1–22.

Last, P.R. & J.D. Stevens. 1994. Sharks and rays of Australia. CSIRO, Canberra. 513 pp.

Metcalfe, J.D. & G.P. Arnold. 1997. Tracking fish with electronic tags. Nature 387: 665–666.

Nakamura, Y. 1993. Vertical and horizontal movements of mature females of *Ommastrephes bartramii* observed by ultrasonic telemetry. pp. 331–336. *In:* T. Okutani, R.K. O'Dor & T. Kubodera (ed.) Recent Advances in Fisheries Biology, Tokai Univ. Press, Tokyo.

Nelson, D.R. 1990. Telemetry studies of sharks: a review, with applications in resource management. pp. 239–256. *In:* H.L. Pratt, S.H. Gruber & T. Taniuchi (ed.) Elasmobranchs as Living Resources: Advances in the Biology, Ecology, Systematics, and the Status of the Fisheries, U.S. Dep. Commer., NOAA Tech. Rep. NMFS 90.

Nelson, D.R., J.N. McKibben, W.S. Strong, Jr., C.G.Lowe, J.A. Sisneros, D.M. Schroeder & R.J. Lavenberg. 1997. An acoustic tracking of a megamouth shark, *Megachasma pelagios*: a crespuscular vertical migrator. Env. Biol. Fish. 49: 389–399.

Nowara, G.B. & T.I. Walker. 1998. Effects of time of solar day, jigging method and jigging depth on catch rates and size of Gould's squid, *Nototodarus gouldi* (McCoy), in southeastern Australian waters. Fisheries Research 34: 279–288.

Olsen, A.M. 1954. The biology, migration and growth rate of the school shark, *Galeorhinus australis* (Macleay) (Carcharhinidae) in south-eastern Australian waters. Aust. J. Mar. Freshw. Res. 5: 353–410.

O'Sullivan, D. & J.M. Cullen. 1983. The food of the squid *Notodarus gouldi* in Bass Strait. Aust. J. Mar. Freshw. Res. 34: 261–285.

Pepperell, J.G. & T.L.O. Davis. 1999. Post-release behaviour of black marlin, *Makaira indica*, caught off the Great Barrier Reef with sportfishing gear. Mar. Biol. (in press).

Punt, A.E. & T.I. Walker. 1998. Stock assessment and risk analysis for the school shark (*Galeorhinus galeus*) resource off southern Australia. Mar. Freshw. Res. 49: 719–731.

Sciarrotta, T.C. & D.R. Nelson. 1997. Diel behavior of the blue shark, *Prionace glauca*, near Santa Catalina Island, California. U.S. Fish. Bull. 75: 519–528.

Stewart, B.S. & R.L. DeLong. 1995. Double migrations of the northern elephant seal, *Mirounga augustirostris*. Journal of Mammology 76: 196–205.

Tricas, T.C. 1979. Relationships of the blue shark, *Prionace glauca*, and its prey species near Santa Catalina Island, California. U.S. Fish. Bull. 77: 175–182.

Welch, D.W. & J.P.Eveson. 1999. An assessment of light-based geoposition estimates from archival tags. Can. J. Fish. Aquat. Sci. 56: 1317–1327.

Wilson, R.P., J.J. Ducamp, W.G. Rees, B.M Culik & K. Niekam. 1992. Estimation of location: global coverage using light intensity. pp. 131–135. *In:* I.G. Priede & S.M. Swift (ed.) Wildlife Telemetry: Remote Monitoring and Tracking of Animals, Ellis Horwood, Chichester.

Wilson, R.P., K. Puetz, J.-B. Charrassin & J. Lage. 1995. Artifacts arising from sampling interval in dive depth studies of marine endotherms. Polar Biology 15: 575–581.

Environmental Biology of Fishes **60**: 299–308, 2001.

Telemetry and satellite tracking of whale sharks, *Rhincodon typus*, in the Sea of Cortez, Mexico, and the north Pacific Ocean

Scott A. Eckert & Brent S. Stewart[a]
Hubbs-Sea World Research Institute, 2595 Ingraham St., San Diego, CA 92109, U.S.A. (e-mail:bstewart@hswri.org)
[a]*Corresponding author; authorship arranged alphabetically*

Received 11 January 1999 Accepted 22 March 2000

Key words: elasmobranchs, satellite telemetry, migration, thermal habitat

Synopsis

We used satellite-linked radio telemetry to document the geographic and vertical movements and thermal habitats of whale sharks in the Sea of Cortez and as they migrated into the north Pacific Ocean. Of 17 sharks tagged between 1994 and 1996, six dispersed widely in the Sea of Cortez during 12–39 days of tracking. Four others left the Sea of Cortez and ranged extensively in the north Pacific Ocean. Indeed, one whale shark migrated to the western north Pacific Ocean, covering over 13 000 km in 37 months of tracking. The sharks generally occupied areas where sea surface water temperatures were between 28 and 32°C, though several ranged to depths of 240 m or deeper where water temperature reached 10°C or colder. Whale sharks may segregate by size and sex, and their movement patterns appear to be related to oceanographic features, such as sea mounts and boundary currents, where primary productivity may be enhanced. These results have important implications for the global conservation of the world's largest yet least known fish. We think that satellite telemetry is a exceptionally promising tool for learning more about the ecology of whale sharks, especially when combined with conventional methods of telemetry and molecular biology.

Introduction

Past studies

Knowledge of the distribution of the world's largest fish, the whale shark, *Rhincodon typus*, is meager and based largely on anecdotal records of sightings, incidental catches in fishing nets and beach cast animals (see syntheses by Wolfson & di Sciara 1981, Wolfson 1986, 1987, Bishop & Abdul-Ghaffar 1993, Beckley et al. 1997, Colman 1997). The species appears to be globally distributed in tropical oceans between 30°N and 35°S latitudes with occasional records as far north as 41°N and south to 36.5°S (Wolfson 1986). Local apparent abundance may be related to seasonal availability of concentrated prey (Taylor 1996, Clark & Nelson 1997, Colman 1997). Whale sharks are thought to be obligate filter-feeding planktivores that primarily eat small crustaceans, yet a few observations suggest that they may also eat algae and small fish on occasion (e.g. Gudger 1941, Silas[1]). Only recently has it been found that whale sharks are obligate lecithotrophic livebearers (cf. Baughman 1955, Joung et al. 1996, Chang et al. 1997). Beyond this, virtually nothing is known of their life history, reproduction or population biology.

Whale sharks are known to aggregate seasonally in the Sea of Cortez, Mexico. The best known sites are Bahía San Luis Gonzaga, Bahía de Los Angeles, Bahía de La Paz, El Bajo, and Banco Gordo (Figure 1), though no predictable pattern among years is yet evident (Gudger 1941, Wolfson & di Sciara 1986, Wolfson 1986, 1987). Such aggregations present unusual opportunities to study some aspects of local whale shark biology (cf. Taylor 1989, 1996, Colman 1997) and to

[1] Silas, E.G. 1986. The whale shark (*Rhiniodon typus* Smith) in Indian coastal waters: is the species endangered or vulnerable? Mar. Fish. Infor. Serv. T & E Ser. 66: 1–37.

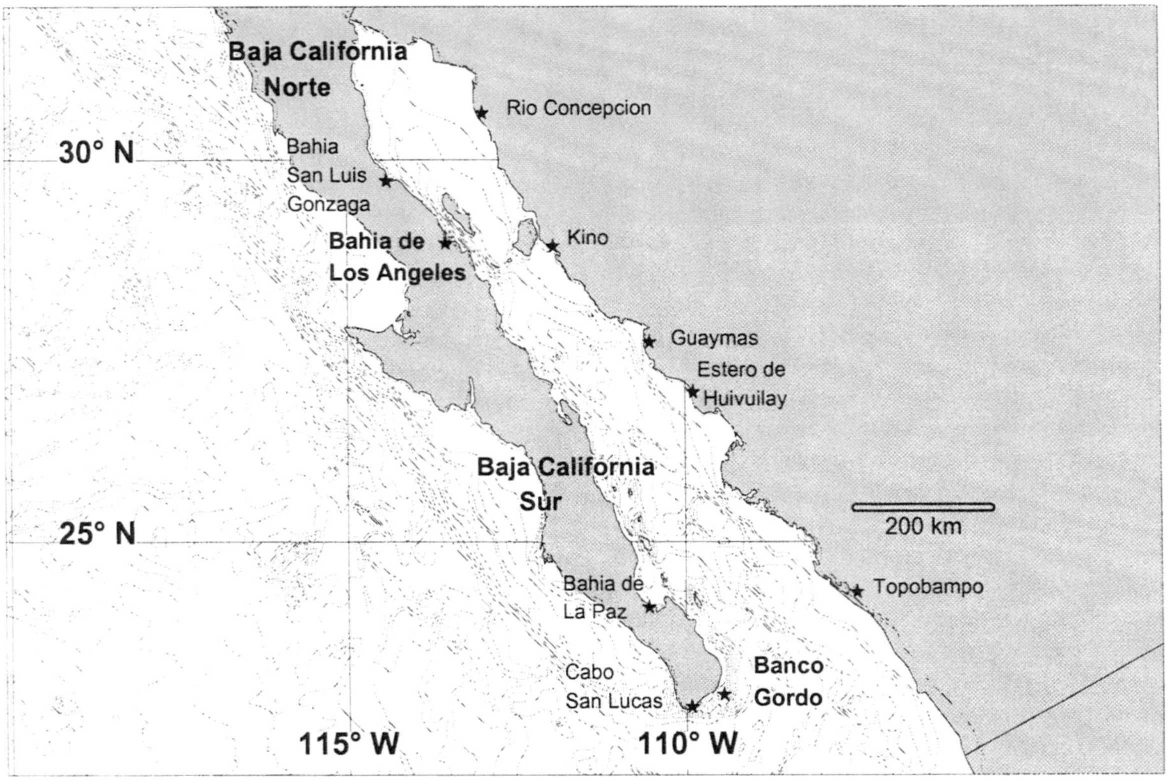

Figure 1. Baja California, the Sea of Cortez, and whale shark tagging sites at Bahía de Los Angeles and Banco Gordo.

allow more detailed study of their ranging behavior using remote sensing techniques. Consequently, to further the long-term interests of Hubbs-Sea World Research Institute in whale shark ecology (e.g., Wolfson & di Sciara 1981, Wolfson 1986, 1987) and owing to our recent applications of satellite-linked radio-telemetry to track the movements and behaviors of a variety of marine vertebrates (e.g., Stewart et al. 1989, 1996, 1998, Heide-Jorgensen et al. 1992, Stewart & DeLong 1995, Eckert & Sarti[2], Stewart 1997), we chose to evaluate the application of radio-telemetry technology to the study of whale sharks in the Sea of Cortez, Baja California, Mexico.

A variety of fishes have been tracked using short-range radio telemetry, though periods of monitoring have generally lasted only a few hours or, occasionally, a few days (e.g., Yuen 1970, Sciarotta & Nelson 1972, Dubsky 1974, Laurs et al. 1977, Nelson et al. 1977, Standora & Nelson 1977, Stasko & Pincock 1977, Carey & Robinson 1981, Carey et al. 1982, Carey & Scharold 1990, Nelson[3], Carey & Clark 1995, Cohen et al. 1999). Satellite-linked tracking and telemetry has become widely used in recent years in studies of terrestrial and marine mammals and various species of birds, but it has only rarely been applied to fishes, and the results have generally been poor (e.g., Priede 1984, Kingman[4], J. Stevens personal communication). One exception is the recent tracking of bluefin tuna, *Thunnus thynnus thynnus*, with timed-release, bouyant transmitters to document gross, short-term movements (Block et al. 1998, Lutcavage et al. 1999).

In the marine environment, satellite telemetry has been most successful in studies of seals and sea lions (pinnipeds), and sea turtles owing to their frequent appearance at the sea surface to breathe which allows

[2] Eckert, S.A. & L.M. Sarti. 1997. Distant fisheries implicated in the loss of the world's largest leatherback nesting population. Marine Turtle Newsletter 78: 2–7.

[3] Nelson, D. R. 1990. Telemetry studies of sharks: a review with applications in resource management. U. S. Nat. Mar. Fish. Serv. Tech. Rept. 90: 245–262.

[4] Kingman, A. 1996. Satellite tracking blue sharks. Shark News 7: 6.

sufficient contact with earth-orbiting satellites for data collection and location determination. Proven techniques for attaching the instruments, by gluing them to the animal's hair or shell or by attaching with a harness, have also been key to reliable short and long term tracking. The infrequent appearance of large fishes at and near the sea-surface means that buoyant satellite transmitters must either be towed behind the animal with a long tether which may allow the transmitter to occasionally break the sea surface often and long enough to make good contact with orbiting satellites (e.g., Priede 1984), or be secured close to the animals body with an in-line corrodible link to allow timed-release of the instrument which can then float to the surface and begin transmitting (e.g., Block et al. 1998, Lutcavage et al. 1999). The former approach may allow the determination of a number of serial locations and thus detailed documentation of migratory routes. The latter, however, may allow only the determination of the location of the animal when the instrument is released relative to where it was attached, but little detail on whereabouts in-between. Moreover, the tracking of the animal is terminated once the transmitter is released. Generally, sufficient opportunities have not been available to test these techniques with large free ranging marine fish to improve the prospects for long-term pelagic tracking, though the miniaturization of transmitter electronics and improved performance of high density batteries are cause for some enthusiasm.

Our goal was to determine whether whale sharks that appear seasonally in some areas of the Sea of Cortez reside there year-round, only appearing near the surface seasonally, or whether they disperse or migrate elsewhere in the Pacific. To accomplish that goal, we tested several methods of configuring and attaching radio transmitters to whale sharks and then document the sharks' geographic and vertical movements remotely using satellite-linked tracking techniques.

Materials and methods

Study area

Whale sharks have been periodically reported in two main areas in the Sea of Cortez, Bahía de Los Angeles in Baja California Norte (BCN) and Banco Gordo in Baja California Sur (BCS) (Figure 1). We observed and tagged sharks in the former location in autumn of 1993, 1994, and 1995 and in the latter area in early spring of 1996. We also made several aerial surveys, using a Cessna 182, along the eastern coast of Baja California to document whale shark abundance there. Moreover, local residents and fishermen, biologists, and tourists alerted us to the appearance of whale sharks in various areas.

Telemetry

We used satellite-linked radio transmitters (platform transmitter terminals, or PTTs) to monitor the vertical and geographic movements of whale sharks and to collect data on the amount of time that sharks spent at various depths and water temperatures. Transmissions from these PTTs were detected and processed by the Argos Data Collection and Location System (Argos DCLS) and then reported to us monthly in digital format.

Briefly, the Argos DCLS consists of two operational satellites that orbit the earth, in near-polar paths, every 101 minutes and which may detect signals from earth-bound transmitters between 6 and 28 times each day depending on the latitude where the PTT is located. If two or more signals are received by a satellite during a single orbit, then a location may be calculated from Doppler shift of successive transmissions. The accuracies of locations vary with several factors but they are often between a few hundred meters to tens of kilometers. The Argos DCLS codes accuracies of locations as either LC1, LC2, or LC3. The predicted accuracies of those locations are all within several kilometers of true. Other locations, primarily determined for tracking of wildlife, are coded as LC0, LCA, and LCB. The accuracies of those locations are not predicted by the Argos DCLS, though previous studies have found that they may often be within several kilometers of true locations (e.g., Argos 1996, Stewart & DeLong 1995, Stewart 1997). Full details of the Argos DCLS may be found elsewhere (Fancy et al.[5], Harris et al.[6], Stewart et al. 1989, Heide-Jorgensen et al. 1992). The key element for our purposes is that a PTT must break the sea surface and transmit while a satellite is overhead for either data to be collected or for a location determined.

We mainly used LC1, LC2, or LC3 locations to describe shark movements, though we also accepted

[5] Fancy, S.G., L.F. Pank, D.C. Douglas, C.H. Curby, G.W. Garner, S.C. Amstrup & W.L. Regelin. 1988. Satellite telemetry: a new tool for wildlife research and management. U.S. Dept. Int., Fish Wildl. Serv. Res. Publ. 172: 1–54.

[6] Harris, R.B., S.G. Fancy, D.C. Douglas, G.W. Garner, S.C. Amstrup, T.R. McCabe & L.F. Pank. 1990. Tracking wildlife by satellite: current systems and performance. U.S. Dept. Int., Fish Wildl. Serv., Fish Wildl. Tech. Rep. 30: 1–52.

LC0 locations if their position relative to the former locations was reasonable relative to prior locations and travel speed. We calculated summary statistics on distance covered and time elapsed between sequential locations to estimate rates of movement.

PTT design and attachment

We used three PTT designs during the study. One was based on an ST-10 transmitter (1/4 watt power output; Telonics, Inc., Mesa Arizona), powered by a 3 volt lithium battery supply (two 2/3A size), encased in electrical potting resin, and then mounted in a float of syntactic foam with a hydrodynamic shape (Type I PTT, Figure 2a). The transmitters were programmed to transmit for 8 h each day. A conductivity circuit prevented transmission when the instrument was submerged. These tactics were used to conserve battery power, extend transmitter life, and ensure spontaneous transmission during brief surface periods if the instrument was submerged for more than 40 s.

A second instrument (Type II PTT, 1/2 W output, Wildlife Computers, Redmond, Washington; Figure 2b) included a microprocessor controller that stored time profiles of depth and water temperature. That instrument and its power supply (four C-cells with a 50 000 transmission capacity) were encased in a buoyant, cylindrical float of syntactic foam (5 cm diameter and 42 cm long). We programmed the microprocessor to measure and store depth and temperature data in a series of intervals which would later yield several frequency histograms for each six-hour time period of each day.

The third instrument was a hybrid of the other two, with a simple ST-10 transmitter that was encased in a Type II float. It was powered by 4 C-cell lithium batteries. Each transmitter float was attached to a 3.0–12.1 m length of either monofilament line (114 kg-test) or braided stainless steel cable (57 kg-test). A swivel was placed in line to prevent kinking. The line was then tethered to a shark using one of several barbed points which was inserted several cm subdermally just behind the first dorsal fin. Three subdermal dart styles were used to attach the tethered buoy. One was a modification of a Floy FII-69 stainless steel dart tag, with the identification streamer replaced by a monofilament tether. Another dart style was a Bandito #108/87 dual wing spear point and the third was a Bandito #21 slip-tip spear point, both commonly used in sport spearfishing.

We located sharks directly from our survey vessel or with the guidance of spotter plane. We then approached

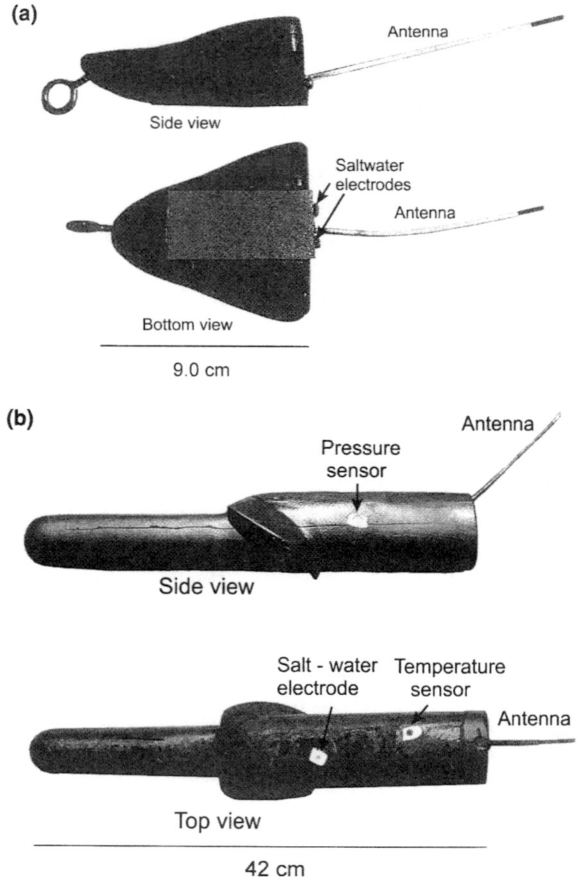

Figure 2. Towed satellite-linked radio transmitter designs used in whale shark tracking: a - type I instrument as viewed from the side (upper) and bottom (lower). b - type II instrument as viewed from the side (upper) and top (lower).

the sharks, by swimming or in a skiff, to within several meters and embedded the subdermal darts 2–4 cm off the dorsal midline near the first dorsal fin using either a 3 m fiberglass spear pole or a Mares Cyrano 850 pneumatic spear gun. The penetration of the darts was limited to a maximum depth of 8 cm by a stop on the spear shaft, as skin thickness is 10–15 cm in the targeted area near the first dorsal fin.

Results

We tagged seven whale sharks (Type I PTT) in Bahía de Los Angeles in September and October 1994 (Table 1). We tagged five more sharks there in early September 1995 (two with Type I PTTs, 3 with Type II PTTs) and then another five further south, at Banco Gordo, in June 1996 (one hybrid Type I PTT, four Type II PTTs).

Geographic movements

No records were obtained for two of the 17 whale sharks tagged. Four of the others were tracked for fewer than 7 days (Table 1). One (#7) of those remained in Bahía de Los Angeles (BLA) for the one day it was tracked, while another (#9) remained there for 3 days until contact was lost. Shark #1 left BLA immediately after it was tagged and moved about 58 km east until contact was lost the following day. Shark #6 remained in BLA for at least 5–7 days until the tag evidently detached and floated ashore where we recovered it 5 days later. Shark #4 remained in BLA for 17 days before contact was lost.

The other ten sharks either moved out of BLA (Figure 3) or out of the Sea of Cortez (Figure 4) and were tracked for 17–1144 days. Four of those sharks (#2, #3, #8, #12) moved into the northern Sea of Cortez until contact was lost, whereastwo (#5, #11) traveled south soon after being tagged (Figure 3). At least four sharks (#10, #13, #14, #15) eventually migrated away from the Sea of Cortez and two of them (#10, #15) traveled substantial distances westward. Indeed, shark #13 covered more than 12 000 km over 37 months (Table 1). Overall, sharks traveled between 23.3 and 12 620 km during 1.1–1144 days of tracking, averaging 4–29 km day^{-1} (Table 1, Figure 5). The weighted mean distance traveled per day was 23.8 ± 0.1 S.E. km (range = 0.1–95.6 km day^{-1}, n = 14). They covered more geographic distance when near the surface (23.9 km day^{-1}), when contact was more frequent, than they did at other times when locations occurred less often (17.5 km day^{-1}; Mann-Whitney 'U' test, $p = 0.02$). Average daily distances traveled appeared to correlate with shark size, though the sample is too small for statistical comparison (Figure 6).

Vertical movements

Data on vertical movements and time spent at various depths and in water of various temperatures were obtained from four whale sharks. During the 24 h prior to contacts with orbiting satellites those sharks spent most (≥80%) of their time at depths less than 10 m (Figure 7a). However, the records also indicate that they also spent time in relatively deep water (Figure 7b).

Table 1. Whale sharks tracked with satellite-linked radio transmitters between 1994 and 1998 in the Sea of Cortez, Mexico, and the North Pacific Ocean (BLA = Bahía de Los Angeles, BG = Banco Gordo)[a].

Shark #	Length (m)	Sex	PTT type[b]	Date tagged	Tagging site	Days tracked[c]	Distance traveled (km)	Mean travel rate (km day^{-1})
1	3.7	F	I	6.10.1994	BLA	1	57.7	—
2	4.6	?	I	22.9.1994	BLA	22	—	—
3	3.0	F	I	19.9.1994	BLA	17	320.3	21.7
4	4.6	?	I	20.9.1994	BLA	17	31.7	3.2
5	4.3	?	I	20.9.1994	BLA	39	818.3	23.5
6	4.0	?	I	23.9.1994	BLA	5–7	23.3	8.4
7	4.0	?	I	7.10.1994	BLA	1	11.2	—
8	6.1	F	I	10.9.1995	BLA	12	404.6	23.6
9	3.7	?	I	10.9.1995	BLA	3	8.5	11.1
10	7.1	?	II	12.9.1995	BLA	1144	12620	17.1
11	3.7	F	II	10.9.1995	BLA	18	46.3	2.0
12	?	?	II	9.9.1995	BLA	28	199.6	—
13	15.0	F	hybrid	20.6.1996	BG	111	2863.6	28.8
14	18.0	F	II	19.6.1996	BG	30	206.8	18.2
15	?	F	II	19.6.1996	BG	665	7762	23.3

[a]Instruments on sharks #1 and #2 transmitted continuously. Those on sharks #3, #4, #5, #6, #7, #8, #9, and #13 operated on a cycle of 8 h on and 16 h off while those on sharks #10, #11, #12, #14, and #15 transmitted on a cycle of 24 h on and 36 h off.

[b]Tether materials were as follows: sharks 1, 2, 6, 7 = 13 m long monofilament tether with a Floy stainless steel dart; sharks 3, 4, 5 = 4 m long monofilament tether with Floy stainless steel dart tip; sharks 8, 11, 12 = 13 m monofilament tether with Bandito 108/87 dart tip; sharks 13, 14, 15 = 15.4 m long stainless steel tether with Bandito 21 dart tip.

[c]The transmitters on sharks 2, 3, 6, 8, 11, 12 are known to have become detached when the anchors pulled out and the transmitters were either recovered or determined to be washed ashore by the telemetry locations and data. The fates of the other transmitters are not known.

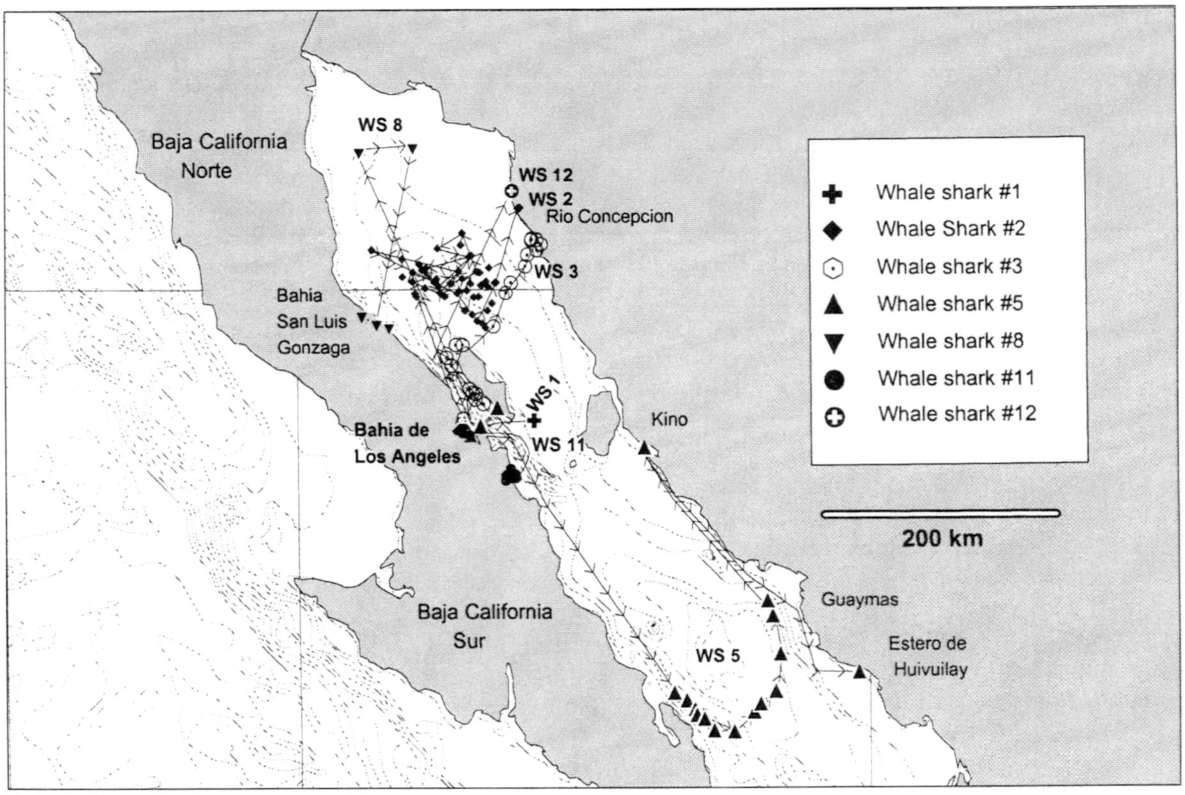

Figure 3. Movements of seven satellite-tagged whale sharks in the Sea of Cortez.

For example, shark #10 spent at least 1% of its time deeper than 240 m. No diel patterns were apparent in vertical movements for any of the sharks.

Thermal habitats

Sharks spent much of their time in areas where sea surface temperatures were between 20 to 32°C. Nonetheless, colder water did not appear to constrain their movements as several spent substantial time in subsurface water of 10°C or cooler (Figure 8). There was no apparent relationship between vertical movements and time of day.

Transmitter performance

The proportion of LC1, LC2, and LC3 locations differed substantially among the three types of instrument. The Type II transmitter, with its cylindrical shaped float and greater power output, performed best (Figure 9). We think that its better performance was primarily due to greater effective power of transmissions, particularly since the hybrid tag, which was housed in the same float as the Type II, performed no better than a Type I transmitter.

Performance of attachment and application method

We cannot easily evaluate the relative performance of the tether materials or the subdermal anchors because we did not systematically match them to the instrument types nor do we know how deeply the anchors penetrated or how well they seated. Nonetheless, instruments with the Bandito 21 anchor appeared to remain on longer (mean = 391 days) than the Bandito 108/87 (mean = 19.3 days) and the Floy (15.4 days) anchors. There were no clear differences in performance among the tethers.

Discussion

This is the first long-term tracking of whale sharks and indeed the first to document long-term

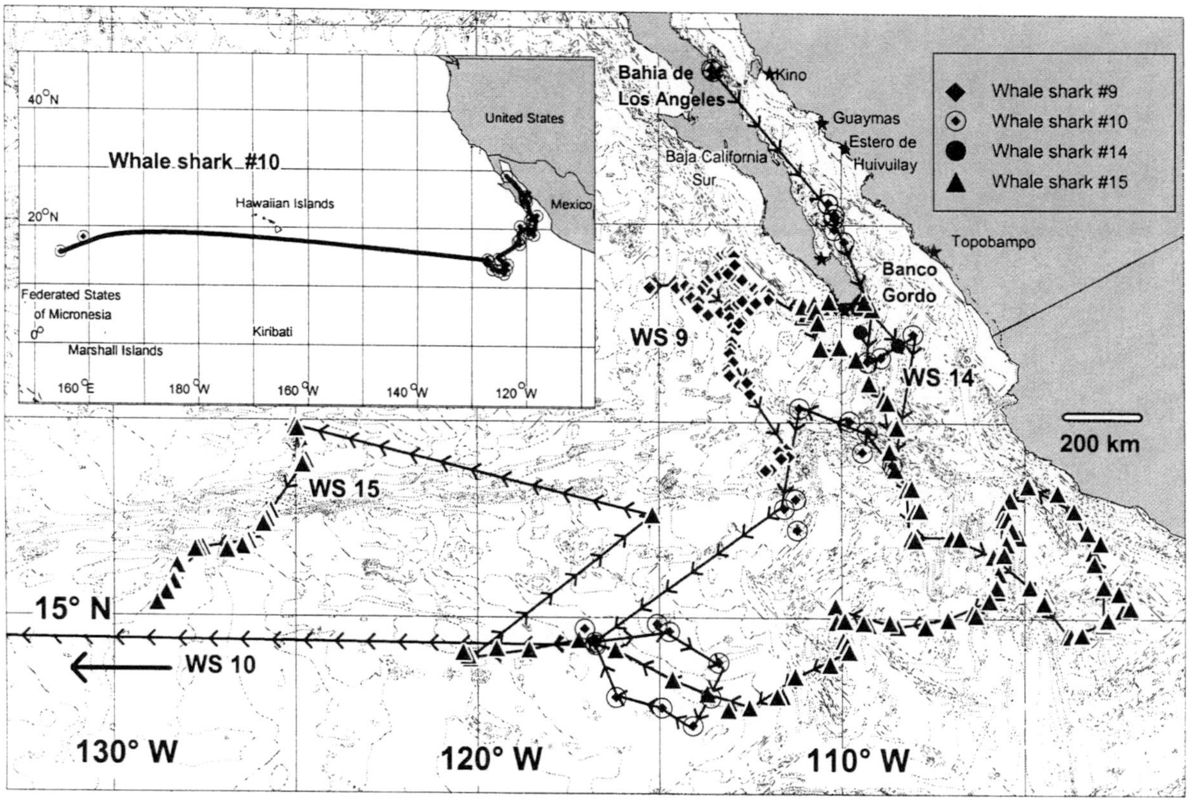

Figure 4. Movements of four satellite-tagged whale sharks that moved out of the Sea of Cortez into the North Pacific Ocean.

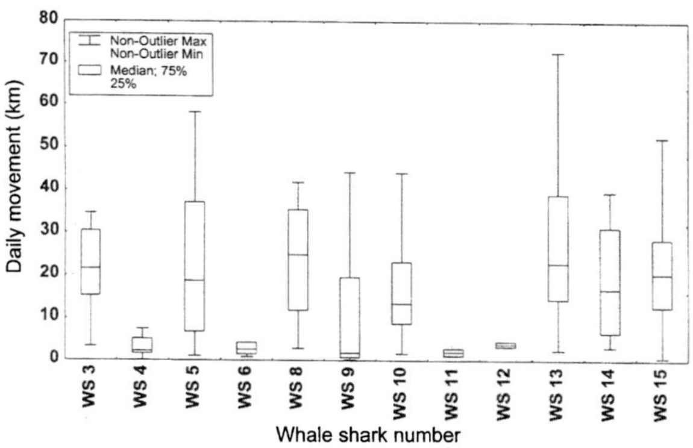

Figure 5. Daily travel distance of 12 whale sharks tagged with satellite-linked radio transmitters in the Sea of Cortez.

movement routes of any fish using satellite telemetry, notwithstanding the recent origin-to-destination tracking of bluefin tuna (Block et al. 1998, Lutcavage et al. 1999). Our results demonstrate the utility of the use of towed satellite-linked radio transmitters to study the ecology of whale sharks and perhaps other large pelagic fishes. Moreover, they indicate rather extensive use of vertical and geographic marine habitats by whale sharks and suggest that whale sharks may range widely, in the North Pacific Ocean at least, and perhaps make migrations that take several years to complete. The tracking of one shark over three years and across the North Pacific Ocean is particularly noteworthy as there has been no prior data to support the hypothesis that whale sharks from local populations (e.g., the eastern and western North Pacific) may mix. Still, there are too few data to determine whether individual whale sharks regularly return to the Sea of Cortez, as they apparently do in waters off western Australia (cf. Taylor 1996). Indeed, our data suggest that they may not, though additional studies are clearly needed to test that hypothesis.

Our data also suggest that whale sharks may segregate by size and sex while in the Sea of Cortez. Virtually all whale sharks that we encountered in the northern Sea of Cortez were less than 4 m long, whereas all of those encountered in the southern reach of the Sea of Cortez were three times that length (Table 1) and all sharks

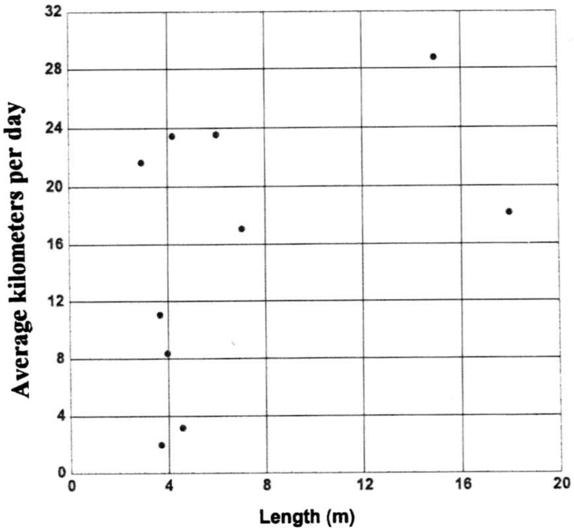

Figure 6. Daily distance traveled versus body length for 12 satellite-tagged whale sharks.

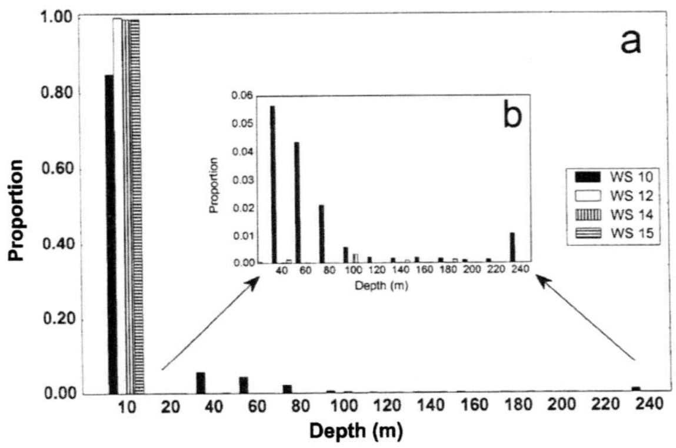

Figure 7. Proportion of time spent at depth for 4 satellite-tagged whale sharks in the Sea of Cortez, Mexico, and the North Pacific Ocean: a – depth profiles show that during the 24 hours preceding radio contact, the majority of time was spend in the upper 10 m of the water column. b – depth profiles show that sharks also spent time in waters substantially deeper than 10 meters and to 240 m and deeper.

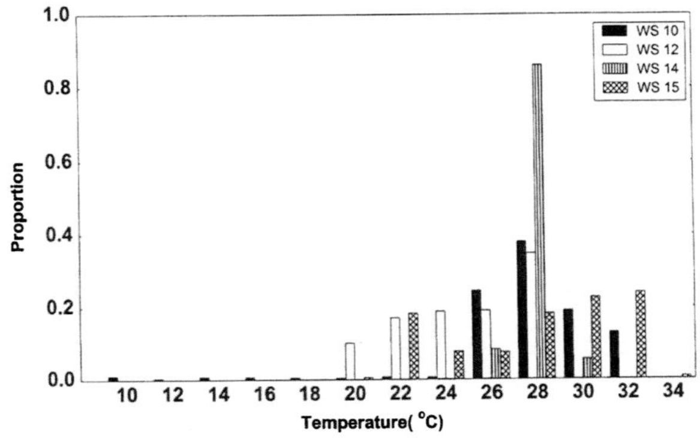

Figure 8. Proportion of time during the 24 hours preceding radio contact that sharks spent in water of various temperature.

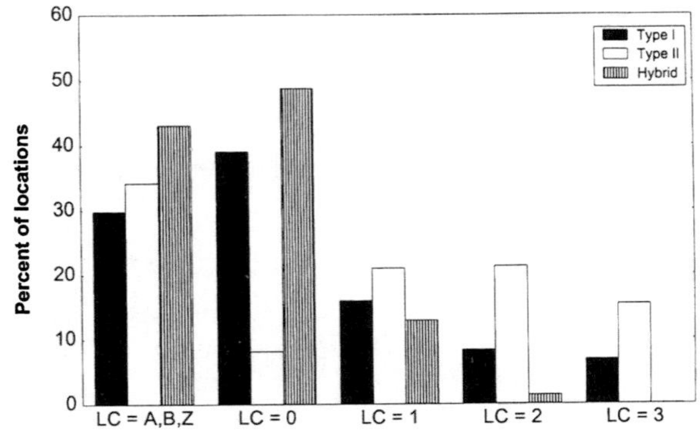

Figure 9. Percentage of locations of various accuracies versus instrument configuration (see Materials and methods for details).

that we found in the latter area were females (most of those appeared to be gravid).

The whale sharks that we tracked appeared to spend long periods at depth punctuated by briefer bouts of time near the surface, when we were able to locate them. However, we think that the data on their vertical movements may be biased overall towards warmer water temperatures and shallower depths, particularly for those sharks that weretracked for longer periods and consequently located less often. This is because the instruments were programmed to store and transmit data in 24 hour blocks. Consequently, if sharks had spent several days or more in deeper water before occasionally approaching the surface, those data would not have been available in the limited instrument memory to be transmitted. In any event, further work using instruments that archive data for longer periods of time and then transmit some or all of it later when the shark surfaces for extended durations or the tags detach and float at the surface (cf. Block et al. 1998, Lutcavage et al. 1999) could resolve this question. Nonetheless, our data do demonstrate that whale sharks spend time, perhaps substantial amounts, at depths greater than 240 m and in water as cold as 10°C.

Our direct experience while in the water with whale sharks indicates that they are strong, powerful swimmers, for short bursts at least. Similarly, the tracking data indicated occasional travel rates of up to 96 km day^{-1} (3.9 km h^{-1}). Overall, however, the tracking data indicate that they are generally slow swimmers and migrators especially compared with other large pelagic sharks. The overall average travel rates that we documented were around 24 km day^{-1} (1 km h^{-1}). These rates compare with 1.5 km h^{-1} for blue sharks (Carey & Scharold 1990), 3.2 km h^{-1} for white sharks (Carey et al. 1982, Strong et al. 1992), 3.6 km h^{-1} for tiger sharks (Tricas et al. 1981), 2.7–3.4 km h^{-1} for basking sharks (Harden Jones 1973, Priede 1984), and 1.2 km h^{-1} for a megamouth shark (Nelson et al. 1997). When we have observed whale sharks in the Sea of Cortez, they appear to swim continuously, though recirculating in small geographic areas while filter-feeding. They also occasionally remain stationary and vertical in the water column while pumping prey-laden water through their gill slits. Thus the tracking data, as indicators of geographic distance covered per hour or day, may not generally reflect true swim speed.

From the movements of sharks outside of the Sea of Cortez, we think that their distribution may be influenced primarily by oceanographic features which correlate with enhanced upwelling and plankton productivity, like the sea mounts near the Revillagigedo Islands and Clipperton Island and along current boundaries in the central north Pacific Ocean.

Future directions

Though our studies have revealed some data on the vertical and geographic movements of whale sharks in the Sea of Cortez and in the North Pacific Ocean, research is clearly needed to document the behaviors of a larger number of individual whale sharks in these regions and in other oceans. Additional information is needed on the relative and seasonal distributions of adult versus juvenile whale sharks and adult females versus males. Moreover, the genetic relationships of putative whale shark populations is an important issue for the global conservation of this poorly known animal. Tracking the habits of these elusive fish using satellite technology presents the most feasible method of further learning, especially when combined with molecular genetic techniques and conventional radio-telemetry.

Because so little is known about whale sharks, virtually any questions about the species' life history, physiology, foraging ecology, demography, and behavior could be posed as hypotheses. Few of those may be actually testable (i.e., truly falsifiable) though, owing to the sharks' cryptic nature and brief seasonal appearances at a few locations. Nonetheless, in further applications of telemetry (especially satellite-linked) and molecular techniques, we think that the following null hypotheses follow directly from the preliminary and suggestive results that we report here and are the most tractable:

(1) Whale sharks are sedentary local residents year round but occur near the sea surface seasonally and live at depth during the others.
(2) The age and sex structure of local whale sharks is reflective of the population at large and the species does not aggregate by age or sex seasonally or during the species life cycle.
(3) There are no inter-region or inter-ocean genetic differences among populations of whale sharks.

Further studies which address these hypotheses, or are even merely descriptive, should contribute substantially to the knowledge of whale sharks and also effectively structure the approach and direction for additional research.

Acknowledgements

We thank P. Daily, R. Davies, Z. Dembisky, T. Dirkson, R. Drinen, D. Jehl, J. Kerivan, J. Kirkpatrick, G. Kooyman, K. Kooyman, F. Murru, A. Resendez, L. Sarti, and D. Simms for assistance in the field. We also thank B. Andrews, J. Goldsmith, D. Kent, M. Shaw and the Animal and Aquarium staff of Sea World of San Diego for supporting various aspects of this study. Special thanks to P. Daily and the crew of the Megalodon for supporting tagging operations in 1996. Finally, we thank B. Bowen, T. Tricas, G. West and an anonymous reviewer for their comments on the manuscript. Support for the studies was provided by Anheuser-Busch Corporation, SeaWorld of Florida, SeaWorld of San Diego, and Hubbs-Sea World Research Institute. Our studies were authorized by research permits from Instituto Nacional de Ecologia (No. 05268 and extension 04336–0918), Instituto Nacional de Pesca (No. 2337, 1974, 1385), La Secretaria de Relaciones Exteriores (in litt. 8 October 1993, 29 July 1995) and approved by the HSWRI Institutional Animal Care and Use Committee.

References cited

Argos. 1996. User's manual. Service Argos, Landover. 176 pp.

Baughman, J.L. 1955. The oviparity of the whale shark, *Rhineodon typus*, with records of this and other fishes in Texas waters. Copeia 1955: 54–55.

Beckley, L.E., G. Cliff, M.J. Smale & L.J.V. Compagno. 1997. Recent strandings and sightings of whale sharks in South Africa. Env. Biol. Fish. 50: 343–348.

Bishop, J.M. & A.R. Abdul-Ghaffar. 1993. Whale shark observations off Kuwait's coast in 1992. J. Fish Biol. 43: 939–940.

Block, B.A., H. Dewar, C. Farwell & E.D. Prince. 1998. A new satellite technology for tracking the movements of Atlantic bluefin tuna. Proc. Nat. Acad. Sci. USA 95: 9384–9389.

Carey, F.G. & E. Clark. 1995. Depth telemetry from the sixgill shark, *Hexanchus griseus*, at Bermuda. Env. Biol. Fish. 42: 7–14.

Carey, F.G. & J.V. Scharold. 1990. Movements of blue sharks (*Prionace glauca*) in depth and course. Mar. Biol. 106: 329–342.

Carey, F.G. & G.H. Robison. 1981. Daily patterns in the activities of swordfish, *Xiphias gladius*, observed by acoustic telemetry. U.S. Fish. Bull. 79: 277–292.

Carey, F.G., J.W. Kanwisher, O. Brazier, G. Gabrielson, J.G. Casey & H.L. Pratt, Jr. 1982. Temperature and activities of a white shark, *Carcharodon carcharias*. Copeia 1982: 254–260.

Chang, W.-B., M.-Y. Leu & L.-S. Fang. 1997. Embryos of the whale shark, *Rhincodon typus*: early growth and size distribution. Copeia 1997: 185–208.

Clark, E. & D.R. Nelson. 1997. Young whale sharks, *Rhincodon typus*, feeding on a copepod bloom near La Paz, Mexico. Env. Biol. Fish. 50: 63–73.

Cohen, M., W.R.C. Beaumont & N.C. Thorp. 1999. Movement and activity patterns of the black piranha. Env. Biol. Fish. 54: 45–52.

Colman, J.G. 1997. A review of the biology and ecology of the whale shark. J. Fish Biol. 51: 1219–1234.

Gudger, E.W. 1941. The food and feeding habits of the whale shark, *Rhineodon typus*. J. Elisha Mitchell Sci. Soc. 57: 57–71.

Harden Jones, F.R. 1973. Tail beat frequency, amplitude, and swimming speed of a shark tracked by sector scanning sonar. J. Cons. Int. Explor. Mer. 35: 95–97.

Heide-Jorgensen, M.P., B.S. Stewart & S. Leatherwood. 1992. Satellite tracking of ringed seals Phoca hispida off northwest Greenland. Ecography 15: 5–61.

Joung, S.-J., C.-T. Chen, E. Clark, S. Uchida & W.Y.P. Huang. 1996. The whale shark, *Rhincodon typus*, is a livebearer: 300 embryos found in one 'megammama' supreme. Env. Biol. Fish. 46: 219–223.

Laurs, R.M., H.S.H. Yuen & J.H. Johnson. 1977. Small-scale movements of albacore, *Thunnus alalunga*, in relation to ocean features as indicated by ultrasonic tracking and oceanographic sampling. U.S. Fish. Bull. 75: 347–355.

Lutcavage, M.E., R.W. Brill, G.B. Skomal, B.C. Chase & P.W. Howey. 1999. Results of a pop-up satellite tagging of spawning class fish in the Gulf of Maine: do North Atlantic bluefin tuna spawn in the mid-Atlantic? Can. J. Fish. Aquat. Sci. 56: 173–177.

Nelson, D.R., J.N. McKibben, W.R. Strong, C.G. Lowe, J.A. Sisneros, D.M. Schroeder & R.J. Lavenberg. 1997. An acoustic tracking of a megamouth shark, *Megachasma pelagios* : a crepuscular vertical migrator. Env. Biol. Fish. 49: 389–399.

Priede, I.G. 1984. A basking shark (*Cetorhinus maximus*) tracked by satellite together with simultaneous remote sensing. Fish. Res. 2: 201–216.

Sciarrotta, T.C. & D.R. Nelson. 1977. Diel behavior of the blue shark, *Prionace glauca*, near Santa Catalina Island, California. U.S. Fish. Bull. 75: 519–528.

Standora, E.A. & D.R. Nelson. 1977. A telemetric study of the behavior of free-swimming Pacific angel sharks, *Squatina californica*. Bull. So. Cal. Acad. Sci. 76: 193–201.

Stasko, A.B. & D.G. Pincock. 1977. Review of underwater biotelemetry, with emphasis on ultrasonic techniques. J. Fish. Res. Board Can. 34: 1261–1285.

Stewart, B.S. 1997. Ontogeny of differential migration and sexual segregation in northern elephant seals. J. Mammal. 78: 1101–1116.

Stewart, B.S. & R.L. DeLong. 1995. Double migrations of the northern elephant seal, *Mirounga angustirostris*. J. Mammal. 76: 196–205.

Stewart, B.S., S. Leatherwood, P.K. Yochem & M.P. Heide-Jorgensen. 1989. Harbor seal tracking and telemetry by satellite. Mar. Mamm. Sci. 5: 361–375.

Stewart, B.S., E.A. Petrov, E.A. Baranov, A. Timonin & M. Ivanov. 1996. Seasonal movements and dive patterns of juvenile Baikal seals, *Phoca sibirica*. Mar. Mamm. Sci. 12: 528–542.

Stewart, B.S., P.K. Yochem, T.S. Gelatt & D.B. Siniff. 1998. Dispersal and habitat use of Weddell seals (*Leptonychotes weddellii*) in the Ross Sea, Antarctica, during their first year of life. N.Z. Nat. Sci. 23: 195.

Strong, W.R., Jr., R.C. Murphy, B.D. Bruce & D.R. Nelson. 1992. Movements and associated observations of bait-attracted white sharks, *Carcharodon carcharias*: a preliminary report. Aust. J. Mar. Freshwater Res. 43: 13–20.

Taylor, G. 1989. Whale sharks of Ningaloo reef, Western Australia: a preliminary study. West. Austr. Nat. 18: 7–12.

Taylor, G. 1996. Seasonal occurrence, distribution and movements of the whale shark, *Rhincodon typus*, at Ningaloo Reef, Western Australia. Mar. Freshwater Res. 47: 637–642.

Tricas, T.C., L.R. Taylor & G. Naftel. 1981. Diel behavior of the tiger shark, *Galeocerdo cuvier*, at French Frigate Shoals, Hawaiian Islands. Copeia 1981: 904–908.

Wolfson, F.H. & G. Notarbartolo di Sciara. 1981. The whale shark, *Rhiniodon typus*, Smith 1828, an annotated bibliography. Atti Soc. Ital. Sci. Nat. Mus. Civ. Stor. Nat. Milano 122: 171–203.

Wolfson, F.H. 1986. Occurrences of the whale shark, *Rincodon typus* Smith. pp. 208–226. *In*: T. Uyeno, R. Aria, T. Taniuchi & K. Matsuura (ed.) Indo-Pacific Fish Biology: Proceedings of the Second International Conference on Indo-Pacific Fishes, Ichthyological Society of Japan, Tokyo.

Wolfson, F.H. 1987. The whale shark, *Rhiniodon typus*, off Baja California, Mexico. Simp. Biol. Mar. 5: 103–108.

Yuen, H.S. 1970. Behavior of skipjack tuna, *Katsuwonus pelamus*, as determined by tracking with ultrasonic devices. J. Fish. Res. Board Can. 27: 2071–2079.

Environmental Biology of Fishes **60:** 309–319, 2001.

Species and subject index

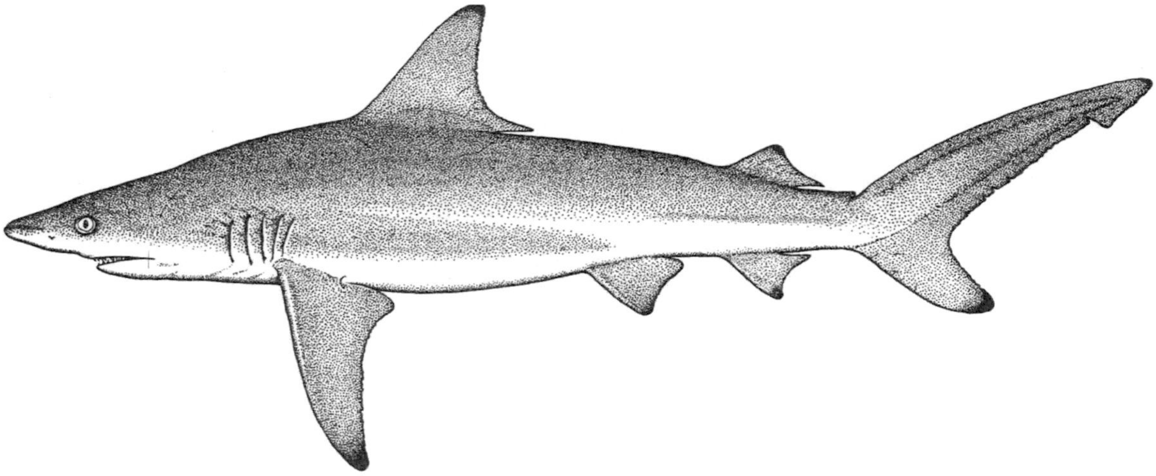

Female of a small blacktip shark, *Carcharhinus limbatus*, 1 m long off the Florida coast (PV).

Female of the porbeagle shark, *Lamna nasus*, 147 cm long, captured in the western north Atlantic (PV).

Kolibri 360 GmbH
Ferdinand-Jühlke-Straße 7
99095 Erfurt, Deutschland
produktsicherheit@kolibri360.de